Recent Results in Cancer Research

Volume 217

Series Editors

Alwin Krämer, German Cancer Research Center, Heidelberg, Germany

Jiade J. Lu, Shanghai Proton and Heavy Ion Center, Shanghai, China

More information about this series at http://www.springer.com/series/392

T.-C. Wu · Mei-Hwei Chang · Kuan-Teh Jeang

Editors

Viruses and Human Cancer

From Basic Science to Clinical Prevention

Second Edition

Editors
T.-C. Wu
School of Medicine
Johns Hopkins University School
of Medicine
Baltimore, MD, USA

Mei-Hwei Chang
Hepatitis Research Center
National Taiwan University Hospital
Taipei, Taiwan

Kuan-Teh Jeang
Molecular Virology Section,
Laboratory of Mole
National Institute of Allergy and Infect
Bethesda, MD, USA

ISSN 0080-0015   ISSN 2197-6767 (electronic)
Recent Results in Cancer Research
ISBN 978-3-030-57364-5   ISBN 978-3-030-57362-1 (eBook)
https://doi.org/10.1007/978-3-030-57362-1

This Springer imprint is published by the registered company Springer Nature Switzerland AG
The registered company address is: Gewerbestrasse 11, 6330 Cham, Switzerland

Contents

An Introduction to Virus Infections and Human Cancer

John T. Schiller and Douglas R. Lowy

1 The Burden of Virus-Induced Cancers

The determination that infection by a specific subset of human viruses is the primary cause of a substantial fraction of human cancers is one of the most important achievements in cancer etiology and intervention. It was recently estimated that a virus infection is the central cause of more than 1,400,000 cancer cases annually, representing approximated 10% of the worldwide cancer burden (Plummer et al. 2016). The widely accepted human oncoviruses are human papillomaviruses (HPV), hepatitis B virus (HBV), hepatitis C virus (HCV), Epstein–Barr virus (EBV), Kaposi's sarcoma-associated herpesvirus (KSHV) (also called human herpesvirus 8), human T-cell lymphotropic virus (HTLV-1), and Merkel cell polyomavirus (MCPyV). This conclusion is based on the cumulative knowledge from a large number of experimental, clinical, and epidemiological studies over the last five decades.

An additional 5% of worldwide cancers, mostly gastric cancer, are attributed to infection by the bacterium *Helicobacter pylori* (Plummer et al. 2016). The number of worldwide incident cases associated with specific virus varies widely, from 640,000 for HPV to 3000 for HTLV (Table 1). Approximately, 85% of the burden of virus-induced cancers is bourn by individuals in the developing regions of the world, with profound implications for translating the knowledge of virus-induced cancers into public health interventions. In addition, some viruses cause more cancers in one sex than the other. Almost 90% of HPV-induced cancers occur in females, while approximately, two-thirds of HBV, HCV, and EBV cancer occur in men (Table 1).

J. T. Schiller (✉) · D. R. Lowy
Laboratory of Cellular Oncology, Center for Cancer Research,
National Cancer Institute, Bethesda, MD, USA
e-mail: schillej@mail.nih.gov

© Springer Nature Switzerland AG 2021
T.-C. Wu et al. (eds.), *Viruses and Human Cancer*, Recent Results
in Cancer Research 217, https://doi.org/10.1007/978-3-030-57362-1_1

1

Table 1 Number of new cancer cases attributable to specific viral infections by gender

Virus	Total	Females	Males
HPV	636,000	570,000	66,000
HBV	420,000	120,000	300,000
HCV	165,000	55,000	110,000
EBV	120,000	40,000	80,000
KSHV	43,000	15,000	29,000
HTLV	2,900	1,200	1,700

Data from (Plummer et al. 2016)

2 The Human Cancers Caused by Viruses

Many different types of cancers are induced by human oncoviruses, and the fraction of each cancer type attributed to a viral infection varies widely (Table 2) (Plummer et al. 2016). HPVs normally infect stratified epithelium and are causally associated with a number of anogenital carcinomas, including cervical, anal, vulvar, vaginal, and vulvar, and also carcinomas of other mucosal epithelium, most notably oropharyngeal. The attributable fraction varies from almost 100% for cervical carcinomas to 4% for oral cavity and laryngeal cancers. Interestingly, the rates of HPV-associated oropharyngeal cancers appear to be substantially increasing in Western Europe and North America, most notably in men (de Martel et al. 2017).

HBV and HCV have a strict tropism for hepatocytes and together are the cause of three-quarters of hepatocellular carcinomas (Petrick and McGlynn 2019) (Table 2). EBV normally infects B lymphocytes and epithelial cells and induces about half of Hodgkin's lymphoma and Burkitt's lymphomas (Saha and Robertson 2019). It is also an etiologic factor in most cases of nasopharyngeal carcinoma (Chen et al. 2019). In addition, EBV-associated gastric cancer is a distinct clinicopathological entity that is present in ∼9% of these cancers (Bae and Kim 2016).

Virtually, all Kaposi's sarcomas are associated with KSHV infection. KSHV is also strongly associated with multicentric Castleman's disease and primary effusion lymphoma, two relatively rare B-cell neoplasms (Katano 2018). HTLV-1 infects lymphocytes and is a central cause of adult T-cell leukemia and lymphoma (Tagaya et al. 2019). MCPyV is commonly detected in normal skin and is responsible for approximately three-quarters of Merkel cell carcinoma, a relatively uncommon skin cancer (Kervarrec et al. 2019).

3 The Diversity of Human Oncoviruses

Human tumor viruses are highly diverse, including viruses with large double-stranded DNA genomes (EBV and KSHV), small double-stranded DNA genomes (HPV, HBV, and MCPyV), positive-sense single-stranded RNA genomes (HCV), and retroviruses (HTLV-1) (Table 3). Some have enveloped virions, specifically

Table 2 Prevalence of viruses in virus-associated cancers

Virus	Cancer	Geographical area	Attributable fraction (%)
HPV	Cervix	World	100
HPV	Penile	World	51
HPV	Anal	World	88
HPV	Vulvar	World	48[a]
HPV	Vaginal	World	78
HPV	Oropharynx	North America	51
HPV	Oropharynx	India	22
HPV	Laryngeal	World	4.6
HBV	Liver	Developing	59[b]
HBV	Liver	Developed	23[b]
HCV	Liver	Developing	33[b]
HCV	Liver	Developed	20[b]
EBV	Hodgkin's lymphoma	Africa	74
EBV	Hodgkin's lymphoma	Asia	56
EBV	Hodgkin's lymphoma	Europe	36
EBV	Burkitt's lymphoma	Sub-Saharan Africa	100
EBV	Burkitt's lymphoma	Other regions	20–30
EBV	Nasopharyngeal carcinoma	High-incidence areas	100
EBV	Nasopharyngeal carcinoma	Low-incidence areas	80
KSHV	Kaposi's sarcoma	World	100
HTLV-1	Adult T-cell leukemia and Lymphoma	World	100
MCPyV	Merkel cell carcinoma	North America	70–80[c]

Data from (Plummer et al. 2016) unless otherwise indicated
[a]Age 15-54 yrs
[b]Data from (de Martel et al. 2012)
[c]Estimate from (Tetzlaff and Nagarajan 2018)

HBV, HCV, EBV and KSHV and HTLV-1, whereas others have naked icosohedral virions, specifically HPV and MCPyV. The carcinogenic mechanisms of oncoviruses also vary widely, as outlined below. However, in all cases, oncogenesis is an uncommon consequence of the normal viral life cycle. Virus-induced cancers almost always arise as monoclonal events from chronic infections, usually many years after the primary infection, indicating that infection is just one component in a multi-step process of carcinogenesis. The exceptional case is KSHV-induced Kaposi's sarcoma, which can arise as a polyclonal tumor within months of infection in immunosuppressed individuals (Cesarman et al. 2019) (also see Chap. 13).

Table 3 Basic features of human oncoviruses

Virus	Genome	Virion structure	Normal tropism	Year isolated (reference)
EBV	Linear 172 kb DS DNA	Enveloped	Epithelium and B cells	1964 (Epstein et al. 1964)
HBV	Circular 3.2 kb partial DS DNA	42 nm enveloped	Hepatocytes	1970 (Dane et al. 1970)
HTLV-1	Linear 9.0 k nt positive-sense RNA	Enveloped	T and B cells	1980 (Poiesz et al. 1980)
HPV16	Circular 7.9 kb DS DNA	55 nm naked Icosahedron	Stratified squamous epithelium	1983 (Dürst et al. 1983)
HCV	Linear 9.6 k nt positive-sense RNA	Enveloped	Hepatocytes	1989 (Choo et al. 1989)
KSHV	Linear 165 kb DS DNA	Enveloped	Oropharyngeal epithelium	1994 (Chang et al. 1994)
MCPyV	Circular 5.4 kb DS DNA	40 nm naked icosahedron	Skin	2008 (Feng et al. 2008)

4 Oncogenic Mechanisms

The oncogenic mechanisms of many tumor viruses, as detailed in later chapters, involve the continued expression of specific viral gene products that regulate proliferative, anti-apoptotic, and/or immune escape activities through an interaction with cellular gene targets. Examples of oncoproteins include E6 and E7 of HPVs, LMP1 of EBV, Tax of HTLV-1, and T antigen of MCPyV (Chaps. 8, and 10 respectively). Although HCV- and HBV-encoded proteins may play a direct role in hepatocarcinogenesis, these viruses may primarily induce cancer more indirectly, as a result of persistent infection causing chronic inflammation and tissue injury (Kanda et al. 2019a, b) (Chaps. 3, 4 and 6). KSHV may act primarily by altering complex cytokine/chemokine networks (Mesri et al. 2010) (Chap. 13). Virally encoded microRNAs, especially those of KSHV and EBV, also play a direct role in carcinogenesis (Wang et al. 2019).

With some viruses, e.g., MCPyV and HPV, malignant progression usually involves mutation and/or insertion of the viral genome into the host DNA, such that it can no longer replicate (Vinokurova et al. 2008; Arora et al. 2012). In addition, the advent of high-throughput sequencing has facilitated the evaluation of the strain variation and cancer risk. Interesting examples have been uncovered for EBV (Kanda et al. 2019a, b) and HPV16 (Mirabello et al. 2016), but the molecular mechanisms that account for these strain differences in carcinogenic potential are not entirely clear.

HIV infection is a strong risk factor for several cancers, particularly cancers that are associated with other virus infections (Vangipuram and Tyring 2019). However, the effect of HIV infection on oncogenesis is thought to be indirect, by inhibiting normal host immune functions that would otherwise control or eliminate the oncovirus infections and/or provide immunosurveillance of emerging tumors. Supporting this hypothesis is the observation that the rates of the same cancers increase in patients with other forms of immunosuppression (Grulich and Vajdic 2015). Other types of retroviruses can induce cancers by insertional mutagenesis in animal models (Fan and Johnson 2011). However, this activity has not been convincingly documented in humans, except in a few patients in experimental gene transfer trials involving delivery of high doses of recombinant retroviral vectors (Romano et al. 2009).

4.1 Viral Infection and Cancer: Establishing Causality

Detecting a virus in a cancer does not establish causality. For instance, the cancer cells might simply be exceptionally susceptible to infection or replication by a particular virus, a scenario that is currently being exploited experimentally in the development of oncolytic virotherapies (Russell and Barber 2018). However, the causal associations between the seven viruses and specific cancers noted above are now convincingly established. They fulfill most, if not all, of the criteria for causality proposed by Sir A. Bradford Hill in the early 1970s (Hill 1971). Multiple epidemiological studies in varying settings have established the strength and consistency of association between infection and cancer for these viruses. Most strikingly, relative risks of over 100 have been calculated for HPV and KSHV infection in the development of cervical carcinoma and Kaposi's sarcoma, respectively, among the highest observed for a cancer risk factor (Moore and Chang 1998; Bosch et al. 2002). In some instances, establishing a strong association required identification of especially oncogenic strains, e.g., HPV16 and 18 among mucosotropic HPVs, or a specific tumor subsets, e.g., oropharyngeal carcinomas among head and neck cancers. Temporality was established by demonstrating that infection proceeds cancer, usually by many years. Integration of the viral gene in the same site in all tumor cells further demonstrated, for some viruses, that the viral infection was an initiating event.

In some cases, the viruses are consistently detected in well-established cancer precursor lesions, as is the case for HPV and high-grade cervical intraepithelial neoplasia (Chap. 8), although in others, such as MCPyV-induced MCC and HPV-associated oropharyngeal cancer, the precursor lesions have not been clearly identified. Demonstrating that, for the most part, populations with higher prevalences of virus infection also had higher incidences of the associated tumor, e.g., HBV and liver cancer, established important dose–response relationships (El-Serag 2012) (Chap. 5). However, these associations are sometimes confounded by high prevalence of the oncovirus in the general population and variability in the exposure to other risk factors. A clear example is the high frequency of EBV infection in the general population, but the induction of EBV-positive Burkitt's lymphoma primarily in areas with a high incidence of malaria (Moormann and Bailey 2016). A large number of laboratory studies established biological plausibility for causality by characterizing the interaction of viral proteins or other viral products with key regulators of proliferation and apoptosis and establishing their immortalizing and transforming activity in vitro and their oncogenic activity in animal models (Chap. 4, 6, 8, 10, 11, and 13). These studies also support the criterion that the associations be in agreement with the current understanding of disease pathogenesis, in this case the molecular biology of carcinogenesis (Mesri et al. 2014). The last Hill criterion, that removing the exposure prevents the disease, has been most convincing demonstrated for HBV and HPV, after introduction of the corresponding vaccine.

4.2 The Importance of Identifying the Viral Etiology for a Cancer

The identification of a virus as a central cause of a specific cancer can have several substantial implications. First, it can provide basic insights into carcinogenic processes, especially the identification of potential cellular targets for diagnoses and interventions that are often relevant to both virus-associated and virus-independent tumors (Mesri et al. 2014). Studies of virally induced carcinogenesis were particularly illuminating in the past decades when the ability to analyze the complexity of host cell genomics and proteomics was much more limited than it is today. For example, the tumor suppressors p53 and pRb were first identified as binding partners of the small DNA tumor viruses in experimental systems and later shown to be targets for several human oncoviruses (Pipas 2019). They are also among the most frequently mutated genes in non-virally induced cancers (Chap. 8).

Second, the presence of the virus can be used in cancer risk assessment and prevention. One example is the increasing use of HPV DNA testing to screen for cervix cancer risk. HPV DNA tests are more sensitive for detection of high-grade premalignant lesions than is the standard Pap test, so intervals between tests can be increased in women who test negative for high-risk HPV DNA in their cervix (Rizzo and Feldman 2018). Another example is HCV screening to identify individuals at high risk of progression to liver cirrhosis and cancer (Chap. 7). HCV screening is now recommended in the USA for all individuals born between 1945 and 1965 (Smith et al. 2012), although the US Prevention Services Task Force has recently made a draft recommendation for widening screening ages to everyone older than 11 years of age (https://www.uspreventiveservicestaskforce.org/Page/Document/draft-research-plan/hepatitis-c-screening1).

Third, viral gene products provide potential targets for therapeutic drugs or therapeutic vaccines for the treatment cancers, precancerous lesions, or chronic infection. There has been substantial research activity in this potentially fruitful area. Although they have not led to viral-based treatment of malignancies, drug studies have had considerable success in the development of antivirals to treat chronic HBV and HCV infection. Pegylated interferon alpha plus a nucleoside/nucleotide analog is currently being used to suppress HBV replication (Chap. 5) and thereby liver cirrhosis and risk of hepatocellular carcinoma (Ren and

Huang 2019). HCV infection can be similarly treated, and a series of direct-acting antiviral drugs targeting the NS3/4A protease and NS5A and NS5B polymerase have been developed that, remarkably, increase sustained virologic response rates to 90-98% (Pradat et al. 2018) (Chap. 7). Treatment of KSHV infection/Kaposi sarcoma centers on reducing the underlying immunosuppression that promotes the disease. Kaposi's sarcoma lesions often regress in HIV-infected individuals after initiation of HAART, but this response is due to reconstitution of the immune system, rather than to direct activity of the drugs against KSHV (Cesarman et al. 2019). No licensed direct-acting antivirals have been developed for HPVs or MCV, despite considerable efforts in the case of HPVs.

Fourth, the knowledge of a viral etiology can serve as the basis of cancer prevention measures. One approach involves behavioral interventions to reduce susceptibility to infection, e.g., limiting exposure to blood products in the case HBV and HCV (Chaps. 5 and 7), limiting number of sexual partners in case of HPV or KSHV, or preventing HTLV-1 transmission by discouraging breast-feeding by infected mothers (Ruff 1994) (Chap. 12).

Alternatively, the identification of human oncoviruses can be used to develop effective vaccines to prevent oncovirus infection. This approach has been successfully implemented for HBV and HPV. HBV prophylactic vaccines were introduced more than thirty years ago (Chap. 5). A dramatic reduction in childhood liver cancers of greater than two-thirds has been documented in Taiwan, a previously high incidence region (Chang et al. 2016). A substantial reduction in adult liver cancer is expected in the near future, as individuals who would have otherwise contracted HBV as infant reach the age of peak cancer incidence. HPV vaccines targeting HPV16 and 18 have been licensed for more than a decade (Schiller and Lowy 2012) (Chap. 9). While substantial reductions in the incidences of HPV-associated cancer are expected in coming years, there has already been a significant reduction in premalignant cervical disease and evidence of herd immunity developing in countries with high vaccination rates (de Sanjose et al. 2019).

There are considerable efforts underway to develop prophylactic and/or therapeutic vaccines against EBV (van Zyl et al. 2019) and HCV (Bailey et al. 2019) (Chap. 7). Commercial development of prophylactic vaccines against these viruses seems possible because there are reasonable non-malignant disease endpoints for initial licensure, mononucleosis in the case of EBV, and liver cirrhosis in the case of HCV. However, specific characteristics of their biology have made the development of effective vaccines challenging. These include multiple entry receptors and viral latency in the case of EBV, and genetic instability in the case of HCV. There has been less effort devoted to developing KSHV and HTLV-1 vaccines because the numbers of worldwide cancers they induce are lower, and these viruses do not appear to be a frequent cause of medically important non-malignant disease, in contrast to EBV and HCV (Schiller and Lowy 2010). They also have proven susceptible to other interventions, specifically reduction in KSHV-induced Kaposi's sarcoma by treating HIV infection and reduction in transmission of HTLV-1 by discouraging infant breast-feeding by infected mothers (Chap. 12). There has been relatively little activity in developing MCV vaccines, perhaps because the natural history of infection is poorly understood, there are no known premalignant lesions or other disease to target, and the cancers are relatively rare.

4.3 The Search for Additional Oncoviruses

Are there other human oncoviruses waiting to be discovered? There are suggestions that viral infections may be associated with several other cancers. For instance, the incidence of non-melanoma skin cancer increases dramatically after immunosuppression, and immunosuppression is associated with clear increases in established virally associated cancers (Grulich and Vajdic 2015). Other cancers that increase less dramatically in immunosuppressed patients and have not been firmly linked to a microbial infection include lung, conjunctival, melanoma, lip, esophageal, laryngeal carcinomas and multiple myeloma. Some epidemiological studies have linked the risk of prostate cancer with sexual activity variables, suggesting involvement of a sexually transmitted infectious agent (Sutcliffe 2010), but its incidence is not increased in immunocompromized individuals (Grulich and Vajdic 2015). Other virus/cancer associations that warrant further investigations include hepatitis delta virus, an HBV-dependent single-strand RNA virus that appears to increase the risk of HCC in HBV-coinfected individuals by threefold (Koh et al. 2019), and BKV in bladder cancer, particularly in immunocompromised patients (Starrett and Buck 2019).

The technologies of high throughput nucleic acid sequencing of entire cellular genomes and transcriptomes, as now applied to a wide variety of human tumors, provide an unprecedented wealth of raw data for the hunt for novel oncoviruses. The discovery of MCV illustrates how this technology can be employed to identify novel human oncoviruses (Feng et al. 2008). However, recent studies that screened for all known viral species have mostly detected established oncoviruses in the tumor collections (Cantalupo et al. 2018). Nevertheless, the possibility remains that highly divergent oncoviruses with undetectable homology to currently known viruses remain to be discovered. However, identification of a viral nucleic acid sequence in a tumor is only the first step. It takes many additional laboratory, clinical, and epidemiological studies to establish that a viral infection is a causal agent in the development of a cancer, as opposed to a passive parasite or simple contaminant of the tumor. Establishment of causality can be particularly difficult in situations where the implicated virus is a common infection in the general population.

It will also be difficult to establish causality if a virus is involved in the initiation of a tumor but not in its maintenance, with the viral genome being lost during progression. Commonly referred to as a "hit and run," it is a plausible but as yet unproven mechanism for human cancers (Schiller and Buck 2011), although there are several well-documented examples of this phenomenon in experimental animal models (Viarisio et al. 2018). Further supporting the possibility of this mechanism is the recent finding that 8% of cervical cancers that are positive for HPV DNA do not express detectable HPV transcripts, functionally equivalent to a hit-and-run (Banister et al. 2017). It may be informative to more closely examine premalignant lesions and cancers in immunocompromised patients, where there may be less selection for eliminating viral gene expression, for further evidence of viruses that may initiate cancers via this mechanism (Starrett and Buck 2019).

5 Conclusions

The discovery and characterization of oncoviruses have been at the forefront of biomedical research over the last several decades. It has provided important insights into basic cell biology and mechanisms of carcinogenesis. In addition, these studies have generated important public health interventions that have the potential to prevent a large number of human cancers. This monograph provides clear evidence that oncoviruses remain a dynamic subject in biomedical science. We expect that future research will generate new and excites insights into the genesis of cancer and effective ways to prevent and treat it.

References

Arora R, Chang Y, Moore PS (2012) MCV and Merkel cell carcinoma: a molecular success story. Curr Opin Virol 2(4):489–498

Bae JM, Kim EH (2016) Epstein-Barr virus and gastric cancer risk: a meta-analysis with meta-regression of case-control studies. J Prev Med Public Health 49(2):97–107

Bailey JR, Barnes E, Cox AL (2019) Approaches, progress, and challenges to hepatitis C vaccine development. Gastroenterology 156(2):418–430

Banister CE, Liu C, Pirisi L, Creek KE, Buckhaults PJ (2017) Identification and characterization of HPV-independent cervical cancers. Oncotarget 8(8):13375–13386

Bosch FX, Lorincz A, Munoz N, Meijer CJ, Shah KV (2002) The causal relation between human papillomavirus and cervical cancer. J Clin Pathol 55(4):244–265

Cantalupo PG, Katz JP, Pipas JM (2018) Viral sequences in human cancer. Virology 513:208–216

Cesarman E, Damania B, Krown SE, Martin J, Bower M, Whitby D (2019) Kaposi sarcoma. Nat Rev Dis Primers 5(1):9

Chang MH, You SL, Chen CJ, Liu CJ, Lai MW, Wu TC, Wu SF, Lee CM, Yang SS, Chu HC, Wang TE, Chen BW, Chuang WL, Soon MS, Lin CY, Chiou ST, Kuo HS, Chen DS, G. Taiwan Hepatoma Study (2016) Long-term effects of hepatitis B immunization of infants in preventing liver cancer. Gastroenterology 151(3):472–480, e471

Chang Y, Cesarman E, Pessin MS, Lee F, Culpepper J, Knowles DM, Moore PS (1994) Identification of herpesvirus-like DNA sequences in AIDS-associated Kaposi's sarcoma. Science 266(5192):1865–1869

Chen YP, Chan ATC, Le QT, Blanchard P, Sun Y, Ma J (2019) Nasopharyngeal carcinoma. Lancet 394(10192):64–80

Choo QL, Kuo G, Weiner AJ, Overby LR, Bradley DW, Houghton M (1989) Isolation of a cDNA clone derived from a blood-borne non-A, non-B viral hepatitis genome. Science 244 (4902):359–362

Dane DS, Cameron CH, Briggs M (1970) Virus-like particles in serum of patients with Australia-antigen-associated hepatitis. Lancet 1(7649):695–698 de Martel C, Ferlay J, Franceschi S, Vignat J, Bray F, Forman D, Plummer M (2012) Global burden of cancers attributable to infections in 2008: a review and synthetic analysis. Lancet Oncol

de Martel C, Plummer M, Vignat J, Franceschi S (2017) Worldwide burden of cancer attributable to HPV by site, country and HPV type. Int J Cancer 141(4):664–670

de Sanjose S, Brotons M, LaMontagne DS, Bruni L (2019) Human papillomavirus vaccine disease impact beyond expectations. Curr Opin Virol 39:16–22

Dürst M, Gissmann L, Ikenberg H, zur Hausen H (1983) A papillomavirus DNA from a cervical carcinoma and its prevalence in cancer biopsy samples from different geographic regions. Proc Natl Acad Sci USA 80:3812–3815

El-Serag HB (2012) Epidemiology of viral hepatitis and hepatocellular carcinoma. Gastroenterology 142(6):1264–1273, e1261

Epstein MA, Achong BG, Barr YM (1964) Virus particles in cultured lymphoblasts from Burkitt's lymphoma. Lancet 1(7335):702–703

Fan H, Johnson C (2011) Insertional oncogenesis by non-acute retroviruses: implications for gene therapy. Viruses 3(4):398–422

Feng H, Shuda M, Chang Y, Moore PS (2008) Clonal integration of a polyomavirus in human Merkel cell carcinoma. Science 319(5866):1096–1100

Grulich AE, Vajdic CM (2015) The epidemiology of cancers in human immunodeficiency virus infection and after organ transplantation. Semin Oncol 42(2):247–257

Hill A (1971) Statistical evidence and inference. In: Principles of medical statistics, 9th edn. Oxford University Press, New York, 309–323

Kanda T, Goto T, Hirotsu Y, Moriyama M, Omata M (2019a) Molecular mechanisms driving progression of liver cirrhosis towards hepatocellular carcinoma in chronic hepatitis B and C infections: a review. Int J Mol Sci 20(6):1358

Kanda T, Yajima M, Ikuta K (2019b) Epstein-Barr virus strain variation and cancer. Cancer Sci 110(4):1132–1139

Katano H (2018) Pathological features of Kaposi's sarcoma-associated herpesvirus infection. Adv Exp Med Biol 1045:357–376

Kervarrec T, Samimi M, Guyetant S, Sarma B, Cheret J, Blanchard E, Berthon P, Schrama D, Houben R, Touze A (2019) Histogenesis of Merkel cell carcinoma: a comprehensive review. Front Oncol 9:451

Koh C, Heller T, Glenn JS (2019) Pathogenesis of and new therapies for hepatitis D. Gastroenterology 156(2), 461–476, e461

Mesri EA, Cesarman E, Boshoff C (2010) Kaposi's sarcoma and its associated herpesvirus. Nat Rev Cancer 10(10):707–719

Mesri EA, Feitelson MA, Munger K (2014) Human viral oncogenesis: a cancer hallmarks analysis. Cell Host Microbe 15(3):266–282

Mirabello L, Yeager M, Cullen M, Boland JF, Chen Z, Wentzensen N, Zhang X, Yu K, Yang Q, Mitchell J, Roberson D, Bass S, Xiao Y, Burdett L, Raine-Bennett T, Lorey T, Castle PE, Burk RD, Schiffman M (2016) HPV16 Sublineage associations with histology-specific cancer risk using HPV whole-genome sequences in 3200 women. J Natl Cancer Inst 108(9):djw100

Moore PS, Chang Y (1998) Kaposi's sarcoma (KS), KS-associated herpesvirus, and the criteria for causality in the age of molecular biology. Am J Epidemiol 147(3):217–221

Moormann AM, Bailey JA (2016) Malaria—how this parasitic infection aids and abets EBV-associated Burkitt lymphomagenesis. Curr Opin Virol 20:78–84

Petrick JL, McGlynn KA (2019) The changing epidemiology of primary liver cancer. Curr Epidemiol Rep 6(2):104–111

Pipas JM (2019) DNA tumor viruses and their contributions to molecular biology. J Virol **93**(9)

Plummer M, de Martel C, Vignat J, Ferlay J, Bray F, Franceschi S (2016) Global burden of cancers attributable to infections in 2012: a synthetic analysis. Lancet Glob Health 4(9):e609–e616

Poiesz BJ, Ruscetti FW, Mier JW, Woods AM, Gallo RC (1980) T-cell lines established from human T-lymphocytic neoplasias by direct response to T-cell growth factor. Proc Natl Acad Sci U S A 77(11):6815–6819

Pradat P, Virlogeux V, Trepo E (2018) Epidemiology and elimination of HCV-related liver disease. Viruses **10**(10)

Ren H, Huang Y (2019) Effects of pegylated interferon-alpha based therapies on functional cure and the risk of hepatocellular carcinoma development in patients with chronic hepatitis B. J Viral Hepat 26(Suppl 1):5–31

Rizzo AE, Feldman S (2018) Update on primary HPV screening for cervical cancer prevention. Curr Probl Cancer 42(5):507–520

Romano G, Marino IR, Pentimalli F, Adamo V, Giordano A (2009) Insertional mutagenesis and development of malignancies induced by integrating gene delivery systems: implications for the design of safer gene-based interventions in patients. Drug News Perspect 22(4):185–196

Ruff AJ (1994) Breastmilk, breastfeeding, and transmission of viruses to the neonate. Semin Perinatol 18(6):510–516

Russell SJ, Barber GN (2018) Oncolytic viruses as antigen-agnostic cancer vaccines. Cancer Cell 33(4):599–605

Saha A, Robertson ES (2019) Mechanisms of B-cell oncogenesis induced by Epstein-Barr virus. J Virol 93(13)

Schiller JT, Buck CB (2011) Cutaneous squamous cell carcinoma: a smoking gun but still no suspects. J Invest Dermatol 131(8):1595–1596

Schiller JT, Lowy DR (2010) Vaccines to prevent infections by oncoviruses. Annu Rev Microbiol 64:23–41

Schiller JT, Lowy DR (2012) Understanding and learning from the success of prophylactic human papillomavirus vaccines. Nat Rev Microbiol 10(10):681–692

Smith BD, Jorgensen C, Zibbell JE, Beckett GA (2012) Centers for disease control and prevention initiatives to prevent hepatitis C virus infection: a selective update. Clin Infect Dis 55(Suppl 1): S49–S53

Starrett GJ, Buck CB (2019) The case for BK polyomavirus as a cause of bladder cancer. Curr Opin Virol 39:8–15

Sutcliffe S (2010) Sexually transmitted infections and risk of prostate cancer: review of historical and emerging hypotheses. Future Oncol 6(8):1289–1311

Tagaya Y, Matsuoka M, Gallo R (2019). 40 years of the human T-cell leukemia virus: past, present, and future. F1000Research 8

Tetzlaff MT, Nagarajan P (2018) Update on Merkel cell carcinoma. Head Neck Pathol 12(1): 31–43

van Zyl DG, Mautner J, Delecluse HJ (2019) Progress in EBV vaccines. Front Oncol 9:104

Vangipuram R, Tyring SK (2019) AIDS-associated malignancies. Cancer Treat Res 177:1–21

Viarisio D, Muller-Decker K, Accardi R, Robitaille A, Durst M, Beer K, Jansen L, Flechten-macher C, Bozza M, Harbottle R, Voegele C, Ardin M, Zavadil J, Caldeira S, Gissmann L, Tommasino M (2018) Beta HPV38 oncoproteins act with a hit-and-run mechanism in ultraviolet radiation-induced skin carcinogenesis in mice. PLoS Pathog 14(1):e1006783

Vinokurova S, Wentzensen N, Kraus I, Klaes R, Driesch C, Melsheimer P, Kisseljov F, Durst M, Schneider A, von Knebel Doeberitz M (2008) Type-dependent integration frequency of human papillomavirus genomes in cervical lesions. Cancer Res 68(1):307–313

Wang M, Gu B, Chen X, Wang Y, Li P, Wang K (2019) The function and therapeutic potential of Epstein-Barr virus-encoded microRNAs in cancer. Mol Ther Nucleic Acids 17:657–668

Epidemiology of Virus Infection and Human Cancer

Chien-Jen Chen, San-Lin You, Wan-Lun Hsu, Hwai-I Yang, Mei-Hsuan Lee, Hui-Chi Chen, Yun-Yuan Chen, Jessica Liu, Hui-Han Hu, Yu-Ju Lin, Yu-Ju Chu, Yen-Tsung Huang, Chun-Ju Chiang, and Yin-Chu Chien

1 Introduction

The International Agency for Research on Cancer (IARC) has comprehensively assessed the carcinogenicity of biological agents to humans based on epidemiological and mechanistic evidence (IARC 2012a; Chen et al. 2014a). Seven viruses including the Epstein–Barr virus (EBV), hepatitis B virus (HBV), hepatitis C virus (HCV), Kaposi's sarcoma herpes virus (KSHV), human immunodeficiency virus, type-1 (HIV-1), human T cell lymphotropic virus, type-1 (HTLV-1), and several types of human papillomavirus (HPV) have been classified as Group 1 human carcinogens as shown in Table 1.

C.-J. Chen (✉) · W.-L. Hsu · Hwai-IYang · H.-C. Chen · Y.-C. Chien
Genomics Research Center, Academia Sinica, 128 Academia Road, Sect. 2, Taipei 115, Taiwan
e-mail: chencj@gate.sinica.edu.tw

S.-L. You
School of Medicine and Big Data Research Centre, Fu-Jen Catholic University, New Taipei, Taiwan

M.-H. Lee · Y.-J. Lin
Institute of Clinical Medicine, National Yang-Ming University, Taipei, Taiwan

Y.-Y. Chen
Taiwan Blood Services Foundation, Taipei, Taiwan

J. Liu
Department of Pediatrics, Perinatal Epidemiology and Health Outcomes Research Unit, Stanford University School of Medicine and Lucile Packard Children's Hospital, Palo Alto, CA, USA

© Springer Nature Switzerland AG 2021
T.-C. Wu et al. (eds.), *Viruses and Human Cancer*, Recent Results in Cancer Research 217, https://doi.org/10.1007/978-3-030-57362-1_2

13

There is sufficient evidence to conclude that EBV causes nasopharyngeal carcinoma, Burkitt's lymphoma, immune suppression-related non-Hodgkin lymphoma, extranodal NK/T cell lymphoma (nasal type), and Hodgkin's lymphoma in humans. The evidence for EBV-caused gastric carcinoma and lymphoepithelioma-like carcinoma is limited. HBV and HCV cause hepatocellular carcinoma with sufficient evidence. The evidence for HCV-caused non-Hodgkin lymphoma, especially B-cell lymphoma, is sufficient, while the evidence for HBV-caused non-Hodgkin lymphoma is limited. There is also limited evidence to conclude that HBV and HCV cause cholangiocarcinoma. The evidence to conclude that HIV-1 causes Kaposi's sarcoma, non-Hodgkin lymphoma, Hodgkin's lymphoma, and cancers of the cervix, anus, and conjunctiva is sufficient. But the evidence for HIV-1 to cause cancers of the vulva, vagina, penis, non-melanoma skin cancer, and hepatocellular carcinoma is limited.

There is sufficient evidence to conclude that HPV-16 causes cancers of the cervix, vulva, vagina, penis, anus, oral cavity, oropharynx, and tonsil; but the evidence for HPV-16 to cause cancer of the larynx is limited. Cervical cancer is caused by several other types of HPV including HPV-18, 31, 33, 35, 39, 45, 51, 52, 56, 58, and 59. The evidence for HPV-26, 30, 34, 53, 66, 67, 68, 69, 70, 73, 82, 85, and 97 to cause cervical cancer is limited. HTLV-1 causes adult T cell leukemia and lymphoma with sufficient evidence. There is sufficient evidence to conclude KSHV causes Kaposi's sarcoma and primary effusion lymphoma, but the evidence for KSHV to cause multicentric Castleman's disease is limited. Based on limited evidence in humans, inadequate evidence in experimental animals, and strong mechanistic evidence in humans, Merkel-cell virus (MCV) was classified as probably carcinogenic to humans by IARC (2012b).

J. Liu
California Perinatal Quality Care Collaborative, Palo Alto, CA, USA

H.-H. Hu
Department of Translational Science, Preclinical Research, PharmaEngine Inc.,
Taipei, Taiwan

Y.-J. Chu
Department of Environmental and Occupational Health Sciences,
University of Washington, Seattle, WA, USA

Y.-T. Huang
Institute of Statistical Sciences, Academia Sinica, Taipei, Taiwan

C.-J. Chiang
Graduate Institute of Epidemiology and Preventive Medicine,
National Taiwan University, Taipei, Taiwan

Table 1 Cancers caused by Group 1 oncogenic viruses with sufficient and limited evidence according to the IARC criteria

Virus	Cancer sites with sufficient evidence	Cancer sites with limited evidence
Epstein–Barr virus (EBV)	Nasopharyngeal carcinoma, Burkitt's lymphoma, immune suppression-related non-Hodgkin lymphoma, extranodal NK/T cell lymphoma (nasal type), Hodgkin's lymphoma	Gastric carcinoma, lympho-epithelioma-like carcinoma
Hepatitis B virus (HBV)	Hepatocellular carcinoma	Cholangiocarcinoma, non-Hodgkin lymphoma
Hepatitis C virus (HCV)	Hepatocellular carcinoma, non-Hodgkin lymphoma	Cholangiocarcinoma
Human immunodeficiency virus, type 1 (HIV-1)	Kaposi's sarcoma, non-Hodgkin lymphoma, Hodgkin's lymphoma, cancers of the cervix, anus, and conjunctiva	Cancers of the vulva, vagina and penis, non-melanoma skin cancer, hepatocellular carcinoma
Human papillomavirus type 16 (HPV-16)	Cancers of the cervix, vulva, vagina, penis, anus, oral cavity, oropharynx, and tonsil	Cancer of the larynx
Human papillomavirus type 18, 31, 33,35, 39, 45, 51, 52, 56, 58, 59 (HPV-18, 31, 33,35, 39, 45, 51, 52, 56, 58, and 59)	Cancer of the cervix	
Human papillomavirus type 26, 30, 34, 53, 66, 67, 68, 69, 70, 73, 82, 85, 97 (HPV-26, 30, 34, 53, 66, 67, 68, 69, 70, 73, 82, 85, and 97)		Cancer of the cervix
Human T cell lymphotrophic virus, type-1 (HTLV-1)	Adult T cell leukemia and lymphoma	
Kaposi's sarcoma herpes virus (KSHV)	Kaposi's sarcoma, primary effusion lymphoma	Multicentric Castleman's disease

The proportion of cancers caused by infectious agents was recently estimated to be more than 20% with wide geographical variations (IARC 2012a). The identification of new cancer sites attributed to these infectious agents means that more cancers are potentially preventable. This chapter will mainly review the epidemiology of oncogenic viruses and their associated cancers.

2 Transmission Routes and Infection Prevalence of Oncogenic Viruses in the World

The transmission routes and global variation in infection prevalence of oncogenic viruses are shown in Table 2 and Fig. 1. EBV is one of the most common viruses in humans, and is highly prevalent throughout the world. Even in remote populations, more than 90% of adults are infected with EBV (IARC 2012a). It is estimated that over 5.5 billion people worldwide are infected with EBV. The virus is primarily transmitted through bodily fluids, particularly saliva, and the age at which primary infection occurs varies significantly. For example, individuals living in over-crowded conditions with poor sanitation are first infected at a younger age than those individuals living in better environments. Two major types of EBV have been

Table 2 Transmission routes and global variations in infection prevalence of oncogenic viruses

Virus	Transmission routes	Areas of highest and lowest prevalence
EBV	Bodily fluids, especially saliva	Highly prevalent throughout the world
HBV	Highly endemic areas: perinatal or child-to-child Low endemic areas: occurs in adulthood through injection drug use, among male homosexuals, though medical procedures and transfusions or hemodialysis	Highest: sub-saharan Africa, Amazon basin, China, Korea, Taiwan, and parts of Southeast Asia Lowest: North and Central America, Australia
HCV	Injection drug use Iatrogenic exposure Less common: perinatal or sexual transmission	Highest: Egypt, China, Mongolia, northern Africa, Pakistan, southern Italy, parts of Japan Lowest: all other areas
HIV-1	Sexual activity Blood contact Mother to child	Highest: Sub-saharan Africa, the Caribbean, Eastern Europe, Central Asia Lowest: Western Europe, parts of East Asia, Australia, Canada, parts of Central America
HPV	Skin-to-skin or skin-to-mucosa contact Sexual activity Less common: perinatal or iatrogenic transmission	Highest: Caribbean, South America, eastern Europe, eastern and western Africa, Australia, Indonesia, and Mongolia Lowest: western Europe, western Asia, USA
HTLV-1	Mother-to-child (such as breastfeeding) Sexual activity Parenteral transmission (such as transfusions)	Highest: Southwest Japan, Sub-Saharan Africa, the Caribbean, South Africa Lowest: East Asia
KSHV	Saliva Less common: prolonged injection drug use, transfusions, transplantation	Highest: Sub-Saharan Africa, Mediterranean region Lowest: Northern Europe, North America, Asia

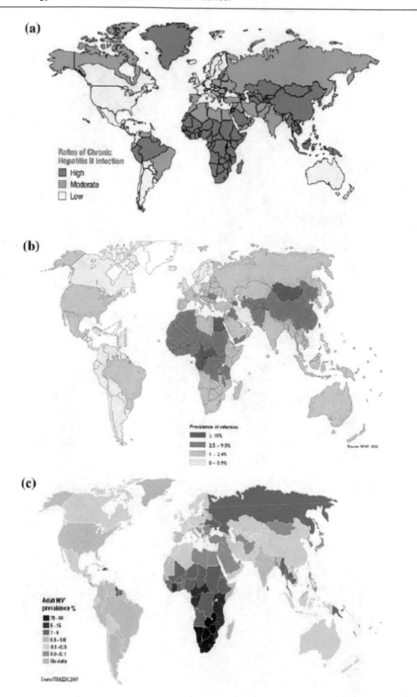

Fig. 1 Estimated prevalence (per 100) of Group 1 oncogenic viruses in the world. **a** HBV, **b** HCV, **c** HIV-1, **d** HPV, **e** HTLV-1, and **f** KSHV

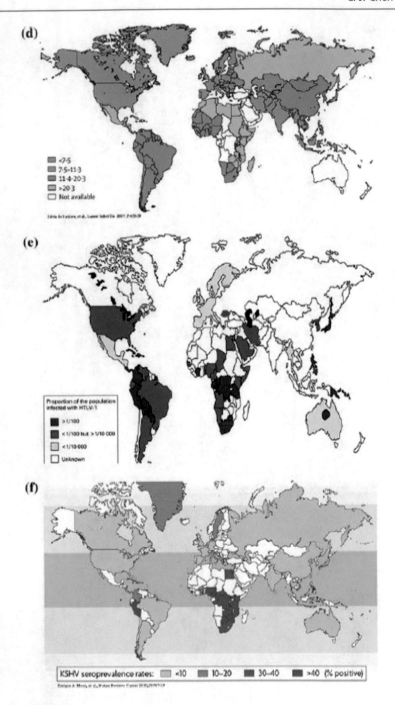

(d)

<7·5
7·5–11·3
11·4–20·3
>20·3
Not available

(e)

Proportion of the population
infected with HTLV-1

>1/100
<1/100 but > 1/10 000
<1/10 000
Unknown

(f)

KSHV seroprevalence rates: <10   10–20   30–40   >40 (% positive)

Fig. 1 (continued)

identified and differ in geographical distribution with EBV-2 being more common in Africa and homosexual men. The role of specific EBV types in the development of different cancers remains unclear. As EBV infection is ubiquitous, the specific geographical distribution of EBV-related malignancies including endemic Burkitt's lymphoma and nasopharyngeal carcinoma is more likely attributable to the variation in the distributions of other cofactors, which may activate EBV replication.

HBV infects more than 2.0 billion people worldwide, with more than 350 million individuals chronically infected (IARC 2012a). There is a wide variation of chronic HBV infection in the world[1] as shown in Table 2 and Fig. 1a. Approximately 45, 43, and 12% of the world population live in areas where the endemicity of chronic HBV infection is high (seroprevalence of hepatitis B surface antigen >8%), medium (2–7%), and low (<2%). The prevalence is highest in sub-Saharan Africa, the Amazon Basin, China, Korea, Taiwan, and several countries in Southeast Asia. In areas of high endemicity, the lifetime risk of HBV infection is more than 60%, with most infections acquired from perinatal and child-to-child transmission, when the risk of chronic infection is greatest. Perinatal (vertical) transmission is predominant in China, Korea, and Taiwan where the seroprevalence of HBeAg in pregnant women is high, while child-to-child (horizontal) transmission is common in sub-Saharan Africa where HBeAg seroprevalence is low in mothers. In areas of medium endemicity, mixed HBV transmission patterns occur in infancy, early childhood, adolescence, and adulthood. In areas of low endemicity, most HBV infections occur in adolescents and young adults through injection drug use, male homosexuality, health care practice, and regular transfusions or hemodialysis.

In addition to the striking geographical variation in seroprevalence of HBsAg in the world, the distribution of the eight genotypes of HBV also varies significantly in different countries (IARC 2012a). Genotype A is prevalent in Europe, Africa, and North America; genotypes B and C are prevalent in East and Southeast Asia; Genotype D is predominant in South Asia, the Middle East, and Mediterranean areas; genotype E is limited to West Africa; genotypes F and G are found in Central and South America; and genotype H is observed in Central America.

HCV infects around 150 million people in the world showing an estimated prevalence of 2.2% (IARC 2012a), with a wide variation in different regions as shown in Table 2 and Fig. 1b. The estimated prevalence of HCV infection (seroprevalence of antibodies against HCV) ranges from <0.1% in the United Kingdom and Scandinavia to 15–20% in Egypt (Alter 2007). A high prevalence of HCV infection is also observed in Mongolia, northern Africa, Pakistan, China, southern Italy, and parts of Japan. At least six major genotypes of HCV have been identified, the geographical distributions of which also vary widely. The response to ribavirin and interferon therapy is better in patients infected with genotypes 2 or 3 than in those infected with genotypes 1 or 4; while the response to direct acting agent therapy is equally good for all genotypes.

---

[1]CDC   http://wwwnc.cdc.gov/travel/yellowbook/2012/chapter-3-infectious-diseases-related-to-travel/hepatitis-b.htm.

Two major transmission routes for HCV have been identified. They are injection drug use and iatrogenic exposures through transfusion, transplantation, and unsafe therapeutic injection. While there has been a large reduction in iatrogenic transmission of HCV since 1990 in developed countries such as Taiwan, Japan and Italy, it continues to be a common source of transmission in low-resource countries where disposable needles tend to be reused. In developed countries, however, injection drug use is the most important transmission route for newly acquired HCV infection. Transmission of HCV through perinatal, sexual, and accidental needle-stick exposures is less efficient than iatrogenic exposure and injection drug use (Lee et al. 2011).

HIV infects an estimated 34 million people worldwide (IARC 2012a; UNAIDS[2] 2012). An estimated 0.8% of all adults aged 15–49 years worldwide are living with HIV, and the burden varies considerably between countries and regions as shown in Table 2 and Fig. 1c. Sub-Saharan Africa remains the most disproportionally affected, with a prevalence of 4.9%. Although the prevalence of HIV infection is nearly 25 times higher in sub-Saharan Africa than it is in Asia, there are still 5 million people in South, Southeast, and East Asia living with HIV infection. After sub-Saharan Africa, other regions heavily affected by HIV are the Caribbean, Eastern Europe and Central Asia, where 1.0% of adults were living with HIV in 2011. In 2011 alone, 2.5 million people, including 0.39 million children, were newly infected with HIV. Since 2001, the annual incidence of HIV infection has fallen in 33 countries, 22 of them in sub-Saharan Africa. However, the incidence of HIV infection is once again rising in Eastern Europe and Central Asia, and new infections are also on the rise in the Middle East and North Africa.

HIV infection is primarily transmitted through three major routes: sexual intercourse, blood contact, and mother-to-child transmission as shown in Table 2. The HIV infectivity is determined by the interaction between agent, host, and environmental factors. The probability of HIV transmission is highest for blood transfusions, followed by mother-to-child transmission, needle sharing, man-to-man sexual transmission, and is the lowest for heterosexual sexual transmission.

HPV infection is highly prevalent throughout the world, and most sexually active individuals will acquire at least one genotype of anogenital HPV infection during their lifetime (IARC 2012a). In a meta-analysis of 157,879 women with normal cytology, the estimated oncogenic HPV DNA point prevalence was reported to be as high as 10%, resulting in an estimate of 600 million infected worldwide (de Sanjose et al. 2007). The point prevalence of HPV infection among women with normal cervical cytology was highest in Caribbean, South America, eastern Europe, eastern and western Africa, Australia, Indonesia, and Mongolia; and lowest in western Europe, western Asia, and the USA demonstrating a large geographical variation as shown in Table 2 and Fig. 1d (Bruni et al. 2019). The estimated point prevalences are highly dynamic, as both incidence and clearance rates are high.

Among 13 known oncogenic HPV types, the most prevalent types include 16, 18, 31, 33, 35, 45, 52, and 58. HPV type 16 is the most common type across all regions, with prevalence ranging from 2.3 to 3.5%. HPV infections are transmitted

---

[2]UNAIDS http://data.unaids.org/pub/epislides/2012/2012_epiupdate_en.pdf.

through direct skin-to-skin or skin-to-mucosa contact. Anogenital HPV types spread mainly through sexual transmission through any type of sexual intercourse in teenagers and young adults. Non-sexual routes, including perinatal and iatrogenic transmissions, account for a minority of HPV infections. The prevalence (%) of HPV type 16 and 18 was increased with severity of cervical lesions. The update prevalence from worldwide metanalysis for women with normal, low-grade lesions (LSIL/CIN-1), high-grad lesions (HSIL/CIN-2/CIN3-/CIS), and cervical cancer was 4.1, 25.8, 51.9 and 69.4%, respectively (Bruni et al. 2019).

Globally, HTLV-1 infects an estimated 15–20 million people (IARC 2012a). HTLV-1 infection is characterized by micro-epidemic hotspots surrounded by low prevalence areas as shown in Table 2 and Fig. 1e (Proietti et al. 2005). The prevalence of HTLV-1 infection ranges from <0.1% in China, Korea, and Taiwan to 20% in Kyushu and Okinawa of Japan. Regions of high endemicity include southwestern Japan, parts of sub-Saharan Africa, the Caribbean Islands, and South Africa. HTLV-1 is primarily transmitted through vertical transmission, sexual transmission, and parenteral transmission. While vertical transmission through breast feeding has a high probability of resulting in mother-to-child infection, in utero infectivity is low due to limited trafficking of HTLV-1-infected lymphocytes across the placenta. The efficiency of sexual transmission of HTLV-1 depends on the proviral load and use of a condom. There has been a significant reduction in parenteral transmission through transfusions due to the sensitive serological examination of blood products. Needle sharing associated with injection drug use is another parenteral route for HTLV-1 transmission.

The prevalence of KSHV has also been shown to have wide geographical variations as shown in Table 2 and Fig. 2f (Dukers and Rezza 2003). Seroprevalence ranges from 2 to 3% in northern Europe to 82% in the Congo (IARC 2012a). Prevalence is generally low (<10%) in northern Europe, the USA, and Asia, elevated in the Mediterranean region (10–30%), and high in sub-Saharan Africa (>50%). KSHV is primarily transmitted via saliva. In countries where KSHV prevalence is high, infection occurs during childhood and increases with age. KSHV may also be transmitted with low efficiency through prolonged injection drug use, blood transfusions, and organ transplantation.

3   Global Variation in the Incidence of Virus-Caused Cancers

Global variations in age-adjusted incidence rates of oncogenic virus-related cancers are shown in Fig. 2 (Ferlay et al. 2018). The age-adjusted incidence rates of nasopharyngeal cancer range from <0.1 in low endemic regions to 9.9 per 100,000 in high endemic areas as shown in Fig. 2a. The highest incidence is seen in southern China, Southeast Asia, and sub-Saharan Africa, and the lowest incidence is seen in Europe, western Africa, and Central America. Interestingly, in different cancer registries throughout the world, individuals of Chinese ethnicity have the

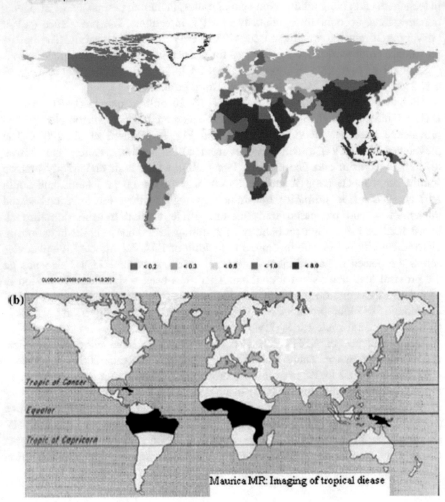

Fig. 2 Age-standardized incidence rate (per 100,000) of virus-caused cancers in the world.
a Nasopharynx, **b** Burkitt lymphoma **c** Liver, **d** Cervix uteri, and **e** Kaposi's sarcoma

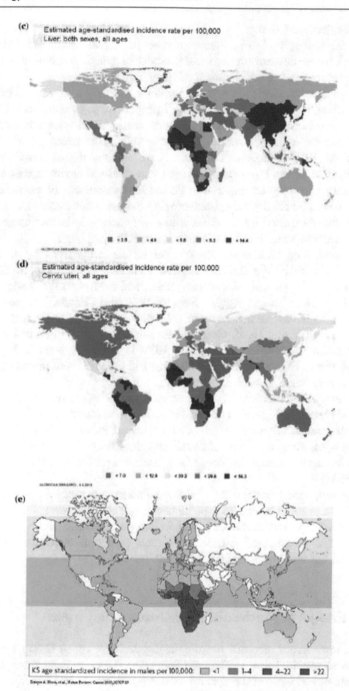

(c) Estimated age-standardised incidence rate per 100,000
Liver: both sexes, all ages

(d) Estimated age-standardised incidence rate per 100,000
Cervix uteri, all ages

(e)

KS age standardized incidence in males per 100,000: ☐ <1 ■ 1–4 ■ 4–22 ■ >22

Fig. 2 (continued)

highest incidence of nasopharyngeal cancer. As EBV infection is ubiquitous in humans, the uniquely high incidence of nasopharyngeal carcinoma among individuals of Chinese descent suggests that lifestyles or genetic susceptibility may play an important role in the development of nasopharyngeal cancer.

The age-adjusted incidence rates of Burkitt's lymphoma are shown in Fig. 2b. Central Africa, equatorial South America, Papua New Guinea, and Caribbean countries are endemic for Burkitt's lymphoma with an incidence rate of 5–35 per 100,000, but the incidence rate of Burkitt lymphoma is relatively low in other countries. As EBV infection is ubiquitous in humans, the extraordinarily high endemicity of Burkitt's lymphoma in Africa suggests local environments or genetic susceptibility may play an important role in the development of endemic Burkitt lymphoma. Despite the limited epidemiological evidence in humans, IARC (2012b) classified malaria caused by infection with *p. falciparum* in holoendemic areas as "probably carcinogenic to humans" to cause Burkitt lymphoma.

The age-adjusted incidence rates of liver cancer range from 1.1 to 93.7 per 100,000 as shown in Fig. 2c. The highest incidence is observed in East Asia, Southeast Asia, Egypt, and sub-Saharan Africa, and the lowest incidence is seen in Europe, the Middle East, Australia, New Zealand, and Canada. The geographical variation in liver cancer incidence is consistent with that of the varying sero-prevalence of HBV and HCV. However, the etiological proportion of liver cancer varies in different countries and regions. HCV is more prevalent in liver cancer patients in Japan, North America and Europe; and HBV is more prevalent in those in Taiwan, Southeast Asia, and China (Chen 2018).

The age-adjusted incidence rates of cervical cancer range from 1.2 to 75.3 per 100,000 as shown in Fig. 2d. The highest incidence is observed in Latin America, South Asia, and sub-Saharan Africa, and the lowest incidence is seen in Europe, North America, Australia, New Zealand, and the Middle East. The geographical variation in cervical cancer incidence is consistent with the varying prevalence of oncogenic HPV.

The age-adjusted incidence rates of Kaposi's sarcoma range from <1.0 to 17.4 per 100,000 as shown in Fig. 2e. The highest incidence is observed in sub-Saharan Africa, while the lowest incidence is seen in Europe, Australia, North America, and East Asia. The geographical variation in Kaposi's sarcoma incidence is consistent with that of the varying prevalence of KSHV.

4 Carcinogenic Mechanisms of Oncogenic Viruses

There are three major mechanisms of carcinogenesis for seven Group 1 oncogenic viruses as shown in Table 3. They are defined as direct, indirect through chronic inflammation, and indirect through immune suppression (IARC 2012a). The direct carcinogens include EBV, HPV, HTLV-1, and KSHV; the indirect carcinogens through chronic inflammation include HBV and HCV; and the indirect carcinogen through immune suppression is HIV-1.

Table 3 Established carcinogenic mechanisms of oncogenic viruses

Mechanism	Group 1 virus (carcinogenic properties)
Direct	EBV (cell proliferation, inhibition of apoptosis, genomic instability, cell migration)
	HPV (immortalization, genomic instability, inhibition of DNA damage response, anti-apoptotic activity)
	HTLV-1 (immortalization and transformation of T cells)
	KSHV (cell proliferation, inhibition of apoptosis, genomic instability, cell migration)
Indirect through chronic inflammation	HBV (inflammation, liver cirrhosis, liver fibrosis)
	HCV (inflammation, liver cirrhosis, liver fibrosis)
Indirect through immune suppression	HIV-1 (immunosuppression)

Direct oncogenic viruses have the following characteristics: (1) The entire or partial viral genome can usually be detected in each cancer cell. (2) The virus can immortalize after the growth of target cells in vitro. (3) The virus expresses several oncogenes that interact with cellular proteins to disrupt cell-cycle checkpoints, inhibit apoptosis, and DNA damage response, cause genomic instability, and induce cell immortalization, transformation, and migration.

Both HBV and HCV cause hepatocellular carcinoma through chronic inflammation, which leads to the production of chemokines, cytokines, and prostaglandins secreted by infected cells and/or inflammatory cells. The chronic inflammation also leads to the production of reactive oxidative species with direct mutagenic effects to deregulate the immune system and promote angiogenesis, which is essential for the neovascularization and survival of tumors.

Individuals infected with HIV-1 have a high risk of cancers caused by another infectious agent. HIV-1 infection, mainly through immunosuppression, leads to increased replication of oncogenic viruses such as EBV and KSHV. Although antiretroviral therapy lowers the risk of many cancers associated with HIV-1, risks remain high worldwide.

5   Lifetime Cumulative Incidence of Virus-Caused Cancers

Some viruses may cause more than one cancer, while some cancers may be caused by more than one virus. However, only a proportion of persons infected by these oncogenic viruses will actually develop specific cancers. The probability of developing specific cancers is determined by the interaction among viral, host, and environmental factors. Table 4 shows the lifetime cumulative incidence, viral factor, host factor, and environmental factors of several virus-caused cancers. The cumulative lifetime (30–75 years old) risk of developing nasopharyngeal carcinoma was 2.2% for men seropositive for IgA antibodies against EBV VCA or antibodies against EBV DNase.

Table 4 Lifetime cumulative incidence and risk predictors of virus-caused cancers

Virus (cancer)	Lifetime incidence	Viral predictors	Host predictors	Environmental predictor
EBV (nasopharyngeal carcinoma)	Men, 2.0%	Elevated serotiter of EBV antibodies, viral load	Male gender; family nasopharyngeal carcinoma history; genetic polymorphisms (*CYP2E1, XRCC1, hOGG1*, HLA, *GABBR1 CLPTM1L/TERT*)	Cantonese salted fish; dietary nitrosamine; wood dust; formaldehyde; tobacco; low intake of plant vitamins, fresh fish, green tea, and coffee
HBV (hepatocellular carcinoma)	Men, 27.4%; women, 8.0%	Persistent infection, viral load, genotype, mutant type, serum HBsAg level	Elder age, male gender, obesity, diabetes, elevated serum levels of androgen, and ALT; family liver cancer history, genetic polymorphisms (DNA repair enzymes, HLA, xenobiotic metabolism enzymes, *NTCP, ADH1B/ALDH2*, telomere length)	Aflatoxins; alcohol; tobacco; low intake of vegetable; low serum level of carotenoids and selenium; HCV co-infection
HCV (hepatocellular carcinoma)	Men, 23.7%; women, 16.7%	Persistent infection, viral load, genotype, mutant type	Elder age, male gender, obesity, diabetes, serum ALT level, family liver cancer history, genetic polymorphisms (*IFNL3*, HLA)	Aflatoxin; alcohol; HBV or HIV-1 co-infection; radiation
HPV (cervical carcinoma)	HPV-16, 34.3%; HPV-52, 23.3%; HPV-58, 33.4%; Any oncogenic HPV, 20.3%	Persistent infection, viral load, genotype, variant type	Elder age, number of pregnancies, family cervical cancer history, serum estrogen level, genetic polymorphisms (DNA repair enzymes, HLA)	Tobacco; immunosuppression, HIV-1 co-infection, contraceptives use, nutrients

Around one-quarter of patients chronically infected with HBV will develop hepatocellular carcinoma. There is a striking difference in the cumulative lifetime incidence of hepatocellular carcinoma between men (27.4%) and women (8.0%) (Huang et al. 2011). The development of HBV-associated hepatocellular carcinoma has been considered as a multistage hepatocarcinogenesis with multifactorial etiology, and involves the interaction of HBV, other virus such as HCV, chemical carcinogens, host characteristics, and genetic susceptibility (Chen et al. 1997; Chen and Chen 2002; Chen and Yang 2011; IARC 2012a; Chen 2018).

Around one-fifth of patients seropositive for antibodies against HCV (anti-HCV) will develop hepatocellular carcinoma. However, the cumulative lifetime incidence of hepatocellular carcinoma shows a less significant gender difference, and is approximately 23.7% for men and 16.7% for women (Huang et al. 2011). The lifetime cumulative risk of hepatocellular carcinoma among anti-HCV seropositives with and without detectable serum HCV RNA level was 24.2 and 3.53%, respectively. Many cofactors are involved in the development of hepatocellular carcinoma in anti-HCV seropositives (Lee et al. 2010, 2014a, b; IARC 2012a).

The cumulative lifetime (30–75 years old) risk of cervical cancer for women who were infected by HPV16, HPV 52, HPV 58, and any Group 1 oncogenic HPV was 34.3%, 23.3%, 33.4%, and 20.3%, respectively. Women with persistent oncogenic HPV infection have a much higher cumulative risk of cervical cancer than those with only transient infection (Chen et al. 2011b).

6 Viral, Host, and Environmental Predictors of Virus-Caused Cancers

As only a proportion of persons infected by oncogenic viruses will actually go on to develop cancers, this strongly suggests the involvement of various risk predictors in the oncogenic process. For example, carcinogenesis would result from the interaction of multiple risk predictors, including viral, host, and environmental factors, as shown in Table 4. Typical viral predictors include various infection markers such as viral load, genotypes, variant types, mutant types, and serotiters of various antigens or antibodies. The host predictors include age, gender, race, anthropometric characteristics, immune status, hormonal level, personal disease history, and family cancer history. Lastly, environmental factors include chemical carcinogens, nutrients, ionizing radiation, immunosuppressive drugs, and co-infections of other infectious agents. The contribution of several additional predictors to the development of virus-associated cancers seems to be substantial, but has not yet been elucidated in detail.

Several predictors for nasopharyngeal carcinoma have been reviewed previously (Chien and Chen 2003). The viral predictors associated with nasopharyngeal carcinoma include the elevated serotiter of antibodies against EBV, including anti-EBV VCA IgA, anti-EBV DNase, and anti-EBNA1, and the elevated serum EBV DNA level (viral load) (Chien et al. 2001; Hsu et al. 2009). Host predictors of nasopharyngeal carcinoma include male gender, family history of nasopharyngeal carcinoma, (Hsu et al. 2011), and genetic polymorphisms of *CYP2E1, XRCC1, hOGG1*, human leukocyte antigens (HLA), *GABBR1*, and a novel locus within *CLPTM1L/TERT* (Hildesheim et al. 1997, 2002; Cho et al. 2003; Hsu et al. 2012a; Bei et al. 2016). Environmental predictors include consumption of Cantonese salted fish, high dietary intake of nitrite and nitrosamine, occupational exposure to wood dust and formaldehyde, long-term tobacco smoking, and low intake of plant vitamins, fresh fish, green tea, and coffee (Ward et al. 2000; Hildesheim et al. 2001; Hsu et al. 2009, 2012b).

Many studies have examined the risk predictors of HBV-caused hepatocellular carcinoma. The viral predictors of HBV-caused hepatocellular carcinoma include positive HBeAg serostatus, elevated serum HBV DNA level (viral load), HBV genotype and mutant types, and elevated serum HBsAg level (Yang et al. 2002, 2008; Chen et al. 2006, 2009a; Lee et al. 2013). Persistently, high HBV DNA levels throughout disease progression also indicate a high risk for hepatocellular carcinoma (Chen et al. 2011a). However, reaching HBV DNA undetectability and HBsAg seroclearance have been shown to significantly decrease the risk for future hepatocellular carcinoma (Liu et al. 2014). Host predictors of HBV-caused hepatocellular carcinoma include elder age, male gender, persistently elevated serum alanine aminotransferase (ALT) level, and family history of hepatocellular carcinoma (Chen et al. 1991, 2011a; Yu et al. 2000a; Yang et al. 2010; Loomba et al. 2013a). In addition, higher parity (Fwu et al. 2009), obesity and diabetes (Chen et al. 2008; Loomba et al. 2013b), elevated serum level of androgen and androgen-related genetic polymorphisms (Yu and Chen 1993; Yu et al. 2000b), and genetic polymorphisms of xenobiotic metabolism enzymes, DNA repair enzymes, *NTCP* (*SLC10A1*), HLA, and telomere length (Chen et al. 1996a; Yu et al. 1995a, 1999a, 2003; Hu et al. 2016; Zeng et al. 2017) are also important host predictors for HBV-caused hepatocellular carcinoma. Genetic polymorphism of *ADH1B* and *ALDH2* are involved in HBV-caused hepatocellular carcinoma through their mediation of habitual alcohol consumption (Liu et al. 2016). Environmental predictors of HBV-caused hepatocellular carcinoma include aflatoxin exposure (Chen et al. 1996b; Wang et al. 1996; Chu et al. 2017), habits of alcohol consumption and tobacco smoking (Chen et al. 1991; Wang et al. 2003; Loomba et al. 2013b), inadequate intake of vegetable, carotenoids and selenium (Yu et al. 1995b, 1999a, b), and co-infection with HCV (Huang et al. 2011). Interestingly, although co-infection with HCV has been shown to result in a higher risk for HCC, it was a sub-additive combined effect, and was also associated with later-onset of HCC, suggesting an antagonistic effect between HBV and HCV (Huang et al. 2011). The mediation effect of HBV and HCV in relation to hepatocellular carcinoma risk and mortality has recently been documented (Huang et al. 2016a, b).

Recently, studies have been able to further clarify viral, host, and environmental predictors of HCV-caused hepatocellular carcinoma. Viral predictors include the elevated serum level of HCV RNA and HCV genotype 1 (Lee et al. 2010, 2014b; Huang et al. 2011). Host predictors include older age, obesity, diabetes, elevated serum ALT level, family history of hepatocellular carcinoma, and genetic polymorphisms of *INFL3* and HLA (Sun et al. 2003; Chen et al. 2008; Lee et al. 2010, 2015, 2018; IARC 2012a). Environmental predictors include aflatoxin exposure, alcohol consumption, tobacco smoking, betel chewing, radiation exposure, and co-infection with HBV or HIV-1 (Sun et al. 2003; Huang et al. 2011; IARC 2012a; Chu et al. 2018).

Viral, host, and environmental factors are associated with oncogenic HPV-caused cervical cancer. Its viral predictors include the persistent infection, elevated viral load, HPV genotypes, and variant types (Chen et al. 2011b, c; Chang et al. 2011; IARC 2012a), while its host predictors include elder age, number of pregnancies, family history of cervical cancer, serum estrogen level, and genetic polymorphisms of DNA repair enzymes and HLA (Chen et al. 2011b; Chuang et al. 2012; IARC 2012a). Environmental predictors of oncogenic HPV-caused cervical cancer include tobacco smoking, immunosuppression, HIV-1 co-infection, use of oral contraceptives, and inadequate intake of micronutrients (IARC 2012a).

7 Risk Calculators of Virus-Caused Cancers

Given that there are many risk predictors for each virus-caused cancer, it would be useful to incorporate them into a risk prediction model or calculator that can predict the cumulative cancer incidence of each cancer. Such risk calculators may provide clinicians with important information for the triage and identification of patients who need intensive treatment, versus those who need only routine follow-up. Moreover, with personalized risk calculations, patients' follow-up intervals, surveillance patterns, and referral strategies can be tailored. Risk calculator is an important clinical tool in the era of precision medicine.

Several risk models/calculators have been developed to predict the cumulative incidence of hepatocellular carcinoma of patients with chronic hepatitis B and C as shown in Table 5 (Yang et al. 2014; Chen et al. 2015; Liu et al. 2015), and only a few calculators were externally validated (Yang et al. 2011, 2016; Lee et al. 2014c; Poh et al. 2016). Most risk prediction models for HBV-caused HCC incorporated a wide range of clinical parameters, were hospital-based cohorts, and did not have satisfactory external validation of the models. In order to perform sufficient external validation of a robust prediction model, the study groups of the IPM model (Han and Ahn 2005), GAG-HCC risk score (Yuen et al. 2009), CUHK Clinical Scoring System (Wong et al. 2010), and REVEAL nomograms (Yang et al. 2010) jointly established the REACH-B risk score (Yang et al. 2011). The REACH-B risk score is a 17-point scoring system incorporating gender, age, serum levels of ALT and HBV DNA, and HBeAg serostatus as its risk predictors. The REACH-B score was first derived from a large community-based cohort of REVEAL-HBV study, then externally validated by a composite hospital-based cohort of IPM, GAG-HCC and CUHK studies. This model was able to predict the 3-, 5-, and 10-year risk of developing HCC with areas under the receiving operating characteristic curve (AUROC) of 0.81, 0.80, and 0.77, respectively, for all chronic hepatitis B patients; and with AUROC of 0.90, 0.78, and 0.81, respectively, for non-cirrhotic patients.

Table 5 Risk calculators for hepatocellular carcinoma caused by HBV and HCV

Risk score (reference)	AUROC (95% confidence interval) for predicting HCC risk by follow-up year and cohort	Risk predictors
IPM (Han and Ahn 2005)	None reported	Gender, Age, Cirrhosis/chronic hepatitis, Chronic HBV/HCV infection, AFP level, ALT level, Alcohol consumption
GAG-HCC Risk score (Yuen et al. 2009)	5-year: 0.88 10-year: 0.89	Gender, Age, Cirrhosis, HBV DNA level, HBVcore promoter mutations
CUHK Clinical scoring system (Wong et al. 2010)	**Internal validation** 5-year: 0.76 (0.66–0.86) 10-year: 0.78 (0.71–0.86)	Age, Cirrhosis, HBV DNA level, Bilirubin level, Albumin level
REVEAL-HBV nomograms (Yang et al. 2010)	**Internal validation** **Model 1** 5-year: 83.1 10-year: 82.1 **Model 2** 5-year: 83.2 10-year: 83.0 **Model 3** 5-year: 83.2 10-year: 83.0	Gender, Age, ALT level, Alcohol consumption, HBV DNA level, HBeAg serostatus, HBV Genotype, Family HCC History
REACH-B Score (Yang et al. 2011)	**External validation (IPM + GAG + HCC/CUHK) for patients without cirrhosis/all patients** 3-year: 0.90 (0.88–0.92)/0.81 (0.79–0.83) 5-year: 0.78 (0.76–0.81)/0.80 (0.78–0.82) 10-year: 0.81 (0.78–0.83)/0.77 (0.75–0.79)	Gender, Age, ALT level, HBV DNA level, HBeAg serostatus
Upgraded REVEAL-HBV nomograms (Lee et al. 2013)	**Internal validation** 5-year: 0.84 10-year: 0.86 15-year: 0.87	Gender, Age, ALT level, HBV DNA level, HBeAg serostatus, HBV Genotype, Family History of HCC, HBsAg level
REVEAL-HBV HCC Risk score (Lin et al. 2013)	**Internal validation by risk model of different combination of risk predictors** (6-year risk) Risk score I: 0.83 Risk score II: 0.89 Risk score III: 0.91	Gender, Age, ALT level, AFP level, HBV DNA level, HBeAg serostatus, AST/ALT ratio, GGT level, Albumin level, Alpha-1 globulin level
REVEAL-HCV HCC Risk Score for Anti-HCV-positives (Lee et al. 2014c)	**Internal (External) validation** 5-year: 0.75 (0.73) 10-year: 0.83 15-year: 0.83	Age, ALT, Cirrhosis, AST/ALT ratio, HCV RNA level, HCV genotype

(continued)

Table 5 (continued)

Risk score (reference)	AUROC (95% confidence interval) for predicting HCC risk by follow-up year and cohort	Risk predictors
REVEAL-HCV HCC Risk score for HCV RNA-positives (Lee et al. 2014c)	**Internal (External) validation** 5-year: 0.65 (0.70) 10-year: 0.77 15-year: 0.73	Age, ALT, Cirrhosis, AST/ALT ratio, HCV RNA level, HCV genotype
RWS-HCC Risk score (Poh et al. 2016)	**External validation by cohort** REACH-B: 0.77 (0.73–0.81) GAG-HCC: 0.83 (0.75–0.91) CUHK: 0.90 (0.86–0.95)	Gender, Age, AFP level, Cirrhosis
REACH-B IIa Score (Yang et al. 2016)	**External validation by ERADICATE-B/CUHK cohort** 3-year: 0.92 (0.82–1.02)/0.85 (0.75–0.95) 5-year: 0.78 (0.70–0.86)/0.82 (0.70–0.93) 10-year: 0.80 (0.76–0.84)/0.78 (0.70–0.87)	Gender, Age, ALT level, HBV DNA level, HBeAg serostatus, HBsAg level
REACH-B IIb Score (Yang et al. 2016)	**External validation by ERADICATE-B/CUHK cohort** 3-year: 0.90 (0.81–1.00)/0.84 (0.76–0.92) 5 year: 0.76 (0.68–0.85)/0.81 (0.71–0.91) 10-year: 0.78 (0.73–0.82)/0.79 (0.72–0.87)	Gender, Age, ALT level, HBeAg serostatus, HBsAg level

With the establishment of quantitative HBsAg levels as an important seromarker in the natural history of chronic hepatitis B, the REVEAL-HBV nomograms and REACH-B risk scores have recently been updated to include this novel risk predictor with original risk predictors in the prediction models. This upgraded REVEAL-HBV nomogram for HCC risk prediction also provided excellent prediction accuracy and discriminatory ability, with an internal validation AUROC of 0.84 and 0.86, respectively, for 5- and 10-year HCC risk prediction (Lee et al. 2013). Predictors included in the derivation of the updated nomogram are gender, age, family history of hepatocellular carcinoma, serum ALT level, HBeAg serostatus, serum levels of HBV DNA and HBsAg, and HBV genotype. Risk scores are assigned for various categories of risk predictors as shown in Table 6. For example, a 60-year-old (risk score = 6) male (risk score = 2) chronic hepatitis B patient who had a family history of hepatocellular carcinoma (risk score = 2), a serum ALT level of 90 IU/L (risk score = 2), a HBeAg-seropositive serostatus, a serum HBV DNA level of 10^7 copies/mL, a serum HBsAg level of 104 IU/mL, and a HBV genotype C infection (risk score = 7) has a sum of risk score of 19. The 5-, 10-, and 15-year cumulative risks of hepatocellular carcinoma by the sum of risk score are shown in the nomogram of Fig. 3. For the male patient with a sum of the risk score as high as 19, his 5-, 10- and 15-year risk of hepatocellular carcinoma

Table 6 Scores assigned to risk predictors of HBV-caused hepatocellular carcinoma

Risk predictor	Risk score
Age	
30–34	0
35–39	1
40–44	2
45–49	3
50–54	4
55–59	5
60–64	6
Gender	
Female	0
Male	2
Family history of hepatocellular carcinoma	
No	0
Yes	2
Serum ALT levels (IU/L)	
<15	0
15–44	1
≥ 45	2
HBeAg/HBV DNA (copies/mL)/HBsAg (IU/mL)/Genotype	
Negative/<10^4/<100/any type	0
Negative/<10^4/100–999/any type	2
Negative/<10^4/≥ 1000/any type	2
Negative/10^4–10^6/<100/any type	3
Negative/10^4–10^6/100–999/any type	3
Negative/10^4–10^6/≥ 1000/any type	4
Negative/$\geq 10^6$/any level/B or B + C	5
Negative/$\geq 10^6$/any level/C	7
Positive/any level/any level/B or B + C	6
Positive/any level/any level/C	7

will be 35, 80, and 90%. In contrast, a 34-year-old woman with no family history of hepatocellular carcinoma, a serum ALT level of 10 IU/L, a HBeAg-negative serostatus, a serum HBV DNA level of 10^3 copies/mL, and a serum HBsAg level of 50 IU/mL (sum of risk score = 0) has 5-, 10-, and 15-year risk of hepatocellular carcinoma of 0.0075, 0.025, and 0.065%.

Serum HBsAg level was included in both REACH-B IIa and IIb risk calculators for the prediction of hepatocellular carcinoma risk in non-cirrhotic patients with chronic HBV infection. The addition of serum HBsAg level to the REACH-B scoring system (REACH-B IIa) or the replacement of serum HBV DNA level by serum HBsAg level (REACH-B IIb) did not alter the accuracy of the scoring systems when externally validated by ERADICATE-B and CUHK studies (Yang et al. 2016).

Fig. 3 Nomogram for predicted 5-, 10-, and 15-year risk of hepatocellular carcinoma by sum of risk score

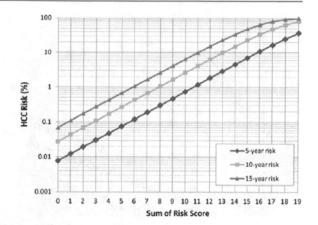

Recently, two risk prediction models for HCV-caused hepatocellular carcinoma were developed and validated as shown in Table 5. The first model utilized age, serum ALT levels, AST/ALT ratio, and a combination of liver cirrhosis, serum HCV RNA level, and HCV genotype to predict the risk of hepatocellular carcinoma among anti-HCV-seropositive patients, while the second model utilized the same variables to predict the risk among patients with detectable serum HCV RNA levels (Lee et al. 2014c). Both models were validated in a high-risk cohort of HCV patients with satisfactory discriminatory ability, showing a 5-year AUROC of 0.73 and 0.70, respectively. These risk prediction models have also been validated using the United States Veterans Affairs database, and the 5-year predictability was acceptable with an AUROC of 0.69 (Matsuda et al. 2014).

The development of the risk calculators needs large-scale prospective cohorts, which have been followed for a long period of time with accurate measurements of risk predictors. Demographical characteristics, viral infection biomarkers, family cancer history, lifestyle variables, and polymorphisms of genetic susceptibility may be incorporated to develop valid and useful cancer risk calculators. Figure 4 shows a nomogram of predicted 1-, 5- and 10-year risk of nasopharyngeal carcinoma derived from a long-term prospective study in Taiwan (Chien et al. 2001). Four risk predictors including family history of nasopharyngeal carcinoma, cumulative pack-years of cigarette smoking, and serotiter of anti-EBV VCA IgA and anti-EBV DNase are incorporated in the model. Based on this 22-point risk score, the risk of nasopharyngeal carcinoma ranges 0.02–5.09% at 1 year, 0.07–13.95% at 5 years, and 0.14–26.20% at 10 years for male adults in Taiwan. The AUROC was as high as 97.9, 87.8 and 90.6% for predicting the 1-, 5- and 10-year risk of nasopharyngeal carcinoma (Hsu et al. 2013).

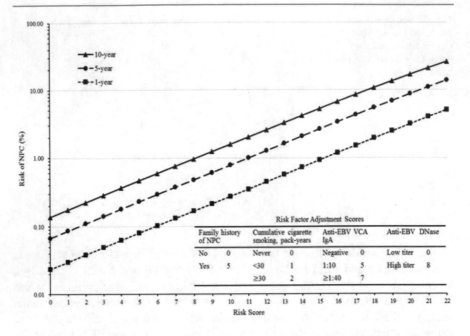

Fig. 4 Nomogram for predicted 1-, 5- and 10-year risk of nasopharyngeal carcinoma by risk score

8 Prevention and Early Detection of Virus-Caused Cancers

The most effective strategy to prevent virus-caused cancers is through the prevention of viral infection through vaccination or transmission interruption, and the antiviral therapy to eliminate oncogenic viruses in human host. Early detection of specific cancers in oncogenic virus-infected individuals through screening may lower the risk of invasive cancer and its fatality. Table 7 shows currently available vaccines or antivirals to prevent or treat patients with oncogenic viral infection. Vaccines are available for the prevention of HBV-caused hepatocellular carcinoma and HPV-caused cervical cancer, while antiviral/antiretroviral therapies are available for the treatment of chronic infection of HBV, HCV, and HIV to prevent hepatocellular carcinoma and Kaposi's sarcoma.

Table 7 Evidence showing the reduction of incidence of virus-caused cancers through the preventive strategy of vaccination, antiviral/antiretroviral therapy, and early detection

Virus	Cancer	Preventive strategy
EBV	Nasopharyngeal carcinoma	Early detection in high-risk group
HBV	Hepatocellular carcinoma	Vaccination and antiviral therapy
HCV	Hepatocellular carcinoma	Antiviral therapy
HIV-1	Kaposi's sarcoma	Antiretroviral therapy
HPV	Cervical cancer	Vaccination and cervical screening

EBV infection is almost ubiquitous in humans, but the incidence of nasopharyngeal carcinoma is low. The prevention of nasopharyngeal carcinoma through the elimination of EBV infection by vaccination or antiviral therapy is not available. However, many seromarkers including various antibodies against EBV-specific antigens and EBV DNA may be used to identify high-risk individuals for the early detection of nasopharyngeal carcinoma. For examples, a novel EBV-based antibody stratification signature has been identified for early detection of nasopharyngeal carcinoma (Coghill et al. 2018). The inclusion of 12 array-identified antibodies in the risk signature resulted in the increase of NPC prediction accuracy of currently used VCAp18/EBNA1 IgA biomarkers, from 82 to 93% in the general Taiwan population and from 78 to 89% in genetically high-risk families.

HBV immunization and antiviral therapy for HBV and HCV may significantly reduce the incidence and mortality of liver cancer (Chen 2018). The national HBV immunization program in Taiwan, which was implemented in 1984, has successfully reduced the incidence of hepatocellular carcinoma in vaccinated birth cohorts (Chang et al. 1997, 2016). The rate ratio of hepatocellular carcinoma incidence between immunized and unimmunized birth cohorts was 0.26, 0.34, 0.37, and 0.42, respectively, for the age group of 6–9, 10–14, 15–19, and 20–26 years. Another study examining 30-year outcomes of the HBV immunization program in Taiwan also showed significant declines in the mortality from infant fulminant hepatitis and chronic liver diseases including cirrhosis, and in the incidence and mortality of hepatocellular carcinoma in the immunized birth cohorts (Chiang et al. 2013). As shown in Table 8, the age-gender-adjusted rate ratio decreased 79% for the mortality and 63% for the incidence of hepatocellular carcinoma in immunized birth cohorts born in 1997–2000 compared to unimmunized birth cohorts born in 1977–1980. In a cohort study of 3.8 million vaccinated neonates in Taiwan, the risk predictors of hepatocellular carcinoma in immunized newborns included maternal HBeAg serostatus and incomplete immunization with hepatitis B immunoglobulin or vaccine (Chien et al. 2014).

Several antivirals have been approved for chronic viral hepatitis therapy. Lamivudine was first approved in 1998 for the treatment of chronic HBV infection, it decreases the hepatocellular carcinoma risk of treated patients significantly with a disadvantage of developing antiviral-resistant YMDD mutants (Liaw et al. 2004). Newly developed antivirals for chronic HBV infection have higher genetic barriers to limit the development of antiviral-resistant strains. A significant decrease in hepatocellular carcinoma incidence among 973 chronic HBV infection patients treated with pegylated interferon or any nucleoside/nucleotide analogues was reported in a USA–Taiwan study, showing treated patients have 77% reduction in hepatocellular carcinoma incidence compared with 4935 untreated patients after adjustment for the REACH-B risk score (Lin et al 2016). In another European study on 1951 adult Caucasian chronic HBV infection patients treated with entecavir or tenofovir, there was a significant decline in annual hepatocellular carcinoma incidence among cirrhosis patients from 3.22% in the first 5 years to 1.57% in 5–10 years after enrollment (Papatheodoridis et al. 2017).

Table 8 Significant reduction in incidence and mortality of hepatocellular carcinoma through National Programs on Hepatitis B Immunization and Chronic Viral Hepatitis Therapy in Taiwan

Birth year	HCC mortality	(Ages 5–29 years)	HCC incidence	(Ages 5–29 years)
Hepatitis B immunization program				
	Rate per 10^5 person-years	Age-sex-adjusted rate ratio (95% CI)	Rate per 10^5 person-years	Age-sex-adjusted rate ratio (95% CI)
1977–1980	0.81	1.00 (reference)	1.14	1.00 (reference)
1981–1984	0.56	0.70 (0.59–0.83)	0.77	0.73 (0.63–0.85)
1985–1988	0.30	0.43 (0.33–0.55)	0.37	0.48 (0.38–0.60)
1989–1992	0.17	0.27 (0.19–0.39)	0.23	0.37 (0.27–0.51)
1993–1996	0.12	0.21 (0.13–0.34)	0.22	0.43 (0.30–0.62)
1997–2000	0.12	0.21 (0.12–0.38)	0.17	0.37 (0.21–0.62)
Calendar year	HCC mortality	(Ages 30–69 years)	HCC incidence	(Ages 30–69 years)
Chronic viral hepatitis therapy program				
	Rate per 10^5 person-years	Age-sex-adjusted rate ratio (95%CI)	Rate per 10^5 person-years	Age-sex-adjusted rate ratio (95% CI)
2000–2003	36.59	1.00 (reference)	54.12	1.00 (reference)
2004–2007	35.77	0.95 (0.93–0.97)	54.79	0.98 (0.96–0.99)
2008–2011	30.21	0.76 (0.75–0.78)	50.77	0.86 (0.85–0.88)
2012–2015	27.44	0.64 (0.62–0.65)	47.55	0.76 (0.74–0.77)

The interferon-based therapy was the standard for the treatment of chronic HCV infection before direct acting agents (DAA) was launched in 2013. HCV genotypes 1 and 4 are less responsive to interferon-based treatment than other genotypes. *INFL3* variants were found to be associated with the efficacy of interferon-based HCV therapy, showing the Asia-Pacific ethnicities have a high frequency of favorable genotypes. In a study of 4639 patients with chronic HCV infection treated with pegylated interferon and ribavirin, the sustained viral response (SVR versus non-SVR) was associated with a significant decline in HCC incidence in cirrhotic (46% reduction) and non-cirrhotic (63% reduction) patients (Lee et al. 2017). DAAs are highly effective for all HCV genotypes without ethnic variation, with low side effects, and convenient with oral intake. In a recent study on 62,354 chronic HCV-infected patients treated with interferon and/or DAAs, SVR versus non-SVR was associated with a significant reduction in HCC in patients treated with DAA only (71% reduction), DAA and interferon (52% reduction), and interferon only (68% reduction) after adjustment for multiple risk factors (Ioannou 2018).

A nationwide chronic viral hepatitis therapy program was implemented in 2003 to treat patients with chronic hepatitis viral infection in Taiwan. A significant reduction in the burden of liver cancer through this program has recently been documented (Chiang et al. 2015). After adjustment for age and gender, there was 36% reduction in the mortality and 24% reduction in the incidence of hepatocellular carcinoma in 2012–2015 compared with the 4-year period before the program launch, i.e., 2000–2003, as shown in Table 8.

Viral hepatitis control has been identified as a sustainable development goal by the United Nations. The global targets for 2030 set by World Health Organization include 90% hepatitis B vaccination coverage, 90% prevention of mother-to-child HBV transmission, 100% blood transfusion and injection safety, 90% diagnosis of HBV and HCV infections, and 80% treatment of eligible patients (WHO 2017). Concerted national and international efforts are in urgent need. The diagnosis and treatment coverage have to be rapidly scaled up through a public health approach to benefit all. Sustainable financing and innovation are also required for the development and delivery of vaccines, diagnostics, and treatments to transform the global hepatitis response.

Many clinical studies demonstrated the efficacy of HPV vaccination in preventing cervical neoplasia, a precursor lesion of cervical cancer. By now, three HPV vaccines now being marketed in many countries throughout the world, a bivalent, a quadrivalent, and a nonavalent vaccine. More than 80 countries have introduced HPV vaccine to protect adolescent girls against cervical cancer. The HPV immunization program in Australia has effectively lowered the incidence of cervical neoplasia in vaccinated female cohorts aged 12–26 years, showing a hazard ratio of 0.53 and 0.73 for cervical high and low-grade lesion, respectively, among vaccinees received three doses (Brotherton et al. 2015). In a modelling study, cervical cancer could be eliminated as a public health problem in Australia within the next 20 years (Hall et al. 2019).

Studies in the USA have also documented the reduced risk of abnormal cervical cytology result among vaccinated females (hazard ratio 0.64 versus unvaccinated females), particularly for those who completed three doses (hazard ratio 0.48) and at age of 11–14 years (hazard ratio 0.36 for 1–2 doses and 0.27 for 3 doses)(Hofstetter et al. 2016). In Canada, women received full vaccination (≥ 3 doses) had an adjusted odds ratio of 0.72 and 0.50 for developing any cervical abnormalities and high-grade lesions, respectively (Kim et al. 2016). In Scotland, women who received three doses of vaccine compared with unvaccinated women had a significant reduction in diagnoses of cervical intraepithelial neoplasia (CIN) 1, CIN 2, and CIN 3 with a relative risk of 0.71, 0.50 and 0.45, respectively (Pollock et al. 2014). Another study among young women in Denmark showed that vaccination with the HPV vaccine is effective in reducing the risk for cervical cancer precursor lesions, despite having only been available since 2006 in Denmark (Baldur-Felskov et al. 2014). Screening for HPV infection in cervical cells and pap smears may aid the early detection of cervical neoplasia and reduce the incidence and mortality of invasive cervical cancer (Chen et al. 2009b, 2014b).

Many studies have also examined the impact of antiretroviral therapy on the incidence of Kaposi's sarcoma in HIV-infected individuals. In a recent review of studies published between 2009 and 2012, the percent reduction in the population-level incidence of Kaposi's sarcoma attributed to antiretroviral therapy among HIV-infected individuals seen in the period when therapy was available, compared to the period when it was not available, ranged from 78 to 95% (Semeere et al. 2012). The reduction in Kaposi's sarcoma incidence was similarly striking among multiple clinic-based cohort studies, which saw reductions of 50–90% since the availability of antiretroviral therapy. However, these results were mostly seen in resource-rich settings. Data from resource-limited settings, where the majority of the burden of Kaposi's sarcoma lies, were limited. More research on the effectiveness of antiretroviral therapy in resource-limited settings is needed.

9   Future Perspectives

Along with the advancement in precision medicine, more and more proteomic and genomic biomarkers associated with the development of virus-caused cancers have been identified. They may be applied to the risk prediction or early detection of cancers. For example, multiple microRNAs have been combined for the diagnosis of hepatocellular carcinoma. However, its efficacy and cost-effectiveness for early diagnosis of hepatocellular carcinoma should be further assessed and compared with those of other methods including abdominal ultrasonography (Chen and Lee 2011).

More importantly, repeated measurements of biomarkers may further improve the risk prediction or early detection of virus-caused cancers (Chen 2005). For example, the trajectory of serum HBV DNA levels has been found to predict long-term risk of hepatocellular carcinoma effectively (Chen et al. 2011a). A recent longitudinal study with serial measurements have identified a novel glycomarker, *Wisteria floribunda* agglutinin-positive human Mac-2-binding protein (WFA$^+$-M2BP), for the long- and short-term prediction of hepatocellular carcinoma (Lin et al. 2018). Patients with increasing changes in serum WFA$^+$-M2BP levels, relative to their baseline levels, had a significantly increased risk. The efficacy of predicting hepatocellular carcinoma for WFA$^+$-M2BP was significantly higher while closer to the disease diagnosis ($p = 0.024$). The AUROC for the risk calculator including the predictors of age, sex, ALT, AFP and WFA$^+$-M2BP was as high as 0.91 for predicting hepatocellular carcinoma occurred within 1 year.

More longitudinal studies with regular follow-up examinations of various biomarkers are in urgent need to identify good molecular targets for the development of preventives, diagnostics, or therapeutics of various virus-caused cancers. A health economic assessment of these biopharmaceuticals in their clinical application.

References

Alter MJ (2007) Epidemiology of hepatitis C virus infection. World J Gastroenterol 13:2436–2441

Brotherton J, Malloya M, Buddb AC, Savillea M, Drennana KT, Gertiga DM (2015) Effectiveness of less than three doses of quadrivalent human papillomavirus vaccine against cervical intraepithelial neoplasia when administered using a standard dose spacing schedule: observational cohort of young women in Australia. Papillomavirus Res 1:59–73

Bruni L, Barrionuevo-Rosas L, Albero G, Aldea M, Serrano B, Mena M, Gómez D, Muñoz J, Bosch FX, de Sanjosé S, ICO Information Centre on HPV and Cancer (2019). Human papillomavirus and related diseases in the world. Summary report (Updated 2019). http://www.hpvcentre.net/statistics/reports/XWX.pdf. Accessed on 22 Jan 2019

Baldur-Felskov B, Dehlendorff C, Munk C, Kjaer S (2014) Early impact of human papillomavirus vaccination on cervical neoplasia—Nationwide follow-up of young Danish women. J Natl Cancer Inst 106(3). https://doi.org/10.1093/jnci/djt460

Bei JX, Su WH, Ng CC, Yu K, Chin YM, Lou PJ, Hsu WL, McKay JD, Chen CJ et al, International Nasopharyngeal Carcinoma (NPC) Genetics Working Group (2016) A GWAS meta-analysis and replication study identifies a novel locus within *CLPTM1L/TERT* associated with nasopharyngeal carcinoma in individuals of Chinese ancestry. Cancer Epidemiol Biomark Prev 25:188–92

Chang MH, Chen CJ, Lai MS, Kong MS, Wu TC, Liang DC, Hsu HM, Shau WY, Chen DS (1997) Taiwan Childhood Hepatoma Study Group. Universal hepatitis B vaccination in Taiwan and the incidence of hepatocellular carcinoma in children. New Engl J Med 336:1855–1859

Chang MH, You SL, Chen CJ, Liu CJ, Lai MW, Wu TC, Wu SF, Lee CM, Yang SS, Chu HC, Wang TE, Chen BW, Chuang WL, Soon MS, Lin CY, Chiou ST, Kuo HS, Chen DS, Taiwan Hepatoma Study Group (2016) Long-term effects of hepatitis B immunization of infants in preventing liver cancer. Gastroenterology 151:472–480

Chang YJ, Chen HC, Lee BH, You SL, Lin CY, Pan MH, Chou YC, Hsieh CY, Chen YM, Chen YJ, Chen CJ (2011) Unique variants of human papillomavirus genotypes 52 and 58 and risk of cervical neoplasia. Int J Cancer 129:965–973

Chen CJ (2005) Time-dependent events in natural history of occult hepatitis B virus infection: Importance of population-based long-term follow-up study with repeated measurements (Editorial). J Hepatology 42:438–440

Chen CJ (2018) Global elimination of viral hepatitis and hepatocellular carcinoma: opportunities and challenges. Gastroenterology 67:595–598

Chen CJ, Chen DS (2002) Interaction of hepatitis B virus, chemical carcinogen and genetic susceptibility: multistage hepatocarcinogenesis with multifactorial etiology (editorial). Hepatology 36:1046–1049

Chen CJ, Lee MH (2011) Early diagnosis of hepatocellular carcinoma by multiple microRNAs: validity, efficacy and cost-effectiveness (editorial). J Clin Oncol 29:4745–4747

Chen CJ, Yang HI (2011) Natural history of chronic hepatitis B REVEALed. J Gastroenterol Hepatol 26:628–638

Chen CJ, Liang KY, Chang AS, Chang YC, Lu SN, Liaw YF, Chang WY, Sheen MC, Lin TM (1991) Effects of hepatitis B virus, alcohol drinking, cigarette smoking and familial tendency on hepatocellular carcinoma. Hepatology 13:398–406

Chen CJ, Yu MW, Liaw YF, Wang LW, Chiamprasert S, Matin F, Hirvonen A, Bell AB, Santella RM (1996a) Chronic hepatitis B carriers with null genotypes of glutathione S-transferase M1 and T1 polymorphisms who are exposed to aflatoxins are at increased risk of hepatocellular carcinoma. Am J Hum Genet 59:128–134

Chen CJ, Wang LY, Lu SN, Wu MH, You SL, Li HP, Zhang YJ, Wang LW, Santella RM (1996b) Elevated aflatoxin exposure and increased risk of hepatocellular carcinoma. Hepatology 24:38–42

Chen CJ, Yu MW, Liaw YF (1997) Epidemiological characteristics and risk factors of hepatocellular carcinoma. J Gastroen Hepatol 12:S294–S308

Chen CJ, Yang HI, Su J, Jen CL, You SL, Lu SN, Huang GT, Iloeje UH (2006) Risk of hepatocellular carcinoma across a biological gradient of serum hepatitis B virus DNA level. JAMA 295:65–73

Chen CL, Yang HI, Yang WS, Liu CJ, Chen PJ, You SL, Wang LY, Sun CA, Lu SN, Chen DS, Chen CJ (2008) Metabolic factors and risk of hepatocellular carcinoma by chronic hepatitis B/C virus infection: a follow-up study in Taiwan. Gastroenterology 135:111–121

Chen CJ, Yang HI, Iloeje UH (2009a) REVEAL-HBV Study Group. Hepatitis B virus DNA levels and outcomes in chronic hepatitis B. Hepatology 49:S72–S84

Chen CJ, Hsu WL, Yang HI, Lee MH, Chen HC, Chien YC, You SL (2014a) Epidemiology of virus infection and human cancer. Recent Results Cancer Res 193:11–32

Chen CJ, Lee MH, Liu J, Yang HI (2015) Hepatocellular carcinoma risk scores: ready to use in 2014? Hepat Oncol 2:1–4

Chen CF, Lee WC, Yang HI, Chang HC, Jen CL, Iloeje UH, Su J, Hsiao KC, Wang LY, You SL, Lu SN, Chen CJ (2011a) Changes in serum levels of HBV DNA and alanine aminotransferase determine risk for hepatocellular carcinoma risk. Gastroenterology 141:1240–1248

Chen HC, You SL, Hsieh CY, Lin CY, Pan MH, Chou YC, Liaw KL, Schiffman M, Hsing AW, Chen CJ (2011b) For CBCSP-HPV Study Group. Prevalence of genotype-specific human papillomavirus infection and cervical neoplasia in Taiwan: a community-based survey of 10,602 women. Int J Cancer 128:1192–1203

Chen HC, Schiffman M, Lin CY, Pan MH, You SL, Chuang LC, Hsieh CY, Liaw KL, Hsing AW, Chen CJ (2011c) CBCSP-HPV Study Group. Persistence of type-specific human papillomavirus infection and increased long-term risk of cervical cancer. J Natl Cancer Inst 103:1387–1396

Chen YY, You SL, Chen CA, Shih LY, Koong SL, Chao KY, Hsiao ML, Hsieh CY, Chen CJ, Force Taiwan Cervical Cancer Screening Task (2009b) Effectiveness of national cervical cancer screening programme in Taiwan: 12-year experience. Br J Cancer 101:174–177

Chen YY, You SL, Koong SL, Liu J, Chen CA, Chen CJ, The Taiwan Cervical Cancer Screening Task Force (2014b) Screening frequency and atypical cells and the prediction of cervical cancer risk. Obs Gyn 123:1003–1011

Chiang CJ, Yang YW, You SL, Lai MS, Chen CJ (2013) Thirty-year outcomes of the national hepatitis B immunization program in Taiwan. JAMA-J Am Med Assoc 310:974–976

Chiang CJ, Yang YW, Chen JD, You SL, Yang HI, Lee MH, Lai MS, Chen CJ (2015) Significant reduction in end-stage liver diseases burden through national viral hepatitis therapy program in Taiwan. Hepatology 61:1154–1162

Chien YC, Chen CJ (2003) Epidemiology and etiology of nasopharyngeal carcinoma: gene-environment interaction. Cancer Rev Asia-Pac 1:1–19

Chien YC, Chen JY, Liu MY, Yang HI, Hsu MM, Chen CJ, Yang CS (2001) Serologic markers of Epstein-Barr virus infection and nasopharyngeal carcinoma in Taiwanese men. New Engl J Med 345:1877–1882

Chien YC, Jan CF, Chiang CJ, Kuo HS, You SL, Chen CJ (2014) Incomplete hepatitis B immunization, maternal carrier status and increased risk of liver diseases: a 20-year cohort study of 3.8 million vaccinees. Hepatology 60:125–132

Cho EY, Hildesheim A, Chen CJ, Hsu MM, Chen IH, Mittl BF, Levine PH, Liu MY, Chen JY, Brinton LA, Cheng YJ, Yang CS (2003) Nasopharyngeal carcinoma and genetic polymorphisms of DNA repair enzymes XRCC1 and hOGG1. Cancer Epidemiol Biomarker Prev 12:1100–1104

Chu YJ, Yang HI, Wu HC, Liu J, Wang LY, Lu SN, Lee MH, Jen CL, You SL, Santella RM, Chen CJ (2017) Elevated serum levels of aflatoxin B_1-albumin adducts increase risk of cirrhosis and cirrhotic hepatocellular carcinoma among chronic hepatitis B virus carriers in Taiwan. Int J Cancer 141:711–720

Chu YJ, Yang HI, Wu HC, Lee MH, Liu J, Wang LY, Lu SN, Jen CL, You SL, Santella RM, Chen CJ (2018) Aflatoxin B_1 exposure increases the risk of hepatocellular carcinoma associated with hepatitis C virus infection or alcohol consumption. Eur J Cancer 94:37–46

Chuang LC, Hu CY, Chen HC, Lin PJ, Lee B, Lin CY, Pan MH, Chou YC, You SL, Hsieh CY, Chen CJ (2012) Associations of human leukocyte antigen class II genotypes with human papillomavirus 18 infection and cervical intraepithelial neoplasia risk. Cancer 118:223–231

Coghill AE, Pfeiffer RM, Proietti C, Hsu WL, Chien YC, Lekieffre L, Krause L, Teng A, Pablo J, Yu KJ, Lou PJ, Wang CP, Liu Z, Chen CJ, Middeldorp J, Mulvenna JP, Bethony J, Hildesheim A, Doolan DL (2018) Identification of a novel, EBV-based antibody risk stratification signature for early detection of nasopharyngeal carcinoma in Taiwan. Clin Cancer Res 24:1305–1314

de Sanjose S, Diaz M, Castellsague X, Clifford G, Bruni L, Munoz N, Bosch FX (2007) Worldwide prevalence and genotype distribution of cervical human papillomavirus DNA in women with normal cytology: a meta-analysis. Lancet Infect Dis 7:453–459

Dukers NH, Rezza G (2003) Human herpesvirus 8 epidemiology: what we do and do not know. AIDS 17:1717–1730

Ferlay J, Colombet M, Soerjomataram I, Mathers C, Parkin DM, Pineros M, Znaor A, Bray F (2018) Estimating the global cancer incidence and mortality in 2018: GLOBOCAN sources and methods. Int J Cancer Epub

Fwu CW, Chien YC, Kirk GD, Nelson KE, You SL, Kuo HS, Feinleib M, Chen CJ (2009) Hepatitis B virus infection and hepatocellular carcinoma among parous Taiwanese women: nationwide cohort study. J Natl Cancer Inst 101:1019–1027

GLOBOCAN. http://gco.iarc.fr/today/online-analysis

Hall MT, Simms KT, Lew J, Smith MA, Brotherton JML, Saville M, Frazer IH, Canfell K (2019) The projected timeframe until cervical cancer elimination in Australia: a modelling study. Lancet Public Health 4:e19–e27

Han KH, Ahn SH (2005) How to predict HCC development in patients with chronic B viral liver disease? Intervirology 48:23–28

Hildesheim A, Anderson LM, Chen CJ, Cheng YJ, Brinton LA, Daly AK, Reed CD, Chen IH, Caporaso NE, Hsu MM, Chen JY, Idle JR, Hoover RN, Yang CS, Chabra SK (1997) CYP2E1 genetic polymorphisms and risk of nasopharyngeal carcinoma in Taiwan. J Natl Cancer Inst 89:1207–1212

Hildesheim A, Dosemeci M, Chan CC, Chen CJ, Cheng YC, Hsu MM, Chen IH, Mittl BF, Sun B, Levine PH, Chen JY, Brinton LA, Yang CS (2001) Occupational exposure to wood, formaldehyde, and solvents and risk of nasopharyngeal carcinoma. Cancer Epidemiol Biomarker Prev 10:1145–1153

Hildesheim A, Apple RJ, Chen CJ, Wang SS, Cheng YC, Klitz W, Mack SJ, Chen IH, Hsu MM, Yang CS, Brinton LA, Levine PH, Erlich HA (2002) Association of HLA class I and II alleles and extended haplotypes with nasopharyngeal carcinoma in Taiwan. J Natl Cancer Inst 94:1780–1789

Hofstetter AM, Ompad DC, Stockwell MS, Rosenthal SL, Soren K (2016) Human papillomavirus vaccination and cervical cytology outcomes among urban low-income minority females. JAMA Pediatr 170:445–452

Hsu WL, Chen JY, Chien YC, Liu MY, You SL, Hsu MM, Yang CS, Chen CJ (2009) Independent effects of EBV and cigarette smoking on nasopharyngeal carcinoma: a 20-year follow-up study on 9,622 males without family history in Taiwan. Cancer Epidemiol Biomark Prev 18:1218–1226

Hsu WL, Yu KJ, Chien YC, Chiang JY, Cheng YJ, Chen JY, Liu MY, Chou SP, You SL, Hsu MM, Lou PJ, Wang CP, Hong JH, Leu YS, Tsai MH, Su MC, Tsai ST, Chao WY, Ger LP, Chen PR, Yang CS, Hildesheim A, Diehl SR, Chen CJ (2011) Familial tendency and risk of nasopharyngeal carcinoma in Taiwan: effects of covariates on risk. Am J Epidemiol 173:292–299

Hsu WL, Pan WH, Chien YC, Yu KJ, Cheng YJ, Chen JY, Liu MY, Hsu MM, Lou PJ, Chen IH, Yang CS, Hildesheim A, Chen CJ (2012a) Lowered risk of nasopharyngeal carcinoma and intake of plant vitamin, fresh fish, green tea and coffee: a case-control study in Taiwan. PLoS ONE 7: e41779

Hsu WL, Tse KP, Liang S, Chien YC, Su WH, Yu KJ, Cheng YJ, Tsang NM, Hsu MM, Chang KP, Chen IH, Chen TI, Yang CS, Golstein AM, Chen CJ, Chang YS, Hildesheim A (2012b) Evaluation of human leukocyte antigen-A (HLA-A), other non-HLA markers on chromosome 6p21 and risk of nasopharyngeal carcinoma. PLoS ONE 7:e42767

Hsu WL, Chien YC, Yu KJ, Chiang CJ, Chen JY, LouPJ, Wang CP, You SL, Diehl SR, Hildesheim A, Chen CJ (2013) A scoring system for the prediction of long-term risk of nasopharyngeal carcinoma. Poster presented at the 6th International nasopharyngeal carcinoma symposium, Istanbul, Kurkey, 20–22 June

Hu HH, Liu J, Lin YL, Luo WS, Chu YJ, Chang CL, Jen CL, Lee MH, Lu SN, Wang LY, You SL, Yang HI, Chen CJ for the REVEAL-HBV Study Group (2016) The rs2296651 (S267F) variant on NTCP (SLC10A1) is inversely associated with chronic hepatitis B and progression to cirrhosis and hepatocellular carcinoma in patients with chronic hepatitis B. Gut 65:1514–1521

Huang YT, Jen CL, Yang HI, Lee MH, Lu SN, Iloeje UH, Chen CJ (2011) REVEAL-HBV/HCV Study Group. Lifetime risk and gender difference of hepatocellular carcinoma among patients affected with chronic hepatitis B and C. J Clin Oncol 29:3643–3650

Huang YT, Yang HI, Liu J, Lee MH, Freeman JR, Chen CJ (2016a) Mediation analysis of hepatitis B and C in relation to hepatocellular carcinoma risk. Epidemiology 27:14–20

Huang YT, Freeman JR, Yang HI, Liu J, Lee MH, Chen CJ (2016b) Mediation effect of hepatitis B and C on mortality. Eur J Epidemiol 31:625–633

IARC (2012a) (International Agency for Research on Cancer) A review of human carcinogens. Part B: biological agent. IARC, Lyon, IARC

IARC (2012b) Carcinogenicity of malaria and of some polyomaviruses. Lancet Oncol 13:339–340

Ioannou GN, Green PK, Berry K (2018) HCVeradication induced by direct-acting antiviral agents reduces the risk of hepatocellular carcinoma. J Hepatol 68:25–32

Kim J, Bell C, Sun M, Kliewer G, Xu L, McInerney M, Svenson LW, Yang H (2016) Effect of human papillomavirus vaccination on cervical cancer screening in Alberta. Can Med Assoc J 188:E281–E288

Lee MH, Yang HI, Lu SN, Jen CL, Yeh SH, Liu CJ, Chen PJ, You SL, Wang LY, Chen WJ, Chen CJ (2010) Hepatitis C virus seromarkers and subsequent risk of hepatocellular carcinoma: long-term predictors from a community-based cohort study. J Clin Oncol 28:4587–4593

Lee MH, Yang HI, Jen CL, Lu SN, Yeh SH, Liu CJ, You SL, Sun CA, Wang LY, Chen WJ, Chen CJ (2011) Community and personal risk factors for hepatitis C virus infection: a survey of 23,820 residents in Taiwan in 1991–2. Gut 60:688–694

Lee MH, Yang HI, Liu J, Batrla-Utermann R, Jen CL, Iloeje UH, Jun J, Lu SN, You SL, Wang LY, Chen CJ, For the REVEAL-HBV Study Group (2013) Prediction models of long-term cirrhosis and hepatocellular carcinoma risk in chronic hepatitis B patients: risk scores integrating host and virus profiles. Hepatology 58:546–554

Lee MH, Yang HI, Yuan Y, L'Italien G, Chen CJ (2014) Epidemiology and natural history of hepatitis C virus infection. World J Gastroenterol 20:9270–9280

Lee MH, Yang HI, Lu SN, Jen CL, You SL, Wang LY, L'Italian G, Chen CJ, Yuan Y, For the REVEAL-HCV Study Group (2014b) Hepatitis C virus genotype 1b increases cumulative lifetime risk of hepatocellular carcinoma. Int J Cancer 135:1119–1126

Lee MH, Lu SN, Yuan Y, Yang HI, Jen CL, You SL, Wang LY, L'Italien G, Chen CJ, For the REVEAL-HCV Study Group (2014c) Development and validation of a clinical scoring system for predicting risk of hepatocellular carcinoma in asymptomatic individuals seropositive for anti-HCV antibodies. PLoS One 9(5):e94760

Lee MH, Yang HI, Lu SN, Lin YJ, Jen CL, Wong KG, Chan SY, Chen LC, Wang LY, L'Italien G, Yuan Y, Chen CJ (2015) Polymorphisms near the IFNL3 gene associated with HCV RNA spontaneous clearance and hepatocellular carcinoma risk. Sci Rep 5:17030

Lee MH, Huang CF, Lai HC, Lin CY, Dai CY, Liu CJ, Wang JH, Huang JF, Su WP, Yang HC, Kee KM, Yeh ML, Chuang PH, Hsu SJ, Huang CI, Kao JT, Chen CC, Chen SH, Jeng WJ, Yang HI, Yuan Y, Lu SN, Sheen IS, Liu CH, Peng CY, Kao JH, Yu ML, Chuang WL, Chen CJ (2017) Clinical efficacy and post-treatment seromarkers associated with the risk of hepatocellular carcinoma among chronic hepatitis C patients. Sci Rep 7:3718

Lee MH, Huang YH, Chen HY, Khor SS, Chang YH, Lin YJ, Jen CL, Lu SN, Yang HI, Nishida N, Sugiyama M, Mizokami M, Yuan Y, L'Italien G, Tokunaga K, Chen CJ, For the REVEAL-HCV Cohort Study (2018) Human leukocyte antigen variants and risk of hepatocellular carcinoma modified by hepatitis C virus genotypes: a genome-wide association study. Hepatology 67:651–661

Liaw YF, Sung JJ, Chow WC, Farrell G, Lee CZ, Yuen H, Tanwandee T, Tao QM, Shue K, Keene ON, Dixon JS, Gray DF, Sabbat J (2004) Lamivudine for patients with chronic hepatitis B and advanced liver disease. N Engl J Med 351:1521–1531

Lin YJ, Lee MH, Yang HI, Jen CL, You SL, Wang LY, Lu SN, Liu J, Chen CJ, For the REVEAL-HBV Study Group (2013) Predictability of liver-related seromarkers for the risk of hepatocellular carcinoma in chronic hepatitis B patients. PLoS One 8:e61448

Lin YJ, Chang CL, Chen LC, Hu HH, Liu J, Korenaga M, Huang YH, Jen CL, Su CY, Nishida N, Sugiyama M, Lu SN, Wang LY, Yuan Y, L'Italien G, Yang HI, Mizokami M, Chen CJ, Lee MH (2018) A Glycomarker for short-term prediction of hepatocellular carcinoma: a longitudinal study with serial measurements. Clin Transl Gastroenterol 9:183

Lin D, Yang HI, Nguyen NH, Hoang J, Kim Y, Vu V, Le A, Chaung K, Nguyen V, Trinh H, Li J, Zhang J, Hsing A, Chen CJ, Nguyen MH (2016) Reduction of chronic hepatitis B hepatocellular carcinoma with antiviral therapy, including low risk patients. Aliment Pharmacol Therap 44:846–855

Liu J, Yang HI, Lee MH, Lu SN, Jen CL, Batrla-Utermann R, Wang LY, You SL, Hsiao CK, Chen PJ, Chen CJ, For the REVEAL-HBV Study Group (2014) Spontaneous seroclearance of hepatitis B seromarkers and subsequent risk of hepatocellular carcinoma. Gut 63:1648–1657

Liu J, Yang HI, Lee MH, Hsu WL, Chen HC, Chen CJ (2015) Epidemiology of virus infection and human cancer. In: Shurin MR, Thanavala Y, Ismail N (eds) Infection and cancer: bi-directorial interactions. Springer, Switzerland, pp 23–48

Liu J, Yang HI, Lee MH, Jen CL, Hu HH, Lu SN, Wang LY, You SL, Chen CJ (2016) Alcohol drinking mediates the association between polymorphisms of ADH1B and ALDH2 and hepatitis B-related hepatocellular carcinoma. Cancer Epdemiol Biomark Prev 25:693–699

Loomba R, Liu J, Yang HI*, Lee MH, Lu SN, Wang LY, Iloeje UH, You SL, Brenner D, Chen CJ, For the REVEAL-HBV Study Group (2013a) Synergistic effects of family history of hepatocellular carcinoma and hepatitis B virus infection on risk for incident hepatocellular carcinoma. Clin Gastroenterol Hepatol 11:1636–1645

Loomba R, Yang HI, Su J, Brenner D, Barrett-Connor E, Iloeje U, Chen CJ (2013) Synergism between obesity and alcohol in increasing the risk of hepatocellular carcinoma: A prospective cohort study. Am J Epidemiol 177:333–342

Matsuda T, McCombs J, Lee MH, Tonnu-Mihara I, L'Italien G, Saab S, Hines P, Yuan Y (2014) External validation of the risk-prediction model for hepatocellular carcinoma [HCC] from the REVEAL-HCV study using data from the U.S. veterans affairs [VA] Health system. In: Paper presented at the EASL, London, UK, 9–13 April

Papatheodoridis GV, Idilman R, Dalekos GN, Buti M, Chi H, van Boemmel F, Calleja JL, Sypsa V, Goulis J, Manolakopoulos S, Loglio A, Siakavellas S, Keskin O, Gatselis N, Hansen BE, Lahretz M, de la Revilla J, Savvidou S, Kourikou A, Vlachogiannakos I, Galanis K, Yurdaydin C, Berg T, Colombo M, Esteban R, Janssen HLA, Lampertico P (2017) The risk of hepatocellular carcinoma decreases after first 5 years of entecavir or tenofovir in Caucasians with chronic hepatitis B. Hepatology 66:1444–1453

Poh Z, Shen L, Yang HI, Seto WK, Wong VW, Lin CY, Goh BG, Chang PJ, Chan HL, Yuen MF, Chen CJ, Tan CK (2016) Real world risk score for hepatocellular carcinoma: development and external validation of a clinically practical predictor (letter to editor). Gut 65:887–888

Pollock KG, Kavanagh K, Potts A, Love J, Cuschieri K, Cubie H, Robrtson C, Cruickshank M, Palmer TJ, Nicoll S, Donaghy M (2014) Reduction of low- and high-grade cervical abnormalities associated with high uptake of the HPV bivalent vaccine in Scotland. Br J Cancer 111:1824–183

Proietti FA, Carneiro-Proietti AB, Catalan-Soares BC, Murphy EL (2005) Global epidemiology of HTLV-1 infection and associated diseases. Oncogene 24:6058–6068

Semeere AS, Busakhala N, Martin JN (2012) Impact of antiretroviral therapy on the incidence of Kaposi's sarcoma in resource-rich and resource-limited settings. Curr Opinion Oncol 24:522–530

Sun CA, Wu DM, Lin CC, Lu SN, You SL, Wang LY, Wu MH, Chen CJ (2003) Incidence and co-factors of hepatitis C virus-related hepatocellular carcinoma: a prospective study of 12,008 men in Taiwan. Am J Epidemiol 157:674–682

UNAIDS (2012) http://data.unaids.org/pub/epislides/2012/2012_epiupdate_en.pdf

Wang LY, Hatch M, Chen CJ, Levin B, You SL, Lu SN, Wu MH, Wu WP, Wang LW, Wang Q, Huang GT, Yang PM, Lee HS, Santella RM (1996) Aflatoxin exposure and the risk of hepatocellular carcinoma in Taiwan. Int J Cancer 67:620–625

Wang LY, You SL, Lu SN, Ho HC, Wu MH, Sun CA, Yang HI, Chen CJ (2003) Risk of hepatocellular carcinoma and habits of alcohol drinking, betel quid chewing, and cigarette smoking: a cohort of 2416 HBsAg-seropositive and 9421 HBsAg-seronegative male residents in Taiwan. Cancer Cause Control 14:241–250

Ward MH, Pan WH, Cheng YJ, Li FH, Brinton LA, Chen CJ, Hsu MM, Chen IH, Levine PH, Yang CS (2000) Dietary exposure to nitrite and nitrosamines and risk of nasopharyngeal cancer in Taiwan. Int J Cancer 86:603–609

WHO (World Health Organization) (2017) Global hepatitis report, 2017. World Health Organization, Geneva

Wong VW, Chan SL, Mo F, Chan TC, Loong HH, Wong GL, Lui YY, Chan AT, Sung JJ, Yeo W, Chan HL, Mok TS (2010) Clinical scoring system to predict hepatocellular carcinoma in chronic hepatitis B carriers. J Clin Oncol 28:1660–1665

Yang HI, Lu SN, You SL, Sun CA, Wang LY, Hsiao K, Chen PJ, Chen DS, Liaw YF, Chen CJ (2002) Hepatitis B e antigen and the risk of hepatocellular carcinoma. New Engl J Med 347:168–174

Yang HI, Yeh SH, Chen PJ, Iloeje UH, Jen CL, Wang LY, Lu SN, You SL, Chen DS, Liaw YF, Chen CJ, REVEAL-HBV Group (2008) Association between Hepatitis B virus genotype and mutants and the risk of hepatocellular carcinoma. J Natl Cancer Inst 100:1134–1143

Yang HI, Sherman M, Su J, Chen PJ, Liaw YF, Iloeje UH, Chen CJ (2010) Nomograms for risk of hepatocellular carcinoma in patients with chronic hepatitis B virus infection. J Clin Oncol 28:2437–2444

Yang HI, Yuen MF, Chan HL, Han KH, Chen PJ, Kim DY, Ahn SH, Chen CJ, Wong VW, Seto WK (2011) Risk estimation for hepatocellular carcinoma in chronic hepatitis B (REACH-B): development and validation of a predictive score. Lancet Oncol 12:568–574

Yang HI, Lee MH, Liu J, Chen CJ (2014) Risk calculators for hepatocellular carcinoma in patients affected with chronic hepatitis B in Asia. World J Gastroenterol 20:6244–6251

Yang HI, Tseng TC, Liu J, Lee MH, Liu CJ, Su TH, Batrla-Utermann R, Chan HL, Kao JH, Chen CJ (2016) Incorporating serum level of hepatitis B surface antigen or omitting level of hepatitis B virus DNA does not affect calculation of risk for hepatocellular carcinoma in patients without cirrhosis. Clin Gastroenterol Hepatol 14:461–468

Yu MW, Chen CJ (1993) Elevated serum testosterone levels and risk of hepatocellular carcinoma. Cancer Res 53:790–794

Yu MW, Gladek-Yarborough A, Chiamprasert S, Santella RM, Liaw YF, Chen CJ (1995a) Cytochrome P-450 2E1 and glutathione S-transferase M1 polymorphisms and susceptibility to hepatocellular carcinoma. Gastroenterology 109:1266–1273

Yu MW, Hsieh HH, Pan WH, Yang CS, Chen CJ (1995b) Vegetable consumption, serum retinol level and risk of hepatocellular carcinoma. Cancer Res 55:1301–1305

Yu MW, Chiu YH, Chiang YC, Chen CH, Lee TH, Santella RM, Chen HD, Liaw YF, Chen CJ (1999a) Plasma carotenoids, glutathione S-transferases M1 and T1 genetic polymorphisms, and risk of hepatocellular carcinoma: independent and interactive effects. Am J Epidemiol 149:621–629

Yu MW, Horng IS, Hsu KH, Chiang YC, Liaw YF, Chen CJ (1999b) Plasma selenium levels and risk of hepatocellular carcinoma among men with chronic hepatitis B virus infection. Am J Epidemiol 150:367–374

Yu MW, Chang HC, Liaw YF, Lin SM, Lee SD, Chen PJ, Hsiao TJ, Lee PH, Chen CJ (2000a) Familial risk of hepatocellular carcinoma among chronic hepatitis B carriers and their relatives. J Natl Cancer Inst 92:1159–1164

Yu MW, Cheng SW, Lin MW, Yang SY, Liaw YF, Chang HC, Hsiao TJ, Lin SM, Lee SD, Chen PJ, Liu CJ, Chen CJ (2000b) Androgen-receptor CAG repeat, plasma testosterone levels, and risk of hepatitis B-related hepatocellular carcinoma. J Natl Cancer Inst 92:2023–2028

Yu MW, Yang SY, Pan IJ, Lin CL, Liu CJ, Liaw YF, Lin SM, Chen PJ, Lee SD, Chen CJ (2003) Polymorphisms in XRCC1 and glutathione S-transferase genes and hepatitis B-related hepatocellular carcinoma. J Natl Cancer Inst 95:1485–1488

Yuen MF, Tanaka Y, Fong DY, Fung J, Wong DK, Yuen JC, But DY, Chan AO, Wong BC, Mizokami M, Lai CL (2009) Independent risk factors and predictive score for the development of hepatocellular carcinoma in chronic hepatitis B. J Hepatology 50(1):80–88

Zeng H, Wu HC, Wang Q, Yang HI, Chen CJ, Santella RM, Shen J (2017) Telomere length and risk of hepatocellular carcinoma: a nested case-control study in the Taiwan Cancer Screening Program cohort. Anticancer Res 37:637–644

Mechanisms of Hepatitis B Virus-Induced Hepatocarcinogenesis

Jiyoung Lee, Kuen-Nan Tsai, and Jing-hsiung James Ou

1 Introduction to Hepatitis B Virus

Hepatitis B virus (HBV) is a hepatotropic virus that can cause severe liver diseases, including acute and chronic hepatitis, cirrhosis, and hepatocellular carcinoma (HCC). There are approximately 250 million people worldwide suffering from chronic HBV infection, resulting in nearly one million deaths annually (Iloeje et al. 2006; Kao and Chen 2002; WHO 2017). Most chronic HBV carriers in endemic areas such as sub-Saharan Africa and China acquired the virus from their carrier mothers early in life, although HBV can also be transmitted via other pathways such as sex or the sharing of injection needles (Milich and Liang 2003; Shin et al. 2016; Ou 1997). Due to the lack of an effective treatment, HBV remains one of the most serious health problems. The infection by HBV contributed to slightly more than 50% of HCC cases worldwide in a 2002 study, rendering it the most important carcinogenic factor of HCC (Parkin 2006). HBV can induce HCC via the induction of chronic liver inflammation, which can cause oxidative DNA damage, liver injury and regeneration, and eventual oncogenic transformation of hepatocytes. Factors like age, gender, duration of infection, alcohol consumption, and the exposure to carcinogens such as aflatoxin can also increase the risk of HCC for HBV patients. In this chapter, we will focus on the direct effect of HBV on hepatocytes in the induction of hepatocarcinogenesis.

J. Lee · K.-N. Tsai · J. J. Ou (✉)
Department of Molecular Microbiology and Immunology,
University of Southern California Keck School of Medicine,
2011 Zonal Avenue, HMR-401, Los Angeles, CA 90033, USA
e-mail: jamesou@usc.edu

© Springer Nature Switzerland AG 2021
T.-C. Wu et al. (eds.), *Viruses and Human Cancer*, Recent Results
in Cancer Research 217, https://doi.org/10.1007/978-3-030-57362-1_3

47

1.1 HBV Genome and Lifecycle

HBV belongs to the *Hepadnaviridae* family (Liang 2009; Tiollais et al. 1985). It is an enveloped virus with a partially double-stranded and circular DNA genome of approximate 3.2 Kb. The HBV genome is remarkably compact and encodes four overlapping genes named C, P, S, and X. The C gene codes for the 21-kDa core protein, which forms the viral core particle that displays the core antigenic determinant (i.e., the core antigen, HBcAg) (Tsai et al. 2018), and a related 25-kDa protein termed the precore protein. The precore protein contains the entire sequence of the core protein plus an amino-terminal extension of 29 amino acids (i.e., the "precore sequence") (Ou 1997; Ou et al. 1986). The first 19 amino acids of the precore protein serve as a signal peptide that directs the precore protein to the endoplasmic reticulum (ER) where it is removed by the signal peptidase to generate the 22-kDa precore protein derivative. This precore protein derivative is subsequently cleaved at its carboxy terminus at multiple sites by the furin-like protease in the Golgi and secreted (Ito et al. 2009). The secreted precore protein derivatives are known as the e antigen (i.e., HBeAg). HBeAg is not essential for HBV replication (Chen et al. 1992; Lamberts et al. 1993), but it has immunomodulatory functions and is important for HBV to establish the persistent infection after the mother-to-child transmission (Milich and Liang 2003; Ou 1997; Tsai et al. 2018; Tian et al. 2016). The P gene codes for the viral DNA polymerase, which is also a reverse transcriptase (Jones and Hu 2013). The S gene codes for the three co-carboxy-terminal envelope proteins known as large (or preS1), middle (or preS2), and small (or major) surface antigen (HBsAg) proteins (Ueda et al. 1991). The X gene codes for the X protein (i.e., HBx), which is a regulatory protein that has diverse functions and can promote hepatocarcinogenesis ((Tang et al. 2006; Kohara et al. 2011; Ng and Lee 2011), also see below). After the infection of hepatocytes, the partially double-stranded HBV genome is delivered into the nucleus where it is repaired to form the covalently closed circular DNA (cccDNA). The cccDNA serves as the template for viral RNA transcription (Guidotti and Chisari 2006). The transcription is unidirectional and regulated by two enhancer elements and four distinct promoters, resulting in the formation of a diverse set of mRNAs that include the 3.5 Kb precore protein mRNA, the 3.5 Kb core protein mRNA, the 2.4 Kb preS1 mRNA, the 2.1 Kb preS2 and major S mRNAs, and the 0.7 Kb X mRNA (Locarnini 2004; Moolla et al. 2002). All of the HBV RNA transcripts are capped, polyadenylated, and terminated at the identical 3' end. The core protein mRNA codes for both the core protein and the viral DNA polymerase. It is also known as the pregenomic RNA (pgRNA) and is packaged by the core protein to form the core particle, in which the pgRNA is converted to the circular and partially double-stranded DNA genome by the viral DNA polymerase that is also packaged. The core particle subsequently interacts with HBsAg embedded in intracellular membranes to form the mature virion, which is then released from infected hepatocytes via a pathway that involves multivesicular bodies (Watanabe et al. 2007; Chou et al. 2015). HBsAg can also be released from cells as empty subviral particles, often in vast excess of infectious viral particles (Gavilanes et al. 1990; Ganem and Prince 2004).

1.2 HBV Genotypes and Clinical Outcomes of Chronic Hepatitis B

Based on the sequence divergence of greater than 8% over the entire viral genome or 4% in the S gene, HBV has been classified into eight well-characterized genotypes A through H (Norder et al. 2004; Stuyver et al. 2000; Arauz-Ruiz et al. 2002; Shi et al. 2012) and two additional genotypes I and J (Tatematsu et al. 2009; Tran et al. 2008). These genotypes have distinct ethnic and geographic distributions. For example, genotype A is prevalent in Sub-Saharan Africa, Western African, and Northern Europe, genotype D is prevalent in Africa, Europe, India, and the Mediterranean region, and genotypes B and C are widespread in Asia (Lin and Kao 2011). HBV genotypes have also been shown to display different pathogenicity and responses to type-I interferon (IFN)-based therapies. Chronic hepatitis B (CHB) patients infected by genotype A or B virus exhibit higher rates of seroclearance of HBeAg and HBV DNA in response to the IFN-α therapy than patients infected by genotypes C or D virus (Erhardt et al. 2005; Kao et al. 2000; Janssen et al. 2005). Furthermore, children chronically infected by genotype A HBV were found to have lower viral load and less severe symptoms than children infected by genotype D HBV (Oommen et al. 2006; Thakur et al. 2002). It has also been reported that differences between HBV genotypes A and D correlated with differences in resistance to the deoxycytidine analog (Zollner et al. 2004) and liver pathogenesis (Verschuere et al. 2005; Yang et al. 2008). Serologically, genotype C tends to be related to a higher incidence of HBeAg positivity and higher viral load than genotype B (Kao 2003). It has also been shown that genotype C is correlated with more severe liver diseases including cirrhosis and HCC, while genotype B is correlated with the development of HCC in young patients with noncirrhotic liver. The biological mechanisms underlying these phenotypic differences among different HBV genotypes, however, remain largely unknown.

1.3 Spliced HBV RNA Variants and HBV Pathogenesis

A series of spliced HBV RNAs (spRNAs), derived from the pgRNA, has been found in the viral particles of CHB patients and in hepatoma cell lines transfected with the HBV genomic DNA (Gunther et al. 1997; Wu et al. 1991; Su et al. 1989). To date, the functions of these spRNAs are still enigmatic because they are not essential for HBV replication. A recent study examined the distribution of spRNA variants extracted from an individual with CHB following liver transplantation and found that the spRNA population altered dynamically over a 15-year period (Betz-Stablein et al. 2016). This finding suggests a possible relationship between spRNAs and HBV pathogenesis and/or persistence. This suggestion is supported by the study of Pol et al. who found that an spRNA-derived protein suppressed TNF-α-stimulated signaling and the infiltration of immune cells into the mouse liver during chronic inflammation (Pol et al. 2015). In addition, the increase of a fraction of HBV-spliced variants in CHB patients was also found to correlate with an impaired response to the IFN-α therapy, due to the inhibition of phosphorylation of IFN-activated *s*ignal transducer and activator of transcription (STAT) 1 and the consequent inhibition of the nuclear translocation of STAT1/STAT2 (Chen et al. 2015). Recent studies also indicated that spRNA-derived proteins could induce specific T-cell responses in HBV-infected patients (Mancini-Bourgine et al. 2007) and interfere with the assembly of HBV core particles and the transcription of HBV RNAs in the cells transfected with HBV DNA (Soussan et al. 2003; Wang et al. 2015). spRNAs also affect host immunity due to their effect on the synthesis of chemokines in hepatocytes, which may contribute to liver immunopathogenesis and immune escape of HBV during chronic HBV infection (Duriez et al. 2017). Furthermore, spRNAs can act as a repressor of HBV RNA transcription and affect the long-term outcomes of CHB diseases and responses to antiviral therapy (Tsai et al. 2015). Overall, the studies on spRNAs indicated that these RNAs might regulate the crosstalk between HBV and its host cells, contributing to HBV persistence and pathogenesis.

1.4   HBV Viral Load and HCC

HCC risk factors for CHB patients include the male gender, age, the serum alanine aminotransferase (ALT) level, the HBeAg level, and the serum HBV DNA level (i.e., viral load), but not the HBsAg level (Tseng et al. 2012). A serum viral DNA level of $\geq 10,000$ copies/mL is a strong risk predictor for HCC, independent of the HBeAg status, the ALT level, and liver cirrhosis (Yang et al. 2002; Iloeje et al. 2007; Chen et al. 2006). For HBeAg-negative patients with viral load of <10,000 copies/mL, the serum HBsAg level of $\geq 1,000$ IU/mL as well as the ALT level and age is a strong risk predictor for HCC (Tseng et al. 2012). High viral load is also a predictor of postoperative recurrence of HCC (Wu et al. 2009; Hung et al. 2008). These findings indicate an important role of HBV viral load in the development and recurrence of HCC.

2   Genetic and Epigenetic Modifications Induced by HBV

Genetic and epigenetic alterations have been found in both viral and host genomes in CHB patients and HBV-related HCC, indicating a critical relationship between these alterations and HBV-induced HCC.

2.1 HBV DNA Integration in Host Chromosomes

More than 80% of HBV-associated HCC had integration of HBV DNA in the host chromosomes (Brechot 2004). The integration of HBV DNA in the host chromosomes of HCC was first reported in 1980 (Brechot et al. 1980; Chakraborty et al. 1980; Edman et al. 1980), and the selection of the integration sites in the host chromosomes was initially thought to be random. However, the studies by whole-genome sequencing of HCC tissues in recent years had identified recurrent integration spots. The integration appears to spread over the entire genome in both tumor and non-tumor tissues, although the integration frequency is higher in tumor tissues than in their paired non-tumor tissues (Zhao et al. 2016). The integration sites in tumor tissues tend to enrich in certain promoter regions (Toh et al. 2013). Recurrent integration within the CpG islands and in the vicinity of the telomere in tumor samples indicates a possible association between the preferential targeting of these sites and the alteration of cellular gene regulation (Zhao et al. 2016). The recurring "hotspot" genes identified in multiple studies include genes encoding the telomerase reverse transcriptase (TERT), myeloid/lymphoid or mixed-lineage leukemia 3 (MLL3), MLL4, cyclin E1 (CCNE1), the protein tyrosine phosphatase receptor type D (PTPRD), unc-5 netrin receptor D (UNC5D), neuregulin 3 (NRG3), catenin delta 2 (CTNND2), the aryl-hydrocarbon receptor repressor (AHRR), SUMO-specific peptidase 5 (SENP5), Rho-associated coiled-coil containing protein kinase 1 (ROCK1), tumor protein 53 (TP53), Axin 1 (AXIN1), AT-rich interactive domain-containing protein 1A (ARID1A), ARID1B, ARID2, catenin beta 1 (CTNNB1), retinoic acid receptor beta (RARB), cyclin A2, fibronectin (FN1), and angiopoietin 1 (ANGPT1) (Zhao et al. 2016; Ding et al. 2012; Jiang et al. 2012; Sung et al. 2012; Wan et al. 2013; Fujimoto et al. 2012; Hai et al. 2014; Li et al. 2013). The PCR analysis using primers specific to human Alu repeat sequences and to HBV DNA to detect viral–host junctions added some more genes to the list: calcium signaling-related genes, 60S ribosomal protein-like encoding genes, platelet-derived growth factor (PDGF), and apoptosis-associated genes (Murakami et al. 2005; Tamori et al. 2005).

The integration of HBV DNA into host chromosomes is not an essential step of the HBV life cycle, and this integration frequently leads to the fragmentation, rearrangement, and disruption of the HBV genome. Rather, the integration of HBV DNA may cause instability of host chromosomes and also enhance the expression of genes related to cancer development, metastasis and angiogenesis, or inactivate tumor suppressor genes to eventually lead to the initiation of carcinogenesis.

As mentioned above, the integration also disrupts the HBV genome, which has a circular structure. A high frequency of the integration occurs in the HBx coding sequence, leading to the production of chimeric transcripts that contain both HBV and host sequences and the expression of C-terminally truncated HBx (Toh et al. 2013; Ou and Rutter 1985; Peng et al. 2005; Ma et al. 2008). C-terminally truncated HBx has been implicated in the induction of oxidative DNA damage (Jung and Kim 2013), expression of MMP-10 (Sze et al. 2013), upregulation of Wnt-5a (Liu et al.

2008), metastasis (Li et al. 2016), cell proliferation (Ma et al. 2008), and increase of the CD133-positive cell subset, which possessed cancer stem cell-like properties (Ng et al. 2016). It may also abrogate the growth-suppressive and apoptotic effects of full-length HBx ((Xu et al. 2007; Tu et al. 2001), also see below). Furthermore, it has been shown to promote hepatocellular proliferation, HCC cell invasion, and metastasis (Sze et al. 2013; Yip et al. 2011). Hybrid transcripts derived from viral and human genes may also be carcinogenic (Ou and Rutter 1985; Lau et al. 2014). The HBx-LINE1 hybrid transcript, for example, was detected in ~20% of HCC patients with a close association with poor prognosis. Approximately 40% of the breakpoints in the HBV genome were located near the 3'-end of the HBx coding sequence and the 5'-end of the precore protein coding sequence while another peak of the breakpoints was detected around nt. 300–500 of the major S protein coding sequence (Zhao et al. 2016). Truncated preS/S proteins had also been detected and found to have gene transactivation functions (Schluter et al. 1994).

2.2 Epigenetic Modifications

2.2.1 DNA Methylation

The term epigenetics refers to all the changes in the chromatin with no change in the DNA sequence. DNA methylation, histone modification, and RNA-associated silencing are three interactive epigenetic changes. Epigenetic changes can be reversed based on physiological conditions. The HBV DNA is subject to epigenetic modifications. The study of DNA methylome, an analysis of genome-wide distribution of DNA methylation, revealed that the HBV genome in liver tumors and cancer cell lines acquired more methylation than in precancerous conditions such as hepatitis and cirrhosis (Fernandez et al. 2009). The study also demonstrated that the methylation of the preS1/preS2/S locus progressed from nearly 0% in hepatic cirrhosis to 11% in chronic hepatitis and 52% in liver tumors. Approximately 70 ~ 80% of the CpG dinucleotides in vertebrate DNA is methylated while such methylation is absent in bacterial DNA (Bachmann and Kopf 2001; Krieg 2002). B cells distinguish between self-DNA and non-self-DNA by their methylation status. The HBV cccDNA is hypermethylated, which may prevent it from being recognized by the innate immune components (Kuss-Duerkop et al. 2018), although its X gene is hypomethylated to allow the steady expression of HBx, which has many regulatory functions to support the HBV lifecycle (Fernandez et al. 2009).

The infection by HBV also affects the methylation of cellular genes. These genes include cell cycle-related genes such as $p16^{INK4A}$ and $p21^{WAF1/CIP1}$, Ras association domain family member 1 (RASSF1A), glutathione S-transferase pi 1 (GSTP1), and cadherin 1 (CDH1) (Rongrui et al. 2014). When compared with non-tumor tissues, the expression of CD82, a tumor suppressor and general suppressor of metastasis, was found to be elevated in HCC tissues (Yu et al. 2014). The introduction of HBV DNA into HepG2 hepatoblastoma cells increased the methylation in the CD82 promoter and suppressed its transcriptional activity (Yu et al. 2014). The hypomethylation of CpG islands in the HBx-transgenic mouse liver led to the downregulation of E- and N-cadherins, which are important for the epithelial-to-mesenchymal transition, Smad6 and Kcp, which are components of the Smad-dependent TGF-β–signaling pathway, and the Wnt-signaling pathway (Lee et al. 2014). These pathways are well known to participate in tumorigenesis. The expression of ankyrin-repeat-containing, SH3-domain-containing, and proline-rich-region containing protein family 1 (ASPP1) and 2 (ASPP2) was downregulated in HCC cell lines and tissues from HCC patients infected by HBV (Zhao et al. 2010). Particularly, a decrease of ASPP2 in HepG2 hepatoblastoma cells was attributable to HBx, which enhanced the recruitment of DNA methyltransferase 1 (DNMT1) and DNMT3a to the ASPP2 promoter for its methylation (Zhao et al. 2010).

The HBx-directed recruitment of DNMT3a and DNMT3b to CpG island 1 within the promoter of metastasis-associated protein 1 (MTA1) activated the expression of MTA1. In this case, CpG island 1 of the MTA1 promoter contains the binding site of the transcriptional repressor p53. The methylation of CpG island 1 reduced the binding of p53, upregulated MTA1 expression, and thus enhanced the invasiveness and the metastasis of HCC (Lee et al. 2012). It had also been reported that HBx-induced upregulation of DNMT3a and DNMT3b led to the methylation of the SOCS-1 promoter and reduced the expression of SOCS-1, a potential tumor suppressor (Fu et al. 2016). Other tumor suppressive genes regulated by HBx-modulated epigenetic events include RASSF1A, procadherin-10 (PCDH10), insulin-like growth factor-binding protein 3 (IGFBP3), and E-cadherin (Ying et al. 2006; Qiu et al. 2014; Fang et al. 2013; Arzumanyan et al. 2012).

2.2.2 Histone Modifications

Post-translational modifications of histones include methylation, acetylation, and phosphorylation. Histone acetylation, which is important for gene regulation, involves the transfer of the acetyl group from acetyl-CoA to the lysine residues in the N-terminal tail of histones. Histone acetylation allows binding of trans-acting factors to activate gene expression and is regulated by histone acetyltransferases (HATs) and histone deacetylases (HDACs). The upregulation of HDACs was observed in HCC and many other human cancers (Weichert 2009), and the suppressive effect of these HDACs on the expression of tumor suppressor genes like $p21^{WAF1/CIP1}$ and $p27^{KIP-1}$ was thought to play an important role in carcinogenesis (Xie et al. 2012). Binding of HBx to methyl-CpG-binding domain protein 2 (MBD2) and transcriptional co-activator CBP/p300 led to upregulation of insulin-like growth factor 2 (IGF-2). HBx facilitated the formation of the MBD2-HBx-CBP/p300 complex, which promoted hypomethylation and acetylation of histone H4 at the IGF-2 promoter, leading to its transcriptional activation (Liu et al. 2015).

2.3 MicroRNAs in HBV-Induced Hepatocarcinogenesis

MicroRNAs are small non-coding RNA molecules involved in gene silencing. Modulation of these microRNAs is implicated in the development of

HBV-associated HCC. HCC patients exhibit a higher oncogenic miR-21 level than healthy individuals (Guo et al. 2017). In addition to miR-21, the upregulation of miR-224, miR-545, miR-374a and the miR-17-92 polycistron that expresses miR-17, 18a, 19a/b, 20a, and 92a (Connolly et al. 2008; Gao et al. 2011; Zhao et al. 2014), and the downregulation of miR-145, miR-199b, Let-7a, and miR-152 were observed in HBV-related HCC (Gao et al. 2011; Wang et al. 2010; Huang et al. 2010; Xu et al. 2014). The HBx-induced upregulation of miR-21 could enhance cell proliferation by inhibition of programmed cell death protein-4 (PDCD4) and PTEN (Damania et al. 2014). It has been proposed that the upregulation of miR-21 by HBx is via the induction of the IL-6 expression followed by the STAT3 activation (Li et al. 2014). Acyl-CoA synthetase long-chain family member 1 (ACSL1) is an important lipid metabolism enzyme in the liver. MiR-205 can directly bind to the 3'-untranslated region (UTR) of the ACSL1 mRNA to regulate its expression (Cui et al. 2014). Cui et al. found a decrease in the level of miR-205 in HCC and a deregulated lipid metabolism as a consequence (Cui et al. 2014). A deregulated lipid metabolism can lead to lipid deposition due to excessive lipid biosynthesis (Tang et al. 2018). A similar increase in ACSL1 and triglyceride was detected in the liver tumors of HBx-transgenic mice, implicating HBx in the suppression of miR-205 (Cui et al. 2014). Wang et al. recently reported that the expression of miR-98, miR-375, miR-335, miR-199a-5p, and miR-22 was altered by HBV infection in patients and that the expression of miR-150, miR-342-3p, miR-663, miR-20b, miR-92a-3p, miR-376c-3p, and miR-92b was altered in HBV-related HCCs (Wang et al. 2017). The biological significance of the alteration of these microRNAs in HBV replication and carcinogenesis, however, is unclear.

2.4 Genetic Variations in HBV DNA

HBV DNA polymerase is error-prone and lacks the proofreading function. The error frequency of the HBV polymerase was estimated to be 6.28×10^{-4} (i.e., one misincorporation in every 1591 nucleotides synthesized) (Park et al. 2003). All four genes in the HBV genome may be mutated (Toh et al. 2013). As mentioned above, the preS1/preS2/S sequence codes for three co-carboxy-terminal envelope proteins termed large (or preS1), middle (or preS2), and small (or major) HBsAg proteins. The large HBsAg functions in the initial binding to the receptor on hepatocytes for HBV to initiate the infection. The function of the middle HBsAg is not clear and its deletion does not affect the replication and maturation of the virus. The small HBsAg is the most abundant HBsAg, and the major hydrophilic region (MHR) located between amino acids 99 and 169 of this protein contains the major antigenic determinant (amino acids 124-147) known as the "a" determinant. This antigenic epitope is recognized by neutralizing antibodies, and mutations in this epitope can escape vaccine-induced immunity (Caligiuri et al. 2016). For example, the substitution mutation at amino acid 145 from glycine to arginine results in immune escape of HBV in patients (Zanetti et al. 1988). Genetic variations within MHR, which reduce the antigenicity of HBsAg and the production of antibody against HBsAg (Aragri et al. 2016), may also be associated with the reactivation of HBV following the immune suppressive therapy (Salpini et al. 2015). In contrast to the mutations at the "a" antigenic determinant, which are associated with the immune escape, mutations in the preS sequence, especially the N-terminal deletion in the preS2 sequence and the C-terminal deletion in the preS1 sequence, are frequently found in the sera and HCC tissues of HCC patients (up to ∼60%) (Pollicino et al. 2014; Fan et al. 2001; Wang et al. 2006).

The large HBsAg protein cannot be secreted and its secretion requires the help from the major HBsAg (Ou and Rutter 1987; Standring et al. 1986). The deletion in the preS sequence may disrupt the interaction between the large HBsAg protein and the major HBsAg protein and lead to its retention in the ER and the induction of ER stress (Pollicino et al. 2014). The ER stress-induced oxidative DNA damage can cause genome instability and ultimately lead to the development of HCC (Wang et al. 2006). Indeed, transgenic mice expressing the large HBsAg with deletions in the preS sequence or carrying the HBV genome with a deletion in the preS2 sequence had been shown to develop hepatic dysplasia and HCC (Wang et al. 2006; Na et al. 2011), confirming the role of preS deletions in HBV carcinogenesis. The development of ground-glass hepatocytes (GGH), which is characterized by abundant and overloaded ER and a liver pathology found in CHB patients, was also found to be associated with the accumulation of the mutated large HBsAg protein in the ER (Su et al. 2014; Chen et al. 2006; Su et al. 2008). The preS mutants had also been shown to enhance anchorage-independent growth of cells, the expression of ER chaperones including GRP78 and GRP94, and the induction of DNA repair gene ogg-1 and inflammatory cytokines (Su et al. 2014; Hung et al. 2004; Hsieh et al. 2004).

Another HBV mutant that is associated with an increased risk of HCC contains a novel mutation in the preS1 sequence. This mutation changes amino acid 4 of the preS1 protein from tryptophan to proline or arginine (W4P/R) and was found to be associated with severe liver diseases including HCC (Lee et al. 2013).

Naturally occurring nucleotide mutations have been identified in all four HBV genes (Caligiuri et al. 2016; Akarca and Lok 1995). One notable mutation is a double-nucleotide mutation of A to T at nucleotide (nt.) 1762 (A1762T) and G to A mutation at nt. 1764 (G1764A). This double mutation, which resides in the basal core promoter (BCP) that controls the transcription of the precore protein and the core protein mRNAs, reduces the precore protein mRNA level and hence the expression of HBeAg by 70% without affecting the core protein mRNA level (Buckwold et al. 1996). This mutant is frequently found in HBeAg-negative patients with chronic hepatitis (Buckwold et al. 1996; Hunt et al. 2000). Further analysis indicated that this double mutation converted a nuclear receptor-binding site in the BCP to an HNF1 transcription factor-binding site and enhanced HBV replication (Buckwold et al. 1996; Li et al. 1999; Locarnini et al. 2003; Parekh et al. 2003; Buckwold et al. 1997). This BCP mutation is often associated with the G to A mutation at nt. 1896 (G1896A) (Ou 1997). The G1986A mutation converts codon 28 of the precore sequence from TGG to the TAG termination codon and abolishes the expression of HBeAg (Locarnini et al. 2003; Kosaka et al. 1991; Liang et al.

1991; Omata et al. 1991). G1896 is located in a highly conserved stem-loop structure termed the epsilon (ε) structure, which is essential for encapsidation initiation of the pgRNA (Chiang et al. 1992; Bartenschlager and Schaller 1992). The G1896A mutation is infrequently observed for HBV genotype A, as G1896 base pairs with C1858, and its mutation to A would destabilize the ε structure for genotype A. In contrast, nt. 1858 is usually T for other HBV genotypes and the G1896A mutation would instead stabilize the ε structure to enhance HBV replication (Li et al. 1993; Lok et al. 1994; Rodriguez-Frias et al. 1995). Both the BCP mutation and the G1896A mutation are associated with a high incidence of HCC (Datta et al. 2012; Croagh et al. 2015; Liu et al. 2006; Kao et al. 2003; Yeh et al. 2010; Iloeje et al. 2012; Lin et al. 2005). Other BCP mutations like T1753C and C1766T either alone or in combination with other BCP mutations and the G1896A mutation brought about similar high carcinogenic consequences (Yang et al. 2008; Kramvis and Kew 1999). These mutations were shown to be associated with the upregulation of S phase kinase-associated protein 2 (SKP2) and the induction of proteasome-dependent degradation of p53 (Yan et al. 2015).

3 Role of HBx in HCC

HBx plays an important role in HBV replication and the loss of its expression significantly impairs HBV replication in HepG2 cells and in mouse models (Xu et al. 2002; Keasler et al. 2009). HBx protein is found in both the nucleus and the cytoplasm with distinctive roles in each locale (Hensel et al. 2017). In the nucleus, HBx stimulates the HBV gene expression and prevents the hypermethylation of the HBV cccDNA (Belloni et al. 2009). HBx in the cytoplasm influences signaling pathways and the activation/inactivation of transcription factors (Elmore et al. 1997; Bouchard et al. 2001). The association of HBx with multiple cellular proteins has been shown to have significant impacts on gene expression, cellular signaling, DNA repair, immune response, and cell proliferation.

3.1 HBx and Gene Expression

HBx does not directly bind DNA, but it can interact with multiple transcription factors to impact the expression of host and viral genes. Its binding partners include RNA polymerase-binding protein (RBP5), transcriptional factor IIB (TFIIB), transcriptional factor IIH (TFIIH), cAMP response element-binding protein (CREB), CREB1-binding protein (CBP)/p300, activating transcription factor 2 (ATF-2), activating protein (AP)-2, and nuclear factor kappa B (NF-κB) (Ali et al. 2014). HBx can also interact with SIRT1 and sequester it from β-catenin, thereby liberating β-catenin, which then transactivates the expression of cancer-promoting genes like cyclin-D1 and c-myc (Srisuttee et al. 2012; Polakis 2012). HBx also interferes with the protein phosphatase 1 (PP1) activity and thereby facilitates the

CREB recruitment to the HBV cccDNA (Cougot et al. 2012). The regulation of gene expression by HBx may also involve epigenetic modifications. HBx restricts the expression of secreted frizzled-related protein SFRP1 and SFRP5 by facilitating the binding of DNMT1 and DNMT3 to the promoters of these two genes for their subsequent hypermethylation (Xie et al. 2014). Indeed, the expression of SFRP1 and SFRP5, members of a family of extracellular glycoproteins, is downregulated in HBV-related HCC (Xie et al. 2014; Liu et al. 2016; Peng et al. 2014). The epigenetic silencing of SFRPs by HBx led to the transactivation of Wnt target genes including c-myc and cyclin D1, thus promoting epithelial–mesenchymal transition (EMT) and hepatocarcinogenesis (Polakis 2012; Xie et al. 2014). HBx also downregulates the expression of two transcription repression factors, SUZ12 and ZNF198. SUZ12 is an essential subunit of the polycomb repressor complex 2 (PRC2), and ZNF198 stabilizes the transcription repressive complex composed of LSD1, Co-REST, and HDAC1. These two transcription repressive complexes are held together by binding the long non-coding RNA HOTAIR. The effect of HBx on SUZ12 and ZNF198 results in the global alteration of chromatin landscape and the derepression of the target genes of PRC2 and LSD1/Co-REST/HDAC1 complexes such as epithelial cell adhesion molecule (EpCAM) (Zhang et al. 2015, 2016), which plays an important role in tumorigenesis and metastasis.

3.2 HBx and Cell Signaling

HBx also contributes to hepatocarcinogenesis by influencing cellular proliferation pathways (Ali et al. 2014). HBx was first reported to stimulate the Ras-Raf-MAP kinase signaling cascade in 1995 (Doria et al. 1995), and subsequently also reported to activate Src family kinases and calcium signaling (Bouchard et al. 2001; Klein and Schneider 1997). HBx could also bind to the C-terminus of the tumor suppressor p53 to inhibit its nuclear translocation and the induction of apoptosis (Elmore et al. 1997). The HBx-induced MTA1/HDAC1 complex had also been shown to stabilized hypoxia-inducible factor-1 alpha (HIF-1α) (Yoo et al. 2008). HIF-1α was highly expressed in HCC patients with poor survival rates (Xie et al. 2008). HBx-induced deacetylation of HIF-1α led to stabilization of this protein that is known to facilitate tumor metastasis, angiogenesis, and malignant transformation (Yoo et al. 2008; Liu et al. 2012). The HBx-induced expression of α-fetoprotein (AFP) and its receptor AFPR also drives Src expression via the activation of PI3K/mTOR pathway, which then promotes the invasion and metastasis of cancer cells (Zhu et al. 2015). Studies also indicated that HBx could activate Notch signaling (Gao et al. 2016; Kongkavitoon et al. 2016; Yang et al. 2017), and in one such study, it was shown that this activation of Notch signaling was due to the induction of Delta-like 4 (Dll4) by HBx via the activation of MEK1/2, PI3K/AKT, and NF-κB pathways (Kongkavitoon et al. 2016). These studies suggested a possible role of the HBx-Dll4-Notch axis in promoting hepatocarcinogenesis.

3.3 HBx and DNA Repair

Cells respond to DNA damage by inactivating cyclin-dependent kinases (CDKs) to stop the cell cycle progression. Failure in such response can lead to mutagenesis, genetic instability, and carcinogenesis. It has been shown that HBx can facilitate the accumulation of DNA damages by interfering with cell cycle control checkpoints and the DNA repair system. Indeed, the expression of HBx could augment sensitivity to hepatocarcinogen diethylnitrosamine or UV light, which causes DNA damage (Wang et al. 1995). HBx has also been shown to decrease the expression of xeroderma pigmentosum complementation group B (XBP) and group D (XBD), thereby interfering with the DNA helicase activity of transcription factor IIH (TFIIH), which is required for DNA repair (Wang et al. 1995). HBx can also bind to the C-terminus of p53 to sequestrate it in the cytoplasm (Kew 2011), and by competing with XPB/D for binding to the C-terminus of p53, HBx interferes with the DNA repair pathway (Qadri et al. 2011). The expression of HBx also reduced the efficacy of transcription-coupled nucleotide excision repair (TCNER), which would remove adducts on the transcribed strand of active genes, in human lymphoblastoid strain TK6 cells (Mathonnet et al. 2004). The expression of HBx in NH32, the isogenic p53-null counterpart of TK6, revealed that the oppressive effect of HBx on TCNER was both p53-dependent and p53-independent. Whether HBx also reduces TCNER in hepatocytes, however, remains to be determined. Interesting structural similarities between HBx and human thymine DNA glycosylase (TDG), an important protein in base excision repair pathway, was also observed (Wang et al. 1995). Although TDG did not directly affect HBV replication, it was proposed that HBx modified or impeded the function of TDG in the DNA repair process, which then led to accumulation of DNA damages and cellular transformation (van de Klundert et al. 2012).

Cell cycle arrest elicited by DNA damage allows cells to repair DNA damage prior to genome replication and cell cycle progression. This prevents DNA damage from being passed down to offspring cells. The unchecked cell cycle progression with the accumulation of genetic errors predisposes cells to carcinogenesis. Cyclin-dependent kinases CDK2 and CDC2 are important regulators of the S phase and the M phase, respectively, of the cell cycle, and their inhibition will induce cell cycle arrest (Satyanarayana and Kaldis 2009; Smits et al. 2000). The cyclin E/CDK2 complex phosphorylates p27^{KIP1} and p21$^{WAF1/CIP1}$, which are cyclin-dependent kinase inhibitors (CKIs), tagging them for degradation for the promotion of cell cycles (Zhu et al. 2005; Nguyen et al. 1999). HBx had been shown to stimulate cell cycle progression, forcing the transit through checkpoint controls and the entry into the S phase, possibly by activating CDK2 and CDC2 (Benn and Schneider 1995). In addition, HBx had also been shown to destabilize p27^{KIP1} (Lee and Kim 2009).

3.4 HBx Mutants

Because the HBx coding sequence overlaps with the BCP, the A1762T and G1764A double mutation within the BCP also changes amino acids 130 and 131 of the HBx sequence from lysine–valine (KV) to methionine–isoleucine (MI) and affects the biological activities of HBx (Ou 1997; Li et al. 1999). In one study, HBx with the K130M mutation was shown to inhibit the p21 expression by repressing the Sp1 transcription factor activity (Kwun and Jang 2004). The BCP/T1753A/T1768A combo mutation also led to the expression of an HBx mutant that decreased the expression of p53 and enhanced the entry of cells into the S phase (Yan et al. 2015). These findings indicated that mutations in HBx might also promote hepatocarcinogenesis via the acceleration of cell cycle progression. The HBx mutant with the proline to serine mutation at amino acid 38 was also frequently found in HBV-infected patients with HCC, suggesting that it may be an independent risk factor for the development of HCC (Muroyama et al. 2006).

4 Conclusions and Future Perspective

In this chapter, we summarize the mechanisms of hepatocarcinogenesis that are induced by HBV infection (Fig. 1). These mechanisms include the genetic and epigenetic alterations in the host genome that may be caused by the insertion of the

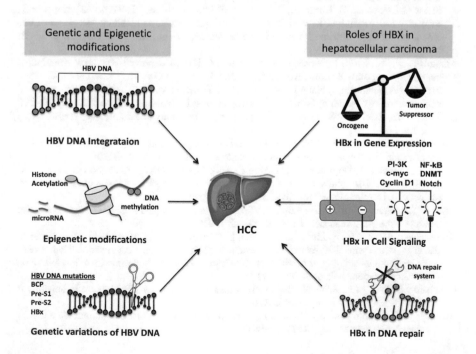

Fig. 1 Illustration of the oncogenic pathways induced by HBV

HBV DNA, the modification of histone acetylation or DNA methylation, or the induction of expression of microRNAs that can sensitize hepatocytes to carcinogenesis. The HBV DNA genome may also be altered by replication errors or its integration into the host chromosomes, leading to the production of variant gene products that include truncated HBx and the large HBsAg protein with deletions in the preS region. The HBx protein also plays a critical role in hepatocarcinogenesis induced by HBV, as it can bind to transcription factors to regulate the expression of oncogenes and tumor suppressor genes. It can also stimulate signaling pathways involved in cell cycle control and inhibit cellular DNA repair to result in the accumulation of genetic mutations in hepatocytes. These changes caused by HBV infection can create a microenvironment favorable to cell proliferation and predispose hepatocytes to oncogenic transformation. Understanding the detailed mechanisms of these alterations will facilitate the development of novel therapeutic interventions for treating HBV-associated HCC.

References

Akarca US, Lok AS (1995) Naturally occurring hepatitis B virus core gene mutations. Hepatology 22:50–60

Ali A, Abdel-Hafiz H, Suhail M, Al-Mars A, Zakaria MK, Fatima K, Ahmad S, Azhar E, Chaudhary A, Qadri I (2014) Hepatitis B virus, HBx mutants and their role in hepatocellular carcinoma. World J Gastroenterol 20:10238–10248

Aragri M, Alteri C, Battisti A, Di Carlo D, Minichini C, Sagnelli C, Bellocchi MC, Pisaturo MA, Starace M, Armenia D, Carioti L, Pollicita M, Salpini R, Sagnelli E, Perno CF, Coppola N, Svicher V (2016) Multiple hepatitis B virus (HBV) quasispecies and immune-escape mutations are present in HBV surface antigen and reverse transcriptase of patients with acute hepatitis B. J Infect Dis 213:1897–1905

Arauz-Ruiz P, Norder H, Robertson BH, Magnius LO (2002) Genotype H: a new Amerindian genotype of hepatitis B virus revealed in Central America. J Gen Virol 83:2059–2073

Arzumanyan A, Friedman T, Kotei E, Ng IO, Lian Z, Feitelson MA (2012) Epigenetic repression of E-cadherin expression by hepatitis B virus x antigen in liver cancer. Oncogene 31:563–572

Bachmann MF, Kopf M (2001) On the role of the innate immunity in autoimmune disease. J Exp Med 193:F47–F50

Bartenschlager R, Schaller H (1992) Hepadnaviral assembly is initiated by polymerase binding to the encapsidation signal in the viral RNA genome. EMBO J 11:3413–3420

Belloni L, Pollicino T, De Nicola F, Guerrieri F, Raffa G, Fanciulli M, Raimondo G, Levrero M (2009) Nuclear HBx binds the HBV minichromosome and modifies the epigenetic regulation of cccDNA function. Proc Natl Acad Sci U S A 106:19975–19979

Benn J, Schneider RJ (1995) Hepatitis B virus HBx protein deregulates cell cycle checkpoint controls. Proc Natl Acad Sci U S A 92:11215–11219

Betz-Stablein BD, Topfer A, Littlejohn M, Yuen L, Colledge D, Sozzi V, Angus P, Thompson A, Revill P, Beerenwinkel N, Warner N, Luciani F (2016) Single-molecule sequencing reveals complex genome variation of hepatitis B virus during 15 years of chronic infection following liver transplantation. J Virol 90:7171–7183

Bouchard MJ, Wang LH, Schneider RJ (2001) Calcium signaling by HBx protein in hepatitis B virus DNA replication. Science 294:2376–2378

Brechot C (2004) Pathogenesis of hepatitis B virus-related hepatocellular carcinoma: old and new paradigms. Gastroenterology 127:S56–S61

Brechot C, Pourcel C, Louise A, Rain B, Tiollais P (1980) Presence of integrated hepatitis B virus DNA sequences in cellular DNA of human hepatocellular carcinoma. Nature 286:533–535

Buckwold VE, Xu Z, Chen M, Yen TS, Ou JH (1996) Effects of a naturally occurring mutation in the hepatitis B virus basal core promoter on precore gene expression and viral replication. J Virol 70:5845–5851

Buckwold VE, Xu Z, Yen TS, Ou JH (1997) Effects of a frequent double-nucleotide basal core promoter mutation and its putative single-nucleotide precursor mutations on hepatitis B virus gene expression and replication. J Gen Virol 78(Pt 8):2055–2065

Caligiuri P, Cerruti R, Icardi G, Bruzzone B (2016) Overview of hepatitis B virus mutations and their implications in the management of infection. World J Gastroenterol 22:145–154

Chakraborty PR, Ruiz-Opazo N, Shouval D, Shafritz DA (1980) Identification of integrated hepatitis B virus DNA and expression of viral RNA in an HBsAg-producing human hepatocellular carcinoma cell line. Nature 286:531–533

Chen CJ, Yang HI, Su J, Jen CL, You SL, Lu SN, Huang GT, Iloeje UH, Group R-HS (2006) Risk of hepatocellular carcinoma across a biological gradient of serum hepatitis B virus DNA level. JAMA 295:65–73

Chen HS, Kew MC, Hornbuckle WE, Tennant BC, Cote PJ, Gerin JL, Purcell RH, Miller RH (1992) The precore gene of the woodchuck hepatitis virus genome is not essential for viral replication in the natural host. J Virol 66:5682–5684

Chen BF, Liu CJ, Jow GM, Chen PJ, Kao JH, Chen DS (2006) High prevalence and mapping of pre-S deletion in hepatitis B virus carriers with progressive liver diseases. Gastroenterology 130:1153–1168

Chen J, Wu M, Wang F, Zhang W, Wang W, Zhang X, Zhang J, Liu Y, Feng Y, Zheng Y, Hu Y, Yuan Z (2015) Hepatitis B virus spliced variants are associated with an impaired response to interferon therapy. Sci Rep 5:16459

Chiang PW, Jeng KS, Hu CP, Chang CM (1992) Characterization of a cis element required for packaging and replication of the human hepatitis B virus. Virology 186:701–711

Chou SF, Tsai ML, Huang JY, Chang YS, Shih C (2015) The dual role of an ESCRT-0 component HGS in HBV transcription and naked capsid secretion. PLoS Pathog 11:e1005123

Connolly E, Melegari M, Landgraf P, Tchaikovskaya T, Tennant BC, Slagle BL, Rogler LE, Zavolan M, Tuschl T, Rogler CE (2008) Elevated expression of the miR-17-92 polycistron and miR-21 in hepadnavirus-associated hepatocellular carcinoma contributes to the malignant phenotype. Am J Pathol 173:856–864

Cougot D, Allemand E, Riviere L, Benhenda S, Duroure K, Levillayer F, Muchardt C, Bucndia MA, Neuveut C (2012) Inhibition of PP1 phosphatase activity by HBx: a mechanism for the activation of hepatitis B virus transcription. Sci Signal 5:ra1

Croagh CM, Desmond PV, Bell SJ (2015) Genotypes and viral variants in chronic hepatitis B: a review of epidemiology and clinical relevance. World J Hepatol 7:289–303

Cui M, Wang Y, Sun B, Xiao Z, Ye L, Zhang X (2014) MiR-205 modulates abnormal lipid metabolism of hepatoma cells via targeting acyl-CoA synthetase long-chain family member 1 (ACSL1) mRNA. Biochem Biophys Res Commun 444:270–275

Damania P, Sen B, Dar SB, Kumar S, Kumari A, Gupta E, Sarin SK, Venugopal SK (2014) Hepatitis B virus induces cell proliferation via HBx-induced microRNA-21 in hepatocellular carcinoma by targeting programmed cell death protein4 (PDCD4) and phosphatase and tensin homologue (PTEN). PLoS ONE 9:e91745

Datta S, Chatterjee S, Veer V, Chakravarty R (2012) Molecular biology of the hepatitis B virus for clinicians. J Clin Exp Hepatol 2:353–365

Ding D, Lou X, Hua D, Yu W, Li L, Wang J, Gao F, Zhao N, Ren G, Li L, Lin B (2012) Recurrent targeted genes of hepatitis B virus in the liver cancer genomes identified by a next-generation sequencing-based approach. PLoS Genet 8:e1003065

Doria M, Klein N, Lucito R, Schneider RJ (1995) The hepatitis B virus HBx protein is a dual specificity cytoplasmic activator of Ras and nuclear activator of transcription factors. EMBO J 14:4747–4757

Duriez M, Mandouri Y, Lekbaby B, Wang H, Schnuriger A, Redelsperger F, Guerrera CI, Lefevre M, Fauveau V, Ahodantin J, Quetier I, Chhuon C, Gourari S, Boissonnas A, Gill U, Kennedy P, Debzi N, Sitterlin D, Maini MK, Kremsdorf D, Soussan P (2017) Alternative splicing of hepatitis B virus: a novel virus/host interaction altering liver immunity. J Hepatol 67:687–699

Edman JC, Gray P, Valenzuela P, Rall LB, Rutter WJ (1980) Integration of hepatitis B virus sequences and their expression in a human hepatoma cell. Nature 286:535–538

Elmore LW, Hancock AR, Chang SF, Wang XW, Chang S, Callahan CP, Geller DA, Will H, Harris CC (1997) Hepatitis B virus X protein and p53 tumor suppressor interactions in the modulation of apoptosis. Proc Natl Acad Sci U S A 94:14707–14712

Erhardt A, Blondin D, Hauck K, Sagir A, Kohnle T, Heintges T, Haussinger D (2005) Response to interferon alfa is hepatitis B virus genotype dependent: genotype A is more sensitive to interferon than genotype D. Gut 54:1009–1013

Fan YF, Lu CC, Chen WC, Yao WJ, Wang HC, Chang TT, Lei HY, Shiau AL, Su IJ (2001) Prevalence and significance of hepatitis B virus (HBV) pre-S mutants in serum and liver at different replicative stages of chronic HBV infection. Hepatology 33:277–286

Fang S, Huang SF, Cao J, Wen YA, Zhang LP, Ren GS (2013) Silencing of PCDH10 in hepatocellular carcinoma via de novo DNA methylation independent of HBV infection or HBX expression. Clin Exp Med 13:127–134

Fernandez AF, Rosales C, Lopez-Nieva P, Grana O, Ballestar E, Ropero S, Espada J, Melo SA, Lujambio A, Fraga MF, Pino I, Javierre B, Carmona FJ, Acquadro F, Steenbergen RD, Snijders PJ, Meijer CJ, Pineau P, Dejean A, Lloveras B, Capella G, Quer J, Buti M, Esteban JI, Allende H, Rodriguez-Frias F, Castellsague X, Minarovits J, Ponce J, Capello D, Gaidano G, Cigudosa JC, Gomez-Lopez G, Pisano DG, Valencia A, Piris MA, Bosch FX, Cahir-McFarland E, Kieff E, Esteller M (2009) The dynamic DNA methylomes of double-stranded DNA viruses associated with human cancer. Genome Res 19:438–451

Fu X, Song X, Li Y, Tan D, Liu G (2016) Hepatitis B virus X protein upregulates DNA methyltransferase 3A/3B and enhances SOCS-1CpG island methylation. Mol Med Rep 13:301–308

Fujimoto A, Totoki Y, Abe T, Boroevich KA, Hosoda F, Nguyen HH, Aoki M, Hosono N, Kubo M, Miya F, Arai Y, Takahashi H, Shirakihara T, Nagasaki M, Shibuya T, Nakano K, Watanabe-Makino K, Tanaka H, Nakamura H, Kusuda J, Ojima H, Shimada K, Okusaka T, Ueno M, Shigekawa Y, Kawakami Y, Arihiro K, Ohdan H, Gotoh K, Ishikawa O, Ariizumi S, Yamamoto M, Yamada T, Chayama K, Kosuge T, Yamaue H, Kamatani N, Miyano S, Nakagama H, Nakamura Y, Tsunoda T, Shibata T, Nakagawa H (2012) Whole-genome sequencing of liver cancers identifies etiological influences on mutation patterns and recurrent mutations in chromatin regulators. Nat Genet 44:760–764

Ganem D, Prince AM (2004) Hepatitis B virus infection–natural history and clinical consequences. N Engl J Med 350:1118–1129

Gao P, Wong CC, Tung EK, Lee JM, Wong CM, Ng IO (2011) Deregulation of microRNA expression occurs early and accumulates in early stages of HBV-associated multistep hepatocarcinogenesis. J Hepatol 54:1177–1184

Gao J, Xiong Y, Wang Y, Wang Y, Zheng G, Xu H (2016) Hepatitis B virus X protein activates Notch signaling by its effects on Notch1 and Notch4 in human hepatocellular carcinoma. Int J Oncol 48:329–337

Gavilanes F, Gomez-Gutierrez J, Aracil M, Gonzalez-Ros JM, Ferragut JA, Guerrero E, Peterson DL (1990) Hepatitis B surface antigen. Role of lipids in maintaining the structural and antigenic properties of protein components. Biochem J 265:857–864

Guidotti LG, Chisari FV (2006) Immunobiology and pathogenesis of viral hepatitis. Annu Rev Pathol 1:23–61

Gunther S, Sommer G, Iwanska A, Will H (1997) Heterogeneity and common features of defective hepatitis B virus genomes derived from spliced pregenomic RNA. Virology 238:363–371

Guo X, Lv X, Lv X, Ma Y, Chen L, Chen Y (2017) Circulating miR-21 serves as a serum biomarker for hepatocellular carcinoma and correlated with distant metastasis. Oncotarget 8:44050–44058

Hai H, Tamori A, Kawada N (2014) Role of hepatitis B virus DNA integration in human hepatocarcinogenesis. World J Gastroenterol 20:6236–6243

Hensel KO, Rendon JC, Navas MC, Rots MG, Postberg J (2017) Virus-host interplay in hepatitis B virus infection and epigenetic treatment strategies. FEBS J 284:3550–3572

Hsieh YH, Su IJ, Wang HC, Chang WW, Lei HY, Lai MD, Chang WT, Huang W (2004) Pre-S mutant surface antigens in chronic hepatitis B virus infection induce oxidative stress and DNA damage. Carcinogenesis 25:2023–2032

Huang J, Wang Y, Guo Y, Sun S (2010) Down-regulated microRNA-152 induces aberrant DNA methylation in hepatitis B virus-related hepatocellular carcinoma by targeting DNA methyltransferase 1. Hepatology 52:60–70

Hung JH, Su IJ, Lei HY, Wang HC, Lin WC, Chang WT, Huang W, Chang WC, Chang YS, Chen CC, Lai MD (2004) Endoplasmic reticulum stress stimulates the expression of cyclooxygenase-2 through activation of NF-kappaB and pp38 mitogen-activated protein kinase. J Biol Chem 279:46384–46392

Hung IF, Poon RT, Lai CL, Fung J, Fan ST, Yuen MF (2008) Recurrence of hepatitis B-related hepatocellular carcinoma is associated with high viral load at the time of resection. Am J Gastroenterol 103:1663–1673

Hunt CM, McGill JM, Allen MI, Condreay LD (2000) Clinical relevance of hepatitis B viral mutations. Hepatology 31:1037–1044

Iloeje UH, Yang HI, Su J, Jen CL, You SL, Chen CJ (2006) Predicting cirrhosis risk based on the level of circulating hepatitis B viral load. Gastroenterology 130:678–686

Iloeje UH, Yang HI, Jen CL, Su J, Wang LY, You SL, Chen CJ (2007) Risk and predictors of mortality associated with chronic hepatitis B infection. Clin Gastroenterol Hepatol 5:921 931

Iloeje UH, Yang HI, Chen CJ (2012) Natural history of chronic hepatitis B: what exactly has REVEAL revealed? Liver Int 32:1333–1341

Ito K, Kim KH, Lok AS, Tong S (2009) Characterization of genotype-specific carboxyl-terminal cleavage sites of hepatitis B virus e antigen precursor and identification of furin as the candidate enzyme. J Virol 83:3507–3517

Janssen HL, van Zonneveld M, Senturk H, Zeuzem S, Akarca US, Cakaloglu Y, Simon C, So TM, Gerken G, de Man RA, Niesters HG, Zondervan P, Hansen B, Schalm SW (2005) Pegylated interferon alfa-2b alone or in combination with lamivudine for HBeAg-positive chronic hepatitis B: a randomised trial. Lancet 365:123–129

Jiang Z, Jhunjhunwala S, Liu J, Haverty PM, Kennemer MI, Guan Y, Lee W, Carnevali P, Stinson J, Johnson S, Diao J, Yeung S, Jubb A, Ye W, Wu TD, Kapadia SB, de Sauvage FJ, Gentleman RC, Stern HM, Seshagiri S, Pant KP, Modrusan Z, Ballinger DG, Zhang Z (2012) The effects of hepatitis B virus integration into the genomes of hepatocellular carcinoma patients. Genome Res 22:593–601

Jones SA, Hu J (2013) Hepatitis B virus reverse transcriptase: diverse functions as classical and emerging targets for antiviral intervention. Emerg Microbes Infect 2:e56

Jung SY, Kim YJ (2013) C-terminal region of HBx is crucial for mitochondrial DNA damage. Cancer Lett 331:76–83

Kao JH (2003) Hepatitis B virus genotypes and hepatocellular carcinoma in Taiwan. Intervirology 46:400–407

Kao JH, Chen DS (2002) Global control of hepatitis B virus infection. Lancet Infect Dis 2:395–403

Kao JH, Wu NH, Chen PJ, Lai MY, Chen DS (2000) Hepatitis B genotypes and the response to interferon therapy. J Hepatol 33:998–1002

Kao JH, Chen PJ, Lai MY, Chen DS (2003) Basal core promoter mutations of hepatitis B virus increase the risk of hepatocellular carcinoma in hepatitis B carriers. Gastroenterology 124:327–334

Keasler VV, Hodgson AJ, Madden CR, Slagle BL (2009) Hepatitis B virus HBx protein localized to the nucleus restores HBx-deficient virus replication in HepG2 cells and in vivo in hydrodynamically-injected mice. Virology 390:122–129

Kew MC (2011) Hepatitis B virus x protein in the pathogenesis of hepatitis B virus-induced hepatocellular carcinoma. J Gastroenterol Hepatol 26(Suppl 1):144–152

Klein NP, Schneider RJ (1997) Activation of Src family kinases by hepatitis B virus HBx protein and coupled signaling to Ras. Mol Cell Biol 17:6427–6436

Kohara K, Munakata T, Kohara M (2011) The pathogenesis of hepatitis C virus induced by viral proteins. Nihon Rinsho 69(Suppl 4):97–102

Kongkavitoon P, Tangkijvanich P, Hirankarn N, Palaga T (2016) Hepatitis B virus HBx activates notch signaling via delta-like 4/Notch1 in hepatocellular carcinoma. PLoS ONE 11:e0146696

Kosaka Y, Takase K, Kojima M, Shimizu M, Inoue K, Yoshiba M, Tanaka S, Akahane Y, Okamoto H, Tsuda F et al (1991) Fulminant hepatitis B: induction by hepatitis B virus mutants defective in the precore region and incapable of encoding e antigen. Gastroenterology 100:1087–1094

Kramvis A, Kew MC (1999) The core promoter of hepatitis B virus. J Viral Hepat 6:415–427

Krieg AM (2002) A role for toll in autoimmunity. Nat Immunol 3:423–424

Kuss-Duerkop SK, Westrich JA, Pyeon D (2018) DNA tumor virus regulation of host DNA methylation and its implications for immune evasion and oncogenesis. Viruses 10

Kwun HJ, Jang KL (2004) Natural variants of hepatitis B virus X protein have differential effects on the expression of cyclin-dependent kinase inhibitor p21 gene. Nucleic Acids Res 32:2202–2213

Lamberts C, Nassal M, Velhagen I, Zentgraf H, Schroder CH (1993) Precore-mediated inhibition of hepatitis B virus progeny DNA synthesis. J Virol 67:3756–3762

Lau CC, Sun T, Ching AK, He M, Li JW, Wong AM, Co NN, Chan AW, Li PS, Lung RW, Tong JH, Lai PB, Chan HL, To KF, Chan TF, Wong N (2014) Viral-human chimeric transcript predisposes risk to liver cancer development and progression. Cancer Cell 25:335–349

Lee J, Kim SS (2009) The function of p27 KIP1 during tumor development. Exp Mol Med 41:765–771

Lee MH, Na H, Na TY, Shin YK, Seong JK, Lee MO (2012) Epigenetic control of metastasis-associated protein 1 gene expression by hepatitis B virus X protein during hepatocarcinogenesis. Oncogenesis 1:e25

Lee SA, Kim KJ, Kim DW, Kim BJ (2013) Male-specific W4P/R mutation in the pre-S1 region of hepatitis B virus, increasing the risk of progression of liver diseases in chronic patients. J Clin Microbiol 51:3928–3936

Lee SM, Lee YG, Bae JB, Choi JK, Tayama C, Hata K, Yun Y, Seong JK, Kim YJ (2014) HBx induces hypomethylation of distal intragenic CpG islands required for active expression of developmental regulators. Proc Natl Acad Sci U S A 111:9555–9560

Li JS, Tong SP, Wen YM, Vitvitski L, Zhang Q, Trepo C (1993) Hepatitis B virus genotype A rarely circulates as an HBe-minus mutant: possible contribution of a single nucleotide in the precore region. J Virol 67:5402–5410

Li J, Buckwold VE, Hon MW, Ou JH (1999) Mechanism of suppression of hepatitis B virus precore RNA transcription by a frequent double mutation. J Virol 73:1239–1244

Li W, Zeng X, Lee NP, Liu X, Chen S, Guo B, Yi S, Zhuang X, Chen F, Wang G, Poon RT, Fan ST, Mao M, Li Y, Li S, Wang J, Jianwang XuX, Jiang H, Zhang X (2013) HIVID: an efficient method to detect HBV integration using low coverage sequencing. Genomics 102:338–344

Li CH, Xu F, Chow S, Feng L, Yin D, Ng TB, Chen Y (2014) Hepatitis B virus X protein promotes hepatocellular carcinoma transformation through interleukin-6 activation of microRNA-21 expression. Eur J Cancer 50:2560–2569

Li W, Li M, Liao D, Lu X, Gu X, Zhang Q, Zhang Z, Li H (2016) Carboxyl-terminal truncated HBx contributes to invasion and metastasis via deregulating metastasis suppressors in hepatocellular carcinoma. Oncotarget 7:55110–55127

Liang TJ (2009) Hepatitis B: the virus and disease. Hepatology 49:S13–S21

Liang TJ, Hasegawa K, Rimon N, Wands JR, Ben-Porath E (1991) A hepatitis B virus mutant associated with an epidemic of fulminant hepatitis. N Engl J Med 324:1705–1709

Lin CL, Kao JH (2011) The clinical implications of hepatitis B virus genotype: Recent advances. J Gastroenterol Hepatol 26(Suppl 1):123–130

Lin CL, Liao LY, Wang CS, Chen PJ, Lai MY, Chen DS, Kao JH (2005) Basal core-promoter mutant of hepatitis B virus and progression of liver disease in hepatitis B e antigen-negative chronic hepatitis B. Liver Int 25:564–570

Liu CJ, Chen BF, Chen PJ, Lai MY, Huang WL, Kao JH, Chen DS (2006) Role of hepatitis B viral load and basal core promoter mutation in hepatocellular carcinoma in hepatitis B carriers. J Infect Dis 193:1258–1265

Liu X, Wang L, Zhang S, Lin J, Zhang S, Feitelson MA, Gao H, Zhu M (2008) Mutations in the C-terminus of the X protein of hepatitis B virus regulate Wnt-5a expression in hepatoma Huh7 cells: cDNA microarray and proteomic analyses. Carcinogenesis 29:1207–1214

Liu W, Shen SM, Zhao XY, Chen GQ (2012) Targeted genes and interacting proteins of hypoxia inducible factor-1. Int J Biochem Mol Biol 3:165–178

Liu XY, Tang SH, Wu SL, Luo YH, Cao MR, Zhou HK, Jiang XW, Shu JC, Bie CQ, Huang SM, Zheng ZH, Gao F (2015) Epigenetic modulation of insulin-like growth factor-II overexpression by hepatitis B virus X protein in hepatocellular carcinoma. Am J Cancer Res 5:956–978

Liu S, Koh SS, Lee CG (2016) Hepatitis B virus X protein and hepatocarcinogenesis. Int J Mol Sci 17

Locarnini S (2004) Molecular virology of hepatitis B virus. Semin Liver Dis 24(Suppl 1):3–10

Locarnini S, McMillan J, Bartholomeusz A (2003) The hepatitis B virus and common mutants. Semin Liver Dis 23:5–20

Lok AS, Akarca U, Greene S (1994) Mutations in the pre-core region of hepatitis B virus serve to enhance the stability of the secondary structure of the pre-genome encapsidation signal. Proc Natl Acad Sci U S A 91:4077–4081

Ma NF, Lau SH, Hu L, Xie D, Wu J, Yang J, Wang Y, Wu MC, Fung J, Bai X, Tzang CH, Fu L, Yang M, Su YA, Guan XY (2008) COOH-terminal truncated HBV X protein plays key role in hepatocarcinogenesis. Clin Cancer Res 14:5061–5068

Mancini-Bourgine M, Bayard F, Soussan P, Deng Q, Lone YC, Kremsdorf D, Michel ML (2007) Hepatitis B virus splice-generated protein induces T-cell responses in HLA-transgenic mice and hepatitis B virus-infected patients. J Virol 81:4963–4972

Mathonnet G, Lachance S, Alaoui-Jamali M, Drobetsky EA (2004) Expression of hepatitis B virus X oncoprotein inhibits transcription-coupled nucleotide excision repair in human cells. Mutat Res 554:305–318

Milich D, Liang TJ (2003) Exploring the biological basis of hepatitis B e antigen in hepatitis B virus infection. Hepatology 38:1075–1086

Moolla N, Kew M, Arbuthnot P (2002) Regulatory elements of hepatitis B virus transcription. J Viral Hepat 9:323–331

Murakami Y, Saigo K, Takashima H, Minami M, Okanoue T, Brechot C, Paterlini-Brechot P (2005) Large scaled analysis of hepatitis B virus (HBV) DNA integration in HBV related hepatocellular carcinomas. Gut 54:1162–1168

Muroyama R, Kato N, Yoshida H, Otsuka M, Moriyama M, Wang Y, Shao RX, Dharel N, Tanaka Y, Ohta M, Tateishi R, Shiina S, Tatsukawa M, Fukai K, Imazeki F, Yokosuka O, Shiratori Y, Omata M (2006) Nucleotide change of codon 38 in the X gene of hepatitis B virus genotype C is associated with an increased risk of hepatocellular carcinoma. J Hepatol 45:805–812

Na B, Huang Z, Wang Q, Qi Z, Tian Y, Lu CC, Yu J, Hanes MA, Kakar S, Huang EJ, Ou JH, Liu L, Yen TS (2011) Transgenic expression of entire hepatitis B virus in mice induces hepatocarcinogenesis independent of chronic liver injury. PLoS ONE 6:e26240

Ng SA, Lee C (2011) Hepatitis B virus X gene and hepatocarcinogenesis. J Gastroenterol 46:974–990

Ng KY, Chai S, Tong M, Guan XY, Lin CH, Ching YP, Xie D, Cheng AS, Ma S (2016) C-terminal truncated hepatitis B virus X protein promotes hepatocellular carcinogenesis through induction of cancer and stem cell-like properties. Oncotarget 7:24005–24017

Nguyen H, Gitig DM, Koff A (1999) Cell-free degradation of p27(kip1), a G1 cyclin-dependent kinase inhibitor, is dependent on CDK2 activity and the proteasome. Mol Cell Biol 19:1190–1201

Norder H, Courouce AM, Coursaget P, Echevarria JM, Lee SD, Mushahwar IK, Robertson BH, Locarnini S, Magnius LO (2004) Genetic diversity of hepatitis B virus strains derived worldwide: genotypes, subgenotypes, and HBsAg subtypes. Intervirology 47:289–309

Omata M, Ehata T, Yokosuka O, Hosoda K, Ohto M (1991) Mutations in the precore region of hepatitis B virus DNA in patients with fulminant and severe hepatitis. N Engl J Med 324:1699–1704

Oommen PT, Wirth S, Wintermeyer P, Gerner P (2006) Relationship between viral load and genotypes of hepatitis B virus in children with chronic hepatitis B. J Pediatr Gastroenterol Nutr 43:342–347

Ou JH (1997) Molecular biology of hepatitis B virus e antigen. J Gastroenterol Hepatol 12:S178–S187

Ou J, Rutter WJ (1985) Hybrid hepatitis B virus-host transcripts in a human hepatoma cell. Proc Natl Acad Sci USA 82:83–87

Ou JH, Rutter WJ (1987) Regulation of secretion of the hepatitis B virus major surface antigen by the preS-1 protein. J Virol 61:782–786

Ou JH, Laub O, Rutter WJ (1986) Hepatitis B virus gene function: the precore region targets the core antigen to cellular membranes and causes the secretion of the e antigen. Proc Natl Acad Sci U S A 83:1578–1582

Parekh S, Zoulim F, Ahn SH, Tsai A, Li J, Kawai S, Khan N, Trepo C, Wands J, Tong S (2003) Genome replication, virion secretion, and e antigen expression of naturally occurring hepatitis B virus core promoter mutants. J Virol 77:6601–6612

Park SG, Kim Y, Park E, Ryu HM, Jung G (2003) Fidelity of hepatitis B virus polymerase. Eur J Biochem 270:2929–2936

Parkin DM (2006) The global health burden of infection-associated cancers in the year 2002. Int J Cancer 118:3030–3044

Peng Z, Zhang Y, Gu W, Wang Z, Li D, Zhang F, Qiu G, Xie K (2005) Integration of the hepatitis B virus X fragment in hepatocellular carcinoma and its effects on the expression of multiple molecules: a key to the cell cycle and apoptosis. Int J Oncol 26:467–473

Peng C, Xiao X, Kang B, He S, Li J (2014) Serum secreted frizzled-related protein 5 levels differentially decrease in patients with hepatitis B virus-associated chronic infection and hepatocellular carcinoma. Oncol Lett 8:1340–1344

Pol JG, Lekbaby B, Redelsperger F, Klamer S, Mandouri Y, Ahodantin J, Bieche I, Lefevre M, Souque P, Charneau P, Gadessaud N, Kremsdorf D, Soussan P (2015) Alternative splicing-regulated protein of hepatitis B virus hacks the TNF-alpha-stimulated signaling pathways and limits the extent of liver inflammation. FASEB J 29:1879–1889

Polakis P (2012) Wnt signaling in cancer. Cold Spring Harb Perspect Biol 4

Pollicino T, Cacciola I, Saffioti F, Raimondo G (2014) Hepatitis B virus PreS/S gene variants: pathobiology and clinical implications. J Hepatol 61:408–417

Qadri I, Fatima K, Abde LHH (2011) Hepatitis B virus X protein impedes the DNA repair via its association with transcription factor, TFIIH. BMC Microbiol 11:48

Qiu X, Zhang L, Lu S, Song Y, Lao Y, Hu J, Fan H (2014) Upregulation of DNMT1 mediated by HBx suppresses RASSF1A expression independent of DNA methylation. Oncol Rep 31:202–208

Rodriguez-Frias F, Buti M, Jardi R, Cotrina M, Viladomiu L, Esteban R, Guardia J (1995) Hepatitis B virus infection: precore mutants and its relation to viral genotypes and core mutations. Hepatology 22:1641–1647

Rongrui L, Na H, Zongfang L, Fanpu J, Shiwen J (2014) Epigenetic mechanism involved in the HBV/HCV-related hepatocellular carcinoma tumorigenesis. Curr Pharm Des 20:1715–1725

Salpini R, Colagrossi L, Bellocchi MC, Surdo M, Becker C, Alteri C, Aragri M, Ricciardi A, Armenia D, Pollicita M, Di Santo F, Carioti L, Louzoun Y, Mastroianni CM, Lichtner M, Paoloni M, Esposito M, D'Amore C, Marrone A, Marignani M, Sarrecchia C, Sarmati L, Andreoni M, Angelico M, Verheyen J, Perno CF, Svicher V (2015) Hepatitis B surface antigen genetic elements critical for immune escape correlate with hepatitis B virus reactivation upon immunosuppression. Hepatology 61:823–833

Satyanarayana A, Kaldis P (2009) A dual role of Cdk2 in DNA damage response. Cell Div 4:9

Schluter V, Meyer M, Hofschneider PH, Koshy R, Caselmann WH (1994) Integrated hepatitis B virus X and 3' truncated preS/S sequences derived from human hepatomas encode functionally active transactivators. Oncogene 9:3335–3344

Shi W, Zhu C, Zheng W, Ling C, Carr MJ, Higgins DG, Zhang Z (2012) Subgenotyping of genotype C hepatitis B virus: correcting misclassifications and identifying a novel subgenotype. PLoS ONE 7:e47271

Shin EC, Sung PS, Park SH (2016) Immune responses and immunopathology in acute and chronic viral hepatitis. Nat Rev Immunol 16:509–523

Smits VA, Klompmaker R, Vallenius T, Rijksen G, Makela TP, Medema RH (2000) p21 inhibits Thr161 phosphorylation of Cdc2 to enforce the G2 DNA damage checkpoint. J Biol Chem 275:30638–30643

Soussan P, Tuveri R, Nalpas B, Garreau F, Zavala F, Masson A, Pol S, Brechot C, Kremsdorf D (2003) The expression of hepatitis B spliced protein (HBSP) encoded by a spliced hepatitis B virus RNA is associated with viral replication and liver fibrosis. J Hepatol 38:343–348

Srisuttee R, Koh SS, Kim SJ, Malilas W, Boonying W, Cho IR, Jhun BH, Ito M, Horio Y, Seto E, Oh S, Chung YH (2012) Hepatitis B virus X (HBX) protein upregulates beta-catenin in a human hepatic cell line by sequestering SIRT1 deacetylase. Oncol Rep 28:276–282

Standring DN, Ou JH, Rutter WJ (1986) Assembly of viral particles in Xenopus oocytes: pre-surface-antigens regulate secretion of the hepatitis B viral surface envelope particle. Proc Natl Acad Sci U S A 83:9338–9342

Stuyver L, De Gendt S, Van Geyt C, Zoulim F, Fried M, Schinazi RF, Rossau R (2000) A new genotype of hepatitis B virus: complete genome and phylogenetic relatedness. J Gen Virol 81:67–74

Su TS, Lai CJ, Huang JL, Lin LH, Yauk YK, Chang CM, Lo SJ, Han SH (1989) Hepatitis B virus transcript produced by RNA splicing. J Virol 63:4011–4018

Su IJ, Wang HC, Wu HC, Huang WY (2008) Ground glass hepatocytes contain pre-S mutants and represent preneoplastic lesions in chronic hepatitis B virus infection. J Gastroenterol Hepatol 23:1169–1174

Su IJ, Wang LH, Hsieh WC, Wu HC, Teng CF, Tsai HW, Huang W (2014) The emerging role of hepatitis B virus pre-S2 deletion mutant proteins in HBV tumorigenesis. J Biomed Sci 21:98

Sung WK, Zheng H, Li S, Chen R, Liu X, Li Y, Lee NP, Lee WH, Ariyaratne PN, Tennakoon C, Mulawadi FH, Wong KF, Liu AM, Poon RT, Fan ST, Chan KL, Gong Z, Hu Y, Lin Z, Wang G, Zhang Q, Barber TD, Chou WC, Aggarwal A, Hao K, Zhou W, Zhang C, Hardwick J, Buser C, Xu J, Kan Z, Dai H, Mao M, Reinhard C, Wang J, Luk JM (2012) Genome-wide survey of recurrent HBV integration in hepatocellular carcinoma. Nat Genet 44:765–769

Sze KM, Chu GK, Lee JM, Ng IO (2013) C-terminal truncated hepatitis B virus x protein is associated with metastasis and enhances invasiveness by C-Jun/matrix metalloproteinase protein 10 activation in hepatocellular carcinoma. Hepatology 57:131–139

Tamori A, Yamanishi Y, Kawashima S, Kanehisa M, Enomoto M, Tanaka H, Kubo S, Shiomi S, Nishiguchi S (2005) Alteration of gene expression in human hepatocellular carcinoma with integrated hepatitis B virus DNA. Clin Cancer Res 11:5821–5826

Tang H, Oishi N, Kaneko S, Murakami S (2006) Molecular functions and biological roles of hepatitis B virus x protein. Cancer Sci 97:977–983

Tang Y, Zhou J, Hooi SC, Jiang YM, Lu GD (2018) Fatty acid activation in carcinogenesis and cancer development: essential roles of long-chain acyl-CoA synthetases. Oncol Lett 16:1390–1396

Tatematsu K, Tanaka Y, Kurbanov F, Sugauchi F, Mano S, Maeshiro T, Nakayoshi T, Wakuta M, Miyakawa Y, Mizokami M (2009) A genetic variant of hepatitis B virus divergent from known human and ape genotypes isolated from a Japanese patient and provisionally assigned to new genotype. J. J Virol 83:10538–10547

Thakur V, Guptan RC, Kazim SN, Malhotra V, Sarin SK (2002) Profile, spectrum and significance of HBV genotypes in chronic liver disease patients in the Indian subcontinent. J Gastroenterol Hepatol 17:165–170

Tian Y, Kuo CF, Akbari O, Ou JH (2016) Maternal-derived hepatitis B virus e antigen alters macrophage function in offspring to drive viral persistence after vertical transmission. Immunity 44:1204–1214

Tiollais P, Pourcel C, Dejean A (1985) The hepatitis B virus. Nature 317:489–495

Toh ST, Jin Y, Liu L, Wang J, Babrzadeh F, Gharizadeh B, Ronaghi M, Toh HC, Chow PK, Chung AY, Ooi LL, Lee CG (2013) Deep sequencing of the hepatitis B virus in hepatocellular carcinoma patients reveals enriched integration events, structural alterations and sequence variations. Carcinogenesis 34:787–798

Tran TT, Trinh TN, Abe K (2008) New complex recombinant genotype of hepatitis B virus identified in Vietnam. J Virol 82:5657–5663

Tsai KN, Chong CL, Chou YC, Huang CC, Wang YL, Wang SW, Chen ML, Chen CH, Chang C (2015) Doubly spliced RNA of hepatitis B virus suppresses viral transcription via TATA-binding protein and induces stress granule assembly. J Virol 89:11406–11419

Tsai KN, Kuo CF, Ou JJ (2018) Mechanisms of hepatitis B virus persistence. Trends Microbiol 26:33–42

Tseng TC, Liu CJ, Yang HC, Su TH, Wang CC, Chen CL, Kuo SF, Liu CH, Chen PJ, Chen DS, Kao JH(2012) High levels of hepatitis B surface antigen increase risk of hepatocellular carcinoma in patients with low HBV load. Gastroenterology 142:1140–1149 e1143; quiz e1113–1144

Tu H, Bonura C, Giannini C, Mouly H, Soussan P, Kew M, Paterlini-Brechot P, Brechot C, Kremsdorf D (2001) Biological impact of natural COOH-terminal deletions of hepatitis B virus X protein in hepatocellular carcinoma tissues. Cancer Res 61:7803–7810

Ueda K, Tsurimoto T, Matsubara K (1991) Three envelope proteins of hepatitis B virus: large S, middle S, and major S proteins needed for the formation of Dane particles. J Virol 65:3521–3529

van de Klundert MA, van Hemert FJ, Zaaijer HL, Kootstra NA (2012) The hepatitis B virus x protein inhibits thymine DNA glycosylase initiated base excision repair. PLoS ONE 7:e48940

Verschuere V, Yap PS, Fevery J (2005) Is HBV genotyping of clinical relevance? Acta Gastroenterol Belg 68:233–236

Wan S, Civan J, Rossi S, Yang H (2013) Profiling HBV integrations in hepatocellular carcinoma. Hepatobiliary Surg Nutr 2:124–126

Wang XW, Yeh H, Schaeffer L, Roy R, Moncollin V, Egly JM, Wang Z, Freidberg EC, Evans MK, Taffe BG et al (1995) p53 modulation of TFIIH-associated nucleotide excision repair activity. Nat Genet 10:188–195

Wang HC, Huang W, Lai MD, Su IJ (2006) Hepatitis B virus pre-S mutants, endoplasmic reticulum stress and hepatocarcinogenesis. Cancer Sci 97:683–688

Wang Y, Lu Y, Toh ST, Sung WK, Tan P, Chow P, Chung AY, Jooi LL, Lee CG (2010) Lethal-7 is down-regulated by the hepatitis B virus x protein and targets signal transducer and activator of transcription 3. J Hepatol 53:57–66

Wang YL, Liou GG, Lin CH, Chen ML, Kuo TM, Tsai KN, Huang CC, Chen YL, Huang LR, Chou YC, Chang C (2015) The inhibitory effect of the hepatitis B virus singly-spliced RNA-encoded p 21.5 protein on HBV nucleocapsid formation. PLoS One 10:e0119625

Wang G, Dong F, Xu Z, Sharma S, Hu X, Chen D, Zhang L, Zhang J, Dong Q (2017) MicroRNA profile in HBV-induced infection and hepatocellular carcinoma. BMC Cancer 17:805

Watanabe T, Sorensen EM, Naito A, Schott M, Kim S, Ahlquist P (2007) Involvement of host cellular multivesicular body functions in hepatitis B virus budding. Proc Natl Acad Sci U S A 104:10205–10210

Weichert W (2009) HDAC expression and clinical prognosis in human malignancies. Cancer Lett 280:168–176

WHO (2017) Global hepatitis report 2017. World Health Organization Geneva

Wu HL, Chen PJ, Tu SJ, Lin MH, Lai MY, Chen DS (1991) Characterization and genetic analysis of alternatively spliced transcripts of hepatitis B virus in infected human liver tissues and transfected HepG2 cells. J Virol 65:1680–1686

Wu JC, Huang YH, Chau GY, Su CW, Lai CR, Lee PC, Huo TI, Sheen IJ, Lee SD, Lui WY (2009) Risk factors for early and late recurrence in hepatitis B-related hepatocellular carcinoma. J Hepatol 51:890–897

Xie H, Song J, Liu K, Ji H, Shen H, Hu S, Yang G, Du Y, Zou X, Jin H, Yan L, Liu J, Fan D (2008) The expression of hypoxia-inducible factor-1alpha in hepatitis B virus-related hepatocellular carcinoma: correlation with patients' prognosis and hepatitis B virus X protein. Dig Dis Sci 53:3225–3233

Xie HJ, Noh JH, Kim JK, Jung KH, Eun JW, Bae HJ, Kim MG, Chang YG, Lee JY, Park H, Nam SW (2012) HDAC1 inactivation induces mitotic defect and caspase-independent autophagic cell death in liver cancer. PLoS ONE 7:e34265

Xie Q, Chen L, Shan X, Shan X, Tang J, Zhou F, Chen Q, Quan H, Nie D, Zhang W, Huang AL, Tang N (2014) Epigenetic silencing of SFRP1 and SFRP5 by hepatitis B virus X protein enhances hepatoma cell tumorigenicity through Wnt signaling pathway. Int J Cancer 135: 635–646

Xu Z, Yen TS, Wu L, Madden CR, Tan W, Slagle BL, Ou JH (2002) Enhancement of hepatitis B virus replication by its X protein in transgenic mice. J Virol 76:2579–2584

Xu R, Zhang X, Zhang W, Fang Y, Zheng S, Yu XF (2007) Association of human APOBEC3 cytidine deaminases with the generation of hepatitis virus B x antigen mutants and hepatocellular carcinoma. Hepatology 46:1810–1820

Xu L, Beckebaum S, Iacob S, Wu G, Kaiser GM, Radtke A, Liu C, Kabar I, Schmidt HH, Zhang X, Lu M, Cicinnati VR (2014) MicroRNA-101 inhibits human hepatocellular carcinoma progression through EZH2 downregulation and increased cytostatic drug sensitivity. J Hepatol 60:590–598

Yan J, Yao Z, Hu K, Zhong Y, Li M, Xiong Z, Deng M (2015) Hepatitis B virus core promoter A1762T/G1764A (TA)/T1753A/T1768A mutations contribute to hepatocarcinogenesis by deregulating Skp2 and P53. Dig Dis Sci 60:1315–1324

Yang HI, Yeh SH, Chen PJ, Iloeje UH, Jen CL, Su J, Wang LY, Lu SN, You SL, Chen DS, Liaw YF, Chen CJ, Group R-HS (2008) Associations between hepatitis B virus genotype and mutants and the risk of hepatocellular carcinoma. J Natl Cancer Inst 100:1134–1143

Yang HI, Lu SN, Liaw YF, You SL, Sun CA, Wang LY, Hsiao CK, Chen PJ, Chen DS, Chen CJ (2002) Hepatitis B e antigen and the risk of hepatocellular carcinoma. N Engl J Med 347:168–174

Yang SL, Ren QG, Zhang T, Pan X, Wen L, Hu JL, Yu C, He QJ (2017) Hepatitis B virus X protein and hypoxiainducible factor-1alpha stimulate Notch gene expression in liver cancer cells. Oncol Rep 37:348–356

Yeh CT, So M, Ng J, Yang HW, Chang ML, Lai MW, Chen TC, Lin CY, Yeh TS, Lee WC (2010) Hepatitis B virus-DNA level and basal core promoter A1762T/G1764A mutation in liver tissue independently predict postoperative survival in hepatocellular carcinoma. Hepatology 52:1922–1933

Ying J, Li H, Seng TJ, Langford C, Srivastava G, Tsao SW, Putti T, Murray P, Chan AT, Tao Q (2006) Functional epigenetics identifies a protocadherin PCDH10 as a candidate tumor suppressor for nasopharyngeal, esophageal and multiple other carcinomas with frequent methylation. Oncogene 25:1070–1080

Yip WK, Cheng AS, Zhu R, Lung RW, Tsang DP, Lau SS, Chen Y, Sung JG, Lai PB, Ng EK, Yu J, Wong N, To KF, Wong VW, Sung JJ, Chan HL (2011) Carboxyl-terminal truncated HBx regulates a distinct microRNA transcription program in hepatocellular carcinoma development. PLoS ONE 6:e22888

Yoo YG, Na TY, Seo HW, Seong JK, Park CK, Shin YK, Lee MO (2008) Hepatitis B virus X protein induces the expression of MTA1 and HDAC1, which enhances hypoxia signaling in hepatocellular carcinoma cells. Oncogene 27:3405–3413

Yu G, Bing Y, Li W, Xia L, Liu Z (2014) Hepatitis B virus inhibits the expression of CD82 through hypermethylation of its promoter in hepatoma cells. Mol Med Rep 10:2580–2586

Zanetti AR, Tanzi E, Manzillo G, Maio G, Sbreglia C, Caporaso N, Thomas H, Zuckerman AJ (1988) Hepatitis B variant in Europe. Lancet 2:1132–1133

Zhang H, Diab A, Fan H, Mani SK, Hullinger R, Merle P, Andrisani O (2015) PLK1 and HOTAIR accelerate proteasomal degradation of SUZ12 and ZNF198 during hepatitis B virus-induced liver carcinogenesis. Cancer Res 75:2363–2374

Zhang H, Xing Z, Mani SK, Bancel B, Durantel D, Zoulim F, Tran EJ, Merle P, Andrisani O (2016) RNA helicase DEAD box protein 5 regulates Polycomb repressive complex 2/Hox transcript antisense intergenic RNA function in hepatitis B virus infection and hepatocarcinogenesis. Hepatology 64:1033–1048

Zhao J, Wu G, Bu F, Lu B, Liang A, Cao L, Tong X, Lu X, Wu M, Guo Y (2010) Epigenetic silence of ankyrin-repeat-containing, SH3-domain-containing, and proline-rich-region- containing protein 1 (ASPP1) and ASPP2 genes promotes tumor growth in hepatitis B virus-positive hepatocellular carcinoma. Hepatology 51:142–153

Zhao Q, Li T, Qi J, Liu J, Qin C (2014) The miR-545/374a cluster encoded in the Ftx lncRNA is overexpressed in HBV-related hepatocellular carcinoma and promotes tumorigenesis and tumor progression. PLoS ONE 9:e109782

Zhao LH, Liu X, Yan HX, Li WY, Zeng X, Yang Y, Zhao J, Liu SP, Zhuang XH, Lin C, Qin CJ, Zhao Y, Pan ZY, Huang G, Liu H, Zhang J, Wang RY, Yang Y, Wen W, Lv GS, Zhang HL, Wu H, Huang S, Wang MD, Tang L, Cao HZ, Wang L, Lee TL, Jiang H, Tan YX, Yuan SX, Hou GJ, Tao QF, Xu QG, Zhang XQ, Wu MC, Xu X, Wang J, Yang HM, Zhou WP, Wang HY (2016) Genomic and oncogenic preference of HBV integration in hepatocellular carcinoma. Nat Commun 7:12992

Zhu H, Nie L, Maki CG (2005) Cdk2-dependent Inhibition of p21 stability via a C-terminal cyclin-binding motif. J Biol Chem 280:29282–29288

Zhu M, Guo J, Li W, Xia H, Lu Y, Dong X, Chen Y, Xie X, Fu S, Li M (2015) HBx induced AFP receptor expressed to activate PI3K/AKT signal to promote expression of Src in liver cells and hepatoma cells. BMC Cancer 15:362

Zollner B, Petersen J, Puchhammer-Stockl E, Kletzmayr J, Sterneck M, Fischer L, Schroter M, Laufs R, Feucht HH (2004) Viral features of lamivudine resistant hepatitis B genotypes A and D. Hepatology 39:42–50

Prevention of Hepatitis B Virus Infection and Liver Cancer

Mei-Hwei Chang

1 Introduction

Hepatocellular carcinoma (HCC) is one of the leading cancer in the world (Parkin et al. 2001). Because of its high fatality (overall ratio of mortality to incidence of 0.93), liver cancer is one of the five most common causes of death from cancer worldwide. According to the 2018 global report from World Health Organization (WHO), liver cancer is the sixth most common cancer and the fourth most common cause of cancer death (WHO 2018). Persistent infection with hepatitis B virus (HBV) or hepatitis C virus (HCV) is associated with approximately 90% of HCC. Evidence from epidemiology, case control study, animal experiments, molecular biology all support the important oncogenic role, either directly or indirectly, of HBV and HCV in HCC. As evidenced by the large population infected with HBV in the developing world, HBV remains the most prevalent oncogenic virus for HCC in humans. HBV is estimated to cause around 55–70% of HCC worldwide, while HCV accounts for around 25% of HCC (Bosch and Ribes 2002). Liver cirrhosis is a common precancerous lesion, accounting for approximately 80% of patients with HCC, including children (Hsu et al. 1983). This sequel usually results from severe liver injury caused by chronic HBV or HCV infection.

The World Health Assembly calls for the elimination of viral hepatitis as a public health threat by 2030 to reduce new infections by 90% and mortality by 65% compared with the 2015 baseline (WHO 2017). Among the proposed strategies to eliminate viral hepatitis, prevention is the most important and cost-effective way to be conducted to achieve the goal.

M.-H. Chang (✉)
Department of Pediatrics, National Taiwan University Hospital, Taipei, Taiwan
e-mail: changmh@ntu.edu.tw

© Springer Nature Switzerland AG 2021
T.-C. Wu et al. (eds.), *Viruses and Human Cancer*, Recent Results
in Cancer Research 217, https://doi.org/10.1007/978-3-030-57362-1_4

2   Disease Burden of HCC

According to the 2018 Cancer, today's report from the International Agency for Research on Cancer (IARC), WHO, liver cancer cases accounts for 4.7% of all new cancer cases and 8.2% of all cancer death cases (IARC, WHO 2018). Approximately, 257 million people (3.5% of the world population) in year 2015 are living with HBV infection. Hepatitis B resulted in 887,000 deaths, mostly from complications (mainly liver cirrhosis and HCC) (WHO 2019). High incidence areas of HCC are mainly in developing regions, such as Eastern and South-Eastern Asia, Middle, and Western Africa. The African (6.1%) and Western Pacific regions (6.2%) had the highest prevalence (WHO 2017). The geographical distribution of the mortality rates is similar to that observed for incidence.

Even in the same country, different ethnic group may have varied incidence of HCC. The annual incidence of HCC in Alaskan Eskimo males was 11.2 per 100,000, five times that of white males in the USA (Heyward et al. 1981). The world geographic distribution of HCC overlaps well with that of the distribution for chronic HBV infection (Beasley 1982). Regions with a high prevalence of HBV infection also have high rates of HCC. HBV causes 60–80% of the primary liver cancer, which accounts for one of the five major cancer deaths particularly in areas highly prevalent for HBV infection, such as Eastern and South-Eastern Asia, the Pacific Rim, and the Northen Africa (Bosch and Ribes 2002; IARC, WHO 2018). The southern parts of Eastern and Central Europe, the Amazon basin, the Middle East, and the Indian subcontinent are also areas with high prevalence of HBV infection and HCC (Lavanchy 2004; WHO 2019).

3   Transmission Routes of Hepatitis B Virus Infection

The age and source of primary HBV infection are important factors affecting the outcome of HBV infection. Maternal serum HBsAg and hepatitis B e antigen (HBeAg) status affect the outcome of HBV infection in their offspring. In Asia and many other endemic areas, before the era of universal HBV immunization, perinatal transmission through HBsAg carrier mothers accounts for 40–50% of HBsAg carriers. Irrespective of the extent of HBsAg carrier rate in the population, around 85–90% of the infants of HBeAg seropositive carrier mothers became HBsAg carriers (Stevens et al. 1975). In endemic areas, HBV infection occurs mainly during infancy and early childhood. In contrast to the infection in adults, HBV infection during early childhood results in a much higher rate of persistent infection and long-term serious complications, such as liver cirrhosis and HCC.

4 Chronic Hepatitis B Virus Infection and Liver Cancer

Liver injury caused by chronic HBV infection is the most important initiation event of hepatocarcinogenesis (Bruix et al. 2004). The role of HBV in tumor formation appears to be complex and may involve both direct and indirect mechanisms of carcinogenesis (Grisham 2001; Villanueva 2007). The outcome of persistent HBV infection is affected by the interaction of host, viral, and environmental factors (Table 1).

4.1 Viral (HBV) Risk Factors for HCC

Chronic HBV infection with persistent positive serum HBsAg is the most important determinant for HCC. A prospective general population study of 22, 707 men in Taiwan showed that the incidence of HCC among subjects with chronic HBV infection is much higher than among non-HBsAg carriers during long-term follow-up. The relative risk is 66. These findings support the hypothesis that HBV has a primary role in the etiology of HCC (Beasley et al. 1981).

HBeAg is a marker of active HBV replication. Chronic HBV-infected subjects with prolonged high HBV replication levels or positive HBeAg after 30 years of age have a higher risk of developing HCC during follow-up. Those HBsAg carriers with persistent seropositive HBeAg have 3–6 times more risk of developing HCC than those with negative serum HBeAg (Yang et al. 2002) (Table 1). Higher HBV

Table 1 Summary of Risk Factors for Progression to HCC in HBV-Infected Individuals

Risk factors	High risk/Low risk	References
Viral factors 1. HBsAg 2. HBeAg in HBsAg-positive persons	Positive/negative = 66/1 Positive/negative = 60/10	Beasley et al. (1981) Yang et al. (2002)
3. HBV DNA level 4. HBV genotype	High $\{[>10^6]/10^5 \sim 10^6/[10^4 \sim <10^5]\}/$ Low $[<10^4]$ copies/ml = 11/9/3/1 [C or D]/[A or B]	Chen et al. (2006) Tseng et al. (2012)
Host factors 1. Age	>40/<40 years = 2–12/1	Chen et al. (2008), Tseng et al. (2012)
2. Age at HBeAg seroconversion	Older (>40 years)/younger (<30 years) = 5/1	Chen et al. (2010)
3. Gender	Male/female = 2–4/1	Ni et al. (1991), Schafer and Sorrell (1999)
4. Family HCC history	Positive/negative = 2–3/1	Turati et al. (2012)
5. Liver cirrhosis	Yes/no = 12/1	Yu et al. (1997)
6. Maternal HBsAg	Positive/negative = 30/1	Chang et al. (2009)
Other factors Smoking Habitual Alcohol	Yes/no = 1–2/1 Yes/no = 1–2/1	Yu et al. (1997), Jee et al. (2004) Yu et al. (1997), Jee et al. (2004)

DNA levels predict higher rates of HCC in those with chronic HBV infection. In comparison to those with serum HBV DNA level $<10^4$ copies/ml, those with greater serum HBV DNA levels [$10^4 \sim <10^5$], [$10^5 \sim 10^6$], or [$>10^6$] copies/ml have a higher risk of HCC [2.7, 8.9, or 10.7] during long-term follow-up (Chen et al. 2006).

There are at least ten genotypes of HBV identified with geographic variation.

Those with HBV genotype C or D infection has a high risk of developing HCC than those infected with genotype A or B HBV (Tseng et al. 2012). In Alaska, those infected with genotype F have a higher risk of HCC than other genotypes (Livingston et al. 2007).

The presence of pre-S mutants carries a high risk of HCC in HBV carriers and was proposed to play a potential role in HBV-related hepatocarcinogensis (Wang et al. 2006). Subjects infected with HBV core promoter mutants were reported to have a higher risk of developing HCC.

4.2   Host Factors for HCC (Table 1)

Older age (>40 yrs) is a risk factor for HCC development (Tseng et al. 2012; Chen et al. 2008). It is very likely due to the accumulation of genetic alterations with gain or loss of genes and liver injury with time during chronic HBV infection. HCC patients are mostly (around 80%) anti-HBe seropositive at diagnosis (Chien et al. 1981). This implies that HCC occurs after long-term HBV infection and liver injury, and that the patients have seroconverted to anti-HBe. Chronic HBV-infected patients with delayed HBeAg seroconversion after age 40 have significantly higher risk of developing HCC (hazard ratio 5.22), in comparison with patients with HBeAg seroconversion before age 30 (Chen et al. 2010).

There is a strong male predominance in HBV-related HCC, with a male to female ratio of $2 \sim 4$:1, even in children (Ni et al. 1991; Schafer and Sorrell 1999), but the mechanisms are not fully understood. The higher activity of androgen pathway functions as a tumor-promoting factor in male hepatocarcinogenesis, and the higher activity of the estrogen pathway functions as a tumor-suppressing factor in female hepatocarcinogenesis (Yeh and Chen 2012). Male predominance of HCC occurs even among young children aged 6–9 years, a possible oncogenic activation of RNA-binding motif on Y chromosome (RBMY) gene may help to explain the male predominance of HCC in children (Chua et al. 2015).

Liver cirrhosis is a pre-cancer lesion for HCC (Yu et al. 1997). Cirrhotic HBV carriers have a 3–8% annual rate of developing HCC. Those with positive HCC family history have a higher risk of HCC in comparison to those without a positive history of HCC. Familial clustering of HCC suggests the role of genetic predisposing factors in addition to the intra-familial transmission of HBV infection (Chang et al. 1984). In a meta-analysis, the pooled relative risk for family history of liver cancer was 2.50 (95% CI, 2.06–3.03) (Turati et al. 2012) (Table 1).

4.3 Maternal Effect

Those with positive maternal serum HBsAg have a 30 times higher risk of developing HCC than those with negative maternal HBsAg (Chang et al. 2009). HBeAg is a soluble antigen produced by HBV. It can cross the placenta barrier from the mother to the infant. Transplacental HBeAg from the mother induces a specific unresponsiveness of helper T cells to HBeAg and HBcAg in neonates born to HBeAg-positive HBsAg carrier mothers (Hsu et al. 1992). This may help to explain why 85–90% of the infants of HBeAg-positive carrier mothers became persistently infected (Beasley et al. 1977).

4.4 Environmental/Life Style Factors

Smoking, habitual alcohol drinking, and in some regions aflatoxin exposure are factors which were related to higher risk of HCC (Yu et al. 1997; Jee et al. 2004; Chen et al. 2008).

5 Strategies of Liver Cancer Prevention

The prognosis of HCC is grave, unless it is detected early and complete resection or ablation is performed. Even in such cases, *de novo* recurrence of HCC is always a problem. Prevention is thus the best way toward the control of HCC. There are three levels of liver cancer prevention, i.e., primary, secondary, and tertiary prevention of liver cancer (Fig. 1).

Primary prevention by universal infant vaccination to block both mother-to-infant and horizontal transmission of HBV infection is the most effective and safe way to prevent HCC. Secondary prevention using antiviral therapy for chronic hepatitis B is aimed at reducing viral replication, liver injury, and fibrosis, shown by the normalization of the liver enzymes, HBeAg clearance, and reduction of HBV DNA levels. Tertiary prevention of HCC using antiviral therapy targeting the subject of successfully treated HCC patients is aimed to prevent the recurrence of HCC.

Other strategies to prevent HCC are also proposed, such as blood and injection safety, prevention of high-risk behavior, changes in environment and/or diet, and liver transplantation for precancerous lesion (e.g., liver cirrhosis) may also be helpful to prevent HBV infection and related liver cancer. In addition, etiology-specific and generic candidate HCC chemoprevention strategies for high-risk subjects, including statins, antidiabetic drugs, selective molecular targeted agents, and dietary and nutritional substances were also reported (Jacobson et al. 1997; Egner et al. 2001; Athuluri-Divakar and Hoshida 2018). Some studies revealed that the use of concomitant medications with statin and nonsteroidal anti-inflammatory drugs (NSAIDs) or aspirin could reduce the risk of HCC

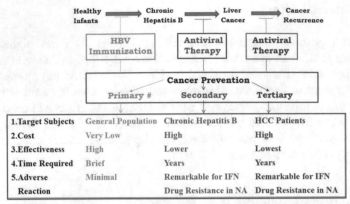

Fig. 1 Strategies for primary, secondary, and tertiary prevention of liver cancer. HBV immunization is the most effective way. For persons who have been infected by hepatitis virus, antiviral therapy may delay or reduce the risk of developing HCC in a minor degree. The effect of other strategies such as chemoprevention and avoidance of risky behavior is still not confirmed and under investigation. HBV = hepatitis B virus; IFN = interferon; NA = nucleos(t)ide analogue

recurrence in patients receiving curative HCC resection, regardless of HBV status (Wu et al. 2012; Lee et al. 2016). However, the effects of those chemoprevention strategies on reducing the risk of HCC recurrence should be further confirmed.

6   Primary Prevention of Hepatitis B Virus Infection by Immunization

6.1   Universal Hepatitis B Vaccination in Infancy

Currently, there are mainly tree strategies of universal immunization programs in the world, depending on the resources and prevalence of HBV infection. (Table 2) In countries with adequate resources, such as the USA, pregnant women are screened for HBsAg but not HBeAg. It is recommended that every infant receives three doses of HBV vaccine. In addition, infants of all HBsAg-positive mothers, regardless of HBeAg status, also receive HBIG within 24 h after birth (Shepard et al. 2006). This strategy saves the cost and the procedure of maternal HBeAg screening but increases the cost of HBIG, which is very expensive.

The first universal hepatitis B vaccination program in the world was launched in Taiwan since July 1984 (Chen et al. 1987). Pregnant women were screened for both serum HBsAg and HBeAg. Infants of HBeAg and HBsAg double positive mothers received HBIG within 24 h after birth (Strategy II).

To save the cost of screening and HBIG, some countries with intermediate/low prevalence of chronic HBV infection or inadequate resources do not screen

Table 2 Current pregnant women screening and universal infant hepatitis B virus (HBV) immunoprophylaxis strategies in different countries and proposed surveillance program for high-risk children with breakthrough infection linked to the specific strategies

Strategy type	Pregnant women screening		Neonatal immunization[a]		
	HBsAg	HBeAg	HBV vaccine	HBIG to children of HBsAg (+)/HBeAg (−) mothers	HBIG to children of HBsAg (+)/HBeAg (+) mothers
I	Yes	No	Yes	Yes	Yes
II[b]	Yes	Yes	Yes	No	Yes
III	No	No	Yes	No	No

[a]Examples of applied countries: Strategy type I: USA, Italy, Korea; Strategy type II: Taiwan, Singapore; Strategy type III: Thailand
[b]In Strategy type II, either simultaneous or sequential HBsAg and HBeAg tests can be applied. All pregnant women are screened for HBsAg and HBeAg at the same time; or all pregnant women are screened for HBsAg, while HBeAg is tested only in those positive for HBsAg; the former strategy is time saving and the latter is budget saving

pregnant women and all infants receive three doses of HBV vaccines without HBIG. Using this strategy, the cost of maternal screening and HBIG can be saved. The efficacy of preventing the infants from chronic infection seems satisfactory (Poovorawan et al. 1989).

6.2 Effect of HBV Vaccination on the Reduction of HBV Infection and Related Complications

HBV vaccine has been part of the WHO global immunization resulting in major declines in acute and chronic HBV infection. Approximately, 90–95% the incidence of chronic HBV infection in children has been reduced in areas where universal HBV vaccination in infancy has been successfully introduced. After the universal vaccination program of HBV, the rate of chronic HBV infection was reduced to approximately one-tenth of that before the vaccination program in the vaccinated infants worldwide. Fulminant or acute hepatitis also has been reduced.

Serial epidemiologic surveys of serum HBV markers were conducted in Taiwan (Hsu et al. 1986; Chen et al. 1996; Ni et al. 2001, 2007, 2016). The HBsAg carrier rate decreased significantly from around 10% before the vaccination program to 0.6–0.7% afterward in vaccinated children younger than 20 years of age. Similar effect has also been observed in many other countries (Whittle et al. 2002; Jang et al. 2001), where universal vaccination programs have been successfully conducted. The HBV vaccination program has indeed reduced both the perinatal and horizontal transmission of HBV worldwide (Da Villa et al. 1995; Whittle et al. 2002). In the reports from Gambia and Korea, universal vaccination programs have also been quite successful. The hepatitis B carrier rate has fallen from 5 to 10% to less than 1% demonstrating that universal vaccination in infancy is more effective than selective immunization for high-risk groups (Montesano 2011).

Worldwide in 2015, the estimated prevalence of HBV infection in children under five years of age was around 1.3%, compared with approximately 4.7% in the pre-vaccination era (WHO 2019). The low incidence of chronic HBV infection in children under five years of age at present can be attributed to the widespread use of hepatitis B vaccine.

6.3 Effect of Liver Cancer Prevention by Immunization Against Hepatitis B Virus Infection (Table 3)

Current therapies for HCC are not satisfactory. Even with early detection and therapy for HCC, recurrence or newly developed HCC is often a troublesome problem during long-term follow-up. Therefore, vaccination is the best way to prevent HBV infection and HCC.

HCC in children is closely related to HBV infection and the characteristics are similar to HCC in adults (Chang et al. 1989). In comparison to most other parts of the world, Taiwan has a high prevalence of HBV infection and HCC in children. Children with HCC in Taiwan are nearly 100% HBsAg seropositive, and most (86%) of them are HBeAg negative. Maternal HBsAg of HCC children are mostly

Table 3 Incidence rates of HCC among children 6–19 years old and adults 20–26 years old, born before versus after universal HBV vaccination program

Age at diagnosis, year	HBV vaccination	Year	Hepatocellular carcinoma		
			No. of HCCs	Incidence rate (per 10^5 person years)	Rate ratio (95% CI)
Taiwan		Birth year			
6–19	No	1963–1984	447	0.57	1.00 (referent)
	Yes	1984–2005	114	0.18	0.31
20–26	No	1956–1984	896	1.33	1.00 (referent)
	Yes	1984–1991	52	0.56	0.42
Khon Kaen	(Thailand)	Birth year			
>5–18	No	Before 1990	15	0.097	1.00 (referent)
	Yes	After 1990	3	0.024	0.25
Alaska	(USA)	Diagnostic year			
<20	No	1969–1984		0.7–2.6	
	Yes	1984–1988		2.9	
	Yes	1989–2008		0.0–1.4	

1. Taiwan-Chang et al. (2016); 2. Thailand-Wichajarn et al. (2008); 3. Alaska- McMahon et al. (2011)

(94%) positive. Most (80%) of the non-tumor portion has liver cirrhosis. Integration of HBV genome into host genome was demonstrated in the HCC tissues in children (Chang et al. 1991). The histological features of HCC in children are very similar to that in adults.

The reduction in HBV infection after the launch of universal hepatitis B vaccination program in July 1984 in Taiwan has had a dramatic effect on the reduction of HCC incidence in children. The rate ratio of the annual incidence of HCC in children and adolescents of 6–19 years old was significantly reduced from 1.00 in those born before the vaccination program to 0.31 in those born afterward (Chang et al. 1997; Chang et al. 2000; Chang et al. 2005; Chang et al. 2009). The cancer prevention effect by the universal HBV vaccine program has been further extended to young adults of 20–26 years old after 27 years of the vaccination program (Chang et al. 2016) (Table 3).

Approximately, 90% of the mothers of the HCC children with known serum HBsAg status were positive for HBsAg. This provides strong evidence of perinatal transmission of maternal HBV as the main route of HBV transmission in HCC children born after the immunization era and was not effectively eliminated by the HBV immunization program (Chang et al 2009).

The incidence of HCC diagnosed during 1985–2007 at Khon Kaen region is significantly lower in Thai children under 18 years old who receive hepatitis B vaccine at birth (year of birth after 1990) than unvaccinated children. The age-standardized incidence rates (ASRs) for liver cancer in children >10 years of age were significantly reduced from 0.88 per million in non-vaccinated to 0.07 per million in vaccinated children (Wichajarn et al. 2008).

Alaska Native people experience the highest rates of acute and chronic HBV infection and HCC in the USA. Universal newborn HBV vaccination coupled with mass screening and immunization of susceptible Alaska Natives has eliminated HCC among Alaska Native children. The incidence of HCC in persons <20 years decreased from 3/100,000 in 1984–1988 to zero in 1995–1999 and no HCC cases have occurred since 1999 (McMahon et al. 2011).

7 Secondary and Tertiary Prevention of Hepatitis B Related HCC

Secondary prevention of HCC uses antiviral agents with either type 1 interferon to induce immune responses or with nucleos(t)ide analogues (NA) to suppress viral replication. Studies have shown that a finite course of conventional interferon-α (IFN) therapy may provide long-term benefit for reducing the progression of liver fibrosis and the development of cirrhosis and HCC (Lin et al. 2007; Miyake et al. 2009; Yang et al. 2009; Wong et al. 2010). Yet significant reduction of HCC was only observed in patients with preexisting cirrhosis and HBeAg seroconverters. Some other meta-analysis revealed inconsistent or no significant reduction of HCC risk after interferon therapy (Lai and Yuen 2013; Cammà et al. 2001; Miyake et al. 2009).

Long-term therapy with nucleos(t)ide analogues may reduce disease progression of chronic hepatitis B, improve fibrosis, and lower the risk of HCC. A multicentered prospective randomized controlled trial of antiviral therapy using the first generation NA (lamivudine) was conducted in Asian patients with HBV-related cirrhosis. HCC was lower (3.9%) in lamivudine-treated patients than in 7.4% of placebo controls after a median follow-up of 32 months ($P = 0.047$) (Liaw et al. 2004). Subsequent meta-analysis confirms that NA treatment reduces but does not eliminate the risk of hepatocellular carcinoma (Thursz et al. 2014).

In patients receiving first-generation NAs with low genetic barrier to drug resistance, the risk of HCC remained higher in patients with resistance-related virological breakthrough than those with sustained virological response (Liaw 2013). Recent studies further illustrated the effect of reducing HBV-related HCC by NAs with high potency and minimal drug resistance [i.e., entecavir (ETV) and tenofovir desoproxial fumarate (TDF)]. Long-term ETV treatment effect was more prominent in patients at higher risk of HCC (with cirrhosis, older age and had more active disease) than younger patients and those without cirrhosis (Hosaka et al. 2013; Su et al. 2016; Wu et al. 2014; Papatheodoridis et al. 2017; Hsu et al. 2018). Suppression of viral replication in non-cirrhosis may reduce the risk of HCC, but since the risk of HCC is not as high as in cirrhosis patients, the magnitude of the risk reduction is less remarkable (Sherman 2013).

For HCC patients who have been treated successfully by surgery, liver transplantation, or local therapy, tertiary prevention of HCC using antiviral therapy against HBV or HCV may potentially prevent late tumor recurrence (Breitenstein et al. 2009). Yet further study is needed to confirm its efficacy.

8 Problems to Be Solved in Liver Cancer Prevention

The risk of developing HCC for vaccinated cohorts was significantly associated with incomplete HBV vaccination and maternal HBsAg or HBeAg seropositivity (Chang et al. 2005; Chang et al. 2009). Failure to prevent HCC by HBV vaccination results mostly from unsuccessful control of HBV transmission from highly infectious mothers. To eradicate HBV infection and its related diseases, we have to overcome the difficulties that hinder the success of universal HBV vaccination.

8.1 Low Coverage Rate of Universal Infant HBV Vaccination

Increasing uptake of HBV vaccine was noted globally year by year. In 2015, global coverage with the three doses of hepatitis B vaccine in infancy reached 84% (WHO 2017). However, in some countries such as in Southeast Asia and Africa, due to inadequate resources, failure to attract national government fund delays the integration of HBV vaccination into the EPI program. Even with integration into the

EPI program, the coverage rate is still inadequate. One main reason is that the parents still have to pay for the HBV vaccines in those countries.

In 1992, WHO recommended that all countries with a high burden of HBV-related diseases should introduce hepatitis B vaccine in their routine infant immunization programs by 1995 and that all countries do so by 1997 (Kane 1996). In 1996, an additional target was added, that is, an 80% reduction in the incidence of new hepatitis B carriers among children worldwide by 2001 (Kane and Brooks 2002). This is particularly urgent in areas where HBV infection and HCC are prevalent.

How to reduce the cost of the vaccine and to increase funds for HBV vaccination to help children of endemic areas with poor economic conditions are important issues to solve for the eradication of HBV infection and its related liver cancer. The Global Alliance for Vaccines and Immunization (GAVI), established in 1999, has contributed in helping the developing countries to increase the coverage of HBV vaccination.

8.2 Poor Compliance Caused by Anxiety to the Adverse Effects of Vaccination or Ignorance

In countries with adequate resources, the ignorance of the parents/guardians or an antivaccine mentality drives some of the people to refuse vaccination. Opposition to vaccination may be reduced by clarification of the vaccine-related side effects. Clarification of this question may help to eliminate opposition to vaccination, which hampers the effort of HBV vaccination and, hence, the goal of eradication of HBV and its related liver diseases (Halsey et al. 1999). Education and propagation of the benefits of HBV vaccination will enhance the motivation of the public and the governments to accept HBV vaccination, even in low prevalence areas.

8.3 Breakthrough Infection or Non-responsiveness in Vacinees

Causes of break through infection or non-responders include high maternal viral load (Lee et al. 1986), intrauterine infection (Tang et al. 1998; Lin et al. 1987), surface gene mutants (Hsu et al. 1999; Hsu et al. 2004; Hsu et al. 2010), poor compliance, genetic hypo-responsiveness, and immune-compromised status. A positive maternal HBsAg serostatus was found in 89% of the HBsAg serospositive subjects born after the launch of the HBV vaccination program in Taiwan (Ni YH et al. 2007). Maternal transmission is the primary reason for breakthrough HBV infection and is the challenge that needs to be addressed in future vaccination programs. Risk factors of failure include a high level of maternal HBV DNA, uterine contraction and placental leakage during the process of delivery, and low level of maternal anti-HBc (Lin et al. 1991; Chang et al. 1996). Mother-to-infant transmission is the major cause of HBV infection among immunized children.

Among 2356 Taiwan children born to HBsAg-positive mothers and identified through prenatal maternal screening, children born to HBeAg-positive mothers are at greatest risk for chronic HBV infection (9.26%), in spite of HBIG injection <24 h after birth and full course of HBV vaccination in infancy (Chen et al. 2012).

Intrauterine HBV infection, though infrequent, is a possible reason for vaccine failure. Although immunoprophylaxis for HBV infection is very successful, still around 2.4% of infants of HBeAg-positive mothers already had detectable HBsAg in the serum at birth or shortly after birth (Tang et al. 1998) and persisted to 12 months of age and later. They become HBsAg carriers despite complete immunoprophylaxis.

The rate of HBsAg gene mutants in HBsAg carriers born after the vaccination program is increasing with time. The rate of HBV surface gene mutation was 7.8%, 17.8%, 28.1%, and 23.8%, respectively, among those seropositive for HBV DNA, at before, and 5, 10, and 15 years after the launch of the HBV vaccination program (Hsu et al. 1999; Hsu et al. 2010). Fortunately, it remained stationary (22.6%) at 20 years after the vaccination program. HBV vaccine covering surface gene mutant proteins is still not urgently needed for routine HBV immunization at present, but careful and continuous monitoring for the surface gene mutants is needed.

Vaccine hypo-responsiveness or non-responsiveness may occur in immuno-compromised host, or in genetically hypo-responders or non-responders to HBV vaccine. Before receiving immunosuppressant or organ transplantation, hepatitis B markers and anti-HBs need to be monitored routinely. Hepatitis B vaccination should be given to those with inadequate anti-HBs levels. A double dose of HBV vaccine can be given to hypo-responders to enhance the vaccine response. Further development of a better vaccine is needed for non-responders to conventional HBV vaccine.

8.4   Problems of Secondary Prevention Using Current Antiviral Therapy

Current antiviral therapies and immune modulating agents did not reach a high sustained response rate. It is difficult to eliminate cccDNA even in those with sustained virologic response; therefore, HBV cannot be eradicated from the hosts in the majority of treated cases with a high viral relapse rate after discontinuation of NAs. Although the newer nucleos(t)ide has much lower resistance rates, the problem of drug resistance still exists after prolonged use. Newer safe therapeutic agents which can permanently eradiate HBV from the host are needed.

The oncogenic process starts early in patients with chronic HBV infection, even in childhood (Chang et al. 1991). A cirrhotic (or severely injured) liver may contain many clones of cells carrying genetic abnormalities and integration of HBV gen-eome into the host that predispose to cancer. Stopping the oncogenic process, by suppressing viral replication, at a late stage, such as in cirrhosis, may reduce or delay, but not eliminate HCC occurrence.

9 Strategies Toward a Successful Control of HBV-Related HCC

Primary prevention by universal vaccination is most cost effective toward a successful control of HBV infection and its complications. Yet currently, there are several problems remained to be solved. The most important strategy is to provide effective primary prevention to every infant for a better control of HBV infection globally, include further increasing the world coverage rates of HBV vaccine, and better methods to act against breakthrough HBV infection/vaccine non-responsiveness. It is extremely important to find ways to reduce the cost of HBV vaccines and to increase funding for HBV vaccination of children living in developing countries endemic of HBV infection. It is particularly urgent in areas where HBV infection and HCC are prevalent.

Increasing efforts are required to eliminate acute and chronic hepatitis B. Due to the competition of other new vaccines, HBV has not captured sufficient attention from policymakers, advocacy groups, or the general public. This is a major challenge for the future (van Herck et al. 2008). It is very important to persuade and support the policy makers of countries that still have no universal HBV vaccination program to establish a program and to encourage the countries which already have a program to increase the coverage rates. A comprehensive public health prevention program should include prevention, detection, and control of HBV infections and its related complications, and evaluation of the effectiveness of prevention activities (Lavanchy 2008).

9.1 Prevention of Breakthrough HBV Infection

Further investigation into the mechanisms of breakthrough HBV infection or non-responders is crucial for setting effective strategies to prevent breakthrough infection of HBV. Current HBV vaccine has induced good immune response and protection against HBV infection in most vaccinees. Yet approximately, 10% breakthrough infection rate occurs in high-risk infants of HBsAg carrier mothers with positive HBeAg and/or high viral load.

Nucleos(t)ide analogue treatment during pregnancy was used attempting to prevent perinatal transmission of HBV infection. A pilot study included 8 highly viremic (HBV DNA $\geq 1.2 \times 10^9$ copies/mL) mothers treated with lamivudine per day during the last month of pregnancy. At 12 months old, 12.5% in the lamivudine group and 28% in the control group were still HBsAg and HBV DNA positive (van Zonneveld et al. 2003). Another clinical trials using lamivudine mg per day were given from 34 weeks of gestation to four weeks after delivery for HBsAg seropoitive highly viremic mothers (Xu et al. 2009) demonstrated a reduction of HBsAg seropositive rate in the infants of the treated group (18%) in comparison to infants of the control group (39%) at week 52. Another study recruited mothers with positive HBeAg and HBV DNA >1.0 $\times$ 10^7 copies/ml.

The incidence of perinatal transmission was lower in the infants that completed follow-up born to the telbivudine-treated mothers than to the controls (0% vs. 8%; $p = 0.002$) (Han et al. 2011).

Tenofovir is considered a preferred choice because of its antiviral potency, more safety data in pregnant women, and lower rates of resistance. A prospective well-controlled trial in Taiwan recruiting pregnant women with HBV DNA ≥ 7.5 log 10 IU/mL has shown a reduction of HBsAg positivity of infants from 10.71 to 1.54% ($P = 0.0481$), with an odds ratio of 0.10 ($P = 0.0434$) in the tenofovir treated group (Chen et al. 2015).

Whether the development of new HBV vaccines against surface antigen gene mutants, and better vaccines for immune-compromised individuals may further reduce the incidence of new HBV infections requires further investigation.

9.2 Screening High-Risk Subjects and Provide Secondary Prevention of HCC

HBsAg carriers are at high risk for HCC. Screening for serum HBsAg is the first step to early detect the high-risk persons for HBV-related HCC screening. With limited resources, the priority target subjects to be screened are illustrated in Fig. 2. They should be screened for HBsAg, and if positive, screening for HCC. Subjects with an HBsAg carrier mother or HBsAg carrier family member(s) are particularly at higher risk of chronic HBV infection and HCC. Screening HBsAg among pregnant women is helpful to give antiviral therapy during third trimester for highly

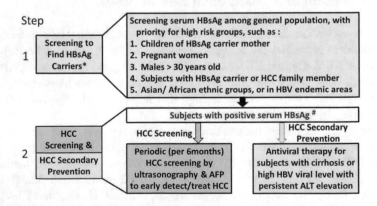

Fig. 2 Screening for HBsAg and secondary prevention of hepatocellular carcinoma. HBsAg carriers are at high risk of developing HCC. So the first step is screening to find HBsAg seropositive persons. *HBsAg carriers are subjects with chronic HBV infection. #Subjects with positive HBsAg, particularly those with special high risk of HCC, i.e., males >40 years, positive HCC family history, cirrhosis, high viral load with persistent abnormal ALT levels, are the priority target groups to receive periodic HCC screening and secondary prevention of HCC

infectious pregnant women to interrupt mother-to-infant transmission. Furthermore, those with positive HBsAg can be followed up regularly to screen or secondarily prevent HCC.

The HCC risk is higher in HBsAg carriers who are males, over 40 years old, with liver cirrhosis, a family history of HCC, or high HBV DNA >10,000 copies/mL (Table 1). For those high-risk subjects, periodic (every fix months) screening of HCC by ultrasonography and alpha-fetoprotein (AFP) is recommended. For those who are living in areas where ultrasound is not readily available, periodic screening with AFP should be considered (Bruix and Sherman 2005).

Secondary prevention of HCC can be considered in high HCC risk patients with chronic HBV infection, such as those with liver cirrhosis, or with high HBV DNA levels and persistent or intermittent abnormal ALT levels. If the future novel antiviral agent(s) is safe and could eradiate HBV, it can be given to patients with chronic HBV infection as early as possible even during childhood (Chang 2013). Potential new therapies including drugs targeting virus (inhibit viral entry, interfere RNA or viral assembly/ encapsidation, or HBsAg production or viral secretion), or targeting host immune system to enhance innate immunity or to restore HBV specific T cell and B cell responses (Kapoor and Kottilil 2014; Serigado et al. 2017; Coffin and Lee 2015; Yang and Bertoletti 2016).

10 Implication in Other Cancer Prevention and Future Prospects

Prevention is the best way to control cancer. Prevention of liver cancer by hepatitis B vaccination is the first successful example of cancer preventive vaccine in human. With the universal hepatitis B vaccination program starting from neonates in most countries in the world, HBV infection and its complications will be further reduced in this century. It is expected that an effective decline in the incidence of HCC in adults will be achieved in the near future. Furthermore, the impact of HBV vaccination on the control of hepatitis B and its related diseases can be extrapolated to other infectious agent-related cancers.

Besides vaccination, addition of hepatitis B immunoglobulin immediately after birth and even antiviral agent during the third trimester of pregnancy to block mother-to-infant transmission of HBV are existing or possible emerging strategies to enhance the prevention efficacy of HBV infection and its related liver cancer. Safe novel antiviral agent with high rate of HBV eradiation is anticipated for better secondary prevention of HCC in patients with chronic HBV infection.

Acknowledgements The work was supported by the Yuanta Foundation, Taiwan, and the National Health Research Institute, Taiwan.

References

Athuluri-Divakar SK, Hoshida Y. (2018) Generic chemoprevention of hepatocellular carcinoma. Ann N Y Acad Sci (17 September, 2018). https://doi.org/10.1111/nyas.13971. Epub ahead of print

Beasley RP (1982) Hepatitis B virus as the etiologic agent in hepatocellular carcinoma: epidemiologic considerations. Hepatology 2:21s–26s

Beasley RP, Hwang LY, Lin CC et al (1981) Hepatocellular carcinoma and hepatitis B virus. A prospective study of 22707 men in Taiwan. Lancet 2:1129–1133

Beasley RP, Trepo C, Stevens CE et al (1977) The e antigen and vertical transmission of hepatitis B surface antigen. Amer J Epidemiol 105:94–98

Braitenstein S, Dimitroulis D, Petrowsky H et al (2009) Systematic review and meta-analysis of interferon after curative treatment of hepatocellular carcinoma in patients with viral hepatitis. Br J Surg 96:975–981

Bosch FX, Ribes J (2002) The epidemiology of primary liver cancer: global epidemiology. In: Tabor E (ed) Viruses and liver cancer. Elsevier Science, Amsterdam, pp 1–16

Bruix J, Boix L, Sala M et al (2004) Focus on hepatocellular carcinoma. Cancer Cell 5:215–219

Bruix J, Sherman M (2005) Management of hepatocellular carcinoma. AASLD Guideline recommendations for HCC screening, Hepatology 42:1208–1236

Cammà C, Giunta M, Andreone P et al (2001) Interferon and prevention of hepatocellular carcinoma in viral cirrhosis: an evidence-based approach. J Hepatol 34:593–602

Chang MH (2013) Paediatrics: Children need optimal management of chronic hepatitis B. Nat Rev Gastroenterol Hepatol. 10:505–506

Chang MH, Chen CJ, Lai MS et al (1997) Universal hepatitis B vaccination in Taiwan and the incidence of hepatocellular carcinoma in children. N Engl J Med 336:1855–1859

Chang MH, Chen DS, Hsu HC et al (1989) Maternal transmission of hepatitis B virus in childhood hepatocellular carcinoma. Cancer 64:2377–2380

Chang MH, Chen PJ, Chen JY et al (1991) Hepatitis B virus integration in hepatitis B virus-related hepatocellularcarcinoma in childhood. Hepatology 13:316–320

Chang MH, Chen TH, Hsu HM et al (2005) Problems in the prevention of childhood hepatocellular carcinoma in the era of universal hepatitis B immunization. Clin Cancer Res 11:7953–7957

Chang MH, Hsu HC, Lee CY et al (1984) Fraternal hepatocellular carcinoma in young children in two families. Cancer 53:1807–1810

Chang MH, Hsu HY, Huang LM et al (1996) The role of transplacental hepatitis B core antibody in the mother-to-infant transmission of hepatitis B virus. J Hepatol 24:674–679

Chang MH, Shau WY, Chen CJ et al (2000) The effect of universal hepatitis B vaccination on hepatocellular carcinoma rates in boys and girls. JAMA 284:3040–3042

Chang MH, You SL, Chen CJ et al (2009) Decreased incidence of hepatocellular carcinoma in hepatitis B vaccinees: a 20-year follow-up study. J Natl Cancer Inst 101:1348–1355

Chang MH, You SL, Chen CJ et al (2016) Long-term effects of hepatitis B immunization of infants in preventing liver cancer. Gastroenterology 151:472–480

Chen CJ, Yang HI, Su J et al (2006) Risk of hepatocellular carcinoma across a biological gradient of serum hepatitis B virus DNA level. JAMA 295:65–73

Chen CL, Yang HI, Yang WS et al (2008) Metabolic factors and risk of hepatocellular carcinoma by chronic hepatitis B/C infection: a follow-up study in Taiwan. Gastroenterology 135:111–121

Chen DS, Hsu NHM, Sung JL et al (1987) A mass vaccination program in Taiwan against hepatitis B virus infection in infants of hepatitis B surface antigen-carrier mothers. JAMA 257:2597–2603

Chen HL, Chang MH, Ni YH et al (1996) Seroepidemiology of hepatitis B virus infection in children: ten years after a hepatitis B mass vaccination program in Taiwan. JAMA 276:906–908

Chen HL, Lin LH, Hu FC et al (2012) Effects of maternal screening and universal immunization to prevent mother-to-infant transmission of HBV. Gastroenterology 142:773–781

Chen HL, Lee CN, Chang CH et al (2015) Efficacy of maternal tenofovir disoproxil fumarate in interrupting mother-to-infant transmission of hepatitis B virus. Hepatology 62:375–386

Chen YC, Chu CM, Liaw YF (2010) Age- specific prognosis following spontaneous hepatitis B e antigen seroconversion in chronic hepatitis B. Hepatology 51:435–444

Chien MC, Tong MJ, Lo KJ et al (1981) Hepatitis B viral markers in patients with primary hepatocellular carcinoma in Taiwan. J Natl Cancer Inst 66:475–479

Coffin CS, Lee SS (2015) New paradigms in hepatitis B management: only diamonds are forever. Br Med Bull 116:79–91

Chua HH, Tsuei DJ, Lee PH et al (2015) RBMY, a novel inhibitor of glycogen synthase kinase 3β, increases tumor stemness and predicts poor prognosis of hepatocellular carcinoma. Hepatology 62:1480–1496

Da Villa G, Picciottoc L, Elia S et al (1995) Hepatitis B vaccination: Universal vaccination of newborn babies and children at 12 years of age versus high risk groups: a comparison in the field. Vaccine 13:1240–1243

Egner PA, Wang JB, Shu YR et al (2001) Chlorophyline intervention reduces aflatoxin-DNA adducts in individuals at high risk for liver cancer. Proc Natl Acad Sci USA 98:14601–14606

Grisham JW (2001) Molecular genetic alterations in primary hepatocellular neoplasms. In: Coleman WB, Tsongalis GJ (eds) The molecular basis of human cancer. Humana Press, Totowa, NJ, pp 269–346

Halsey NA, Duclos P, van Damme P et al (1999) Hepatitis B vaccine and central nervous system demyelinating diseases. Pediatrc Infect Dis J 18:23–24

Han GR, Cao MK, Zhao W et al (2011) A prospective and open-label study for the efficacy and safety of telbivudine in pregnancy for the prevention of perinatal transmission of hepatitis B virus infection. J Hepatol 55:1215–1221

Heyward WL, Lanier AP, Bender TR et al (1981) Primary hepatocellular carcinoma in Alaskan Natives, 1969–1979. Int J Cancer 28:47–50

Hosaka T, Suzuki F, Kobayashi M et al (2013) Long-term entecavir treatment reduces hepatocellular carcinoma incidence in patients with hepatitis B virus infection. Hepatology 58:98–107

Hsu HC, Lin WS, Tsai MJ (1983) Hepatitis B surface antigen and hepatocellular carcinoma in Taiwan. With special reference to types and localization of HBsAg in the tumor cells. Cancer 52:1825–1832

Hsu HY, Chang MH, Chen DS et al (1986) Baseline seroepidemiology of hepatitis B virus infection in children in Taipei, 1984: A study just before mass hepatitis B vaccination program in Taiwan. J Med Virol 18:301–307

Hsu HY, Chang MH, Hsieh KH et al (1992) Cellular immune response to hepatitis B core antigen in maternal-infant transmission of hepatitis B virus. Hepatology 15:770–776

Hsu HY, Chang MH, Liaw SH et al (1999) Changes of hepatitis B surface variants in carrier children before and after universal vaccination in Taiwan. Hepatology 30:1312–1317

Hsu HY, Chang MH, Ni YH et al (2010) Twenty-year trends in the emergence of hepatitis B surface antigen variants in children and adolescents after universal vaccination in Taiwan. J Infect Dis 201:1192–1200

Hsu YC, Ho HJ, Lee TY et al (2018) Temporal trend and risk determinants of hepatocellular carcinoma in chronic hepatitis B patients on entecavir or tenofovir. J Viral Hepat. 25:543–551

Hsu HY, Chang MH, Ni YH, Chen HL (2004) Survey of hepatitis B surface variant infection in children 15 years after a nationwide vaccination program in Taiwan. Gut 53:1499–1503

International Agency for Research on Cancer, World Health Organization (2018) Cancer today. http://gco.iarc.fr/today/fact-sheets-cancers. Accessed December 29, 2018

Jacobson LP, Zhang BC, Shu YR et al (1997) Oltipratz chemoprevention trial in Qidong, People's Republic of China: study design and clinical outcomes. Cancer Epidemiol Biomarkers Prev 6:257–265

Jang MK, Lee JY, Lee JH et al (2001) Seroepidemiology of HBV infection in South Korea, 1995 through 1999. Korean J Intern Med 16:153–159

Jee SH, Ohrr H, Sull JW et al (2004) Cigarette smoking, alcohol drinking, hepatitis B, and risk for hepatocellular carcinoma in Korea. J Natl Cancer Inst 96:1851–1855

Kane MA (1996) Global status of hepatitis B immunization. Lancet 348:696

Kane MA, Brooks A (2002) New immunization initiatives and progress toward the global control of hepatitis B. Current Opinion in Infectious Diseases 15:465–469

Kapoor R, Kottilil S (2014) Strategies to eliminate HBV infection. Future Virol 9:565–585

Lai CL, Yuen MF (2013) Prevention of hepatitis B virus-related hepatocellular carcinoma with antiviral therapy. Hepatology 57:399–408

Lavanchy D (2004) Hepatitis B virus epidemiology, disease burden, treatment, and current and emerging prevention and control measures. J Viral Hepatitis 11:97–107

Lavanchy D (2008) Chronic viral hepatitis as a public health issue in the world. Best Pract Res Clin Gastroenterol 22:991–1008

Lee PC, Yeh CM, Hu YW et al (2016) Antiplatelet therapy is associated with a better prognosis for patients with hepatitis B virus-related hepatocellular carcinoma after liver resection. Ann Surg Oncol 23(Suppl 5):874–883

Lee SD, Lo KJ, Wu JC et al (1986) Prevention of maternal-infant hepatitis B virus transmission by immunization: the role of serum hepatitis B virus DNA. Hepatology 6:369–373

Liaw YF, Sung JJ, Chow WC et al (2004) Lamivudine for patients with chronic hepatitis B and advanced liver disease. N Engl J Med 351:1521–1531

Liaw YF (2013) Impact of therapy on the long-term outcome of chronic hepatitis B. Clin Liver Dis 17:413–423

Lin HH, Chang MH, Chen DS et al (1991) Early predictor of the efficacy of immunoprophylaxis against perinatal hepatitis B transmission: Analysis of prophylaxis failure. Vaccine 9:457–460

Lin HH, Lee TY, Chen DS et al (1987) Transplacental leakage of HBeAg-positive maternal blood as the most likely route in causing intrauterine infection with hepatitis intrauterine infection with hepatitis B virus. J Pediatr 111:877–881

Lin SM, Yu ML, Lee CM et al (2007) Interferon therapy in HBeAg positive chronic hepatitis reduces progression to cirrhosis and hepatocellular carcinoma. J Hepatol 46:45–52

Livingston SE, Simonetti J, McMahon B et al (2007) Hepatitis B virus genotypes in Alaska native people with hepatocellular carcinoma: preponderance of genotype F. J Infect Dis 195(5–11):1

McMahon BJ, Bulkow LR, Singleton RJ et al (2011) Elimination of hepatocellular carcinoma and acute hepatitis B in children 25 Years after a hepatitis B newborn and catch-up immunization program. Hepatology 54:801–807

Miyake Y, Kobashi H, Yamamoto K (2009) Meta-analysis: the effect of interferon on development of hepatocellular carcinoma in patients with chronic hepatitis B virus infection. J Gastroenterol 44:470–475

Montesano R (2011) Preventing primary liver cancer: the HBV vaccination project in the Gambia (West Africa). Environmental Health 10(Suppl 1):S6

Ni YH, Chang MH, Hsu HY et al (1991) Hepatocellular carcinoma in childhood. Clinical manifestations and prognosis. Cancer 68:1737–1741

Ni YH, Chang MH, Huang LM et al (2001) Hepatitis B virus infection in children and adolescents in an hyperendemic area: 15 years after universal hepatitis B vaccination. Ann Intern Med 135:796–800

Ni YH, Huang LM, Chang MH et al (2007) Two decades of universal hepatitis B vaccination in Taiwan: Impact and implication for future strategies. Gastroenterology 132:1287–1293

Ni YH, Chang MH, Jan CF et al (2016) Continuing decrease in hepatitis B virus infection 30 years after initiation of infant vaccination program in Taiwan. Clin Gastroenterol Hepatol 14:1324–1330

Papatheodoridis GV, Idilman R, Dalekos GN et al (2017) The risk of hepatocellular carcinoma decreases after the first 5 years of entecavir or tenofovir in caucasians with chronic hepatitis B. Hepatology 66:1444–1453

Parkin DM, Bray F, Ferlay J et al (2001) Estimating the world cancer burden: Globocan 2000. Int J Cancer 94:153–156

Poovorawan Y, Sanpavat S, Pongpunlert W et al (1989) Protective efficacy of a recombinant DNA hepatitis B vaccine in neonates of HBe antigen-positive mothers. JAMA 261:3278–3281

Schafer DF, Sorrell MF (1999) Hepatocellular carcinoma. Lancet 353:1253–1257

Serigado JM, Izzy M, Kalia H (2017) Novel therapies and potential therapeutic targets in the management of chronic hepatitis B. Eur J Gastroenterol Hepatol 29:987–993

Shepard CW, Simard EP, Finelli L et al (2006) Hepatitis B virus infection: epidemiology and vaccination. Epidemiol Rev 28:112–125

Sherman M (2013) Does hepatitis B treatment reduce the incidence of hepatocellular carcinoma? Hepatology 58:18–20

Stevens CE, Beasley RP, Tsui J et al (1975) Vertical transmission of hepatitis B antigen in Taiwan. N Engl J Med 292:771–774

Su TH, Hu TH, Chen CY et al (2016) C-TEAM study group and the Taiwan Liver diseases consortium. Four-year entecavir therapy reduces hepatocellular carcinoma, cirrhotic events and mortality in chronic hepatitis B patients. Liver Int. 36:1755–1764

Tang JR, Hsu HY, Lin HH et al (1998) Hepatitis B surface antigenemia at birth: a longterm follow-up study. J Pediatr 133:374–377

Thursz MR (2014) Treating chronic hepatitis with antiviral drugs to prevent liver cancer. In: Steward Bernard W, Wild Christopher P (eds) World cancer report 2014. WHO, IARC, pp 587–593

Tseng TC, Liu CJ, Yang HC et al (2012) High levels of hepatitis B surface antigen increase risk of hepatocellular carcinoma in patients with low HBV load. Gastroenteorlogy 142:1140–1149

Turati F, Edefonti V, Talamini R et al (2012) Family history of liver cancer and hepatocellular carcinoma. Hepatology 55:1416–1425

Van Herck K, Vorsters A, Van Damme P (2008) Prevention of viral hepatitis (B and C) reassessed. Best Pract Res Clin Gastroenterol 22:1009–1029

van Zonneveld M, van Nunen AB, Niesters HG et al (2003) Lamivudine treatment during pregnancy to prevent perinatal transmission of hepatitis B virus infection. J Viral Hepat 10:294–297

Villanueva A, Newell P, Chiang DY et al (2007) Genomics and signaling pathways in hepatocellular carcinoma. Semin Liver Dis 27:55–76

Wang HC, Huang W, Lai MD et al (2006) Hepatitis B virus pre-S mutants, endoplasmic reticulum stress and hepatocarcinogenesis. Cancer Sci 97:683–688

Whittle H, Jaffar S, Wansbrough M et al (2002) Observational study of vaccine efficacy 14 years after trial of hepatitis B vaccination in Gambian children. BMJ 325:569–573

Wichajarn K, Kosalaraksa P, Wiangnon S (2008) Incidence of hepatocellular carcinoma in children in Khon Kaen before and after national hepatitis B vaccine program. Asian Pac J Cancer Prev 9:507–509

Wong GL, Yiu KK, Wong VW et al (2010) (2010) Metaanalysis: reduction in hepatic events following interferon-alfa therapy of chronic hepatitis B. Aliment Pharmacol Ther 32:1059–1068

World Health Organization (2017) Global hepatitis report 2017. Geneva, Switzerland

World Health Organization (2019). (https://www.who.int/news-room/fact-sheets/detail/hepatitis-b. Accessed January, 2019

Wu CY, Chen YJ, Ho HJ et al (2012) (2012) Association between nucleoside analogues and risk of hepatitis B virus-related hepatocellular carcinoma recurrence following liver resection. JAMA 308:1906–1914

Wu CY, Lin JT, Ho HJ et al (2014) Association of nucleos(t)ide analogue therapy with reduced risk of hepatocellular carcinoma in patients with chronic hepatitis B: a nationwide cohort study. Gastroenterology 147:143–151

Xu WM, Cui YT, Wang L et al (2009) Lamivudine in late pregnancy to prevent perinatal transmission of hepatitis B virus infection: a multicentre, randomized, double-blind, placebo-controlled study. J Viral Hepat 16:94–103

Yang HI, Lu SN, Liaw YF et al (2002) Hepatitis B e antigen and the risks of hepatocellular carcinoma. N Engl J Med 347:168–174

Yang N, Bertoletti A (2016) Advances in therapeutics for chronic hepatitis B. Hepatol Int 10:277–285

Yang YF, Zhao W, Zhong YD et al (2009) Interferon therapy in chronic hepatitis B reduces progression to cirrhosis and hepatocellular carcinoma: a meta-analysis. J Viral Hepat. 16:265–271

Yeh SH, Chen PJ (2012) Gender disparity of hepatocellular carcinoma: the roles of sex hormones. Oncology 78(suppl 1):172–179

Yu MW, Hsu FC, Sheen IS et al (1997) Prospective study of hepatocellular carcinoma and liver cirrhosis in asymptomatic chronic hepatitis B virus carriers. Am J Epidemiol 145:1039–1047

The Oncogenic Role of Hepatitis C Virus

Kazuhiko Koike and Takeya Tsutsumi

1 Introduction

Worldwide, about 170 million people are persistently infected with hepatitis C virus (HCV), which induces chronic hepatitis, cirrhosis, and eventually hepatocellular carcinoma (HCC) (Saito et al. 1990). Owing to the recent advances in HCV research, particularly the identification of the JFH-1 strain (Wakita et al. 2005), the HCV lifecycle has been elucidated. Accordingly, direct-acting antivirals (DAAs) against HCV were developed. DAAs can eliminate HCV efficiently and safely, and almost all HCV-infected patients achieve a sustained viral response (SVR). However, HCC can develop even in patients who achieved an SVR, albeit at a lower frequency than in untreated patients. This post-SVR HCC is an important problem in clinical practice.

As mechanisms of hepatocarcinogenesis by HCV, DNA damage induced by cytokines and oxidative stress by chronic inflammation, or mutations of genomic DNA induced by repeated cellular destruction and regeneration, have been considered. In fact, an elevated level of serum alanine aminotransferase (reflecting hepatitis activity) and a lower platelet count (reflecting progression of fibrosis) are predictive of HCC development. However, accumulating in vitro data suggest that the core protein, which constitutes the HCV particle, affects cell proliferation, transcription, and apoptosis, suggesting that HCV itself may be carcinogenic (Koike 2007). Furthermore, we and other groups have reported that transgenic mice

K. Koike (✉)
Department of Gastroenterology, Graduate School of Medicine,
The University of Tokyo, Tokyo, Japan
e-mail: kkoike-tky@umin.ac.jp

T. Tsutsumi
Division of Infectious Diseases, Advanced Clinical Research Center, Institute of Medical Science, The University of Tokyo, Tokyo, Japan
e-mail: kkoike-tky@umin.ac.jp

© Springer Nature Switzerland AG 2021
T.-C. Wu et al. (eds.), *Viruses and Human Cancer*, Recent Results
in Cancer Research 217, https://doi.org/10.1007/978-3-030-57362-1_5

harboring the HCV core protein gene developed HCC in the absence of hepatic inflammation (Moriya et al. 1998; Machida et al. 2006; Lerat et al. 2002; Naas et al. 2005). These data indicate that HCV, and particularly its core protein, promote hepatocarcinogenesis by modulating the gene expression and functions of host cells involved in processes necessary for malignant transformation, *i.e.*, HCV itself has oncogenic activity. Also, hepatic steatosis, accumulation of oxidative stress, and insulin resistance, which are frequent in HCV-infected patients, also occur in HCV core-transgenic mice, and mitochondrial dysfunction might contribute to these effects. Therefore, knowledge of the mechanism underlying HCC development in persistent HCV infection is needed.

2 Hepatitis C Virus and Viral Proteins

HCV is an enveloped RNA virus belonging to the family *Flaviviridae*, and contains a positive-sense, single-stranded RNA genome of approximately 9600 nucleotides (nt) within the nucleocapsid (Houghton et al. 1991). The genome consists of a large open reading frame (ORF) encoding a polyprotein of approximately 3010 amino acids (aa) (Fig. 1). The ORF is contiguous to highly conserved untranslated regions (UTR) at the 5′- and 3′-termini. The complete 5′–UTR consists of 341 nt and acts as an internal ribosomal entry site. This promotes translation of the RNA genome using a cap-independent mechanism rather than ribosome scanning from the 5′–end of a capped molecule.

The polyprotein is processed by cellular and viral proteases to generate the viral structural and non-structural proteins. The structural proteins, which are encoded by the NH_2-terminal quarter of the genome, include the core protein and the envelope proteins, E1 and E2. E2 has an alternative form, E2-p7, which is reportedly

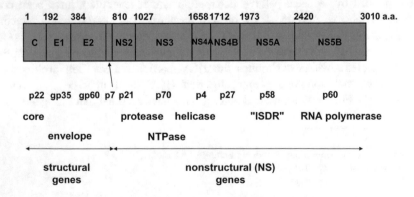

Fig. 1 Structure of the HCV genome. The HCV genome encodes a polyprotein of 3,010 aa, which is processed to structural and nonstructural proteins by cellular or viral proteases. One of the structural proteins, the core protein, has shown a variety of characteristics in vitro and in vivo. ISDR, interferon sensitivity-determining region

associated with virion assembly and release (Joyce et al. 2009, Atoom et al. 2014). NS2, NS3, NS4A, NS4B, NS5A, and NS5B are the non-structural proteins of HCV coded in the remaining portion of the genome. These include serine protease (NS3/4A), NTPase/helicase (NS3), and RNA-dependent RNA polymerase (NS5B).

The core protein of HCV occupies residues 1–191 of the precursor polyprotein and is cleaved between the core and E1 proteins by a host signal peptidase. The C-terminal membrane anchor of the core protein is further processed by a host signal-peptide peptidase (Moradpour et al. 2007). The mature core protein is estimated to consist of 177–179 aa and has a high level of homology among HCV genotypes. The HCV core protein possesses a hydrophilic N-terminal region (domain 1; residues 1–117) followed by a hydrophobic region (domain 2) from residues 118 to 170. Domain 1 is rich in basic residues and is implicated in RNA binding and homo-oligomerization. The amphipathic helices I and II (residues 119–136 and 148–164, respectively) in domain 2 are involved in the association of HCV core protein with lipids (Boulant et al. 2006). In addition, the region spanning residues 112–152 is associated with the membranes of the endoplasmic reticulum (ER) and mitochondria (Suzuki et al. 2005). The core protein is also localized to the nucleus (Miyamoto et al. 2007; Shirakura et al. 2007) and binds to the nuclear proteasome activator 28γ (PA28γ)/REGγ, resulting in its PA28γ-dependent degradation (Moriishi et al. 2003). Autophagy is involved in the degradation of organelles and elimination of microorganisms; disruption of autophagy leads to disorders involving protein deposition. Replication of HCV RNA induces autophagy in a strain-dependent manner, suggesting that HCV harnesses autophagy to prevent cell death and dysfunction of autophagy is implicated in the genotype-specific pathogenesis of HCV (Taguwa et al. 2011).

3 Possible Role of HCV in Hepatocarcinogenesis

The mechanism underlying hepatocarcinogenesis in HCV infection is unclear, despite the fact that nearly 80% of patients with HCC in Japan and 30% of those worldwide (Perz et al. 2006) are persistently infected with HCV (Kiyosawa et al. 1990; Saito et al. 1990; Yotsuyanagi et al. 2000). These lines of evidence prompted us to evaluate the role of HCV in hepatocarcinogenesis. HCV–induced inflammation leads to necrosis of hepatocytes; their subsequent regeneration enhances genetic aberrations in host cells, the accumulation of which leads to HCC. This hypothesis presupposes indirect involvement of HCV in HCC via hepatic inflammation. This poses the question: can inflammation alone explain the high incidence (90% over 15 years) or multicentric nature of HCC in HCV-infected patients?

The putative indirect role of HCV must be weighed against the rarity of HCC in patients with autoimmune hepatitis in which severe inflammation in the liver persists, even after the development of cirrhosis. Therefore, viral proteins may induce neoplasia. This possibility was evaluated by introducing HCV genes into hepatocytes, but the result was negative. This is likely because of the weak carcinogenic activity of HCV, which takes a long time to manifest. Indeed, HCC development in HCV-infected individuals requires 30–40 years. Humanized immunocompromised mice harboring human hepatocytes support HCV replication, but this does not induce HCC. Therefore, investigations of the carcinogenetic activity of HCV in vivo have used transgenic mice.

4 In Vivo Oncogenic Activity of HCV Core Protein in Mice

A major issue regarding the pathogenesis of HCV-associated liver lesions is the direct pathological effects of HCV proteins. Although several strategies have been applied, the relationship between HCV proteins and disease phenotype remains unclear. For this purpose, several lines of mice transgenic for HCV cDNA have been established (Table 1); some carry the entire coding region of the HCV genome (Lerat et al. 2002), the core region only (Machida et al. 2006; Moriya et al. 1997), the envelope region only (Koike et al. 1995; Pasquinelli et al. 1997), the core and envelope regions (Lerat et al. 2002; Naas et al. 2005), and the core to NS2 regions (Wakita et al. 1998). Although mRNA from the NS region of HCV cDNA has been detected in the liver of such transgenic mice (Honda et al. 1999; Lerat et al. 2002), HCV NS proteins have not. The reason for this is unclear but may be because the HCV NS proteins are harmful to mouse development. If so, establishment of mouse strains that produce the HCV proteins at low levels may be feasible.

We have engineered transgenic mouse lines carrying the HCV genome by introducing cDNA of HCV genotype 1b (Moriya et al. 1997; Moriya et al. 1998). The four transgenic mouse lines carry the core gene, envelope genes, NS genes, or only NS5A gene under the same transcriptional regulatory element. Among them, only transgenic mice carrying the core gene developed HCC in two independent lineages (Moriya et al. 1998). The envelope gene-transgenic mice did not develop HCC, despite high levels of the E1 and E2 proteins (Koike et al. 1995; Koike et al. 1997). The transgenic mice carrying the NS genes or NS5A gene also did not develop HCC.

The core gene-transgenic mice, early in life, develop hepatic steatosis, a histologic characteristic of chronic hepatitis C, along with lymphoid follicle formation and bile duct damage (Bach et al. 1992). Thus, the core gene-transgenic mice recapitulate chronic hepatitis C. Notably, significant hepatic inflammation is not observed in these mice. Late in life, the core gene-transgenic mice develop HCC. The development of steatosis and HCC is reproduced in other HCV-transgenic mouse lines, which harbor the entire HCV genome or its structural genes, including the core gene (Lerat et al. 2002; Machida et al. 2006; Naas et al. 2005). Therefore, the HCV core protein per se has oncogenic potential in vivo. In fact, the core protein modulates intracellular signaling pathways, including mitogen-activated protein kinase in vivo (Tsutsumi et al. 2002b; Tsutsumi et al. 2003), which promotes the proliferation of hepatocytes. Further investigation of the core-transgenic mice revealed that the HCV core protein exerts several effects (see below) and may play a role in hepatocarcinogenesis.

Table 1 Consequences to the expression HCV proteins in mice

HCV gene	Genotype	Promoter	Protein expression	Phenotypes	References
Core	1b	HBV	Similar to patients	Steatosis, HCC, insulin resistance, oxidative stress	Moriya (1997, 1998) Moriishi (2003, 2007) Shintani (2004) Miyamoto (2007)
Core	1b	EF-1a	Similar to patients	Steatosis, adenoma, HCC, oxidative stress	Machida (2006)
E1–E2	1b	HBV	Abundant	None in the liver	Koike (1995), Koike (1997)
Core-E1–E2	1b	Albumin	Similar to patients	Steatosis, HCC, oxidative stress	Lerat (2003)
Core-E1–E2	1a	CMV	Similar to patients	Steatosis, HCC	Naas (2005)
Structural proteins	1b	MHC	Low in the liver	Hepatitis	Honda (1999)
Entire polyprotein	1b	Albumin	Only mRNA detectable	Steatosis, HCC	Lerat (2003)
Entire polyprotein	1a	A1-anti-trypsin		Steatosis, intrahepatic T cell recruitment	Alonzi (2004)
NS3/4A	1a	MUP		None (modulation of immunity)	Frelin (2006)
NS5A	1a	apoE		None (resistance to TNF)	Majumder (2002)

HBV, hepatitis B virus; EF, elongation factor; MUP, major urinary protein; Alb, albumin; CMV, cytomegalovirus; MHC, major histocompatibility complex; AT, anti-trypsin; apo E, apolipoprotein E

5 Induction of Oxidative Stress via Mitochondria by HCV

Augmentation of oxidative stress is implicated in the pathogenesis of liver disease in HCV–infected patients (Farinati et al. 1995). Reactive oxygen species (ROS) are endogenous oxygen-containing molecules formed as normal products during aerobic metabolism. ROS can induce genetic mutations as well as chromosomal alterations and thus contribute to carcinogenesis (Fujita et al. 2008; Kato et al. 2001). While the HCV core protein is localized predominantly to the cytoplasm in association with lipid droplets, it is also present in the nucleus and mitochondria (Moriya et al. 1998). Mitochondria are a major source of ROS and HCV induces oxidative stress in vivo as well as in vitro by localizing to mitochondria and disrupting their function. Hepatic ROS production is increased in the HCV core-transgenic mice at an early age, which is compensated for by upregulation of catalase and reduced synthesis of glutathione. However, in older mice, the compensatory effect is inadequate, leading to ROS accumulation in hepatocytes (Moriya et al. 2001). NS5A also induces ER stress and increases Ca efflux, leading to enhanced Ca influx into mitochondria and an increased ROS level (Tardif et al. 2002). Induction of oxidative stress is also observed in vitro in HCV-replicating cells such as subgenomic-replicon cells and JFH-1 cells (Boudreau et al. 2009). In addition, oxidative stress is enhanced in the liver of chimeric mice harboring HCV-infected human hepatocytes (Joyce et al. 2009). Furthermore, patients with chronic hepatitis C have increased oxidative DNA damage in hepatocytes and peripheral leukocytes (Fujita et al. 2008; Yen et al. 2012). These data suggest that HCV directly contributes to hepatocarcinogenesis by inducing oxidative stress, which triggers DNA damage. Also, the main site of HCV-induced ROS production is mitochondria. Indeed, proteomic profiling of biopsy specimens from HCV-infected human livers with advanced fibrosis revealed impairment of both key mitochondrial processes, including fatty acid oxidation, and the response to oxidative stress (Diamond et al. 2007). The mechanism underlying the HCV-induced increased ROS production by mitochondria has been investigated.

HCV core protein is localized to, and induces morphological changes of, mitochondria. In addition, HCV core protein suppresses the activity of complex I in the mitochondrial respiratory chain (Korenaga et al. 2005; Piccoli et al. 2007). This suppression is mediated in part by the direct interaction of HCV core protein with a mitochondrial protein, prohibitin. Prohibitin is a ubiquitously expressed and highly conserved protein that plays the predominant role in inhibiting cell-cycle progression and cellular proliferation by attenuating DNA synthesis (Mishra et al. 2005). It is localized to the nucleus and interacts with transcription factors vital for cell-cycle progression. Mitochondrial prohibitin acts as a chaperone by stabilizing newly synthesized mitochondrial proteins (Nijtmans et al. 2000). By two-dimensional polyacrylamide gel electrophoresis (2D-PAGE) of mitochondria isolated from HepG2 cells stably expressing the HCV core protein, prohibitin was found to be upregulated. We found that the interaction between prohibitin and mitochondria-encoded subunit II of COX is suppressed in core-expressing cells (Tsutsumi et al. 2009). This suggests that HCV core protein disrupts the formation and function of the mitochondrial respiratory chain by interacting with prohibitin and suppressing its function as a chaperone, leading to increased oxidative stress. Indeed, suppression of prohibitin function results in increased ROS production and mice lacking intrahepatic expression of prohibitin exhibits ROS accumulation and HCC development (Theiss et al. 2007; Ko et al. 2010). HCV core protein is also associated with mitophagy, mitochondrion-specific autophagy. Under normal conditions, mitochondria with morphological or functional abnormalities are rapidly removed by mitophagy, but abnormal mitochondria accumulate in the presence of the HCV core protein. This suppressive effect may be due to interaction of the HCV core protein with Parkin, a promoter of mitophagy. Parkin is an E3 ubiquitin ligase predominantly localized to the cytoplasm, but translocates to mitochondria in response to their depolarization owing to activation of PTEN-induced putative kinase 1 (PINK1). In mitochondria, PINK1 ubiquitinates itself and outer mitochondrial proteins, thereby priming mitophagy. However, in the presence of HCV core protein, Parkin is retained in the cytoplasm, leading to suppression of mitophagy (Hara et al. 2014). Other factors associated with mito-phagy, such as the 'mitophagy receptors' Bnip3 and Nix, may also be targeted by HCV core protein. In any case, the accumulation of abnormal mitochondria caused by the suppression of mitophagy further increases oxidative stress, which con-tributes to DNA damage, finally leading to the development of HCC.

6   Effect of HCV on Iron Metabolism

As discussed above, chronic hepatitis C is characterized by increased oxidative stress. Iron accumulation in the liver aggravates oxidative stress, as indicated by increased levels of DNA adducts in the liver (Farinati et al. 1995). In addition, iron accumulates in the liver of HCV core-transgenic mice (Moriya et al. 2010). Therefore, the contribution of abnormal iron metabolism by HCV to hepatocar-cinogenesis is focused, and in fact, the risk of HCC development is higher in HCV-infected patients with elevated hepatic iron accumulation. As a possible molecular mechanism, HCV core protein modulates the expression of heme-oxygenase-1 (HO-1), a key factor in iron metabolism. HO-1 catalyzes the initial and rate-limiting reaction in heme catabolism and cleaves pro-oxidant heme to biliverdin, which in mammals is converted to bilirubin; both biliverdin and bilirubin have antioxidant activity (Stocker et al. 1987). HO-1 has been suggested to be an important antioxidant in the presence of glutathione depletion (Oguro et al. 1998). Thus, HO-1 is an endogenous protective mechanism against oxidative stress, and particularly iron overload. Also, in cultured cells and transgenic mice, HCV decreases the expression of hepcidin, a protein that suppresses iron absorption from the gastrointestinal tract, leading to increased absorption and subsequent accumu-lation of iron (Nishina et al. 2008; Miura et al. 2008). Therefore, HCV-induced abnormalities of iron metabolism contribute to ROS accumulation. This notion is supported by the fact that phlebotomy decreases the incidence of HCC in HCV-infected patients.

7   Induction of Hepatic Steatosis and Insulin Resistance by HCV

In patients with chronic hepatitis C, hepatic steatosis and diabetes mellitus are more frequent comorbidities, as compared to patients with other chronic liver diseases. The grade of hepatic steatosis correlates with the intrahepatic HCV load, and hepatic steatosis and insulin resistance are improved by elimination of HCV.

Furthermore, hepatic steatosis and insulin resistance occur in HCV core-transgenic mice at an early age in the absence of inflammation, suggesting direct involvement of HCV. Several mechanisms of intrahepatic lipid accumulation by HCV have been proposed (Fig. 2)—upregulation of intrahepatic fatty acid synthesis via the activation of sterol regulatory element binding protein-1 (SREBP1), decreased consumption of fatty acids due to disruption of mitochondrial function, increased import of fatty acids due to insulin resistance, and reduced export of very low-density lipoprotein (VLDL) due to downregulation of the activity of microsomal triglyceride protein (MTP) (Perlemuter et al. 2002). Hepatic steatosis is beneficial to HCV because lipid droplets in hepatocytes are indispensable for viral replication (Miyanari et al. 2007). Insulin resistance is caused by functional suppression of insulin receptor substrate-1, a key factor in the intracellular insulin signaling pathway, due to upregulation of tumor necrosis factor-α and suppressor of cytokine signaling (Tsutsumi et al. 2002b; Shintani et al. 2004; Miyoshi et al. 2005).

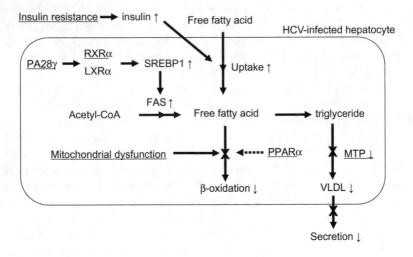

Fig. 2 Molecular mechanisms of HCV-induced intrahepatic lipid accumulation. HCV, and particularly the core protein, affects several pathways associated with lipid metabolism and induces hepatic steatosis. Underlining indicates cellular proteins or processes affected by HCV. First, the core protein induces insulin resistance, promoting the peripheral release and hepatic uptake of fatty acids. Second, the core protein suppresses the activity of MTP, inhibiting the secretion of VLDL from the liver, resulting in an increased hepatic triglyceride level. Third, the transcription factor, SREBP-1c, is upregulated by the core protein, resulting in increased production of triglycerides. Finally, impaired β-oxidation of fatty acids due to mitochondrial dysfunction induced by the core protein leads to the accumulation of fatty acids. PA28g, proteasome activator 28g; RXRα, retinoid X receptor; LXRα, liver X receptor; SREBP1, sterol regulatory element binding protein-1; FAS, fatty acid synthase; PPARα peroxisome proliferator-activated receptor-α; MTP, microsomal triglyceride transfer protein; VLDL, very low–density lipoprotein

In patients with chronic hepatitis C, the presence of hepatic steatosis or insulin resistance is independently associated with a poor response to interferon-based therapy and progression to fibrosis, and with HCC development in patients with liver cirrhosis. DNA damage accompanied by lipid-induced chronic inflammation and ROS production, and insulin-induced cell proliferation, are possible mechanisms of hepatocarcinogenesis. The HCV core protein activates the nuclear receptors retinoid X receptor-α (RXRα) and peroxisome proliferator-activated receptor-α (PPARα) (Tsutsumi et al. 2002a). PPARα forms a heterodimer with RXRα to modulate the expression of genes related to lipid metabolism. The fact that HCV core-transgenic mice lacking the PPARα gene (core-transgenic/PPARα-knockout mice) do not develop hepatic steatosis and HCC suggests that these nuclear receptors play important roles in HCV-induced hepatocarcinogenesis (Tanaka et al. 2008).

8 Interaction of HCV Core Protein with Host Proteins

HCV proteins, particularly the core protein, are associated with intracellular signaling, transcription, transformation, apoptosis, and autophagy. However, most of the data are from in vitro cell-culture studies, and the results differed according to the cell line or expression system used. Therefore, whether the data reflect the situation in HCV-infected patients is unclear, so demonstrating the effect of HCV in vivo is vital, for which transgenic mice are useful.

Proteasome activator 28γ (PA28γ) interacts with the HCV core protein in vitro in normal hepatocytes of HCV core-transgenic mice (Moriishi et al. 2003). PA28γ is a well-conserved, proteasome-associated protein that mediates the degradation of host proteins by binding to, and regulating the activity of, the 20S proteasome, although the mechanism is unclear. Overexpression of PA28γ promotes degradation of the HCV core protein. In contrast, nuclear accumulation of the HCV core protein occurs in PA28γ-knockout hepatocytes, suggesting degradation by the PA28γ-proteasome system. PA28γ-knockout mice have an almost normal phenotype without pathological changes, but core-transgenic/PA28γ-knockout mice do not develop hepatic steatosis, unlike young core-transgenic mice (Moriishi et al. 2007). The HCV core protein promotes the binding of heterodimers of liver X receptor-α (LXRα) and RXRα to LXR-responsive element, which activates SREBP1c expression, but this effect is downregulated in the absence of PA28γ. Furthermore, the increased accumulation of ROS and the insulin resistance in core-transgenic mice are absent in core-transgenic/PA28γ-knockout mice and, surprisingly, core-transgenic/PA28γ-knockout mice do not develop HCC. These findings suggest that PA28γ plays an important role in the induction of HCC by the HCV core protein. Also, functional activation of PA28γ by interaction with the core protein may induce hepatocarcinogenesis because PA28γ expression is upregulated in several cancers (Chen et al. 2013; Okamura et al. 2003). Given that the development of HCC is prevented in its absence, PA28γ may be a novel therapeutic target.

9 Conclusion

The results of HCV mouse studies indicate that the HCV core protein has carcinogenic activity in vivo; thus, HCV has hepatic oncogenic potential. In transgenic mice, HCV proteins, particularly the core protein, induce hepatic steatosis, mitochondrial dysfunction, insulin resistance, and ROS accumulation, and the interactions of these mechanisms result in HCC development (Fig. 3).

Accumulation of a complete set of cellular genetic aberrations is required for the development of neoplasia, such as colorectal cancer (Kinzler and Vogelstein 1996). Mutations in the APC gene (inactivation), in the K-*ras* gene (activation), and in the p53 gene (inactivation) accumulate, resulting in the development of colorectal cancer. This theory, Vogelstein-type carcinogenesis, has been extended to other cancers. The induction of HCC by HCV core protein suggests an alternative mechanism of hepatocarcinogenesis. The HCV core protein may enable some of the steps in hepatocarcinogenesis to be skipped, leading to the development of HCC even in the absence of the set of genetic aberrations required for carcinogenesis (Fig. 4). Such non–Vogelstein induction of HCC may explain the unusual events in HCV carriers (Koike 2005).

Due to the remarkable progress in therapies for HCV infection, almost all HCV-infected patients achieve an SVR, but some cannot eliminate HCV due to the progression of liver fibrosis or drug-resistance mutations. Furthermore, HCC can develop in patients who have achieved an SVR. Therefore, it is important to develop therapeutic strategies to prevent and cure HCC. For this purpose, the

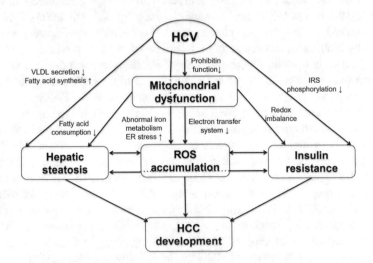

Fig. 3 Molecular mechanism of HCV-induced hepatocarcinogenesis. HCV, and particularly its core protein, impairs several cellular pathways and induces mitochondrial dysfunction, hepatic steatosis, insulin resistance, and ROS accumulation. The interactions of these mechanisms lead to hepatocarcinogenesis

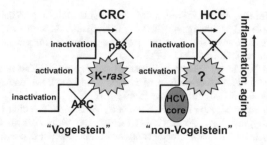

Fig. 4 The role of HCV in hepatocarcinogenesis. Multiple steps are required to induce cancer. Hepatocarcinogenesis requires the accumulation of genetic mutations in hepatocytes. The HCV core protein may enable some of the required steps to be skipped. The effect of the core protein would be one-step up in the stairway to HCC, even in the absence of a set of genetic aberrations necessary for carcinogenesis. Such a non-Vogelstein mechanism may explain the atypical modes of hepatocarcinogenesis in the presence of HCV, such as the very high incidence and multicentric nature of HCC. CRC, colorectal cancer; HCC, hepatocellular carcinoma; APC, adenomatous polyposis coli

above-mentioned mechanisms of HCV-induced hepatocarcinogenesis are useful, and compounds targeting mitochondria, nuclear receptors, or PA28γ may be promising candidate anti–hepatocarcinogenic agents.

References

Alonzi T, Agrati C, Costabile B, Cicchini C, Amicone L, Cavallari C, Rocca CD, Folgori A, Fipaldini C, Poccia F, Monica NL, Tripodi M (2004) Steatosis and intrahepatic lymphocyte recruitment in hepatitis C virus transgenic mice. J Gen Virol 85:1509–1520

Atoom AM, Taylor NG, Russell RS (2014) The elusive function of the hepatitis C virus p7 protein. Virology 462–463:377–387

Bach N, Thung SN, Schaffner F (1992) The histological features of chronic hepatitis C and autoimmune chronic hepatitis: a comparative analysis. Hepatology 15:572–577

Boudreau HE, Emerson SU, Korzeniowska A, Jendrysik MA, Leto TL (2009) Hepatitis C virus (HCV) proteins induce NADPH oxidase 4 expression in a transforming growth factor beta-dependent manner: a new contributor to HCV-induced oxidative stress. J Virol 83 (24):12934–12946

Boulant S, Montserret R, Hope RG, Ratinier M, Targett-Adams P, Lavergne JP et al (2006) Structural determinants that target the hepatitis C virus core protein to lipid droplets. J Biol Chem 281:22236–22247

Chen D, Yang X, Huang L, Chi P (2013) The expression and clinical significance of PA28 γ in colorectal cancer. J Investig Med 61:1192–1196

Diamond DL, Jacobs JM, Paeper B, Proll SC, Gritsenko MA, Carithers RL Jr et al (2007) Proteomic profiling of human liver biopsies: hepatitis C virus-induced fibrosis and mitochondrial dysfunction. Hepatology 46:649–657

Farinati F, Cardin R, De Maria N, Della Libera G, Marafin C, Lecis E, Burra P, Floreani A, Cecchetto A, Naccarato R (1995) Iron storage, lipid peroxidation and glutathione turnover in chronic anti-HCV positive hepatitis. J Hepatol 22:449–456

Frelin L, Brenndörfer ED, Ahlén G, Weiland M, Hultgren C, Alheim M, Glaumann H, Rozell B, Milich DR, Bode JG, Sällberg M (2006) The hepatitis C virus and immune evasion: non-structural 3/4A transgenic mice are resistant to lethal tumour necrosis factor alpha mediated liver disease. Gut 55:1475–1483

Fujita N, Sugimoto R, Ma N, Tanaka H, Iwasa M, Kobayashi Y, Kawanishi S, Watanabe S, Kaito M, Takei Y (2008) Comparison of hepatic oxidative DNA damage in patients with chronic hepatitis B and C. J Viral Hepat 15:498–507

Hara Y, Yanatori I, Ikeda M, Kiyokage E, Nishina S, Tomiyama Y, Toida K, Kishi F, Kato N, Imamura M, Chayama K, Hino K (2014) Hepatitis C virus core protein suppresses mitophagy by interacting with parkin in the context of mitochondrial depolarization. Am J Pathol 184:3026–3039

Honda A, Arai Y, Hirota N, Sato T, Ikegaki J, Koizumi T, Hatano M, Kohara M, Moriyama T, Imawari M, Shimotohno K, Tokuhisa T (1999) Hepatitis C virus structural proteins induce liver cell injury in transgenic mice. J Med Virol 59:281–289

Houghton M, Weiner A, Han J, Kuo G, Choo QL (1991) Molecular biology of hepatitis C viruses. Implications for diagnosis, development and control of viral diseases. Hepatology 14:381–388

Joyce MA, Walters KA, Lamb SE, Yeh MM, Zhu LF, Kneteman N, Doyle JS, Katze MG, Tyrrell DL (2009) HCV induces oxidative and ER stress, and sensitizes infected cells to apoptosis in SCID/Alb-uPA mice. PLoS Pathog 5:e1000291

Kato J, Kobune M, Nakamura T, Kuroiwa G, Takada K, Takimoto R, Sato Y, Fujikawa K, Takahashi M, Takayama T, Ikeda T, Niitsu Y (2001) Normalization of elevated hepatic 8-hydroxy-2′-deoxyguanosine levels in chronic hepatitis C patients by phlebotomy and low iron diet. Cancer Res 61:8697–8702

Kinzler KW, Vogelstein B (1996) Lessons from hereditary colorectal cancer. Cell 87:159–170

Kiyosawa K, Sodeyama T, Tanaka E, Gibo Y, Yoshizawa K, Nakano Y et al (1990) Interrelationship of blood transfusion, non-A, non-B hepatitis and hepatocellular carcinoma: analysis by detection of antibody to hepatitis C virus. Hepatology 12:671–675

Ko KS, Tomasi ML, Iglesias-Ara A, French BA, French SW, Ramani K, Lozano JJ, Oh P, He L, Stiles BL, Li TW, Yang H, Martínez-Chantar ML, Mato JM, Lu SC (2010) Liver-specific deletion of prohibitin 1 results in spontaneous liver injury, fibrosis, and hepatocellular carcinoma in mice. Hepatology 52:2096–2108

Koike K (2005) Molecular basis of hepatitis C virus-associated hepatocarcinogenesis: lessons from animal model studies. Clin Gastroenterol Hepatol 3:S132–S135

Koike K (2007) Hepatitis C virus contributes to hepatocarcinogenesis by modulating metabolic and intracellular signaling pathways. J Gastroenterol Hepatol 22(Suppl 1):S108–S111

Koike K, Moriya K, Ishibashi K, Matsuura Y, Suzuki T, Saito I et al (1995) Expression of hepatitis C virus envelope proteins in transgenic mice. J Gen Virol 76:3031–3038

Koike K, Moriya K, Yotsuyanagi H, Shintani Y, Fujie H, Ishibashi K et al (1997) Sialadenitis resembling Sjögren's syndrome in mice transgenic for hepatitis C virus envelope genes. Proc Natl Acad Sci USA 94:233–236

Korenaga M, Wang T, Li Y, Showalter LA, Chan T, Sun J, Weinman SA (2005) Hepatitis C virus core protein inhibits mitochondrial electron transport and increases reactive oxygen species (ROS) production. J Biol Chem 280:37481–37488

Lerat H, Honda M, Beard MR, Loesch K, Sun J, Yang Y et al (2002) Steatosis and liver cancer in transgenic mice expressing the structural and nonstructural proteins of hepatitis C virus. Gastroenterology 122:352–365

Machida K, Cheng KT, Lai CK, Jeng KS, Sung VM, Lai MM (2006) Hepatitis C virus triggers mitochondrial permeability transition with production of reactive oxygen species, leading to DNA damage and STAT3 activation. J Virol 80:7199–7207

Majumder M, Ghosh AK, Steele R, Zhou XY, Phillips NJ, Ray R, Ray RB (2002) Hepatitis C virus NS5A protein impairs TNF-mediated hepatic apoptosis, but not by an anti-FAS antibody, in transgenic mice. Virology 294:94–105

Mishra S, Murphy LC, Nyomba BL, Murphy LJ (2005) Prohibitin: a potential target for new therapeutics. Trends Mol Med 11:192–197

Miura K, Taura K, Kodama Y, Schnabl B, Brenner DA (2008) Hepatitis C virus-induced oxidative stress suppresses hepcidin expression through increased histone deacetylase activity. Hepatology 48:1420–1429

Miyamoto H, Moriishi K, Moriya K, Murata S, Tanaka K, Suzuki T et al (2007) Hepatitis C virus core protein induces insulin resistance through a PA28γ-dependent pathway. J Virol 81:1727–1735

Miyanari Y, Atsuzawa K, Usuda N, Watashi K, Hishiki T, Zayas M, Bartenschlager R, Wakita T, Hijikata M, Shimotohno K (2007) The lipid droplet is an important organelle for hepatitis C virus production. Nat Cell Biol 9:1089–1097

Miyoshi H, Fujie H, Shintani Y, Tsutsumi T, Shinzawa S, Makuuchi M, Kokudo N, Matsuura Y, Suzuki T, Miyamura T, Moriya K, Koike K (2005) Hepatitis C virus core protein exerts an inhibitory effect on suppressor of cytokine signaling (SOCS)-1 gene expression. J Hepatol 43:757–763

Moradpour D, Penin F, Rice CM (2007) Replication of hepatitis C virus. Nat Rev Microbiol 5:453–463

Moriishi K, Okabayashi T, Nakai K, Moriya K, Koike K, Murata S et al (2003) Proteasome activator PA28γ-dependent nuclear retention and degradation of hepatitis C virus core protein. J Virol 77:10237–10249

Moriishi K, Mochizuki R, Moriya K, Miyamoto H, Mori Y, Abe T et al (2007) Critical role of PA28g in hepatitis C virus-associated steatogenesis and hepatocarcinogenesis. Proc Natl Acad Sci USA 104:1661–1666

Moriya K, Yotsuyanagi H, Shintani Y, Fujie H, Ishibashi K, Matsuura Y et al (1997) Hepatitis C virus core protein induces hepatic steatosis in transgenic mice. J Gen Virol 78:1527–1531

Moriya K, Fujie H, Shintani Y, Yotsuyanagi H, Tsutsumi T, Matsuura Y et al (1998) Hepatitis C virus core protein induces hepatocellular carcinoma in transgenic mice. Nat Med 4:1065–1068

Moriya K, Nakagawa K, Santa T, Shintani Y, Fujie H, Miyoshi H et al (2001) Oxidative stress in the absence of inflammation in a mouse model for hepatitis C virus-associated hepatocarcinogenesis. Cancer Res 61:4365–4370

Moriya K, Miyoshi H, Shinzawa S, Tsutsumi T, Fujie H, Goto K, Shintani Y, Yotsuyanagi H, Koike K (2010) Hepatitis C virus core protein compromises iron-induced activation of antioxidants in mice and HepG2 cells. J Med Virol 82:776–792

Naas T, Ghorbani M, Alvarez-Maya I, Lapner M, Kothary R, De Repentigny Y et al (2005) Characterization of liver histopathology in a transgenic mouse model expressing genotype 1a hepatitis C virus core and envelope proteins 1 and 2. J Gen Virol 86:2185–2196

Nijtmans LG, de Jong L, Artal Sanz M, Coates PJ, Berden JA, Back JW et al (2000) Prohibitins act as a membrane-bound chaperone for the stabilization of mitochondrial proteins. EMBO J 19:2444–2451

Nishina S, Hino K, Korenaga M, Vecchi C, Pietrangelo A, Mizukami Y, Furutani T, Sakai A, Okuda M, Hidaka I, Okita K, Sakaida I (2008) Hepatitis C virus-induced reactive oxygen species raise hepatic iron level in mice by reducing hepcidin transcription. Gastroenterology 134:226–238

Oguro T, Hayashi M, Nakajo S, Numazawa S, Yoshida T (1998) The expression of heme oxygenase-1 gene responded to oxidative stress produced by phorone, a glutathione depletor, in the rat liver; the relevance to activation of c-jun n-terminal kinase. J Pharmacol Exp Ther 287:773–778

Okamura T, Taniguchi S, Ohkura T, Yoshida A, Shimizu H, Sakai M, Maeta H, Fukui H, Ueta Y, Hisatome I, Shigemasa C (2003) Abnormally high expression of proteasome activator-gamma in thyroid neoplasm. J Clin Endocrinol Metab 88:1374–1383

Pasquinelli C, Shoenberger JM, Chung J et al (1997) Hepatitis C virus core and E2 protein expression in transgenic mice. Hepatology 25:719–727

Perlemuter G, Sabile A, Letteron P, Vona G, Topilco A, Koike K et al (2002) Hepatitis C virus core protein inhibits microsomal triglyceride transfer protein activity and very low density lipoprotein secretion: a model of viral-related steatosis. FASEB J 16:185–194

Perz JF, Armstrong GL, Farrington LA, Hutin YJ, Bell BP (2006) The contributions of hepatitis B virus and hepatitis C virus infections to cirrhosis and primary liver cancer worldwide. J Hepatol 45:529–538

Piccoli C, Scrima R, Quarato G, D'Aprile A, Ripoli M, Lecce L et al (2007) Hepatitis C virus protein expression causes calcium-mediated mitochondrial bioenergetic dysfunction and nitro-oxidative stress. Hepatology 46:58–65

Saito I, Miyamura T, Ohbayashi A, Harada H, Katayama T, Kikuchi S et al (1990) Hepatitis C virus infection is associated with the development of hepatocellular carcinoma. Proc Natl Acad Sci USA 87:6547–6549

Shintani Y, Fujie H, Miyoshi H, Tsutsumi T, Kimura S, Moriya K et al (2004) Hepatitis C virus and diabetes: direct involvement of the virus in the development of insulin resistance. Gastroenterology 126:840–848

Shirakura M, Murakami K, Ichimura T, Suzuki R, Shimoji T, Fukuda K et al (2007) E6AP ubiquitin ligase mediates ubiquitylation and degradation of hepatitis C virus core protein. J Virol 81:1174–1185

Suzuki R, Sakamoto S, Tsutsumi T, Rikimaru A, Tanaka K, Shimoike T et al (2005) Molecular determinants for subcellular localization of hepatitis C virus core protein. J Virol 79:1271–1281

Stocker R, Yamamoto Y, McDonagh AF, Glazer AN, Ames BN (1987) Bilirubin is an antioxidant of possible physiological importance. Science 235:1043–1046

Taguwa S, Kambara H, Fujita N, Noda T, Yoshimori T, Koike K, Moriishi K, Matsuura Y (2011) Dysfunction of autophagy participates in vacuole formation and cell death in cells replicating hepatitis C virus. J Virol 85:13185–13194

Tanaka N, Moriya K, Kiyosawa K, Koike K, Gonzalez FJ, Aoyama T (2008) PPARalpha activation is essential for HCV core protein-induced hepatic steatosis and hepatocellular carcinoma in mice. J Clin Invest 118:683–694

Tardif KD, Mori K, Siddiqui A (2002) Hepatitis C virus subgenomic replicons induce endoplasmic reticulum stress activating an intracellular signaling pathway. J Virol 76:7453–7459

Theiss AL, Idell RD, Srinivasan S, Klapproth JM, Jones DP, Merlin D et al (2007) Prohibitin protects against oxidative stress in intestinal epithelial cells. FASEB J 21:197–206

Tsutsumi T, Suzuki T, Shimoike T, Suzuki R, Moriya K, Shintani Y et al (2002a) Interaction of hepatitis C virus core protein with retinoid X receptor alpha modulates its transcriptional activity. Hepatology 35:937–946

Tsutsumi T, Suzuki T, Moriya K, Yotsuyanagi H, Shintani Y, Fujie H et al (2002b) Intrahepatic cytokine expression and AP-1 activation in mice transgenic for hepatitis C virus core protein. Virology 304:415–424

Tsutsumi T, Suzuki T, Moriya K, Shintani Y, Fujie H, Miyoshi H et al (2003) Hepatitis C virus core protein activates ERK and p38 MAPK in cooperation with ethanol in transgenic mice. Hepatology 38:820–828

Tsutsumi T, Matsuda M, Aizaki H, Moriya K, Miyoshi H, Fujie H, Shintani Y, Yotsuyanagi H, Miyamura T, Suzuki T, Koike K (2009) Proteomics analysis of mitochondrial proteins reveals overexpression of a mitochondrial protein chaperone, prohibitin, in cells expressing hepatitis C virus core protein. Hepatology 50:378–386

Wakita T, Taya C, Katsume A et al (1998) Efficient conditional transgene expression in hepatitis C virus cDNA transgenic mice mediated by the Cre/loxP system. J Biol Chem 273:9001–9006

Wakita T, Pietschmann T, Kato T, Date T, Miyamoto M, Zhao Z, Murthy K, Habermann A, Kräusslich HG, Mizokami M, Bartenschlager R, Liang TJ (2005) Production of infectious hepatitis C virus in tissue culture from a cloned viral genome. Nat Med 11:791–796

Yen HH, Shih KL, Lin TT, Su WW, Soon MS, Liu CS (2012) Decreased mitochondrial deoxyribonucleic acid and increased oxidative damage in chronic hepatitis C. World J Gastroenterol 18:5084–5089

Yotsuyanagi H, Shintani Y, Moriya K, Fujie H, Tsutsumi T, Kato T et al (2000) Virological analysis of non-B, non-C hepatocellular carcinoma in Japan: frequent involvement of hepatitis B virus. J Infect Dis 181:1920–1928

Prevention of Hepatitis C Virus Infection and Liver Cancer

E. J. Lim and J. Torresi

1 Introduction

Hepatocellular carcinoma (HCC) is the fifth most prevalent cancer and the second leading cause of cancer-related death worldwide (Lafaro et al. 2015). Incidence of HCC is highest in the less developed countries of Asia and Africa, with China accounting for over 50% of cases (Petrick et al. 2016). However, the rates of HCC have been increasing rapidly in America, Europe and Australia (Petrick et al. 2016), where the majority of HCC cases are attributable to chronic hepatitis C (HCV) infection (Choo et al. 2016). Of the more than 42,000 new cases of HCC in the United States in 2018 (http://seer.cancer.gov), an estimated 50–60% are related to HCV (El-Serag and Kanwal 2014). Although a strong association exists between chronic HCV and the development of HCC, the precise mechanisms by which HCV infection promotes HCC are uncertain. There is, however, good evidence to suggest that eradication of the virus in both cirrhotic and noncirrhotic HCV-infected patients reduces the subsequent risk of developing HCC (Morgan et al. 2013).

E. J. Lim
Department of Gastroenterology, Austin Hospital, Heidelberg, Victoria 3084, Australia

Department of Medicine, The University of Melbourne, Austin Hospital, Heidelberg, Victoria 3084, Australia

J. Torresi (✉)
Department of Microbiology and Immunology, The University of Melbourne, Parkville, VIC, Australia
e-mail: josepht@unimelb.edu.au

© Springer Nature Switzerland AG 2021
T.-C. Wu et al. (eds.), *Viruses and Human Cancer*, Recent Results in Cancer Research 217, https://doi.org/10.1007/978-3-030-57362-1_6

2 Hepatitis C Infection

It is estimated that over 71 million individuals are chronically infected with HCV worldwide (Blach et al. 2017). HCV is transmitted via infected blood. Currently, in Western countries, acquisition of HCV occurs primarily through intravenous drug use and tattoos (Razali et al. 2007), whereas in Asia and Africa, infection mainly occurs through the use of contaminated blood products and medical instruments. In contrast to these modes, HCV transmission via sexual and perinatal routes is infrequent (Razali et al. 2007).

Following infection with HCV, up to 80% of patients are unable to clear the virus, resulting in chronic infection which may ultimately progress to cirrhosis in approximately 20% of individuals (Westbrook and Dusheiko 2014). The natural history of HCV infection is summarized in Fig. 1. The majority of chronically infected patients remain asymptomatic for many years, thus delaying both the diagnosis and treatment. It is often not until the development of complications of cirrhosis such as hepatic decompensation and HCC that these patients present to medical care, and around 15% of patients with chronic HCV will develop HCC (Rein et al. 2011).

3 Hepatitis C Virology

HCV belongs to the genus *Hepacivirus* in the *Flaviviridae* family (Forns and Bukh 1999). It has a single-stranded linear RNA genome of approximately 9600 nucleotides that encodes a large polyprotein of approximately 3000 amino acids (Bartenschlager et al. 2011). The structure of the HCV genome and functions of the various viral proteins are summarized in Fig. 2. HCV exists as six major genotypes with genotype 1 being the dominant genotype in the United States, Europe and Australia (Messina et al. 2015). Genotype 2 is primarily found in West Africa, genotype 3 is more prevalent in India and Southeast Asia, whereas genotype 4 is more commonly seen in the Middle East and Africa (Gower et al. 2014). Each

Fig. 1 Natural history of HCV infection. Adapted from Alter (1995)

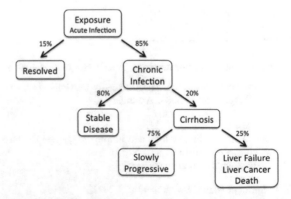

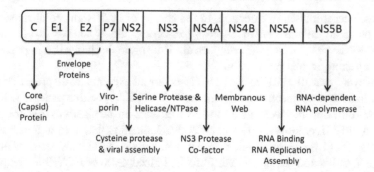

Fig. 2 HCV genome structure and functions of viral proteins. Adapted from Bartenschlager et al. (2011)

genotype contains multiple subtypes according to viral sequencing, identified with lower case letters (a, b, c, etc.). The HCV genotype may influence treatment response, severity of liver disease, and risk of HCC development, but the data are contradictory. A meta-analysis found that genotype 1b HCV infection was associated with a doubling of the risk of HCC development compared to infection with other genotypes (Raimondi et al. 2009). However, in a more recent cohort study of 100,000 patients, those chronically infected with genotype 3 HCV were found to have an 80% greater risk of HCC compared with genotype 1 chronic infection (Kanwal et al. 2014).

Like other RNA viruses, the RNA polymerase protein of HCV lacks proof-reading ability and as a consequence, replication of the viral genome is error-prone resulting in a high mutation rate. The result is great genetic heterogeneity that leads to the evolution of diverse viral quasi-species. This viral diversity interferes with the development of effective host humoral immune responses against the virus thereby promoting viral persistence within infected individuals (Forns and Bukh 1999).

4 Hepatitis C and Associated Risk Factors for HCC Development

HCV is recognized as a major cause of HCC globally. In a large population-based prospective study, infection with HCV conferred a 20-fold increased risk of developing HCC compared to HCV-negative individuals (Sun et al. 2003). In another large prospective cohort study, HCV infection conferred a cumulative lifetime risk of HCC of 24% in men and 17% in women (Huang et al. 2011). This strong association between chronic HCV and HCC has been noted since the early 1980s, when the virus was known as non-A, non-B hepatitis (Kiyosawa et al. 1984). Almost all HCV-related HCCs occur in the setting of established cirrhosis, with cirrhosis itself being a strong independent risk factor for developing HCC (Tsukuma et al. 1993). Consequently, HCC develops many years (often 2–3 decades) after initial HCV infection. In the setting of HCV-induced cirrhosis, HCC develops at an annual rate of 1–5% (Westbrook and Dusheiko 2014), with an 11.5% 4-year risk of HCC (Serfaty et al. 1998), emphasizing the importance of HCC screening in this population.

However, not all HCV-related HCCs occur in patients with pre-existing cirrhosis. In a large prospective study, about 17% of HCV-positive patients with HCC were not cirrhotic but were noted to have at least an Ishak fibrosis score of 3 or more on serial liver biopsies (Lok et al. 2009), indicating that even in the absence of cirrhosis, HCC tends to develop in HCV-infected individuals with established chronic hepatitis and advanced liver fibrosis. Indeed, 20% of HCV-infected patients with Metavir fibrosis stage 3 (noncirrhotic) have been noted to develop HCC (Alkofer et al. 2011). More recently, a small retrospective study noted that 10% of HCCs developed in HCV-infected patients who only had a Metavir fibrosis stage of F0 to F2 (Lewis et al. 2013).

Risk factors that may contribute to the progression of HCV-associated liver disease to cirrhosis and HCC include concurrent alcohol consumption, older age at the time of infection, male gender, coinfection with HIV or hepatitis B, immunosuppression, associated insulin resistance or non-alcoholic steatohepatitis, and a higher degree of inflammation and fibrosis on liver biopsy (Chen and Morgan 2006).

Significant alcohol intake of >40 g alcohol/day in women and >60 g of alcohol/day in men for more than 5 years is associated with a 2- to 3-fold increased risk of cirrhosis and decompensated liver disease in HCV-infected individuals (Wiley et al. 1998). Furthermore, the risk of developing HCC is doubled in HCV-infected individuals who consume >60 g of alcohol/day compared to those consuming <60 g/day (Donato et al. 2002). Also, the presence of chronic HCV infection has been associated with more advanced liver disease and increased mortality in alcoholic individuals compared to alcoholic patients with chronic hepatitis B infection (Mendenhall et al. 1991).

Age of infection is also an independent risk factor for the development of more severe liver disease in chronic HCV. After controlling for duration of HCV infection, patients who acquire HCV infection at an older age (>40 years old) are significantly more likely to progress to advanced liver fibrosis than individuals infected at a younger age (Poynard et al. 1997). The incidence of HCC is up to 29 times higher for individuals who become infected with HCV after 39 years of age compared to those infected before the age of 19 years (Pradat et al. 2007). HCV-infected men have a 2–4 fold higher risk of developing HCC than HCV-infected women (El-Serag and Kanwal 2014).

In the setting of HIV–HCV coinfection, a low CD4 count is associated with higher HCV viral loads as well as increased hepatic fibrosis and accelerated progression to cirrhosis (Di Martino et al. 2001). Other causes of immunosuppression, such as organ transplantation, have also been associated with more rapid liver fibrosis progression (Berenguer et al. 2000). Individuals coinfected with HCV and hepatitis B virus (HBV), also have a higher risk of HCC development (Shi et al. 2005). Kruse et al. showed that individuals with HCV/HBV-coinfection but without detectable HBV DNA had a risk of HCC equivalent to HCV-monoinfected patients.

However, in coinfected patients with detectable HBV replication, the risk of HCC was doubled (Kruse et al. 2014).

Curiously, daily coffee consumption may have a protective effect on HCV-induced HCC (Wakai et al. 2007). HCV-infected individuals with HCC who drank at least one cup of coffee a day were noted to have a 69% lower mortality compared to individuals who did not drink coffee (Kurozawa et al. 2005).

4.1 Hepatitis C Viral Hepatocarcinogenesis

How HCV infection results in the development of HCC is not entirely clear. There is evidence to suggest that HCV may interact with various intracellular signal transduction pathways or affect epigenetic changes thereby altering hepatocyte physiology to directly promote malignant transformation. HCV core protein has been strongly implicated in hepatocarcinogenesis. Core protein expression has been noted to promote HCC development in transgenic mice in the absence of hepatic inflammation or fibrosis indicating that viral proteins themselves may directly have a carcinogenic effect (Moriya et al. 1998). Core protein is noted to interact with the mitogen-activated protein (MAP) kinase signalling pathway (Hayashi et al. 2000) and upregulate mTOR (Tholey and Ahn 2015), thereby promoting cell proliferation. Core protein has also been shown to inhibit p53 tumour suppressor protein allowing the proliferation of genetically aberrant hepatocytes (Kao et al. 2004).

The non-structural proteins of HCV have also been implicated in hepatocarcinogenesis. NS3 (Ishido and Hotta 1998) and NS5A (Lan et al. 2002) have been shown to bind to p53, perhaps inhibiting p53-mediated apoptosis of malignant cells. The NS5A protein was noted to activate the beta-catenin signalling pathway (Park et al. 2009) and mTOR pathway (Peng et al. 2010) to promote cell proliferation. While the NS5B protein has been found to downregulate retinoblastoma (Rb) protein, a key tumour suppressor protein that regulates cellular response to DNA damage, by targeting Rb for proteasome degradation (Munakata et al. 2005).

Also, it has been shown that the tumour suppressor gene p16INK4A in tumour tissue resected from the livers of HCV-infected patients with HCCs is hypermethylated, and this results in the inactivation of p16INK4A, a feature not seen in non-HCV-associated HCCs (Li et al. 2004).

Second, immune-mediated liver inflammation and the promotion of apoptosis of HCV-infected hepatocytes result in a compensatory stimulation of cell proliferation to replace dead hepatocytes. The increased cell turn-over permits the accumulation of genetic mutations within hepatocytes, and this together with the surrounding inflammatory liver milieu promotes HCC development (Hino et al. 2002; Budhu and Wang 2006). Indeed, inflammation-mediated hepatocyte apoptosis with compensatory cell proliferation was shown to enhance HCC development in mice (Qiu et al. 2011), and increased inflammatory activity within the liver parenchyma has been associated with a poorer prognosis in patients with HCV-associated HCC (Maki et al. 2007).

5 Prevention of HCC in Patients with Hepatitis C-Induced Cirrhosis

The incidence of HCV-related HCC continues to rise worldwide because of the high number of individuals with chronic HCV infection, the presence of associated co-morbidities, and the longer survival of patients with advanced liver disease as a result of improved management of the complications of liver failure. A recent meta-analysis showed that the successful clearance of HCV in patients with advanced liver disease was associated with a reduction in the risk of subsequent HCC development from 17.8 to 4.2%, indicating that despite already having cirrhosis, successful antiviral therapy of HCV-infected patients will still reduce the future risk of HCC (Morgan et al. 2013). However, as the risk of HCC is not completely eliminated despite achieving viral eradication, ongoing HCC screening is still indicated in patients with cirrhosis (Aleman et al. 2013). There appears to be no benefit of therapy on reducing HCC risk if viral eradication was not successful. A systematic review not only found that SVR was related to a reduced incidence of liver failure and HCC, but successful viral eradication may lead to cirrhosis regression (Ng and Saab 2011).

In a meta-analysis of patients with HCV-induced cirrhosis who develop HCC, it was found that after curative treatment of HCC via local ablative therapy or surgical resection, successful eradication of HCV with interferon therapy was associated with a reduced risk of HCC recurrence from 61 to 35% (Singal et al. 2010). Furthermore, successful treatment with pegylated interferon and ribavirin was associated with an improved hepatic functional reserve and increased survival (96 vs. 61% at 3 years) in this cohort (Ishikawa et al. 2012). These studies indicate that antiviral therapy is also useful in the secondary prevention of HCC in HCV-induced cirrhosis. In patients undergoing liver transplantation for HCV-associated HCC, interferon therapy for recurrent HCV post-liver transplant was found to decrease subsequent HCC recurrence from 27 to 4% (Kohli et al. 2012). More recently, the newer direct-acting antiviral (DAA) drugs have also been shown to be highly efficacious in clearing HCV post-liver transplantation, with a 98.5% success rate reported regardless of viral genotype and fibrosis stage (Pyrsopoulos et al. 2018). DAA regimens have been shown to be highly effective in eradicating post-liver transplant HCV with minimal side effects and a durable response on long-term follow-up (Beinhardt et al. 2018).

6 Prevention of Cirrhosis and HCC in Patients with Hepatitis C-Induced Chronic Hepatitis

The potential long-term benefits of successful antiviral therapy of HCV-infected patients with chronic hepatitis include the normalization of serum transaminase levels with resultant reduction in hepatic necroinflammation and fibrosis, the improvement in health-related quality of life, and the reduction in HCC risk, all of which enhance patient survival (Patel et al. 2006). We know that HCC primarily develops in HCV-infected patients with cirrhosis. In order to reduce the risk of developing HCV-related HCC, the aim would be to provide treatment to eradicate HCV prior to the development of cirrhosis. Indeed, studies have shown that successful viral clearance with antiviral therapy results in clinical and histological improvement in the vast majority of patients (Marcellin et al. 1997), with an associated reduction in the risk of subsequent development of cirrhosis and HCC (Pradat et al. 2007).

In a large multicentre European study of largely noncirrhotic (89%) HCV-infected patients, successful viral eradication was associated with the progression to cirrhosis in only 2.3% of patients with no patients developing HCC, whereas a failure to achieve viral clearance was associated with progression to cirrhosis in 20% of patients and development of HCC in 4.2% (Pradat et al. 2007). In another study of HCV-infected patients, the majority of whom (90%) did not have cirrhosis, successful viral clearance with interferon therapy was associated with a 10-fold reduction in the risk of subsequent HCC development from 2.31 to 0.24%/year (Maruoka et al. 2012). The meta-analysis by Morgan et al. concluded that achieving sustained virological response (SVR), defined as having an undetectable HCV RNA 12 weeks post-completion of antiviral therapy, resulted in a 4.6% absolute reduction in HCC development regardless of fibrosis stage (Morgan et al. 2013). Indeed, attaining SVR is associated with a 54% reduction in all-cause mortality in HCV-infected individuals (Backus et al. 2011). In patients with no evidence of significant hepatic fibrosis, SVR is thought to be associated with an improvement in liver histology back to normal and a reduction in the risk of HCC back to population levels. As such, these patients may be discharged from ongoing surveillance (Westbrook and Dusheiko 2014).

7 Antiviral Treatment of Hepatitis C

A vaccine for the prevention of HCV is not yet available, and therefore the only effective means to prevent the development of liver cirrhosis and HCC is with antiviral therapy. The aim of such treatment is to eradicate the virus, resulting in the clearance of HCV from the body, and thereby halt the progression of liver injury and fibrosis. On long-term follow-up of patients achieving SVR, more than 99% remained HCV negative, essentially equating SVR with a cure for HCV infection (Simmons et al. 2016).

7.1 Pegylated Interferon Alfa and Ribavirin

For many years, the only available treatment for chronic HCV infection was the combination of pegylated interferon alfa and ribavirin. Interferon alfa induced the transcription of genes involved in cell cycle regulation, apoptosis, and promotion of an antiviral state within hepatocytes (Thomas et al. 2003). Besides the direct effect against virally-infected host cells, interferon alfa is also involved in modulating the immune system by enhancing the CD8+ cytotoxic T cell response against infected cells, as well as promoting the proliferation of B cells to augment the production of antibodies against HCV (Thomas et al. 2003). Treatment was frequently associated with a myriad of side effects resulting in patient morbidity (Chevaliez and Pawlotsky 2007) and despite a prolonged treatment duration of 48 weeks, viral clearance was only achieved in 52% of patients with genotype 1 HCV (Hadziyannis et al. 2004). These regimens are no longer recommended for HCV treatment in the US (Chung et al. 2015) or Europe (Liver 2018).

7.2 Direct-Acting Antiviral Agents

The recent advent of DAA drugs has vastly improved the landscape of antiviral therapy for chronic HCV and has provided a more effective approach to the long-term prevention of HCV-associated HCC. Indeed, combinations of these oral agents are now considered the standard of care for HCV treatment (Chung et al. 2015). These DAAs targets and inhibit specific HCV proteins, including the viral protease (NS3/4A protein), polymerase (NS5B protein), and the NS5A protein, thereby disrupting the HCV replication cycle (Poordad and Dieterich 2012). These DAAs have resulted in significant improvements in achieving SVR, resulting in the rapid normalization of hepatic inflammation and regression of liver fibrosis (Tada et al. 2017) (Table 1).

The first DAAs to be released were the protease inhibitors telaprevir (Incivek, Vertex) and boceprevir (Victrelis, Merck Sharp and Dohme) in 2011. Both were only effective in patients infected with genotype 1 HCV and were unable to affect SVR alone. As such, they had to be added to pegylated interferon alfa and ribavirin treatment (Hezode et al. 2014), thereby increasing the number and severity of side effects but also augmenting the SVR rates in both treatment naive and treatment-experienced patients. Telaprevir increased SVR rates from 44 to 75% compared to standard therapy alone and boceprevir enhanced SVR rates from 38 to 66% compared to pegylated interferon and ribavirin alone. Common side effects of telaprevir therapy include skin rash, anaemia, and gastrointestinal symptoms, while common side effects of boceprevir include anaemia and dysgeusia (Jacobson et al. 2011; Poordad et al. 2011).

It soon became apparent that single DAA agents were not only insufficiently potent to achieve SVR on their own, but single-agent use quickly led to the development of antiviral resistance due to the rapid selection of drug-resistant HCV variants (Conteduca et al. 2014). The answer came from utilizing regimens consisting of combinations of DAAs with different mechanisms of action in order to prevent the development of antiviral resistance (Pawlotsky 2016). Furthermore, by utilizing multiple DAAs together, treatment duration has generally been shortened to 12 weeks for most regimens regardless of HCV genotype, degree of hepatic

Table 1 Summary of the outcomes of Phase 3 clinical trials with Zepatier, Epclusa, Vosevi and Maviret

DAA regimen	Population	Treatment history	Treatment	Endpoint/Outcome SVR12	References
Zepatier GPV/EBV					
C-EDGE-TN	Gt 1a, 1b, 4 NC or C	TN	GPV/EBV 12w	92% Gt 1a 99% Gt 1b 100% Gt 4 97% noncirrhotic 94% cirrhotic	Zeuzem et al. (2015)
C-EDGE TE	GT14,6, NC or C ± HIV	TE-PR	GPV/EBV ± 12w GPV/EBV ± 16w	12w: 92% Gt 1a 100% Gt 1b 16w: 94% Gt 1a 98% Gt 1b	Kwo et al. (2017)
C-SURFER	GT1, NC or C ± CRD	TN	GPV/EBV 12w	99.1%	Bruchfeld et al. (2017)
C-EDGE COINFECTION	GT1, NC ± HIV	TN	GFV/EBV 12w	97% Gt 1a 96% Gt 1b 96% Gt 4	Rockstroh et al. (2015)
Epclusa SOF/VEL					
ASTRAL-1	GT 1,2,4,5,6 NC or C	TN (68%) or TE (32%)	SOF/VEL 12w	99% Overall 98% Gt 1a 99% Gt 1b 100% Gt 2 100% Gt 4 97% Gt 5 100% Gt 6	Feld et al. (2015)

(continued)

Table 1 (continued)

DAA regimen	Population	Treatment history	Treatment	Endpoint/Outcome SVR12	References
ASTRAL-2 & 3	Gt 2,3 NC or C	TN TE-PR	SOF/VEL 12w or SOF/RBV 12w	99% Gt 2, SOF/VEL 95% Gt 3, SOF/VEL 94% Gt 2, SOF/RBV 80% Gt 3, SOF/RBV	Foster et al. (2015)
Vosevi **SOF/VEL/VOX**					
POLARIS-1	Gt 1,2,3,4,5,6 NC or C	TE-DAA	SOF/VEL/VOX 12w	96% Overall 96% Gt 1a 100% Gt 1b 100% Gt 2 95 Gt 3 91 Gt 4 100 Gt 5 100 Gt 6	Bourliere et al. (2017)
POLARIS-4	Gt 1,2,3,4 NC or C	TE-DAA	SOF/VEL/VOX 12w or SOF/VEL 12w	98% Overall S/V/V 98% Gt 1a S/V/V 96% Gt 1b S/V/V 100% Gt 2 S/V/V 96% Gt 3 S/V/V 100% Gt 4 S/V/V 90% Overall S/V 89% Gt 1a S/V	Bourliere et al. (2017)

(continued)

Table 1 (continued)

DAA regimen	Population	Treatment history	Treatment	Endpoint/Outcome SVR12	References
				95% Gt 1b S/V 97% Gt 2 S/V 85% Gt 3 S/V	
Maviret (GLE/PIB)					
ENDURANCE-1	GT1, NC ± HIV	TN or TE-PRS	GLE/PIB 12w GLE/PIB 8w	99.7% 99.1%	Zeuzem et al. (2018)
ENDURANCE-2	GT2, NC	TN or TE-PRS	GLE/PIB 12w	[a]99.5% 100%	Asselah et al. (2018)
ENDURANCE-3	GT3, NC	TN	GLE/PIB 8w GLE/PIB 12w DCV/SOF 12w	94.9% 95.3% 96.5%	Zeuzem et al. (2018)
ENDURANCE-4	GT4,5,6, NC	TN or TE-PRS	GLE/PIB 12w	99% overall	Asselah et al. (2018)
EXPEDITION-4	GT1-6, NC or C ± CRD	TN or TE-PRS	GLE/PIB 12w	98%	Gane et al. (2017)
ENDURANCE-5,6	GT4,5,6 NC or C	TN or TE-PRS	GLE/PIB 8w GLE/PIB 12w	Overall, 97.6% 99% 89% (cirrhotic)	Asselah et al. (2019)

[a]DAA-treatment-experienced cohort; *RCTs* Randomized control trials; *PRS* Pegylated interferon, ribavirin and/or sofosbuvir; *NC* Noncirrhotic; *C* Cirrhotic; *TE* Treatment-experienced; *TN* Treatment naive; *CRD* Chronic renal disease; *DCV* Daclatasvir; *SOF* Sofosbuvir; *GLE/PIB* Glecaprevir/Pibrentasvir; *GPV/EBV* Grazoprevir/Elbasvir; *SOF/VEL* Sofosbuvir/Velpatasvir; *SOF/VELVOX* Sofosbuvir/Velpatasvir/Voxilaprevir; *RBV* Ribavirin

fibrosis, or previous antiviral treatment failure (Table 1). These all-oral regimens are also far better tolerated than interferon-containing regimens.

DAA drugs are relatively safe and well-tolerated and there are few absolute contraindications to therapy, with the main issue being drug interactions. Hence it is essential to determine what medications the patient is on, taking into account herbal and over-the-counter drugs, and check for interactions prior to commencement of DAA therapy. Drugs with the potential to cause toxicity should be ceased or switched to another drug in the same class that does not have interaction for the duration of DAA therapy. The University of Liverpool website (www.hep-druginteractions.org) is a useful resource for the screening of potential drug interactions with DAAs. Specific DAA regimens are discussed below.

8 Zepatier® (Elbasvir 50 mg + Grazoprevir 100 mg, Merck & Co.)

Zepatier is a once-daily, single-tablet regimen combining elbasvir and grazoprevir. This treatment is only effective for genotype 1 and 4 HCV infection (Zeuzem et al. 2015). Elbasvir is a potent inhibitor of the NS5A protein which is crucial for viral RNA replication and assembly (Coburn et al. 2013). Grazoprevir is a potent inhibitor of the NS3/4A protease, required for the proteolytic cleavage of the viral polyprotein into the mature proteins which are essential for HCV replication. Grazoprevir has been shown to be effective against common NS3 resistance-associated substitutions (Summa et al. 2012). Although 12-week treatment with Zepatier is highly efficacious in both treatment naive and treatment-experienced patients with a reported SVR rate of over 95% (Zeuzem et al. 2015), in certain difficult to treat cohorts such as treatment-experienced patients or treatment naive patients with pre-existing NS5A resistance-associated substitutions, the recommendation is to extend therapy to 16 weeks and add weight-based ribavirin (Kwo et al. 2017). This is because the presence of NS5A resistance-associated substitutions may result in resistance to elbasvir, thereby reducing the efficacy of this regimen. However, Lawitz et al. showed that by extending treatment duration and adding ribavirin, SVR rates in this cohort was increased from 91 to 100% (Lawitz et al. 2015). In patients with advanced chronic renal disease and infected with HCV genotype 1, a 12-week treatment course of Zepatier has been shown to be safe and produce an SVR of 99% (Bruchfeld et al. 2017). Similarly, in treatment-naive HIV-coinfected patients, Zepatier has been shown to produce an SVR of 96% (Rockstroh et al. 2015). Zepatier is generally well-tolerated with only a small proportion of patients experiencing headaches (17%), fatigue (15%) or nausea (9%) (Zeuzem et al. 2015), however, when combined with ribivirin, these adverse effects were more frequent, together with an increased rate of anaemia (8%) and other cytopenias (Forns et al. 2015), necessitating ribavirin dose-reduction in affected patients. Zepatier can be used safely in patients with renal impairment even up to stage 5 chronic kidney disease on haemodialysis (Roth et al. 2015), but it is contraindicated in patients with Child-Pugh B or C cirrhosis as it contains a protease inhibitor (Vermehren et al. 2018).

9 Epclusa® (Sofosbuvir 400 mg + Velpatasvir 100 mg, Gilead Sciences)

Epclusa is a single-tablet, daily dose regimen consisting of velpatasvir and sofosbuvir. The nucleoside-analogue sofosbuvir binds to and inhibits the active site of the HCV NS5B polymerase, the structure of which is highly conserved across all HCV genotypes (Buhler and Bartenschlager 2012), thus giving sofosbuvir pan-genotypic efficacy with a high drug resistance barrier (Sofia et al. 2010). Velpatasvir is a second-generation NS5A inhibitor which also has pan-genotypic potency, effective even in patients with detected NS5A resistance-associated substitutions, and has a high barrier to resistance (Cheng et al. 2013). Hence, Epclusa is the first of the pan-genotypic DAA regimens effective against all 6 HCV genotypes, with SVR rates of between 97 and 100% achieved after 12 weeks of treatment regardless of the presence of cirrhosis (99% SVR overall) (Feld et al. 2015). Epclusa is well-tolerated with no statistically significant difference seen between the treatment and placebo groups in the commonly reported side effects of headache, fatigue and nausea, however, cytopaenias were reported in 1% of patients receiving treatment (Feld et al. 2015). This regimen is approved for use in decompensated cirrhosis, usually in combination with ribavirin if tolerated (Vermehren et al. 2018), where the addition of ribavirin to 12-week treatment with Epclusa was noted to increase the SVR rate in this cohort from 83 to 94% (Curry et al. 2015). Addition of weight-based ribavirin should also be considered in treatment-experienced genotype 3 patients as this cohort is known to have a suboptimal response to Epclusa alone (Foster et al. 2015). As Sofosbuvir is mainly renally-excreted, its use is contraindicated in patients with a glomerular filtration rate of <30 ml/min, and Epclusa is not recommended for patients with severe renal impairment.

10 Vosevi® (Sofosbuvir 400 mg + Velpatasvir 100 mg + Voxilaprevir 100 mg, Gilead Sciences)

This single daily tablet fixed-dose regimen adds voxilaprevir, to the NS5B inhibitor/NS5A inhibitor combination of Epclusa. Voxilaprevir is a new generation pan-genotypic reversible inhibitor of the HCV NS3/4A protease with enhanced activity against the common NS3 resistance-associated substitutions (Taylor et al. 2015). This 12-week regimen was shown to be effective in patients who failed to respond to prior NS5A inhibitor-containing DAA treatment, achieving SVR in 91–100% of patients with genotype 1–6 HCV (Bourliere et al. 2017). In treatment-experienced patients not previously exposed to an NS5A inhibitor, Vosevi achieved SVR in 98% of genotype 1–3 patients, whereas Epclusa achieved SVR in 90% of these patients (Bourliere et al. 2017), indicating that the addition of a protease inhibitor to Epclusa improved viral clearance in this cohort. Commonly reported side effects to Vosevi treatment include headache, nausea, diarrhoea and fatigue, but no treatment cessation resulted from adverse events. Currently, the triple DAA combination of Vosevi is recommended as salvage therapy after failure of dual DAA regimens, especially in the setting of previous NS5A inhibitor use, rather than first-line therapy, with data showing that Vosevi was effective regardless of the presence of resistance-associated substitutions prior to commencing therapy obviating the need for pre-treatment resistance testing (Bourliere et al. 2017). With 80% of sofosbuvir being renally-excreted, Vosevi is not recommended in patients with severe chronic kidney disease. As Vosevi contains a protease inhibitor, its use is contraindicated in decompensated cirrhosis (Vermehren et al. 2018).

11   Maviret® (Glecaprevir 100 mg + Pibrentasvir 40 mg, AbbVie)

Maviret is a single-tablet once-daily regimen consisting of glecaprevir and pibrentasvir effective against all 6 HCV genotypes with reported SVR rates of 95–99% across all genotypes with 12-week therapy (Asselah et al. 2018; Zeuzem et al. 2018). Glecaprevir is a new generation NS3/4A protease inhibitor which has high antiviral potency across all HCV genotypes and does not seem to be as prone to the development of resistance as the previous generation protease inhibitors (Ng et al. 2018). Pibrentasvir is a new generation NS5A inhibitor with potent pan-genotypic activity and efficacy against the common NS5A resistance-associated substitutions (Ng et al. 2017). The presence of cirrhosis did not adversely impact treatment efficacy with Maviret with reported SVR rates of 98–100% in cirrhotic patients with genotype 1–6 HCV, with no patient developing hepatic decompensation on treatment (Forns et al. 2017; Wyles et al. 2017). However, in treatment-experienced patients, especially with prior NS5A inhibitor exposure, extension of Maviret therapy from 12 to 16 weeks may be beneficial. In treatment-experienced genotype 3-infected patients, SVR rates of 91 and 95% were reported for 12 and 16-week treatment, respectively (Wyles et al. 2017). While NS5A-treatment-experienced genotype 1-infected patients showed SVR rates for 12 and 16-week treatment of 88 and 94% respectively (Poordad et al. 2018). Maviret has also been shown to produce high SVR rates in patients infected with genotypes 5 and 6 (Asselah et al. 2019). In this phase, 3b study patients without cirrhosis received 8 weeks of Maviret while patients with compensated cirrhosis received 12 weeks of treatment. Patients were either treatment naive or had prior treatment with pegylated interferon with or without ribavirin or sofosbuvir. Overall, an SVR12 was achieved in 97.6% of patients(Asselah et al. 2019).

Maviret treatment is well-tolerated with commonly occurring adverse events of headache and fatigue not statistically different between the Maviret treatment group and placebo group (Asselah et al. 2018). Maviret has been shown to be safe in severe chronic kidney disease as both glecaprevir and pibrentasvir have negligible renal excretion (Gane et al. 2017). As Maviret contains a protease inhibitor, its use is contraindicated in decompensated cirrhosis (Vermehren et al. 2018).

12   Risk of HCC Development with DAA Therapy

Initial reports raised concern that HCV eradication via DAA treatment may in fact increase the risk of developing HCC. Conti et al. found that in their cohort of HCV cirrhotic patients, the majority of whom did not have a history of previous HCC, 8% were found to have developed HCC within 24 weeks of completing DAA therapy (Conti et al. 2016). Another small study of HCV-infected individuals treated with DAA therapy that specifically excluding patients with previous HCC found a HCC incidence rate of 9% developing within 6 months of DAA therapy (Ravi et al. 2017). These rates were higher than the 1–5% annual reported rate of HCC development in cirrhotic patients with untreated HCV (Westbrook and Dusheiko 2014). The rate of HCC recurrence was also noted to be high after the DAA treatment of HCV. Reig and colleagues noted in their cohort of HCV cirrhotics who had previous curative treatment of HCC, 28% developed evidence of HCC recurrence within 6 months of DAA treatment completion (Reig et al. 2016). In patients who received liver transplantation for HCV-associated HCC, pre-transplant DAA therapy was also associated with an increased rate of HCC recurrence within the transplanted liver (Yang et al. 2016). This promotion of tumourogenesis may be due to immune dysregulation caused by DAA therapy (Meissner et al. 2016), with the rapid viral clearance effected by DAAs perhaps resulting in impaired immune surveillance of tumour cells (Nault and Colombo 2016). It was noted that this increased risk of HCC development was not seen with interferon-based treatments (van der Meer et al. 2012), and interferon therapy has been shown to reduce the risk of HCC recurrence (Singal et al. 2010) purportedly due to the immune stimulant properties of interferon. However, it must be noted that due to the propensity of interferon to cause hepatic decompensation, interferon-based therapy was only used in well-compensated cirrhotics, unlike DAA therapy which may be used safely in more advanced patients, who are innately at higher risk of developing HCC.

However, subsequent large studies and meta-analyses have refuted these claims. In a large retrospective cohort study of 22,500 patients treated with DAAs, achieving SVR was associated with a significant reduction in HCC risk, with the annual HCC incidence rate falling from 3.5 to 0.9% compared to patients who did not achieve SVR (Kanwal et al. 2017). A meta-analysis including over 13,000 patients across 41 studies showed no statistical difference in the rate of HCC development following SVR with DAA regimens compared to interferon-based treatment (Waziry et al. 2017). Furthermore, Tsai et al. noted that in HCV-infected patients post-curative treatment of HCC, treatment with pegylated interferon and ribavirin was also associated with a significance rate of HCC recurrence (22.9%) within 6 months after antiviral therapy (Tsai et al. 2017), indicating the predilection of HCC to redevelop early regardless of the type of antiviral therapy used. Another meta-analysis analyzing more than 31,000 HCV-infected individuals showed that achieving SVR with DAA therapy reduced the risk of HCC at all stages of hepatic fibrosis (hazard ratio 0.24, $p < 0.001$) (Morgan et al. 2013). A large retrospective study involving over 62,000 patients in the US not only showed that DAA therapy for HCV was not associated with an increased risk of HCC development compared to interferon-based therapy, but achieving SVR with DAA therapy also decreased the subsequent risk of HCC by 71%, and subsequent HCC development was related to advanced disease stage rather than treatment used (Ioannou et al. 2018). To answer the question of post-liver transplant HCC recurrence, a study of HCV-infected individuals who received DAA therapy prior to undergoing liver transplantation for HCC was performed and only 8.5% of patients developed tumour recurrence at 24 months post-transplant (Donato et al. 2017).

13   Risk of Hepatitis B Reactivation and Flare with DAA Therapy

In patients infected with both HCV and HBV, the HCV viral replication often dominates resulting in a low to undetectable HBV viral load (Yu et al. 2015). DAA therapy for HCV has been found to promote HBV flares by rapidly removing the suppressive effect of HCV on HBV replication while having no antiviral activity against HBV. In one study using all-oral DAA therapy in patients coinfected with HCV and HBV, on-treatment development of hepatitis occurred in 30% of HBsAg-positive patients, with 10% progressing to liver failure (Wang et al. 2017), but this did not adversely affect HCV SVR rates. Even in patients with previous HBV infection (HBcAb positive but HBsAg negative and HBV viral load unde-tected), DAA treatment for HCV has been associated with on-therapy HBV reac-tivation resulting in significant hepatitic flares. Reactivation causing fulminant hepatitis requiring liver transplantation has even been reported (Ende et al. 2015).

The most recent AASLD guidelines (hcvguidelines.org) and EASL guidelines (Liver 2018) both recommend screening for HBV status with HBsAg, HBsAb and HBcAb prior to the initiation of DAA therapy. Patients with a detectable HBV viral load are at risk of a hepatitis flare on DAA treatment and should receive prophy-lactic HBV nucleoside-analogue therapy which should be continued for a further 12 weeks after cessation of HCV DAA therapy. Patients with previous HBV infection (HBcAb positive only) should be monitored for the loss of HBsAb or detection of HBsAg while on DAA therapy, indicating the presence of HBV reactivation.

14 Prevention of the Acquisition of HCV in High-Risk Patients

14.1 Preventative Vaccines

Despite HCV being discovered over 20 years ago, there is currently still no effective vaccine to prevent HCV infection, and the development of a preventative vaccine remains an area of intense research. Recent advances in the treatment of HCV with DAAs have significantly improved SVR rates. However, these treatments will not prevent re-infection particularly in high-risk populations where re-infection rates of up 30% have been reported (Bate et al. 2010; Sacks-Davis et al. 2013; Midgard et al. 2016).

Simulation models of hepatitis C dynamics in high-risk populations have all predicted that the introduction of a vaccine, even with modest efficacy, will have a significant effect on reducing the incidence of HCV. Moreover, vaccination after successful treatment with DAAs is also predicted to be as effective at reducing HCV prevalence as vaccinating an equivalent number of people who inject drugs (PWID) in the community (Scott et al. 2015).

A vaccine producing sterilizing immunity is not required in order to achieve HCV elimination. In a US study, a vaccine with an efficacy of 80% and a high vaccination rate of 1% per month targeted to high-risk individuals is predicted to reduce the incidence of HCV from 13.5 to 2.3% per person-years 30 years after vaccine introduction. Even a vaccine of modest (65%) efficacy and vaccination coverage of 0.6% per month would produce a fall in the incidence of chronic HCV to 2.9% after 30 years (Hahn et al. 2009). A UK study showed that by achieving annual vaccination rates of 162, 77 and 44 per 1000 people who inject drugs (PWID) for low (50% protection for 5 years), moderate (70% protection for 10 years), and high (90% protection for 20 years) vaccine efficacies resulted in a halving of chronic HCV prevalence over a 40 year period (Stone et al. 2016). The introduction of DAAs is not a reason to overlook the potential benefit of a vaccine. The introduction of a vaccine of 60–90% efficacy in the era of DAAs is predicted to significantly reduce HCV prevalence especially in populations with high (50%) to very high (75%) chronic HCV prevalence (Scott et al. 2015). A preventative vaccine in a combined approach with DAAs and harm minimization strategies will be the only way to enable us to fulfill the goal of eliminating HCV as a global health burden.

We know that individuals who spontaneously clear HCV infection develop a strong and broadly cross-reactive CD4+ and CD8+ T cell responses against HCV core and non-structural proteins NS3, NS4 and NS5 (Lauer et al. 2004) as well as the production of cross-reactive neutralizing antibodies (NAb) (Pestka et al. 2007). The early induction of broad NAb is associated with control of viraemia and protection against HCV infection in chimpanzees, humanized uPa-SCID liver chimeric mice and humans (Dorner et al. 2011; Osburn et al. 2014). A strong NAb response is essential for a protective HCV vaccine. The importance of both CD4+

and CD8+ T cell responses in clearance of and protection against HCV has been borne out by numerous studies (Smyk-Pearson et al. 2006; Bharadwaj et al. 2009; Schulze Zur Wiesch et al. 2012; Swadling et al. 2014). Strong and broad HCV-specific T cell responses are important in the spontaneous clearance of HCV (Smyk-Pearson et al. 2006; Schulze Zur Wiesch et al. 2012). In contrast to persistent infection, spontaneous resolution of HCV infection has been temporally linked to the appearance of strong, long-lived, polyfunctional CD4+ and CD8+ T cell responses that are directed against multiple HCV antigens (Lechner et al. 2000). The role of T cell responses in HCV control is further reinforced by studies showing persistence of HCV in chimpanzees after viral challenge following the depletion of both CD4+ and CD8+ T cells (Grakoui et al. 2003).

As such, a preventative HCV vaccine would need to reliably generate all these responses against the various genotypes and quasi-species of HCV in inoculated individuals. Few vaccine strategies other than live attenuated viruses or virus-like particles (VLP) are likely to fulfill these criteria.

14.2   Recombinant Adenoviral and MVA Vaccines for HCV

Several HCV containing vaccine candidates that predominantly result in the production of HCV-specific T cell responses have now been described. These have included recombinant adenoviral and modified vaccinia Ankara (MVA), DNA and VLP vaccines in various prime-boost approaches (Folgori et al. 2006; Mikkelsen et al. 2011; Barnes et al. 2012; Swadling et al. 2014; Kumar et al. 2016).

Several studies have now been reported using recombinant adenoviral vectors encoding the non-structural proteins of HCV to produce live attenuated vaccines capable of producing CD4+ and CD8+ T-specific responses. These vaccines have been tested in various animal models and some have also progressed to clinical trials in humans. In early studies in mice, immunization with a replication-deficient recombinant adenovirus encoding HCV NS3 protein resulted in the production of strong HCV-specific T cell responses. This vaccine also resulted in protection against the recombinant vaccinia virus expressing the HCV NS3 and this protective response was correlated to CD8+ T-specific responses (Mikkelsen et al. 2011). A recombinant adenovirus vaccine encoding the NS3 gene of HCV has also been shown to induce strong CD8+ T cell responses in chimpanzees and this vaccine also resulted in the protection against challenge with a heterologous virus and the development of acute hepatitis in chimpanzees (Folgori et al. 2006). These studies demonstrated that recombinant adenoviral vaccines have the potential to prevent HCV in humans.

The earlier studies in primates paved the way for studies of recombinant adenoviral HCV vaccines in humans. In one of the first studies in healthy human volunteers, a vaccine consisting of recombinant human adenovirus 6 (Ad6) and chimpanzee adenovirus 3 (ChAd3) encoding the NS3-5B genes of genotype 1B of HCV produced CD4+ and CD8+ T cell responses against homologous and heterologous HCV non-structural proteins (Barnes et al. 2012). The vaccine also produced polyfunctional memory CD4+ and CD8+ T cells in the vaccine recipients (Barnes et al. 2012). In a subsequent study, ChAd3 and MVA vectors encoding the NS3, NS4, NS5A, and NS5B proteins of HCV genotype 1b were tested in a prime-boost strategy in human volunteers. Vaccination with the ChAd3 vaccine followed by boosting with the MVA vaccine produced HCV-specific polyfunctional CD8+ and CD4+ T cell responses against homologous and heterologous HCV antigens together with long-lived memory T cell responses (Swadling et al. 2014).

A more recent study investigated immune responses in healthy human volunteers following vaccination with replication-defective ChAd3 and ChAd6 vectors followed by boosting with recombinant MVA vaccines delivering both HCV non-structural and HIV-1 conserved immunogens simultaneously (Hartnell et al. 2018). The co-administration of both HCV and HIV vaccines produced strong and broad polyfunctional CD4+ and CD8+ T cells responses that were similar to the responses produced using the different regimens alone. The immune responses were also maintained for up to 34 weeks after vaccination with HCV non-structural and HIV-1 conserved immunogens simultaneously. These vaccines, however, do not produce neutralizing antibody responses which are a central requirement for protection against HCV.

14.3 Recombinant Protein-Based Vaccines for HCV

The efficacy of an HCV vaccine will reside in its ability to produce broad NAbs (brNAb) in addition to CD4+ and CD8+ T cell responses. Recombinant protein and virus-like particle (VLP) based vaccines have been shown to produce brNAb responses. The vaccination of chimpanzees with recombinant E1 and E2 proteins produced in mammalian cells has been shown to prevent the development of persistent infection after homologous or heterologous virus challenge. This recombinant HCV E1E2 vaccine adjuvanted with MF59C has also been shown to be safe and immunogenic in humans resulting in the production of NAb and CD4$^+$ T cell responses (Frey et al. 2010). Furthermore, immunization of human volunteers with recombinant gpE1/E2 (HCV genotype 1a) resulted in the production of broad cross-neutralizing antibody responses (Law et al. 2013). It has also been shown that vaccination of mice and macaques with a genotype 1a HCV E2 glycoprotein and retroviral Gag pseudotypic particle vaccine produces high-titre NAb responses (Garrone et al. 2011).

Recombinant protein vaccines are also able to produce T cell responses against multiple antigenic targets. The co-administration of a recombinant HCV core, E1, E2, and NS3 protein vaccine in mice and African green monkeys has been shown to induce strong core and NS3 specific T cell responses. Furthermore, immune mice controlled viremia after challenge with a vaccinia virus expressing HCV structural proteins (Martinez-Donato et al. 2014).

14.4 Recombinant Virus-like Particle (VLP) Based Vaccines for HCV

Hepatitis C virus-like particles (HCV VLPs) have been shown to produce NAb and T cell responses in a number of animal models (Chua et al. 2012; Beaumont et al. 2013; Kumar et al. 2016; Earnest-Silveira et al. 2016a, b). HCV-specific NAbs recognize tertiary or quaternary structures (Giang et al. 2012) this makes VLPs attractive as a potential vaccine for HCV as VLPs present conformational epitopes in their native state (Garrone et al. 2011; Chua et al. 2012; Beaumont et al. 2013). HCV VLPs also produce stronger cytotoxic T cell responses in mice compared to DNA vaccines encoding HCV core, E1 and E2 (Murata et al. 2003).

Insect cell-derived VLPs expressing the core, E1 and E2 structural proteins of genotype 1a HCV have been shown to produce broad HCV-specific immune responses (Baumert et al. 1998, 1999; Murata et al. 2003; Steinmann et al. 2004).

Vaccination of mice with insect cell-derived HCV VLPs resulted in cross-neutralizing antibody responses against the HCV structural proteins (Baumert et al. 1998). Furthermore, vaccination of HLA-A2.1 transgenic and BALB/c mice with HCV VLPs produced strong humoral and HCV core-specific CD4+ and CD8+ T cell responses and protection against recombinant vaccinia virus expressing the HCV structural proteins (Baumert et al. 1998, 1999; Lechmann et al. 2001; Murata et al. 2003; Steinmann et al. 2004). The importance of CD4+ and CD8+ T cells in producing protective responses against vaccinia-HCV has also been shown in adoptive transfer experiments (Murata et al. 2003). In addition, the HCV VLPs were able to stimulate the maturation of human dendritic cells (Barth et al. 2005). The immunogenicity of HCV VLPs and the effects of novel adjuvants were further tested in a nonhuman primate model (Jeong et al. 2004). Baboons were immunized with HCV VLPs adjuvanted with AS01B developed HCV-specific antibody, CD4+ and CD8+ T cell responses (Jeong et al. 2004). A subsequent study of HCVVLPs in the chimpanzee showed that this vaccine produced HCV-specific CD4+ and CD8+ T cell responses and prevent progression to persistent infection following the HCV challenge (Elmowalid et al. 2007).

Genotype 1a HCV VLPs have also been produced in human hepatocyte-derived cells. These HCV VLPs possess the biochemical, biophysical properties and morphological characteristic of HCV virions (Gastaminza et al. 2010; Catanese et al. 2013; Earnest-Silveira et al. 2016a, b; Collett et al. 2019) and have been shown to produce NAb and HCV-specific T cell responses in mice (Chua et al. 2012; Earnest-Silveira et al. 2016a, b). In addition, the HCV VLPs bound neutralizing human monoclonal antibodies (HuMAbs) targeting conserved antigenic domain B and D epitopes of the E2 protein (Keck et al. 2004, 2008, 2011; Fauvelle et al. 2016; Keck et al. 2016). A genotype 3a HCV VLP vaccine has also been shown to produce broad humoral and cellular immune responses in mice (Kumar et al. 2016). An advance in the approach of HCV VLP vaccines has been the development of a quadrivalent genotype 1a/1b/2a/3a HCV VLP vaccine (Earnest-Silveira et al. 2016a, b). This vaccine has been shown to produce brNAb, memory B cell and T cell responses in mice and pigs (Christiansen et al. 2018a, b, 2019).

Vaccination of mice with recombinant retrovirus-based virus-like particles (retroVLPs) made of Gag of murine leukemia virus and pseudotyped with HCV E1 and E2 envelope glycoproteins produces strong homologous HCV-specific NAb and T cell responses (Huret et al. 2013). Also, the co-administration of this vaccine with retroVLPs displaying NS3 has been shown to produce strong HCV NS3 and E1E2 specific T cell responses (Huret et al. 2013). In a prime-boost immunization series using retroVLPs and a recombinant serotype 5 adenovirus (rAd5) expressing HCV-E1/E2 envelope glycoprotein (rAdE1E2) mice were primed with rAdE1E2 followed by boosting with retroVLPs. This approach resulted in stronger E2-specific antibody responses than retroVLPs alone (Desjardins et al. 2009; Garrone et al. 2011). The prime-boost strategy also produced cross-neutralizing HCV NAb against five genotypes of HCV (Garrone et al. 2011).

In an alternative approach, a chimeric HBs-HCV VLP vaccine containing E1–E2 heterodimers of genotype 1a HCV has been shown to produce cross-NAb responses against heterologous HCV genotypes (Patient et al. 2009; Beaumont et al. 2013, 2016; Beaumont and Roingeard 2015). The immunogenicity of the chimeric HBs-HCV particles was assessed in rabbits and shown to produce strong HCV E1 and E2 specific antibody responses and HCV neutralizing antibody responses against HCV genotypes 1a, 1b, 2a and 3 (Beaumont et al. 2013). The immunogenicity of the chimeric vaccine and the strength of the HCV E2 responses was not affected by pre-existing immunity to HBsAg (Beaumont and Roingeard 2015).

Modified HBsAg particles by carrying HCV-specific B and T cell epitopes in the 'a' determinant of the HBs protein have also been shown to produce HCV-specific immune responses in mice (Netter et al. 2001, 2003; Woo et al. 2006; Haqshenas et al. 2007; Vietheer et al. 2007). Vaccination of mice with a combination of particles carrying different HCV E2 HVR1 epitopes resulted in a stronger antibody response than vaccination with the individual particles (Netter et al. 2001). Finally, the presence of pre-existing anti-HBs antibody has been shown to have no effect on the production of anti-HVR1 antibody, suggesting that the vaccine could be used in individuals who have previously been vaccinated against HBV (Netter et al. 2001).

14.5 Public Health Measures

As a safe and effective preventative vaccine remains elusive, strategies to reduce HCV transmission among individuals at high risk of acquiring the virus should be employed. In particular, harm reduction measures to reduce unsafe injecting practices amongst intravenous drug users, such as behavioural interventions, the access to sterile needles and syringes, and the management of substance abuse, have been shown to reduce the risk of HCV infection by about 75% (Hagan et al. 2011). The universal screening of blood donors is also important to prevent transmission via contaminated blood products, as is adhering to universal precautions and strict needle-stick protocols within healthcare facilities.

15 Conclusions

HCV is currently a serious global health concern, chronically infecting about 3% of the world's population, leading to chronic hepatitis, cirrhosis, liver failure and HCC, thereby causing significant morbidity and mortality. Furthermore, HCV infection is a major cause of HCC in the Western world, with the majority of HCCs developing in the setting of cirrhosis. The new highly potent DAA combinations are able to cure HCV in the vast majority of infected patients, regardless of viral genotype, viral load, or fibrosis stage. Thus, it is imperative to identify at-risk individuals and provide antiviral therapy prior to the development of established cirrhosis in order to reduce the risk of subsequent HCC. Even after the development of cirrhosis, successful HCV clearance is still associated with reduced HCC risk. Preventative and therapeutic vaccines against HCV remain an area of ongoing research and hopefully, an effective vaccine will be available in the future.

References

Aleman S, Rahbin N, Weiland O, Davidsdottir L, Hedenstierna M, Rose N, Verbaan H, Stal P, Carlsson T, Norrgren H, Ekbom A, Granath F, Hultcrantz R (2013) A risk for hepatocellular carcinoma persists long-term after sustained virologic response in patients with hepatitis C-associated liver cirrhosis. Clin Infect Dis 57(2):230–236

Alkofer B, Lepennec V, Chiche L (2011) Hepatocellular cancer in the non-cirrhotic liver. J Visceral Surg 148(1):3–10

Alter MJ (1995) Epidemiology of hepatitis C in the West. Semin Liver Dis 15(1):5–14

Asselah T, Kowdley KV, Zadeikis N, Wang S, Hassanein T, Horsmans Y, Colombo M, Calinas F, Aguilar H, de Ledinghen V, Mantry PS, Hezode C, Marinho RT, Agarwal K, Nevens F, Elkhashab M, Kort J, Liu R, Ng TI, Krishnan P, Lin C-W, Mensa FJ (2018) Efficacy of glecaprevir/pibrentasvir for 8 or 12 weeks in patients with hepatitis C virus genotype 2, 4, 5, or 6 infection without cirrhosis. Clin Gastroenterol Hepatol 16(3):417–426

Asselah T, Lee SS, Yao BB, Nguyen T, Wong F, Mahomed A, Lim SG, Abergel A, Sasadeusz J, Gane E, Zadeikis N, Schnell G, Zhang Z, Porcalla A, Mensa FJ, Nguyen K (2019) Efficacy and safety of glecaprevir/pibrentasvir in patients with chronic hepatitis C virus genotype 5 or 6 infection (ENDURANCE-5,6): an open-label, multicentre, phase 3b trial. Lancet Gastroenterol Hepatol 4(1):45–51

Backus LI, Boothroyd DB, Phillips BR, Belperio P, Halloran J, Mole LA (2011) A Sustained virologic response reduces risk of all-cause mortality in patients with hepatitis C. Clin Gastroenterol Hepatol 9(6):509–U145

Barnes E, Folgori A, Capone S, Swadling L, Aston S, Kurioka A, Meyer J, Huddart R, Smith K, Townsend R, Brown A, Antrobus R, Ammendola V, Naddeo M, O'Hara G, Willberg C, Harrison A, Grazioli F, Esposito ML, Siani L, Traboni C, Oo Y, Adams D, Hill A, Colloca S, Nicosia A, Cortese R, Klenerman P (2012) Novel adenovirus-based vaccines induce broad and sustained T cell responses to HCV in man. Sci Transl Med 4(115):115ra111

Bartenschlager R, Penin F, Lohmann V, Andre P (2011) Assembly of infectious hepatitis C virus particles. Trends Microbiol 19(2):95–103

Barth H, Ulsenheimer A, Pape GR, Diepolder HM, Hoffman M, Neumann-Haefelin C, Thimme R, Henneke P, Klein RS, Paranhos-Baccala G, Depla E, Liang TJ, Blum H, Baumert TF (2005) Uptake and presentation of hepatitis C virus-like particles by human dendritic cells. Blood 105 (9):3605–3614

Bate JP, Colman AJ, Frost PJ, Shaw DR, Harley HA (2010) High prevalence of late relapse and reinfection in prisoners treated for chronic hepatitis C. J Gastroenterol Hepatol 25(7):1276–1280

Baumert TF, Ito S, Wong DT, Liang TJ (1998) Hepatitis C virus structural proteins assemble into viruslike particles in insect cells. J Virol 72(5):3827–3836

Baumert TF, Vergalla J, Satoi J, Thomson M, Lechmann M, Herion D, Greenberg HB, Ito S, Liang TJ (1999) Hepatitis C virus-like particles synthesized in insect cells as a potential vaccine candidate. Gastroenterology 117(6):1397–1407

Beaumont E, Patient R, Hourioux C, Dimier-Poisson I, Roingeard P (2013) Chimeric hepatitis B virus/hepatitis C virus envelope proteins elicit broadly neutralizing antibodies and constitute a potential bivalent prophylactic vaccine. Hepatology 57(4):1303–1313

Beaumont E, Roch E, Chopin L, Roingeard P (2016) Hepatitis C virus E1 and E2 proteins used as separate immunogens induce neutralizing antibodies with additive properties. PLoS ONE 11 (3):e0151626

Beaumont E, Roingeard P (2015) Chimeric hepatitis B virus (HBV)/hepatitis C virus (HCV) subviral envelope particles induce efficient anti-HCV antibody production in animals pre-immunized with HBV vaccine. Vaccine 33(8):973–976

Beinhardt S, Al-Zoairy R, Kozbial K, Stattermayer AF, Maieron A, Stauber R, Strasser M, Zoller H, Graziadei I, Rasoul-Rockenschaub S, Trauner M, Ferenci P, Hofer H (2018) Long-term follow-up of ribavirin-free DAA-based treatment in HCV recurrence after orthotopic liver transplantation. Liver International 38(7):1188–1197

Berenguer M, Ferrell L, Watson J, Prieto M, Kim M, Rayon M, Cordoba J, Herola A, Ascher N, Mir J, Berenguer J, Wright TL (2000) HCV-related fibrosis progression following liver transplantation: increase in recent years. J Hepatol 32(4):673–684

Bharadwaj M, Thammanichanond D, Aitken CK, Moneer S, Drummer HE, Lilly SL, Holdsworth R, Bowden SD, Jackson DC, Hellard M, Torresi J, McCluskey J (2009) TCD8 response in diverse outcomes of recurrent exposure to hepatitis C virus. Immunol Cell Biol 87 (6):464–472

Blach S, Zeuzem S, Manns M, Altraif I, Duberg AS, Muljono DH, Waked I, Alavian SM, Lee MH, Negro F, Abaalkhail F, Abdou A, Abdulla M, Abou Rached A, Aho I, Akarca U, Al Ghazzawi I, Al Kaabi S, Al Lawati F, Al Namaani K, Al Serkal Y, Al-Busafi SA, Al-Dabal L, Aleman S, Alghamdi AS, Aljumah AA, Al-Romaihi HE, Andersson MI, Arendt V, Arkkila P, Assiri AM, Baatarkhuu O, Bane A, Ben-Ari Z, Bergin C, Bessone F, Bihl F, Bizri AR, Blachier M, Blasco AJ, Mello CEB, Bruggmann P, Brunton CR, Calinas F, Chan HLY, Chaudhry A, Cheinquer H, Chen CJ, Chien RN, Choi MS, Christensen PB, Chuang WL, Chulanov V, Cisneros L, Clausen MR, Cramp ME, Craxi A, Croes EA, Dalgard O, Daruich JR, de Ledinghen V, Dore GJ, El-Sayed MH, Ergor G, Esmat G, Estes C, Falconer K, Farag E, Ferraz MLG, Ferreira PR, Flisiak R, Frankova S, Gamkrelidze I, Gane E, Garcia-Samaniego J, Khan AG, Gountas I, Goldis A, Gottfredsson M, Grebely J, Gschwantler M, Pessoa MG, Gunter J, Hajarizadeh B, Hajelssedig O, Hamid S, Hamoudi W, Hatzakis A, Himatt SM, Hofer H, Hrstic I, Hui YT, Hunyady B, Idilman R, Jafri W, Jahis R, Janjua NZ, Jarcuska P, Jeruma A, Jonasson JG, Kamel Y, Kao JH, Kaymakoglu S, Kershenobich D, Khamis J, Kim YS, Kondili L, Koutoubi Z, Krajden M, Krarup H, Lai MS, Laleman W, Lao WC, Lavanchy D, Lazaro P, Leleu H, Lesi O, Lesmana LA, Li M, Liakina V, Lim YS, Luksic B, Mahomed A, Maimets M, Makara M, Malu AO, Marinho RT, Marotta P, Mauss S, Memon MS, Correa MCM, Mendez-Sanchez N, Merat S, Metwally AM, Mohamed R, Moreno C, Mourad FH, Mullhaupt B, Murphy K, Nde H, Njouom R, Nonkovic D, Norris S, Obekpa S, Oguche S, Olafsson S, Oltman M, Omede O, Omuemu C, Opare-Sem O, Ovrehus ALH, Owusu-Ofori S, Oyunsuren TS, Papatheodoridis G, Pasini K, Peltekian KM, Phillips RO, Pimenov N, Poustchi H, Prabdial-Sing N, Qureshi H, Ramji A, Razavi-Shearer D, Razavi-Shearer K, Redae B, Reesink HW, Ridruejo E, Robbins S, Roberts LR, Roberts SK, Rosenberg WM, Roudot-Thoraval F, Ryder SD, Safadi R, Sagalova O, Salupere R, Sanai FM, Avila JFS,

Saraswat V, Sarmento-Castro R, Sarrazin C, Schmelzer JD, Schreter I, Seguin-Devaux C, Shah SR, Sharara AI, Sharma M, Shevaldin A, Shiha GE, Sievert W, Sonderup M, Souliotis K, Speiciene D, Sperl J, Starkel P, Stauber RE, Stedman C, Struck D, Su TH, Sypsa V, Tan SS, Tanaka J, Thompson AJ, Tolmane I, Tomasiewicz K, Valantinas J, Van Damme P, van der Meer AJ, van Thiel I, Van Vlierberghe H, Vince A, Vogel W, Wedemeyer H, Weis N, Wong VWS, Yaghi C, Yosry A, Yuen MF, Yunihastuti E, Yusuf A, Zuckerman E, Razavi H, Polaris Observ HCVC (2017) Global prevalence and genotype distribution of hepatitis C virus infection in 2015: a modelling study. Lancet Gastroenterol Hepatol 2(3):161–176

Bourliere M, Gordon SC, Flamm SL, Cooper CL, Ramji A, Tong M, Ravendhran N, Vierling JM, Tran TT, Pianko S, Bansal MB, Ledinghen VD, Hyland RH, Stamm LM, Dvory-Sobol H, Svarovskaia E, Zhang J, Huang KC, Subramanian GM, Brainard DM, McHutchison JG, Verna EC, Buggisch P, Landis CS, Younes ZH, Curry MP, Strasser SI, Schiff ER, Reddy KR, Manns MP, Kowdley KV, Zeuzem S, Polaris and P.-Investigators (2017) Sofosbuvir, velpatasvir, and voxilaprevir for previously treated HCV infection. New Engl J Med 376 (22):2134–2146

Bruchfeld A, Roth D, Martin P, Nelson DR, Pol S, Londono MC, Monsour H Jr, Silva M, Hwang P, Arduino JM, Robertson M, Nguyen BY, Wahl J, Barr E, Greaves W (2017) Elbasvir plus grazoprevir in patients with hepatitis C virus infection and stage 4-5 chronic kidney disease: clinical, virological, and health-related quality-of-life outcomes from a phase 3, multicentre, randomised, double-blind, placebo-controlled trial. Lancet Gastroenterol Hepatol 2 (8):585–594

Budhu A, Wang XW (2006) The role of cytokines in hepatocellular carcinoma. J Leukoc Biol 80 (6):1197–1213

Buhler S, Bartenschlager R (2012) New targets for antiviral therapy of chronic hepatitis C. Liver Int: Off J Int Assoc Study Liver 32(Suppl 1):9–16

Catanese MT, Uryu K, Kopp M, Edwards TJ, Andrus L, Rice WJ, Silvestry M, Kuhn RJ, Rice CM (2013) Ultrastructural analysis of hepatitis C virus particles. Proc Natl Acad Sci USA 110 (23):9505–9510

Chen SL, Morgan TR (2006) The natural history of hepatitis C virus (HCV) infection. Int J Med Sci 3(2):47–52

Cheng G, Yu M, Peng B, Lee YJ, Trejo-Martin A, Gong R, Bush C, Worth A, Nash M, Chan K, Yang H, Beran R, Tian Y, Perry J, Taylor J, Yang C, Paulson M, Delaney W, Link JO (2013) GS-5816, a second generation HCV NS5A inhibitor with potent antiviral activity, broad genotypic coverage and a high resistance barrier. J Hepatol 58:S484–S485

Chevaliez S, Pawlotsky J-M (2007) Hepatitis C virus: virology, diagnosis and management of antiviral therapy. World J Gastroenterol 13(17):2461–2466

Choo SP, Tan WL, Goh BKP, Tai WM, Zhu AX (2016) Comparison of hepatocellular carcinoma in Eastern versus Western populations. Cancer 122(22):3430–3446

Christiansen D, Earnest-Silveira L, Chua B, Boo I, Drummer HE, Grubor-Bauk B, Gowans EJ, Jackson DC, Torresi J (2018a) Antibody responses to a quadrivalent hepatitis C viral-like particle vaccine adjuvanted with toll-like receptor 2 agonists. Viral Immunol 31(4):338–343

Christiansen D, Earnest-Silveira L, Chua B, Meuleman P, Boo I, Grubor-Bauk B, Jackson DC, Keck ZY, Foung SKH, Drummer HE, Gowans EJ, Torresi J (2018b) Immunological responses following administration of a genotype 1a/1b/2/3a quadrivalent HCV VLP vaccine. Sci Rep 8 (1):6483

Christiansen D, Earnest-Silveira L, Grubor-Bauk B, Wijesundara DDK, Boo I, Ramsland R, Vincan E, Drummer HE, Gowans EJ, Torresi J (2019) Pre-clinical evaluation of a quadrivalent HCV VLP vaccine in pigs following microneedle delivery. Scientific Reports 9(1):1–3

Chua BY, Johnson D, Tan A, Earnest-Silveira L, Sekiya T, Chin R, Torresi J, Jackson DC (2012) Hepatitis C VLPs delivered to dendritic cells by a TLR2 targeting lipopeptide results in enhanced antibody and cell-mediated responses. PLoS ONE 7(10):e47492

Chung RT, Davis GL, Jensen DM, Masur H, Saag MS, Thomas DL, Aronsohn AI, Charlton MR, Feld JJ, Fontana RJ, Ghany MG, Godofsky EW, Graham CS, Kim AY, Kiser JJ, Kottilil S, Marks KM, Martin P, Mitruka K, Morgan TR, Naggie S, Raymond D, Reau NS, Schooley RT, Sherman KE, Sulkowski MS, Vargas HE, Ward JW, Wyles DL, Panel AIHG (2015) Hepatitis C guidance: AASLD-IDSA recommendations for testing, managing, and treating adults infected with hepatitis C virus. Hepatology 62(3):932–954

Coburn CA, Meinke PT, Chang W, Fandozzi CM, Graham DJ, Hu B, Huang Q, Kargman S, Kozlowski J, Liu R, McCauley JA, Nomeir AA, Soll RM, Vacca JP, Wang D, Wu H, Zhong B, Olsen DB, Ludmerer SW (2013) Discovery of MK-8742: an HCV NS5A inhibitor with broad genotype activity. Chem Med Chem 8(12):1930–1940

Collett S, Torresi J, Earnest-Silveira L, Christiansen D, Elbourne A, Ramsland PA (2019) Probing and pressing surfaces of hepatitis C virus-like particles. J Colloid Interface Sci 545:259–268

Conteduca V, Sansonno D, Russi S, Pavone F, Dammacco F (2014) Therapy of chronic hepatitis C virus infection in the era of direct-acting and host-targeting antiviral agents. J Infect 68(1):1–20

Conti F, Buonfiglioli F, Scuteri A, Crespi C, Bolondi L, Caraceni P, Foschi FG, Lenzi M, Mazzella G, Verucchi G, Andreone P, Brillanti S (2016) Early occurrence and recurrence of hepatocellular carcinoma in HCV-related cirrhosis treated with direct-acting antivirals. J Hepatol 65(4):727–733

Curry MP, O'Leary JG, Bzowej N, Muir AJ, Korenblat KM, Fenkel JM, Reddy KR, Lawitz E, Flamm SL, Schiano T, Teperman L, Fontana R, Schiff E, Fried M, Doehle B, An D, McNally J, Osinusi A, Brainard DM, McHutchison JG, Brown RS Jr, Charlton M, Investigators A- (2015) Sofosbuvir and velpatasvir for HCV in patients with decompensated cirrhosis. N Engl J Med 373(27):2618–2628

Desjardins D, Huret C, Dalba C, Kreppel F, Kochanek S, Cosset FL, Tangy F, Klatzmann D, Bellier B (2009) Recombinant retrovirus-like particle forming DNA vaccines in prime-boost immunization and their use for hepatitis C virus vaccine development. J Gene Med 11(4):313–325

Di Martino V, Rufat P, Boyer N, Renard P, Degos F, Martinot-Peignoux M, Matheron S, Le Moing V, Vachon F, Degott C, Valla D, Marcellin P (2001) The influence of human immunodeficiency virus coinfection on chronic hepatitis C in injection drug users: a long-term retrospective cohort study. Hepatol (Baltimore, Md) 34(6):1193–1199

Donato F, Tagger A, Gelatti U, Parrinello G, Boffetta P, Albertini A, Decarli A, Trevisi P, Ribero ML, Martelli C, Porru S, Nardi G (2002) Alcohol and hepatocellular carcinoma: the effect of lifetime intake and hepatitis virus infections in men and women. Am J Epidemiol 155 (4):323–331

Donato MF, Invernizzi F, Rossi G, Iavarone M (2017) Interferon-free therapy of hepatitis C during wait list and post-transplant risk of hepatocellular carcinoma recurrence. J Hepatol 67(6):1355–1356

Dorner M, Horwitz JA, Robbins JB, Barry WT, Feng Q, Mu K, Jones CT, Schoggins JW, Catanese MT, Burton DR, Law M, Rice CM, Ploss A (2011) A genetically humanized mouse model for hepatitis C virus infection. Nature 474(7350):208–211

Earnest-Silveira L, Christiansen D, Herrmann S, Ralph SA, Das S, Gowans EJ, Torresi J (2016a) Large scale production of a mammalian cell derived quadrivalent hepatitis C virus like particle vaccine. J Virol Methods 236:87–92

Earnest-Silveira L, Chua B, Chin R, Christiansen D, Johnson D, Herrmann S, Ralph SA, Vercauteren K, Mesalam A, Meuleman P, Das S, Boo I, Drummer H, Bock CT, Gowans EJ, Jackson DC, Torresi J (2016b) Characterization of a hepatitis C virus-like particle vaccine produced in a human hepatocyte-derived cell line. J Gener Virol 97(8):1865–1876

El-Serag HB, Kanwal F (2014) Epidemiology of hepatocellular carcinoma in the united states: where are we? where do we go? Hepatology 60(5):1767–1775

Elmowalid GA, Qiao M, Jeong SH, Borg BB, Baumert TF, Sapp RK, Hu Z, Murthy K, Liang TJ (2007) Immunization with hepatitis C virus-like particles results in control of hepatitis C virus infection in chimpanzees. Proc Natl Acad Sci USA 104(20):8427–8432

132   E. J. Lim and J. Torresi

Ende AR, Kim NH, Yeh MM, Harper J, Landis CS (2015) Fulminant hepatitis B reactivation leading to liver transplantation in a patient with chronic hepatitis C treated with simeprevir and sofosbuvir: a case report. J Med Case Rep 9:164
Fauvelle C, Colpitts CC, Keck ZY, Pierce BG, Foung SK, Baumert TF (2016) Hepatitis C virus vaccine candidates inducing protective neutralizing antibodies. Expert Rev Vaccines 15 (12):1535–1544
Feld JJ, Jacobson IM, Hezode C, Asselah T, Ruane PJ, Gruener N, Abergel A, Mangia A, Lai CL, Chan HLY, Mazzotta F, Moreno C, Yoshida E, Shafran SD, Towner WJ, Tran TT, McNally J, Osinusi A, Svarovskaia E, Zhu Y, Brainard DM, McHutchison JG, Agarwal K, Zeuzem S, Investigators A- (2015) Sofosbuvir and velpatasvir for HCV genotype 1, 2, 4, 5, and 6 infection. N Engl J Med 373(27):2599–2607
Folgori A, Capone S, Ruggeri L, Meola A, Sporeno E, Ercole BB, Pezzanera M, Tafi R, Arcuri M, Fattori E, Lahm A, Luzzago A, Vitelli A, Colloca S, Cortese R, Nicosia A (2006) A T-cell HCV vaccine eliciting effective immunity against heterologous virus challenge in chimpanzees. Nat Med 12(2):190–197
Forns X, Bukh J (1999) The molecular biology of hepatitis C virus. Genotypes and quasispecies. Clin Liver Dis 3(4):693–716, vii
Forns X, Gordon SC, Zuckerman E, Lawitz E, Calleja JL, Hofer H, Gilbert C, Palcza J, Howe AYM, DiNubile MJ, Robertson MN, Wahl J, Barr E, Buti M (2015) Grazoprevir and elbasvir plus ribavirin for chronic HCV genotype-1 infection after failure of combination therapy containing a direct-acting antiviral agent. J Hepatol 63(3):564–572
Forns X, Lee SS, Valdes J, Lens S, Ghalib R, Aguilar H, Felizarta F, Hassanein T, Hinrichsen H, Rincon D, Morillas R, Zeuzem S, Horsmans Y, Nelson DR, Yu Y, Krishnan P, Lin C-W, Kort JJ, Mensa FJ (2017) Glecaprevir plus pibrentasvir for chronic hepatitis C virus genotype 1, 2, 4, 5, or 6 infection in adults with compensated cirrhosis (EXPEDITION-1): a single-arm, open-label, multicentre phase 3 trial. Lancet Infect Dis 17(10):1062–1068
Foster GR, Afdhal N, Roberts SK, Braeu N, Gane EJ, Pianko S, Lawitz E, Thompson A, Shiffman ML, Cooper C, Towner WJ, Conway B, Ruane P, Bourliere M, Asselah T, Berg T, Zeuzem S, Rosenberg W, Agarwal K, Stedman CAM, Mo H, Dvory-Sobol H, Han L, Wang J, McNally J, Osinusi A, Brainard DM, McHutchison JG, Mazzotta F, Tran TT, Gordon SC, Patel K, Reau N, Mangia A, Sulkowski M, Astral and A.-. Investigators (2015) Sofosbuvir and velpatasvir for HCV genotype 2 and 3 infection. New Engl J Med 373(27):2608–2617
Frey SE, Houghton M, Coates S, Abrignani S, Chien D, Rosa D, Pileri P, Ray R, Di Bisceglie AM, Rinella P, Hill H, Wolff MC, Schultze V, Han JH, Scharschmidt B, Belshe RB (2010) Safety and immunogenicity of HCV E1E2 vaccine adjuvanted with MF59 administered to healthy adults. Vaccine 28(38):6367–6373
Gane E, Lawitz E, Pugatch D, Papatheodoridis G, Brau N, Brown A, Pol S, Leroy V, Persico M, Moreno C, Colombo M, Yoshida EM, Nelson DR, Collins C, Lei Y, Kosloski M, Mensa FJ (2017) Glecaprevir and pibrentasvir in patients with HCV and severe renal impairment. N Engl J Med 377(15):1448–1455
Garrone P, Fluckiger AC, Mangeot PE, Gauthier E, Dupeyrot-Lacas P, Mancip J, Cangialosi A, Du Chene I, LeGrand R, Mangeot I, Lavillette D, Bellier B, Cosset FL, Tangy F, Klatzmann D, Dalba C (2011) A prime-boost strategy using virus-like particles pseudotyped for HCV proteins triggers broadly neutralizing antibodies in macaques. Sci Transl Med 3(94):94ra71
Gastaminza P, Dryden KA, Boyd B, Wood MR, Law M, Yeager M, Chisari FV (2010) Ultrastructural and biophysical characterization of hepatitis C virus particles produced in cell culture. J Virol 84(21):10999–11009
Giang E, Dorner M, Prentoe JC, Dreux M, Evans MJ, Bukh J, Rice CM, Ploss A, Burton DR, Law M (2012) Human broadly neutralizing antibodies to the envelope glycoprotein complex of hepatitis C virus. Proc Natl Acad Sci USA 109(16):6205–6210
Gower E, Estes C, Blach S, Razavi-Shearer K, Razavi H (2014) Global epidemiology and genotype distribution of the hepatitis C virus infection. J Hepatol 61:S45–S57

Grakoui A, Shoukry NH, Woollard DJ, Han JH, Hanson HL, Ghrayeb J, Murthy KK, Rice CM, Walker CM (2003) HCV persistence and immune evasion in the absence of memory T cell help. Science 302(5645):659–662

Hadziyannis SJ, Sette H Jr, Morgan TR, Balan V, Diago M, Marcellin P, Ramadori G, Bodenheimer H Jr, Bernstein D, Rizzetto M, Zeuzem S, Pockros PJ, Lin A, Ackrill AM, P. I. S. Group (2004) Peginterferon-alpha2a and ribavirin combination therapy in chronic hepatitis C: a randomized study of treatment duration and ribavirin dose. Ann Int Med 140(5):346–355

Hagan H, Pouget ER, Des Jarlais DC (2011) A systematic review and meta-analysis of interventions to prevent hepatitis C virus infection in people who inject drugs. J Infect Dis 204 (1):74–83

Hahn JA, Wylie D, Dill J, Sanchez MS, Lloyd-Smith JO, Page-Shafer K, Getz WM (2009) Potential impact of vaccination on the hepatitis C virus epidemic in injection drug users. Epidemics 1(1):47–57

Haqshenas G, Dong X, Netter H, Torresi J, Gowans EJ (2007) A chimeric GB virus B encoding the hepatitis C virus hypervariable region 1 is infectious in vivo. J Gen Virol 88(Pt 3):895–902

Hartnell F, Brown A, Capone S, Kopycinski J, Bliss C, Makvandi-Nejad S, Swadling L, Ghaffari E, Cicconi P, Del Sorbo M, Sbrocchi R, Esposito I, Vassilev V, Marriott P, Gardiner CM, Bannan C, Bergin C, Hoffmann M, Turner B, Nicosia A, Folgori A, Hanke T, Barnes E, Dorrell L (2018) A novel vaccine strategy employing serologically different chimpanzee adenoviral vectors for the prevention of HIV-1 and HCV coinfection. Front Immunol 9:3175

Hayashi J, Aoki H, Kajino K, Moriyama M, Arakawa Y, Hino O (2000) Hepatitis C virus core protein activates the MAPK/ERK cascade synergistically with tumor promoter TPA, but not with epidermal growth factor or transforming growth factor alpha. Hepatology (Baltimore, Md) 32(5):958–961

Hezode C, Fontaine H, Dorival C, Zoulim F, Larrey D, Canva V, De Ledinghen V, Poynard T, Samuel D, Bourliere M, Alric L, Raabe J-J, Zarski J-P, Marcellin P, Riachi G, Bernard P-H, Loustaud-Ratti V, Chazouilleres O, Abergel A, Guyader D, Metivier S, Tran A, Di Martino V, Causse X, Dao T, Lucidarme D, Portal I, Cacoub P, Gournay J, Grando-Lemaire V, Hillon P, Attali P, Fontanges T, Rosa I, Petrov-Sanchez V, Barthe Y, Pawlotsky J-M, Pol S, Carrat F, Bronowicki J-P, Grp CS (2014) Effectiveness of telaprevir or boceprevir in treatment-experienced patients with HCV genotype 1 infection and cirrhosis. Gastroenterology 147(1):132–U235

Hino O, Kajino K, Umeda T, Arakawa Y (2002) Understanding the hypercarcinogenic state in chronic hepatitis: a clue to the prevention of human hepatocellular carcinoma. J Gastroenterol 37(11):883–887

Huang YT, Jen CL, Yang HI, Lee MH, Su J, Lu SN, Iloeje UH, Chen CJ (2011) Lifetime risk and sex difference of hepatocellular carcinoma among patients with chronic hepatitis B and C. J Clin Oncol 29(27):3643–3650

Huret C, Desjardins D, Miyalou M, Levacher B, Amadoudji Zin M, Bonduelle O, Combadiere B, Dalba C, Klatzmann D, Bellier B (2013) Recombinant retrovirus-derived virus-like particle-based vaccines induce hepatitis C virus-specific cellular and neutralizing immune responses in mice. Vaccine 31(11):1540–1547

Ioannou GN, Green PK, Berry K (2018) HCV eradication induced by direct-acting antiviral agents reduces the risk of hepatocellular carcinoma. J Hepatol 68(1):25–32

Ishido S, Hotta H (1998) Complex formation of the nonstructural protein 3 of hepatitis C virus with the p53 tumor suppressor. FEBS Lett 438(3):258–262

Ishikawa T, Higuchi K, Kubota T, Seki K-I, Honma T, Yoshida T, Kamimura T (2012) Combination PEG-IFN a-2b/ribavirin therapy following treatment of hepatitis C virus-associated hepatocellular carcinoma is capable of improving hepatic functional reserve and survival. Hepatogastroenterology 59(114):529–532

Jacobson IM, McHutchison JG, Dusheiko G, Di Bisceglie AM, Reddy KR, Bzowej NH, Marcellin P, Muir AJ, Ferenci P, Flisiak R, George J, Rizzetto M, Shouval D, Sola R, Terg RA, Yoshida EM, Adda N, Bengtsson L, Sankoh AJ, Kieffer TL, George S, Kauffman RS, Zeuzem S, Team AS (2011) Telaprevir for previously untreated chronic hepatitis C virus infection. N Engl J Med 364(25):2405–2416

Jeong SH, Qiao M, Nascimbeni M, Hu Z, Rehermann B, Murthy K, Liang TJ (2004) Immunization with hepatitis C virus-like particles induces humoral and cellular immune responses in nonhuman primates. J Virol 78(13):6995–7003

Kanwal F, Kramer J, Asch SM, Chayanupatkul M, Cao YM, El-Serag HB (2017) Risk of hepatocellular cancer in HCV patients treated with direct-acting antiviral agents. Gastroenterology 153(4):996

Kanwal F, Kramer JR, Ilyas J, Duan ZG, El-Serag HB (2014) HCV genotype 3 is associated with an increased risk of cirrhosis and hepatocellular cancer in a national sample of US veterans with HCV. Hepatology 60(1):98–105

Kao CF, Chen SY, Chen JY, Lee YHW (2004) Modulation of p53 transcription regulatory activity and post-translational modification by hepatitis C virus core protein. Oncogene 23(14):2472–2483

Keck Z, Op De Beeck A, Hadlock KG, Xia J, Li T-K, Dubuisson J, Foung SK (2004) Hepatitis C virus E2 has three immunogenic domains containing conformational epitopes with distinct propertires and biological functions. J Virol 78(17):9224–9232

Keck ZY, Li TK, Xia J, Gal-Tanamy M, Olson O, Li SH, Patel AH, Ball JK, Lemon SM, Foung SK (2008) Definition of a conserved immunodominant domain on hepatitis C virus E2 glycoprotein by neutralizing human monoclonal antibodies. J Virol 82(12):6061–6066

Keck ZY, Saha A, Xia J, Wang Y, Lau P, Krey T, Rey FA, Foung SK (2011) Mapping a region of hepatitis C virus E2 that is responsible for escape from neutralizing antibodies and a core CD81-binding region that does not tolerate neutralization escape mutations. J Virol 85 (20):10451–10463

Keck ZY, Wang Y, Lau P, Lund G, Rangarajan S, Fauvelle C, Liao GC, Holtsberg FW, Warfield KL, Aman MJ, Pierce BG, Fuerst TR, Bailey JR, Baumert TF, Mariuzza RA, Kneteman NM, Foung SK (2016) Affinity maturation of a broadly neutralizing human monoclonal antibody that prevents acute hepatitis C virus infection in mice. Hepatology 64 (6):1922–1933

Kiyosawa K, Akahane Y, Nagata A, Furuta S (1984) Hepatocellular carcinoma after non-A, non-B posttransfusion hepatitis. Am J Gastroenterol 79(10):777–781

Kohli V, Singhal A, Elliott L, Jalil S (2012) Antiviral therapy for recurrent hepatitis C reduces recurrence of hepatocellular carcinoma following liver transplantation. Transp Int: Off J Eur Soc Organ Transp 25(2):192–200

Kruse RL, Kramer JR, Tyson GL, Duan ZG, Chen L, El-Serag HB, Kanwal F (2014) Clinical outcomes of hepatitis B virus coinfection in a united states cohort of hepatitis C virus-infected patients. Hepatology 60(6):1871–1878

Kumar A, Das S, Mullick R, Lahiri P, Tatineni R, Goswami D, Bhat P, Torresi J, Gowans EJ, Karande AA (2016) Immune responses against hepatitis C virus genotype 3a virus-like particles in mice: a novel VLP prime-adenovirus boost strategy. Vaccine 34(8):1115–1125

Kurozawa Y, Ogimoto I, Shibata A, Nose T, Yoshimura T, Suzuki H, Sakata R, Fujita Y, Ichikawa S, Iwai N, Tamakoshi A, Grp JS (2005) Coffee and risk of death from hepatocellular carcinoma in a large cohort study in Japan. Br J Cancer 93(5):607–610

Kwo P, Gane EJ, Peng CY, Pearlman B, Vierling JM, Serfaty L, Buti M, Shafran S, Stryszak P, Lin L, Gress J, Black S, Dutko FJ, Robertson M, Wahl J, Lupinacci L, Barr E, Haber B (2017) Effectiveness of elbasvir and grazoprevir combination, with or without ribavirin, for treatment-experienced patients with chronic hepatitis C infection. Gastroenterology 152(1):164

Lafaro KJ, Demirjian AN, Pawlik TM (2015) Epidemiology of hepatocellular carcinoma. Surg Oncol Clin N Am 24(1):1

Lan KH, Sheu ML, Hwang SJ, Yen SH, Chen SY, Wu JC, Wang YJ, Kato N, Omata M, Chang FY, Lee SD (2002) HCVNS5A interacts with p53 and inhibits p53-mediated apoptosis. Oncogene 21(31):4801–4811

Lauer GM, Barnes E, Lucas M, Timm J, Ouchi K, Kim AY, Day CL, Robbins GK, Casson DR, Reiser M, Dusheiko G, Allen TM, Chung RT, Walker BD, Klenerman P (2004) High resolution analysis of cellular immune responses in resolved and persistent hepatitis C virus infection. Gastroenterology 127(3):924–936

Law JL, Chen C, Wong J, Hockman D, Santer DM, Frey SE, Belshe RB, Wakita T, Bukh J, Jones CT, Rice CM, Abrignani S, Tyrrell DL, Houghton M (2013) A hepatitis C virus (HCV) vaccine comprising envelope glycoproteins gpE1/gpE2 derived from a single isolate elicits broad cross-genotype neutralizing antibodies in humans. PLoS ONE 8(3):e59776

Lawitz E, Gane E, Pearlman B, Tam E, Ghesquiere W, Guyader D, Alric L, Bronowicki JP, Lester L, Sievert W, Ghalib R, Balart L, Sund F, Lagging M, Dutko F, Shaughnessy M, Hwang P, Howe AYM, Wahl J, Robertson M, Barr E, Haber B (2015) Efficacy and safety of 12 weeks versus 18 weeks of treatment with grazoprevir (MK-5172) and elbasvir (MK-8742) with or without ribavirin for hepatitis C virus genotype 1 infection in previously untreated patients with cirrhosis and patients with previous null response with or without cirrhosis (C-WORTHY): a randomised, open-label phase 2 trial. Lancet 385(9973):1075–1086

Lechmann M, Murata K, Satoi J, Vergalla J, Baumert TF, Liang TJ (2001) Hepatitis C virus-like particles induce virus-specific humoral and cellular immune responses in mice. Hepatology 34 (2):417–423

Lechner F, Wong DK, Dunbar PR, Chapman R, Chung RT, Dohrenwend P, Robbins G, Phillips R, Klenerman P, Walker BD (2000) Analysis of successful immune responses in persons infected with hepatitis C virus. J Exp Med 191(9):1499–1512

Lewis S, Roayaie S, Ward SC, Shyknevsky I, Jibara G, Taouli B (2013) Hepatocellular carcinoma in chronic hepatitis C in the absence of advanced fibrosis or cirrhosis. Am J Roentgenol 200(6): W610–W616

Li X, Hui A-M, Sun L, Hasegawa K, Torzilli G, Minagawa M, Takayama T, Makuuchi M (2004) p16INK4A hypermethylation is associated with hepatitis virus infection, age, and gender in hepatocellular carcinoma. Clin Cancer Res: Off J Am Assoc Cancer Res 10(22):7484–7489

Liver, EAS (2018). EASL recommendations on treatment of hepatitis C 2018. J Hepatol 69 (2):461–511

Lok AS, Seeff LB, Morgan TR, di Bisceglie AM, Sterling RK, Curto TM, Everson GT, Lindsay KL, Lee WM, Bonkovsky HL, Dienstag JL, Ghany MG, Morishima C, Goodman ZD, H.-C. T. Group (2009) Incidence of hepatocellular carcinoma and associated risk factors in hepatitis C-related advanced liver disease. Gastroenterology 136(1):138–148

Maki A, Kono H, Gupta M, Asakawa M, Suzuki T, Matsuda M, Fujii H, Rusyn I (2007) Predictive power of biomarkers of oxidative stress and inflammation in patients with hepatitis C virus-associated hepatocellular carcinoma. Ann Surg Oncol 14(3):1182–1190

Marcellin P, Boyer N, Gervais A, Martinot M, Pouteau M, Castelnau C, Kilani A, Areias J, Auperin A, Benhamou JP, Degott C, Erlinger S (1997) Long-term histologic improvement and loss of detectable intrahepatic HCV RNA in patients with chronic hepatitis C and sustained response to interferon-alpha therapy. Ann Intern Med 127(10):875–881

Martinez-Donato G, Amador-Canizares Y, Alvarez-Lajonchere L, Guerra I, Perez A, Dubuisson J, Wychowsk C, Musacchio A, Aguilar D, Duenas-Carrera S (2014) Neutralizing antibodies and broad, functional T cell immune response following immunization with hepatitis C virus proteins-based vaccine formulation. Vaccine 32(15):1720–1726

Maruoka D, Imazeki F, Arai M, Kanda T, Fujiwara K, Yokosuka O (2012) Long-term cohort study of chronic hepatitis C according to interferon efficacy. J Gastroenterol Hepatol 27 (2):291–299

Meissner EG, Kohli A, Virtaneva K, Sturdevant D, Martens C, Porcella SF, McHutchison JG, Masur H, Kottilil S (2016) Achieving sustained virologic response after interferon-free hepatitis C virus treatment correlates with hepatic interferon gene expression changes independent of cirrhosis. J Viral Hepatitis 23(7):496–505

Mendenhall CL, Seeff L, Diehl AM, Ghosn SJ, French SW, Gartside PS, Rouster SD, Buskell-Bales Z, Grossman CJ, Roselle GA (1991) Antibodies to hepatitis B virus and hepatitis C virus in alcoholic hepatitis and cirrhosis: their prevalence and clinical relevance. The VA Cooperative Study Group (No. 119). Hepatology (Baltimore, Md) 14(4 Pt 1):581–589

Messina JP, Humphreys I, Flaxman A, Brown A, Cooke GS, Pybus OG, Barnes E (2015) Global distribution and prevalence of hepatitis C virus genotypes. Hepatology 61(1):77–87

Midgard H, Bjoro B, Maeland A, Konopski Z, Kileng H, Damas JK, Paulsen J, Heggelund L, Sandvei PK, Ringstad JO, Karlsen LN, Stene-Johansen K, Pettersson JH, Dorenberg DH, Dalgard O (2016) Hepatitis C reinfection after sustained virological response. J Hepatol 64 (5):1020–1026

Mikkelsen M, Holst PJ, Bukh J, Thomsen AR, Christensen JP (2011) Enhanced and sustained CD8+ T cell responses with an adenoviral vector-based hepatitis C virus vaccine encoding NS3 linked to the MHC class II chaperone protein invariant chain. J Immunol 186(4):2355–2364

Morgan RL, Baack B, Smith BD, Yartel A, Pitasi M, Falck-Ytter Y (2013) Eradication of hepatitis C virus infection and the development of hepatocellular carcinoma a meta-analysis of observational studies. Ann Int Med 158(5):329

Moriya K, Fujie H, Shintani Y, Yotsuyanagi H, Tsutsumi T, Ishibashi K, Matsuura Y, Kimura S, Miyamura T, Koike K (1998) The core protein of hepatitis C virus induces hepatocellular carcinoma in transgenic mice. Nat Med 4(9):1065–1067

Munakata T, Nakamura M, Liang YQ, Li K, Lemon SM (2005) Down-regulation of the retinoblastoma tumor suppressor by the hepatitis C virus NS5B RNA-dependent RNA polymerase. Proc Natl Acad Sci USA 102(50):18159–18164

Murata K, Lechmann M, Qiao M, Gunji T, Alter HJ, Liang TJ (2003) Immunization with hepatitis C virus-like particles protects mice from recombinant hepatitis C virus-vaccinia infection. Proc Natl Acad Sci U S A 100(11):6753–6758

Nault JC, Colombo M (2016) Hepatocellular carcinoma and direct acting antiviral treatments: controversy after the revolution. J Hepatol 65(4):663–665

Netter HJ, Macnaughton TB, Woo WP, Tindle R, Gowans EJ (2001) Antigenicity and immunogenicity of novel chimeric hepatitis B surface antigen particles with exposed hepatitis C virus epitopes. J Virol 75(5):2130–2141

Netter HJ, Woo WP, Tindle R, Macfarlan RI, Gowans EJ (2003) Immunogenicity of recombinant HBsAg/HCV particles in mice pre-immunised with hepatitis B virus-specific vaccine. Vaccine 21(21–22):2692–2697

Ng TI, Krishnan P, Pilot-Matias T, Kati W, Schnell G, Beyer J, Reisch T, Lu L, Dekhtyar T, Irvin M, Tripathi R, Maring C, Randolph JT, Wagner R, Collins C (2017). In vitro antiviral activity and resistance profile of the next-generation hepatitis C virus NS5A inhibitor pibrentasvir. Antimicrobial Agents Chemotherapy 61(5)

Ng TI, Tripathi R, Reisch T, Lu L, Middleton T, Hopkins TA, Pithawalla R, Irvin M, Dekhtyar T, Krishnan P, Schnell G, Beyer J, McDaniel KF, Ma J, Wang G, Jiang L-J, Or YS, Kempf D, Pilot-Matias T, Collins C (2018) In vitro antiviral activity and resistance profile of the next-generation hepatitis C virus NS3/4A protease inhibitor glecaprevir. Antimicrobial Agents Chemotherapy 62(1)

Ng V, Saab S (2011) Effects of a sustained virologic response on outcomes of patients with chronic hepatitis C. Clin Gastroenterol Hepatol 9(11):923–930

Osburn WO, Snider AE, Wells BL, Latanich R, Bailey JR, Thomas DL, Cox AL, Ray SC (2014) Clearance of hepatitis C infection is associated with the early appearance of broad neutralizing antibody responses. Hepatology 59(6):2140–2151

Park CY, Choi SH, Kang SM, Kang JI, Ahn BY, Kim H, Jung G, Choi KY, Hwang SB (2009) Nonstructural 5A protein activates beta-catenin signaling cascades: implication of hepatitis C virus-induced liver pathogenesis. J Hepatol 51(5):853–864

Patel K, Muir AJ, McHutchison JG (2006) Diagnosis and treatment of chronic hepatitis C infection. BMJ (Clin Res Ed) 332(7548):1013–1017

Patient R, Hourioux C, Vaudin P, Pages JC, Roingeard P (2009) Chimeric hepatitis B and C viruses envelope proteins can form subviral particles: implications for the design of new vaccine strategies. New Biotechnol 25(4):226–234

Pawlotsky JM (2016) Hepatitis C virus resistance to direct-acting antiviral drugs in interferon-free regimens. Gastroenterology 151(1):70–86

Peng L, Liang DY, Tong WY, Li JH, Yuan ZH (2010) Hepatitis C virus NS5A Activates the mammalian target of rapamycin (mTOR) pathway, contributing to cell survival by disrupting the interaction between FK506-binding protein 38 (FKBP38) and mTOR. J Biol Chem 285 (27):20870–20881

Pestka JM, Zeisel MB, Blaser E, Schurmann P, Bartosch B, Cosset F-L, Patel AH, Meisel H, Baumert J, Viazov S, Rispeter K, Blum HE, Roggendorf M, Baumert TF (2007) Rapid induction of virus-neutralizing antibodies and viral clearance in a single-source outbreak of hepatitis C. Proc Natl Acad Sci USA 104(14):6025–6030

Petrick JL, Braunlin M, Laversanne M, Valery PC, Bray F, McGlynn KA (2016) International trends in liver cancer incidence, overall and by histologic subtype, 1978–2007. Int J Cancer 139(7):1534–1545

Poordad F, Dieterich D (2012) Treating hepatitis C: current standard of care and emerging direct-acting antiviral agents. J Viral Hepatitis 19(7):449–464

Poordad F, McCone J Jr, Bacon BR, Bruno S, Manns MP, Sulkowski MS, Jacobson IM, Reddy KR, Goodman ZD, Boparai N, DiNubile MJ, Sniukiene V, Brass CA, Albrecht JK, Bronowicki J-P, Investigators S (2011) Boceprevir for untreated chronic HCV genotype 1 infection. N Engl J Med 364(13):1195–1206

Poordad F, Pol S, Asatryan A, Buti M, Shaw D, Hezode C, Felizarta F, Reindollar RW, Gordon SC, Pianko S, Fried MW, Bernstein DE, Gallant J, Lin C-W, Lei Y, Ng TI, Krishnan P, Kopecky-Bromberg S, Kort J, Mensa FJ (2018) Glecaprevir/Pibrentasvir in patients with hepatitis C virus genotype 1 or 4 and past direct-acting antiviral treatment failure. Hepatology 67(4):1253–1260

Poynard T, Bedossa P, Opolon P (1997) Natural history of liver fibrosis progression in patients with chronic hepatitis C. The OBSVIRC, METAVIR, CLINIVIR, and DOSVIRC groups. Lancet 349(9055):825–832

Pradat P, Tillmann HL, Sauleda S, Braconier JH, Saracco G, Thursz M, Goldin R, Winkler R, Alberti A, Esteban JI, Hadziyannis S, Rizzetto M, Thomas H, Manns MP, Trepo C, Grp H (2007) Long-term follow-up of the hepatitis CHENCORE cohort: response to therapy and occurrence of liver-related complications. J Viral Hepatitis 14(8):556–563

Pyrsopoulos N, Trilianos P, Lingiah VA, Fung P, Punnoose M (2018) The safety and efficacy of ledipasvir/sofosbuvir with or without ribavirin in the treatment of orthotopic liver transplant recipients with recurrent hepatitis C: real-world data. Eur J Gastroenterol Hepatol 30(7):761–765

Qiu W, Wang XW, Leibowitz B, Yang WC, Zhang L, Yu J (2011) PUMA-mediated apoptosis drives chemical hepatocarcinogenesis in mice. Hepatology 54(4):1249–1258

Raimondi S, Bruno S, Mondelli MU, Maisonneuve P (2009) Hepatitis C virus genotype 1b as a risk factor for hepatocellular carcinoma development: a meta-analysis. J Hepatol 50(6):1142–1154

Ravi S, Axley P, Jones D, Kodali S, Simpson H, McGuire BM, Singal AK (2017) Unusually high rates of hepatocellular carcinoma after treatment with direct-acting antiviral therapy for hepatitis C related cirrhosis. Gastroenterology 152(4):911–912

Razali K, Thein HH, Bell J, Cooper-Stanbury M, Dolan K, Dore G, George J, Kaldor J, Karvelas M, Li J, Maher L, McGregor S, Hellard M, Poeder F, Quaine J, Stewart K, Tyrrell H,

Weltman M, Westcott O, Wodak A, Law M (2007) Modelling the hepatitis C virus epidemic in Australia. Drug Alcohol Depend 91(2–3):228–235

Reig M, Marino Z, Perello C, Inarrairaegui M, Ribeiro A, Lens S, Diaz A, Vilana R, Darnell A, Varela M, Sangro B, Calleja JL, Forns X, Bruix J (2016) Unexpected high rate of early tumor recurrence in patients with HCV-related HCC undergoing interferon-free therapy. J Hepatol 65 (4):719–726

Rein DB, Wittenborn JS, Weinbaum CM, Sabin M, Smith BD, Lesesne SB (2011) Forecasting the morbidity and mortality associated with prevalent cases of pre-cirrhotic chronic hepatitis C in the United States. Diges Liver Dis 43(1):66–72

Rockstroh JK, Nelson M, Katlama C, Lalezari J, Mallolas J, Bloch M, Matthews GV, Saag MS, Zamor PJ, Orkin C, Gress J, Klopfer S, Shaughnessy M, Wahl J, Nguyen BY, Barr E, Platt HL, Robertson MN, Sulkowski M (2015) Efficacy and safety of grazoprevir (MK-5172) and elbasvir (MK-8742) in patients with hepatitis C virus and HIV co-infection (C-EDGE CO-INFECTION): a non-randomised, open-label trial. Lancet HIV 2(8):e319–e327

Roth D, Nelson DR, Bruchfeld A, Liapakis A, Silva M, Monsour H, Martin P, Pol S, Londono MC, Hassanein T, Zamor PJ, Zuckerman E, Wan S, Jackson B, Nguyen BY, Robertson M, Barr E, Wahl J, Greaves W (2015) Grazoprevir plus elbasvir in treatment-naive and treatment-experienced patients with hepatitis C virus genotype 1 infection and stage 4–5 chronic kidney disease (the C-SURFER study): a combination phase 3 study. Lancet 386 (10003):1537–1545

Sacks-Davis R, Aitken CK, Higgs P, Spelman T, Pedrana AE, Bowden S, Bharadwaj M, Nivarthi UK, Suppiah V, George J, Grebely J, Drummer HE, Hellard M (2013) High rates of hepatitis C virus reinfection and spontaneous clearance of reinfection in people who inject drugs: a prospective cohort study. PLoS ONE 8(11):e80216

Schulze Zur Wiesch J, Ciuffreda D, Lewis-Ximenez L, Kasprowicz V, Nolan BE, Streeck H, Aneja J, Reyor LL, Allen TM, Lohse AW, McGovern B, Chung RT, Kwok WW, Kim AY, Lauer GM (2012) Broadly directed virus-specific CD4+ T cell responses are primed during acute hepatitis C infection, but rapidly disappear from human blood with viral persistence. J Exp Med 209(1) 61–75

Scott N, McBryde E, Vickerman P, Martin NK, Stone J, Drummer H, Hellard M (2015) The role of a hepatitis C virus vaccine: modelling the benefits alongside direct-acting antiviral treatments. BMC Med 13:198

Serfaty L, Aumaitre H, Chazouilleres O, Bonnand AM, Rosmorduc O, Poupon RE, Poupon R (1998) Determinants of outcome of compensated hepatitis C virus-related cirrhosis. Hepatology (Baltimore, Md) 27(5):1435–1440

Shi J, Zhu L, Liu S, Xie WF (2005) A meta-analysis of case-control studies on the combined effect of hepatitis B and C virus infections in causing hepatocellular carcinoma in China. Br J Cancer 92(3):607–612

Simmons B, Saleem J, Hill A, Riley RD, Cooke GS (2016) Risk of late relapse or reinfection with hepatitis C virus after achieving a sustained virological response: a systematic review and meta-analysis. Clin Infect Dis 62(6):683–694

Singal AK, Freeman DH, Anand BS (2010) Meta-analysis: interferon improves outcomes following ablation or resection of hepatocellular carcinoma. Aliment Pharmacol Ther 32 (7):851–858

Smyk-Pearson S, Tester IA, Lezotte D, Sasaki AW, Lewinsohn DM, Rosen HR (2006) Differential antigenic hierarchy associated with spontaneous Recovery from hepatitis C virus infection: implications for vaccine design. J Infect Dis 194:454–463

Sofia MJ, Bao D, Chang W, Du J, Nagarathnam D, Rachakonda S, Reddy PG, Ross BS, Wang P, Zhang H-R, Bansal S, Espiritu C, Keilman M, Lam AM, Steuer HMM, Niu C, Otto MJ, Furman PA (2010) Discovery of a beta-D-2′-Deoxy-2′-alpha-fluoro-2′-beta-C-methyluridine nucleotide prodrug (PSI-7977) for the treatment of hepatitis C virus. J Med Chem 53 (19):7202–7218

Steinmann D, Barth H, Gissler B, Schurmann P, Adah MI, Gerlach JT, Pape GR, Depla E, Jacobs D, Maertens G, Patel AH, Inchauspe G, Liang TJ, Blum HE, Baumert TF (2004) Inhibition of hepatitis C virus-like particle binding to target cells by antiviral antibodies in acute and chronic hepatitis C. J Virol 78(17):9030–9040

Stone J, Martin NK, Hickman M, Hellard M, Scott N, McBryde E, Drummer H, Vickerman P (2016) The potential impact of a hepatitis C vaccine for people who inject drugs: is a vaccine needed in the age of direct-acting antivirals? PLoS ONE 11(5):e0156213

Summa V, Ludmerer SW, McCauley JA, Fandozzi C, Burlein C, Claudio G, Coleman PJ, DiMuzio JM, Ferrara M, Di Filippo M, Gates AT, Graham DJ, Harper S, Hazuda DJ, McHale C, Monteagudo E, Pucci V, Rowley M, Rudd MT, Soriano A, Stahlhut MW, Vacca JP, Olsen DB, Liverton NJ, Carroll SS (2012) MK-5172, a selective inhibitor of hepatitis C virus NS3/4a protease with broad activity across genotypes and resistant variants. Antimicrob Agents Chemother 56(8):4161–4167

Sun C-A, Wu D-M, Lin C-C, Lu S-N, You S-L, Wang L-Y, Wu M-H, Chen C-J (2003) Incidence and cofactors of hepatitis C virus-related hepatocellular carcinoma: a prospective study of 12,008 men in Taiwan. Am J Epidemiol 157(8):674–682

Swadling L, Capone S, Antrobus RD, Brown A, Richardson R, Newell EW, Halliday J, Kelly C, Bowen D, Fergusson J, Kurioka A, Ammendola V, Del Sorbo M, Grazioli F, Esposito ML, Siani L, Traboni C, Hill A, Colloca S, Davis M, Nicosia A, Cortese R, Folgori A, Klenerman P, Barnes E (2014) A human vaccine strategy based on chimpanzee adenoviral and MVA vectors that primes, boosts, and sustains functional HCV-specific T cell memory. Science translational medicine 6(261):261ra153

Tada T, Kumada T, Toyoda H, Mizuno K, Sone Y, Kataoka S, Hashinokuchi S (2017) Improvement of liver stiffness in patients with hepatitis C virus infection who received direct-acting antiviral therapy and achieved sustained virological response. J Gastroenterol Hepatol 32(12):1982–1988

Taylor JG, Appleby T, Barauskas O, Chen X, Dvory-Sobol H, Gong R, Lee J, Nejati E, Schultz B, Wang Y, Yang C, Yu M, Zipfel S, Chan K (2015) Preclinical profile of the PAN-genotypic HCV NS3/4A protease inhibitor GS-9857. J Hepatol 62:S681–S681

Tholey DM, Ahn J (2015) Impact of hepatitis C virus infection on hepatocellular carcinoma. Gastroenterol Clin North Am 44(4):761

Thomas H, Foster G, Platis D (2003) Mechanisms of action of interferon and nucleoside analogues. J Hepatol 39(Suppl 1):S93–S98

Tsai P-C, Huang C-F, Yu M-L (2017) Unexpected early tumor recurrence in patients with hepatitis C virus-related hepatocellular carcinoma undergoing interferon-free therapy: Issue of the interval between HCC treatment and antiviral therapy. J Hepatol 66(2):464

Tsukuma H, Hiyama T, Tanaka S, Nakao M, Yabuuchi T, Kitamura T, Nakanishi K, Fujimoto I, Inoue A, Yamazaki H, Kawashima T (1993) Risk-factors for hepatocellular-carcinoma among patients with chronic liver-disease. N Engl J Med 328(25):1797–1801

van der Meer AJ, Veldt BJ, Feld JJ, Wedemeyer H, Dufour JF, Lammert F, Duarte-Rojo A, Heathcote EJ, Manns MP, Kuske L, Zeuzem S, Hofmann WP, de Knegt RJ, Hansen BE, Janssen HLA (2012) Association between sustained virological response and all-cause mortality among patients with chronic hepatitis C and advanced hepatic fibrosis. Jama-Journal of the American Medical Association 308(24):2584–2593

Vermehren J, Park JS, Jacobson I, Zeuzem S (2018) Challenges and perspectives of direct antivirals for the treatment of hepatitis C virus infection. J Hepatol

Vietheer PT, Boo I, Drummer HE, Netter HJ (2007) Immunizations with chimeric hepatitis B virus-like particles to induce potential anti-hepatitis C virus neutralizing antibodies. Antivir Ther 12(4):477–487

Wakai K, Kurozawa Y, Shibata A, Fujita Y, Kotani K, Ogimoto I, Naito M, Nishio K, Suzuki H, Yoshimura T, Tamakoshi A, Grp JS (2007) Liver cancer risk, coffee, and hepatitis C virus infection: a nested case-control study in Japan. Br J Cancer 97(3):426–428

Wang C, Ji D, Chen J, Shao Q, Li B, Liu JL, Wu V, Wong A, Wang YD, Zhang XY, Lu L, Wong C, Tsang S, Zhang Z, Sun J, Hou JL, Chen GF, Lau G (2017) Hepatitis due to reactivation of hepatitis B virus in endemic areas among patients with hepatitis C treated with direct-acting antiviral agents. Clin Gastroenterol Hepatol 15(1):132–136

Waziry R, Hajarizadeh B, Grebely J, Amin J, Law M, Danta M, George J, Dore GJ (2017) Hepatocellular carcinoma risk following direct-acting antiviral HCV therapy: a systematic review, meta-analyses, and meta-regression. J Hepatol 67(6):1204–1212

Westbrook RH, Dusheiko G (2014) Natural history of hepatitis C. J Hepatol 61:S58–S68

Wiley TE, McCarthy M, Breidi L, Layden TJ (1998) Impact of alcohol on the histological and clinical progression of hepatitis C infection. Hepatology (Baltimore, Md.) 28(3):805–809

Woo WP, Doan T, Herd KA, Netter HJ, Tindle RW (2006) Hepatitis B surface antigen vector delivers protective cytotoxic T-lymphocyte responses to disease-relevant foreign epitopes. J Virol 80(8):3975–3984

Wyles D, Poordad F, Wang S, Alric L, Felizarta F, Kwo PY, Maliakkal B, Agarwal K, Hassanein T, Weilert F, Lee SS, Kort J, Lovell SS, Liu R, Lin C-W, Pilot-Matias T, Krishnan P, Mensa FJ (2017) Glecaprevir/pibrentasvir for hepatitis C virus genotype 3 patients with cirrhosis and/or prior treatment experience: a partially randomized phase 3 clinical trial. Hepatology (Baltimore, Md.)

Yang JD, Aqel BA, Pungpapong S, Gores GJ, Roberts LR, Leise MD (2016) Direct acting antiviral therapy and tumor recurrence after liver transplantation for hepatitis C-associated hepatocellular carcinoma. J Hepatol 65(4):859–860

Yu G, Chi XM, Wu RH, Wang XM, Gao XZ, Kong F, Feng XW, Gao YD, Huang XX, Jin JL, Qi Y, Tu ZK, Sun B, Zhong J, Pan Y, Niu JQ (2015) Replication inhibition of hepatitis B virus and hepatitis C virus in co-infected patients in chinese population. Plos One 10(9)

Zeuzem S, Foster GR, Wang S, Asatryan A, Gane E, Feld JJ, Asselah T, Bourliere M, Ruane PJ, Wedemeyer H, Pol S, Flisiak R, Poordad F, Chuang WL, Stedman CA, Flamm S, Kwo P, Dore GJ, Sepulveda-Arzola G, Roberts SK, Soto-Malave R, Kaita K, Puoti M, Vierling J, Tam E, Vargas HE, Bruck R, Fuster F, Paik SW, Felizarta F, Kort J, Fu B, Liu R, Ng TI, Pilot-Matias T, Lin CW, Trinh R, Mensa FJ (2018) Glecaprevir-Pibrentasvir for 8 or 12 weeks in HCV genotype 1 or 3 infection. N Engl J Med 378(4):354–369

Zeuzem S, Ghalib R, Reddy KR, Pockros PJ, Ben Ari Z, Zhao Y, Brown DD, Wan SY, DiNubile MJ, Nguyen BY, Robertson MN, Wahl J, Barr E, Butterton JR (2015) Grazoprevir-elbasvir combination therapy for treatment-naive cirrhotic and noncirrhotic patients with chronic hepatitis C virus genotype 1, 4, or 6 infection a randomized trial. Ann Intern Med 163(1):1–13

High-Risk Human Papillomaviruses and DNA Repair

Kavi Mehta and Laimonis Laimins

1 Introduction

Papillomaviruses (PVs) are non-enveloped DNA viruses that are the causative agents of benign and malignant epithelial lesions. In the 1930s, investigators found that filtered extracts of papillomas from cottontail rabbits could establish new infections in uninfected rabbits (Zhou et al. 2013; Lowy 2007). While some rabbits cleared these lesions, others developed squamous cell carcinomas. In the 1940s, Peyton Rous determined that treating cottontail rabbit papillomavirus lesions with coal tar resulted in the rapid appearance of carcinomas suggesting that other factors besides viral infection could influence progression (Zhou et al. 2013; Lowy 2007). Furthermore, extracts from papillomas of the mouth of rabbits were not able to produce lesions in genitalia indicating that these viruses exhibited tissue tropism. Subsequently, bovine papillomavirus 1 was found to induce large lesions in cows and became a main focus of study for decades due to the ease in harvesting virions from these lesions and BPV's ability to transform mouse fibroblasts (Lowy 2007; Kawai and Akira 2011). Around the turn of the century, it was observed that cutaneous warts from human hands were not transmissible to the genitalia; however, it was not until the seventies that it was determined that different papillomavirus types were responsible and exhibited tissue tropism. While some papillomavirus induced warts remained benign, others had the ability to progress to squamous cell carcinoma. These findings preceded the realization that cervical neoplasias not only resembled viral papillomas, but contained high amounts of human papillomavirus DNAs (HPV) (Zhou et al. 2013; Lowy 2007).

K. Mehta · L. Laimins (✉)
Department of Microbiology-Immunology, Northwestern University,
Feinberg School of Medicine, Chicago, IL, USA
e-mail: l-laimins@northwestern.edu

© Springer Nature Switzerland AG 2021
T.-C. Wu et al. (eds.), *Viruses and Human Cancer*, Recent Results
in Cancer Research 217, https://doi.org/10.1007/978-3-030-57362-1_7

Genital human papillomaviruses (HPVs) are spread primarily through sexual contact, and their life cycles are intimately linked to the differentiation of squamous epithelia. Over 75% of sexually active individuals have been infected with genital HPVs at some point in their lives (Pivarcsi et al. 2004; Castellsagué 2008). Over 200 types of HPVs have been identified and approximately 40 infect the genital tract. Of these, 12 are considered high-risk types and include HPV 16, 18, 31, 33, 35, 39, 45, 51, 52, 56, 58, and 59. Low-risk HPV types (e.g., HPV 6 or 11) induce benign lesions such as genital warts and laryngeal papillomas, while high-risk types are the etiological agents of cervical cancer, anogenital cancers, as well as many head and neck squamous cell carcinomas (HNSCCs). High-risk types are responsible for over 7% of all cancers worldwide (Zhou et al. 2013; Lowy 2007; Forman et al. 2012), and over 99% of cases of cervical cancer are HPV-associated (Lowy 2007; Kawai and Akira 2011; Moody and Laimins 2008). In the US, half of the approximately 10,000 women diagnosed with cervical cancer each year will die from this disease (Pivarcsi et al. 2004; Castellsagué 2008; Parkin and Bray 2006). Recent studies suggest that high-risk HPVs are responsible for over 60% of oropharyngeal cancers, and this number has been increasing over the last decade (Chaturvedi 2012; Robinson et al. 2013; Schiffman et al. 2016). Three prophylactic HPV vaccines have been developed that target both high-risk and low-risk types. The vaccines consist of multivalent virus-like particles from specific high-risk and common low-risk HPV types. The bivalent version protects against HPV 16 and 18 infections, the quadrivalent version targets HPV 16, 18, 11, and 6, while the nanovalent version is directed against HPV 6, 11, 16, 18, 31, 33, 45, 52, and 58. These vaccines are highly effective in blocking initial infection by HPV, but have no effect on pre-existing infections (Prue et al. 2017; Lu et al. 2011). There are no effective treatments outside of surgery or cryotherapy for treating existing HPV lesions, and therefore, it remains critical to understand the viral life cycle to uncover how viral infection progresses to malignancy and to identify new therapeutics. One of the host pathways that HPV hijacks is the DNA damage response (DDR) which is required for HPV's differentiation-dependent amplification. Before describing the DDR and its role in HPV infection, it is first important to discuss the multiple factors that regulate the viral life cycle.

2   Genome Organization

HPV genomes consist of double-stranded circular DNAs that are approximately 8 kB is size and encode for between 6 and 8 open-reading frames. Viral transcription takes place from a single DNA strand, and early gene expression is regulated in large part by alternative splicing (Fig. 1). Upon infection, viral genomes establish themselves as nuclear extrachromosomal elements or episomes, and maintenance of these elements is necessary for viral genome replication and persistence. A region 500–1000 base pairs upstream of the early coding sequences that is alternatively referred to as the upstream regulatory region (URR), the long

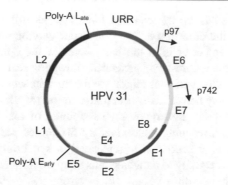

Fig. 1 Schematic representation of the HPV 31 genome. Circular map of HPV 31 genome identifying the upstream regulatory region (URR), which is a non-coding region that contains binding sites for multiple transcription factors and the origin of replication. The genome contains both an early (p97) and late promoter (p742), along with early and late polyadenylation sites. The late promoter is activated upon epithelial differentiation. The early genes include, E1, E2, E6, E7, E1^E4, E5, and E8^E2C. The two late capsid proteins are encoded by the L1 and L2 genes

control region (LCR), or the non-coding region (NCR), regulates early transcription and contains the viral origin of replication (McBride et al. 2012; Mighty and Laimins 2013). Early (E for early) transcription precedes productive viral replication and directs expression of polycistronic messages that encode for E1, E2, E1^E4, E5, E8^E2C, E6, and E7. The E1 and E2 proteins are DNA binding factors that bind to sequences in the URR and help to recruit cellular replication factors to viral origins. E2 also helps regulate early gene expression while E8^E2C acts primarily as a repressor. E6 and E7 are the two viral oncoproteins that control cell-cycle progression and allow for cells to remain active in the cell cycle upon differentiation to allow for genome amplification in suprabasal layers. E1^E4 and E5 regulate late viral events (Moody and Laimins 2008). Early viral transcripts are translated using a "leaky scanning," mechanism where the first open-reading frame is translated at a high rate while the sequential ORFs are translated at lower rates. This mechanism contributes to modulation of the levels of viral proteins, so that E6 and E7 are translated at high rates and others such as E5 are translated at very low levels (Remm et al. 1999). Upon differentiation, E1^E4, E5, L1, and L2 are expressed from the late viral promoter (p742) located in the E7 ORF (Moody and Laimins 2008; Beglin et al. 2009). Late viral genes L1 and L2 encode the major and minor capsid proteins of HPV and are critical for viral entry and egress. In many HPV-induced cancers, high-risk HPV genomes are found integrated into host DNAs. Integration leads to disruption of E2 expression which results in increased expression of the HPV encoded oncogenes E6 and E7.

Keratinocyte differentiation, HPV replication, and oncogenesis: The life cycle of human papillomavirus is closely linked to epithelial differentiation. Differentiated normal keratinocytes are divided into four distinct layers: the *stratum basale*, the *stratum spinosum*, the *stratum granulosum*, and the *stratum corneum*.

The basal layer is made up of cells that have not yet differentiated and remain proliferative, while the other three layers represent varying degrees of differentiation. Genital HPVs infect cells in the basal layer of the epithelium that become exposed through microabrasions generated through sexual activity (Fig. 2). Following entry, the viral genomes migrate to the nucleus where they associate with PML bodies which may help initiate viral transcription (McBride 2008). Within the nucleus, the virus rapidly undergoes several rounds of amplification using host replication machinery, reaching approximately 50 copies per cell. In persistently infected basal cells, viral genomes are replicated in synchrony with cellular replication and distributed equally to daughter cells at approximately 50–100 copies per cell. Evidence from two-dimensional gel electrophoresis suggests that genome replication during maintenance may occur through the formation of theta-structures (Flores and Lambert 1997). Cells in the basal layer can remain infected for years as they evade immune surveillance and provide a repository for production of new viruses. After cell division, one daughter cell moves away from the basal layer and begins the process of differentiation. Normal keratinocytes exit the cell cycle upon differentiation, but viral infection prevents cell-cycle exit, locking cells in G1 initially, but subsequently pushing cells to re-renter S/G2 where the genomes are amplified to approximately 1000 copies per cell. Coincident with amplification the capsid proteins, L1 and L2 are synthesized which self-assemble into icosahedrons and package viral DNA. Newly synthesized viruses are then released from cells in the *stratum corneum*.

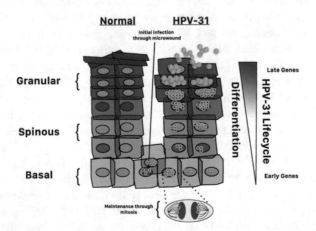

Fig. 2 HPV life cycle is intimately linked to epithelial differentiation. HPV infects cells in the basal layer that are exposed through microwounds and establishes its genomes at low copy number in these cells while expressing early genes. The HPV copy number in basal cells is maintained constant through cell division. Upon differentiation, one daughter cell moves away from the basal layer and begins differentiating leading to activation of the late promoter, amplification of the HPV genome, and subsequent assembly of progeny virions

Progression to malignancy often occurs in cells in which viral genomes have integrated into host chromosomes, and this leads to increased expression of the E6 and E7 oncoproteins. Little is known about what influences integration and how E6 and E7 contribute to this process though both can induce genetic instability in cells. In many cases of HPV-associated malignancies, HPV integrates at random fragile sites into the host genome retaining expression of only E6 and E7. Integration may contribute to genetic instability through the generation of double-stranded breaks and the activation of the ataxia telangiectasia mutated pathway (ATM) leading to mutations in the host genome (Kadaja et al. 2009).

3 Viral Oncoproteins

E6 and E7: The E6 and E7 proteins play major roles in manipulating the cell cycle in HPV-infected cells. Although present in low-risk HPV types, only high-risk E6 and E7 exhibit transformation activity. Although E6 and E7 are able individually to transform NIH-3T3 mouse fibroblasts, both are required for efficient immortalization of keratinocytes (Howley and Lowy 2007).

A major function of high-risk E7 is to bind and inhibit the activity of the retinoblastoma family of proteins, including pRb, p107, and p130. High-risk E7 proteins are approximately17 kDa in size and contain three conserved domains (CR1, CR2, and CR3) that share extensive homologies with adenovirus E1A (Zhang et al. 2006). High-risk E7 binds pRb ten-times more efficiently than the E7 of low-risk types (Münger et al. 1989). The CR1 and CR2 domains are both implicated in pRb binding and degradation with CR2 containing a conserved LXCXE motif. CR3 contains two zinc-finger domains that interact with the C-terminal domain of pRb and may play a role in stabilizing binding (Liu et al. 2006). E7 binds pRb and targets it for ubiquitin-mediated proteosomal degradation (Boyer et al. 1996). This degradation permits constitutive activation of the transcription factor E2F1 that regulates G1-S transition in both undifferentiated and differentiated cells (Moody and Laimins 2010). E2F normally associates with the Rb family members in G1, which represses its activation function and is released upon phosphorylation of Rb by CDK kinases in a cell-cycle dependent manner. E2F factors are bound to the promoters of genes that are normally expressed in S-phase, and release from Rb binding leads to activation of expression. p130 and p107 are also targets of E7 binding and act similar to Rb (Zhang et al. 2006). Interestingly, E7 also induces genetic instability by causing missegregation of chromosomes in a manner independent of its pRb binding through the degradation of the cyclin-dependent kinase inhibitor p21 as well as through alterations in centrosomes (Duensing and Münger 2003). In addition to its role in pRb family binding and degradation, E7 regulates E2Fs activity through its interactions with histone deacetylases (HDACS). HDACS remove acetyl groups from histones, and this results in heterochromatinization and repression of a DNA locus. E7 binds HDACS through the CR3, zinc-finger binding domain (Longworth and Laimins 2004).

As a by-product of altering cell-cycle progression, E7 expression leads to enhanced levels of p53 that can lead to apoptosis (Demers et al. 1994). To counteract the effects of elevated p53 levels, the high-risk E6 proteins have evolved to degrade p53. A major role of high-risk E6 in the HPV life cycle is to bind p53 and recruit it into a complex with the E3-ubiquitin ligase E6 associated protein (E6AP), which ubiquitinates p53, targeting it for proteosomal degradation. E6 can also bind the coactivators p300/CBP preventing p53 acetylation which also inhibits its transcription activation ability (Zimmermann et al. 1999; Gu et al. 1997). Another major target of p53 is p21, and E6 induced degradation of p53 prevents p21 mediated G1/S or G2/M checkpoint arrest, as well as apoptosis. This allows for HPV to replicate during an aberrant cell cycle. Recent observations demonstrate that HPV genomes containing knockout mutations in E6 are deficient in viral maintenance and knockdown of p53 restores episomal maintenance (Lorenz et al. 2013). Other studies have demonstrated that E6's role in viral maintenance and cell proliferation also relies on its PDZ binding motif (Lee and Laimins 2004; Nicolaides et al. 2011). It has been shown that E6 interacts with several PDZ domain containing proteins, including Dlg, MAGI-1/2/3, MUPP1, and Scribble all of which E6 targets for proteosomal degradation (Massimi et al. 2004). These proteins are thought to organize complexes of proteins that contribute to tumor suppressor activity. Another critical activity of E6 is the activation of htert, the catalytic subunit of telomerase. This is important for keratinocyte immortalization through direct interaction of E6 and E6AP with the htert promoter (Liu et al. 2009). Increased expression of either htert or c-myc independent of E6 expression readily duplicates E6's effects on htert, demonstrating its importance in immortalization (Galloway et al. 1998),(Liu et al. 2007). Recent studies have outlined a role for high-risk HPV E6 in regulating chromatin modifiers such as Tip60 and SET7 which inhibit transcriptional regulation by p53 indicating that E6 can control p53 expression at various stages (Vande Pol and Klingelhutz 2013). The high-risk E6 and E7 are sufficient to immortalize keratinocytes through the binding and degrading of p53 and pRb and the upregulation of telomerase activity. It is important to note, however, that in low-risk HPV types, immortalization does not occur due to E6 and E7's lower affinities for p53 and Rb, and E6's inability to activate htert expression (Van Doorslaer and Burk 2012).

E5: The HPV E5 protein is expressed in both early and late stages of the viral life cycle and associates with the Golgi apparatus, endoplasmic reticulum, as well as endosomal and nuclear membranes (Conrad et al. 1993; Hausen zur 2000; Ashrafi et al. 2005). Expression of E5 in mouse fibroblasts and human keratinocytes results in EGF-dependent proliferation suggesting that E5 may serve as a regulator of epidermal growth factor receptor (EGFR) activity, analogous to BPV E5's regulation of platelet-derived growth factor beta (PDGFR-β) that leads to transformation (Petti and Dimaio 1994; Straight et al. 1995). Transgenic mice expressing HPV 16 E5 under the control of the K14 promoter develop large tumors suggesting that E5 can act as an oncoprotein (Genther Williams et al. 2005). E5 also enhances the efficiency of E6 and E7 transformation but exhibits no transformation activity when expressed by itself (Valle and Banks 1995). Studies in COS cells demonstrate that

HPV 16 E5 is capable of interacting with a variety of transmembrane receptor proteins including EGFR and PDGFR-β. Furthermore, E5 is able to stabilize endosome acidification and cell surface EGFR expression, in turn stabilizing EGFR at endosomes upon stimulation with ligand (Straight et al. 1995; Hwang et al. 1995). EGFR stabilization may be associated with oncogenic activity due to enhanced EGFR-mediated mitogenic activity. Interestingly, HPV-31 genomes with E5 knockouts lose the ability to activate late viral transcription and amplification suggesting that E5 may contribute to regulation of late viral functions (Fehrmann et al. 2003; Genther et al. 2003). Recent studies indicate other targets for E5 including the ER-associated lipoprotein A4 that is linked to Akt-mediated proliferative capacity, and Bap-31 an ER-associated membrane complex shuttling protein, that among other functions shuttles MHC-I proteins through the ER. Both A4 and Bap31 are regulated by E5 leading to enhanced cellular proliferation (Regan and Laimins 2008; Halavaty et al. 2014).

4 The DNA Damage Response

The host DNA damage response (DDR) has evolved to repair single-stranded DNA (ssDNA) and double-stranded DNA (dsDNA) breaks as well as inter- and instra-strand cross-links induced by external damaging agents such as radiation or replication errors (Fradet-Turcotte et al. 2016). The DDR repairs thousands of lesions per cell per day and is vitally important for cell-cycle progression as well as to maintain genetic fidelity (Fradet-Turcotte et al. 2016; Wallace and Galloway 2014). If these lesions are not repaired, cells accumulate damage which leads to genetic instability that contributes to progression to malignancy. The PI3K-related protein kinases Ataxia-telengectasia mutated (ATM) and Ataxia-telengectasia Rad3 related (ATR), and DNA-dependent protein kinase (DNA-PK) are activated in response to different types of DNA breaks resulting in the activation of downstream signaling pathways. This activation leads to cell-cycle checkpoint arrest, repair, or apoptosis in the case of damage that cannot be repaired. ATR and ATM cross-talk as they share many downstream substrates while knockdown or mutation of either results in aberrant signaling and an impaired checkpoint response. Many cancers lack appropriate G1 checkpoint control and rely on ATR and ATM to fix DNA lesions that frequently arise. These pathways are also amplified in many cancers (Rundle et al. 2017).

The ATM pathway is activated in response to double-stranded breaks (DSBs) leading to the autophophosrylation of the ATM kinase at serine 1981. The autophosphorylation of ATM then recruits the MRN complex that is made up of Nijmegen Breakage Syndrome 1 (NBS1) Rad50 and Mre11 to the DSB. ATM also phosphorylates the modified histone H2AX at serine 139 known as γ-H2AX which bind to regions surrounding the lesion further amplifying the DDR. γ-H2AX in turn recruits mediator of DNA damage checkpoint protein 1 (MDC1) which is responsible for recruiting the ubiquitin ligases RNF8 and RNF168 to the DNA lesion as well as the non-homologous end-joining (NHEJ) and homologous recombination (HR) repair factors such as 53BP1 and BRCA1 to the lesion (Uckelmann and Sixma 2017; Mattiroli et al. 2012; Doil et al. 2009). 53BP1 and BRCA1 recruitment mediates the choice between which of the two repair pathways fixes the lesion (Fradet-Turcotte et al. 2016). NHEJ repair is the major repair mechanism in mammalian cells and occurs without extensive processing and the presence of the sister chromatid. p53BP1 is recruited to the ends of DSBs along with the Ku70-Ku80 complex that tethers the ends together (Fradet-Turcotte et al. 2016; Chapman et al. 2012). The ends are then ligated in an error-prone manner by DNA ligase-IV and X-ray repair cross-complementing protein 4 (XRCC4) (Chapman et al. 2012; Burma and Chen 2004). This process occurs primarily in the G1 phase.

Homologous recombination occurs in S/G2 phases as it requires the presence of the sister chromatid for homology-based repair and strand invasion. In S/G2 phases BRCA1, CtBP-interacting protein (CtIP) and the MRN complex are recruited to the DSB following ATM activation to start DNA end resection which is a necessary step for HR. 5′–3′ end resection by the exonuclease Exo1, Dna2 nuclease, and the helicase Sgs1 are required for the resection and establishment of ssDNA. This is followed by the recruitment of HR factors and sister strand invasion (Niu et al. 2010). The ssDNA is then coated with the ssDNA stabilizing heterotrimeric replication protein A (RPA). BRCA1 recruitment to the DSB promotes recruitment of BRCA2 through the BRCA1-BRCA2 partner and localizer of BRCA2 (PALB2) (Rohini Roy JCSNP 2012). This is followed by displacement of RPA from the ssDNA and recruitment of Rad51 which results in its polymerization and formation of Rad51 nucleofilaments that search for sister chromatid homology leading to repair. If the damage is irreparable, apoptosis is induced through the action of BRCA1 (Fradet-Turcotte et al. 2016; Yuan et al. 1999; Chen et al. 1998).

Although ATM has a major role in DNA repair, it also plays a role in regulating cell-cycle checkpoints in efforts to maintain integrity. Cell-cycle checkpoints provide cells with a means to delay replication to allow for repair or apoptosis and are regulated by an extensive signaling cascade. ATM's role in cell-cycle checkpoint control was first realized when it was found that ATM was required for p53 activation. p53 is the classical tumor suppressor that is important for regulating the G1/S-phase checkpoint (Derheimer and Kastan 2010). Upon DNA damage, activated ATM phosphorylates p53, mouse double minute 2 homolog (MDM2), as well as checkpoint Kinase 2 (CHK2) leading to activation of the kinase inhibitor p21. This results in the inhibition of cyclin-E/CDK complexes and halts the G1/S transition. ATM signaling also plays roles in S-phase and intra-S-phase arrest by activating several proteins including structural maintenance of chromosomes protein 1 (SMC-1), NBS1, Fanconi anemia group D2 protein (FANCD2), and CHK2.

ATR responds to ssDNA breaks induced by radiation, genotoxic stress, depleted nucleotide pools, as well as replication stress due to stalled replication forks (Cimprich and Cortez 2008; Saldivar et al. 2017). ATR is recruited to ssDNA, i.e., that has been recognized by RPA with its partner ATR-interacting protein (ATRIP) which contains an RPA binding domain. This also leads to the further recruitment of the proliferating cell nuclear antigen (PCNA) like Rad9-Rad1-Hus1 (9-1-1) complex that loads onto junctional dsDNA near the RPA-coated DNAs. The 9-1-1-complex recruits DNA topoisomerase 2-binding protein 1 (TOPBP1) which interacts with ATR and regulates its activation.

ATR activation leads to the phosphorylation of checkpoint protein 1 (CHK1) at two different sites, which is important for the regulation of cell-cycle checkpoints. This phosphorylation is mediated by Claspin which interacts with Rad17 part of the 9-1-1 clamp (Cimprich and Cortez 2008). Activation of CHK1 leads to dissociation from chromatin as well as signaling to the CDC25 phosphatases which prevent cell entry into mitosis (Smits et al. 2006). ATR also plays a role in coordinating origin firing and stabilizing stalled replication forks making it essential for cell viability even from very early embryonic stages (Shechter et al. 2004).

5 The DNA Damage Response and HPV

High-risk HPVs constitutively activate both the ATM and ATR DNA damage repair pathways, and this is necessary for productive replication in differentiating cells. The HPV E7 and E1 proteins can independently activate the ATM and ATR pathways. The E7 protein acts through the innate immune regulator STAT-5 which in turn activates ATM through its effects on the acetyltranferase Tip60. At the same time STAT-5 activates ATR through increased expression of the TopBP1 protein. Tip60 must acetylate ATM prior to its activation by phosphorylation and TopBP1 forms complexes with ATR resulting in its phosphorylation and recruitment to sites of DNA breaks. One by-product of this activation is the induction of genetic instability and as well as viral genome integration which may contribute to the development of virally induced cancers.

High-risk HPVs activate ATM and ATR pathways in both undifferentiated and differentiated cells. Importantly, small molecule inhibitors of either ATM or ATR prevent the differentiation-dependent amplification of HPV (Moody and Laimins 2009; Hong et al. 2015; Edwards et al. 2013). Inhibition of ATM or CHK2 prevents HPV amplification, and this results in reduced levels of Cdc25c which in turn regulates transition into G2, which is when HPV genome amplification occurs (Moody and Laimins 2009). HPV genomes are recruited to nuclear replication centers that contain many members of the DDR including, TopBP1, pATM, γ-H2AX, pCHK1, pCHK2, NBS1, BRCA1, pATR, Rad51, and FANCD2 (Kadaja et al. 2009; Wallace and Galloway 2014; Satsuka et al. 2015; Spriggs and Laimins 2017). In addition, shRNA knockdown studies have shown that Rad51, BRCA1, and NBS1 are required for differentiation-dependent amplification (Anacker et al. 2014). Interestingly, FANCD2, ATR and TopBP1, NBS1, and SMC1 are also required for efficient viral maintenance in undifferentiated cells (Hong et al. 2015; Spriggs and Laimins 2017; Anacker et al. 2014; Mehta et al. 2015). Many of these factors have been shown to bind to HPV genomes using ChIP analyses (Anacker et al. 2014; Gillespie et al. 2012).

Constitutive activation of the cell cycle by E6 and E7 leads to the depletion of free nucleotide pools, replication stress, and genomic instability due to the multiple origin firing and stalled replication forks (Bester et al. 2011). Recent studies indicate that high-risk E6 and E7 induce DNA breaks in both viral and cellular DNAs. Interestingly, homologous recombination repair factors are then preferentially recruited to viral DNAs to repair the breaks (Mehta and Laimins 2018). This preferential recruitment of DDR factors such as RAD51 and BRCA1 results in rapid repair of breaks and genome amplification that occurs in G2. This is a novel mechanism that links viral induction of DNA breaks to amplification of viral genomes as well as genetic instability in host chromsomes.

6 Summary

Human papillomaviruses are the causative agents of many anogenital cancers including almost all cervical cancers. In addition, over 60% of oropharyngeal cancers are associated with infection by high-risk HPVs, and the numbers are increasing rapidly in Western countries. While prophylactic vaccines have been developed that are highly effective in blocking initial infections, they have no effect

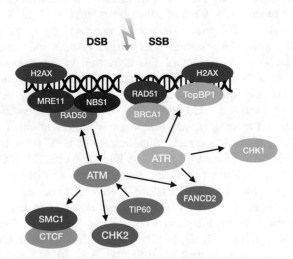

Fig. 3 Schematic of the DNA damage response pathway factors demonstrated to play a role in the HPV life cycle. The ATM pathway is activated in response to double-stranded breaks while ATR responds to single-strand DNA breaks. Activation of these pathways leads to checkpoint signaling, repair, and/or apoptosis. HPV nuclear replication foci contain members of these pathways including TopBP1, pATM, H2AX, pCHK1, CHK2, NBS1, BRCA1, ATR, Rad51, and FANCD2. Homologous recombination repair factors, Rad51, BRCA1, and NBS1 are all required for differentiation-dependent amplification of HPV. FANCD2, ATR, TopBP1, NBS1, and SMC1 are required for efficient maintenance of viral episomes in undifferentiated cells

on existing virally caused lesions. The life cycle of high-risk HPVs is dependent upon differentiation and activation of host DNA repair pathways. This includes both ATM and ATR pathways whose activation is required for both productive replication as well as stable maintenance of episomes. HPV proteins activate the DNA damage repair pathways through induction of DNA breaks in both cellular and viral DNAs. The preferential repair of breaks in viral genomes results in productive replication of HPV DNAs while at the same time promoting genetic instability in host chromosomes (Fig. 3).

Acknowledgements This work was supported by grants from the NCI to LAL (RO1CA059655 and RO1 CA142861). KM was supported by the Cellular and Molecular Basis of Disease Training Program (T32 NIH T32 GM08061).

References

Anacker DC, Gautam D, Gillespie KA et al (2014) Productive replication of human papillomavirus 31 requires DNA repair factor Nbs1. J Virol 88:8528–8544. https://doi.org/10.1128/JVI.00517-14

Ashrafi GH, Haghshenas MR, Marchetti B et al (2005) E5 protein of human papillomavirus type 16 selectively downregulates surface HLA class I. Int J Cancer 113:276–283. https://doi.org/10.1002/ijc.20558

Beglin M, Melar-New M, Laimins L (2009) Human papillomaviruses and the interferon response. J Interferon Cytokine Res 29:629–635. https://doi.org/10.1089/jir.2009.0075

Bester AC, Roniger M, Oren YS et al (2011) Nucleotide deficiency promotes genomic instability in early stages of cancer development. Cell 145:435–446. https://doi.org/10.1016/j.cell.2011.03.044

Boyer SN, Wazer DE, Band V (1996) E7 protein of human papilloma virus-16 induces degradation of retinoblastoma protein through the ubiquitin-proteasome pathway. Cancer Res 56:4620–4624

Burma S, Chen DJ (2004) Role of DNA–PK in the cellular response to DNA double-strand breaks. DNA Repair (Amst.)

Castellsagué X (2008) Natural history and epidemiology of HPV infection and cervical cancer. Gynecol Oncol

Chapman JR, Taylor M, Boulton SJ (2012) Playing the end game: DNA double-strand break repair pathway choice—science direct. Mol Cell

Chaturvedi AK (2012) Epidemiology and clinical aspects of HPV in head and neck cancers. Head Neck Pathol 6(Suppl 1):S16–S24. https://doi.org/10.1007/s12105-012-0377-0

Chen J, Silver DP, Walpita D et al (1998) Stable interaction between the products of the BRCA1 and BRCA2 tumor suppressor genes in mitotic and meiotic cells. Mol Cell 2:317–328

Cimprich KA, Cortez D (2008) ATR: an essential regulator of genome integrity. Nat Rev Mol Cell Biol 9:616–627. https://doi.org/10.1091/mbc.12.5.1199

Conrad M, Bubb VJ, Schlegel R (1993) The human papillomavirus type 6 and 16 E5 proteins are membrane-associated proteins which associate with the 16-kilodalton pore-forming protein. J Virol

Demers GW, Halbert CL, Galloway DA (1994) Elevated wild-type p53 protein levels in human epithelial cell lines immortalized by the human papillomavirus type 16 E7 gene. Virology 198:169–174. https://doi.org/10.1006/viro.1994.1019

Derheimer FA, Kastan MB (2010) Multiple roles of ATM in monitoring and maintaining DNA integrity. FEBS Lett 584:3675–3681. https://doi.org/10.1016/j.cmet.2006.10.002

Doil C, Mailand N, Bekker-Jensen S et al (2009) RNF168 binds and amplifies ubiquitin conjugates on damaged chromosomes to allow accumulation of repair proteins. Cell 136:435–446. https://doi.org/10.1016/j.cell.2008.12.041

Duensing S, Münger K (2003) Human papillomavirus type 16 E7 oncoprotein can induce abnormal centrosome duplication through a mechanism independent of inactivation of retinoblastoma protein family members. J Virol 77:12331–12335

Edwards TG, Helmus MJ, Koeller K et al (2013) Human papillomavirus episome stability is reduced by aphidicolin and controlled by DNA damage response pathways. J Virol 87:3979–3989. https://doi.org/10.1128/JVI.03473-12

Fehrmann F, Klumpp DJ, Laimins LA (2003) Human papillomavirus type 31 E5 protein supports cell cycle progression and activates late viral functions upon epithelial differentiation. J Virol 77:2819–2831

Flores ER, Lambert PF (1997) Evidence for a switch in the mode of human papillomavirus type 16 DNA replication during the viral life cycle. J Virol 71:7167–7179

Forman D, de Martel C, Lacey CJ et al (2012) Global burden of human papillomavirus and related diseases. Vaccine 30(Suppl 5):F12–F23. https://doi.org/10.1016/j.vaccine.2012.07.055

Fradet-Turcotte A, Sitz J, Grapton D, Orthwein A (2016) BRCA2 functions: from DNA repair to replication fork stabilization. Endocr Relat Cancer 23:T1–T17. https://doi.org/10.1530/ERC-16-0297

Galloway DA, Kiyono T, Foster SA et al (1998) Both Rb/p16INK4a inactivation and telomerase activity are required to immortalize human epithelial cells. Nature 396:84–88. https://doi.org/10.1038/23962

Genther SM, Sterling S, Duensing S et al (2003) Quantitative role of the human papillomavirus type 16 E5 gene during the productive stage of the viral life cycle. J Virol 77:2832–2842

Genther Williams SM, Disbrow GL, Schlegel R et al (2005) Requirement of epidermal growth factor receptor for hyperplasia induced by E5, a high-risk human papillomavirus oncogene. Cancer Res 65:6534–6542. https://doi.org/10.1158/0008-5472.CAN-05-0083

Gillespie KA, Mehta KP, Laimins LA, Moody CA (2012) Human papillomaviruses recruit cellular DNA repair and homologous recombination factors to viral replication centers. J Virol 86:9520–9526. https://doi.org/10.1128/JVI.00247-12

Gu W, Shi XL, Roeder RG (1997) Synergistic activation of transcription by CBP and p53. Nature 387:819–823. https://doi.org/10.1038/42972

Halavaty KK, Regan J, Mehta K, Laimins L (2014) Human papillomavirus E5 oncoproteins bind the A4 endoplasmic reticulum protein to regulate proliferative ability upon differentiation. Virology

Hausen zur H (2000) Papillomaviruses causing cancer: evasion from host-cell control in early events in carcinogenesis. J Nat Cancer Inst

Hong S, Cheng S, Iovane A, Laimins LA (2015) STAT-5 regulates transcription of the topoisomerase IIβ-binding protein 1 (TopBP1) gene to activate the ATR pathway and promote human papillomavirus replication. mBio 6:e02006–15. https://doi.org/10.1128/mbio.02006-15

Howley PM, Lowy DR (2007) Papillomaviruses, 5 edn., pp. 2299–2354

Hwang ES, Nottoli T, Dimaio D (1995) The HPV16 E5 protein: expression, detection, and stable complex formation with transmembrane proteins in COS cells. Virology

Kadaja M, Isok-Paas H, Laos T et al (2009) Mechanism of genomic instability in cells infected with the high-risk human papillomaviruses. PLoS Pathog 5:e1000397. https://doi.org/10.1371/journal.ppat.1000397

Kawai T, Akira S (2011) Toll-like receptors and their crosstalk with other innate receptors in infection and immunity. Immunity

Lee C, Laimins LA (2004) Role of the PDZ domain-binding motif of the oncoprotein E6 in the pathogenesis of human papillomavirus type 31. J Virol 78:12366–12377. https://doi.org/10.1128/JVI.78.22.12366-12377.2004

Liu X, Clements A, Zhao K, Marmorstein R (2006) Structure of the human papillomavirus E7 oncoprotein and its mechanism for inactivation of the retinoblastoma tumor suppressor. J Biol Chem 281:578–586. https://doi.org/10.1074/jbc.M508455200

Liu X, Dakic A, Zhang Y et al (2009) HPV E6 protein interacts physically and functionally with the cellular telomerase complex. Proc Natl Acad Sci USA 106:18780–18785. https://doi.org/10.1073/pnas.0906357106

Liu X, Disbrow GL, Yuan H et al (2007) Myc and human papillomavirus type 16 E7 genes cooperate to immortalize human keratinocytes. J Virol 81:12689–12695. https://doi.org/10.1128/JVI.00669-07

Longworth MS, Laimins LA (2004) The binding of histone deacetylases and the integrity of zinc finger-like motifs of the E7 protein are essential for the life cycle of human papillomavirus type 31. J Virol 78:3533–3541

Lorenz LD, Rivera Cardona J, Lambert PF (2013) Inactivation of p53 rescues the maintenance of high risk HPV DNA genomes deficient in expression of E6. PLoS Pathog 9:e1003717. https://doi.org/10.1371/journal.ppat.1003717

Lowy DR (2007) History of papillomavirus research. The Papillomaviruses. Springer, US, Boston, MA, pp 13–28

Lu B, Kumar A, Castellsagué X, Giuliano AR (2011) Efficacy and safety of prophylactic vaccines against cervical HPV infection and diseases among women: a systematic review and meta-analysis. BMC Infect. Dis. 11(1):13. https://doi.org/10.1186/1471-2334-11-13

Massimi P, Gammoh N, Thomas M, Banks L (2004) HPV E6 specifically targets different cellular pools of its PDZ domain-containing tumour suppressor substrates for proteasome-mediated degradation. Oncogene 23:8033–8039. https://doi.org/10.1038/sj.onc.1207977

Mattiroli F, Vissers JHA, van Dijk WJ et al (2012) RNF168 ubiquitinates K13-15 on H2A/H2AX to drive DNA damage signaling. Cell 150:1182–1195. https://doi.org/10.1016/j.cell.2012.08.005

McBride AA (2008) Replication and partitioning of papillomavirus genomes. Adv Virus Res 72:155–205. https://doi.org/10.1016/S0065-3527(08)00404-1

McBride AA, Sakakibara N, Stepp WH, Jang MK (2012) Hitchhiking on host chromatin: how papillomaviruses persist. Biochimica et Biophysica Acta (BBA) - Gene Regulatory Mechanisms 1819:820–825. https://doi.org/10.1016/j.bbagrm.2012.01.011

Mehta K, Gunasekharan V, Satsuka A. Laimins L (2015) Human papillomaviruses activate and recruit SMC1 cohesin proteins for the differentiation-dependent life cycle through association with CTCF insulators 1–25. https://doi.org/10.1371/journal.ppat.1004763&domain=pdf

Mehta K, Laimins L (2018) Human papillomaviruses preferentially recruit DNA repair factors to viral genomes for rapid repair and amplification. mBio 9:e00064–18. https://doi.org/10.1038/nprot.2008.73

Mighty KK, Laimins LA (2013) The role of human papillomaviruses in oncogenesis. Viruses and human cancer. Springer, Berlin Heidelberg, Berlin, Heidelberg, pp 135–148

Moody CA, Laimins LA (2008) The life cycle of human papillomaviruses. Springer, US, New York, NY, pp 75–104

Moody CA, Laimins LA (2009) Human papillomaviruses activate the ATM DNA damage pathway for viral genome amplification upon differentiation. PLoS Pathog 5:e1000605. https://doi.org/10.1371/journal.ppat.1000605.s004

Moody CA, Laimins LA (2010) Human papillomavirus oncoproteins: pathways to transformation. Nat Publishing Group 10:550–560. https://doi.org/10.1038/nrc2886

Münger K, Werness BA, Dyson N et al (1989) Complex formation of human papillomavirus E7 proteins with the retinoblastoma tumor suppressor gene product. EMBO J 8:4099–4105

Nicolaides L, Davy C, Raj K et al (2011) Stabilization of HPV16 E6 protein by PDZ proteins, and potential implications for genome maintenance. Virology 414:137–145. https://doi.org/10.1016/j.virol.2011.03.017

Niu H, Chung W-H, Zhu Z et al (2010) Mechanism of the ATP-dependent DNA end-resection machinery from saccharomyces cerevisiae. Nature 467:108–111. https://doi.org/10.1038/nature09318

Parkin DM, Bray F (2006) Chapter 2: THE burden of HPV-related cancers. Vaccine 24(Suppl 3): S3–11–25. https://doi.org/10.1016/j.vaccine.2006.05.111

Petti L, Dimaio D (1994) Specific interaction between the bovine papillomavirus E5 transforming protein and the beta receptor for platelet-derived growth factor in stably transformed and acutely transfected cells. J Virol

Pivarcsi A, Koreck A, Bodai L et al (2004) Differentiation-regulated expression of toll-like receptors 2 and 4 in HaCaT keratinocytes. Arch Dermatol Res 296:120–124. https://doi.org/10.1007/s00403-004-0475-2

Prue G, Lawler M, Baker P, Warnakulasuriya S (2017) Human papillomavirus (HPV): making the case for 'Immunisation for All'. Oral Dis 23:726–730. https://doi.org/10.1111/odi.12562

Regan JA, Laimins LA (2008) Bap31 Is a novel target of the human papillomavirus E5 protein. J Virol 82:10042–10051. https://doi.org/10.1128/JVI.01240-08

Remm M, Remm A, Ustav M (1999) Human papillomavirus type 18 E1 protein is translated from polycistronic mRNA by a discontinuous scanning mechanism. J Virol 73:3062–3070

Robinson M, Suh Y-E, Paleri V et al (2013) Oncogenic human papillomavirus-associated nasopharyngeal carcinoma: an observational study of correlation with ethnicity, histological subtype and outcome in a UK population. Infect Agents Cancer 8:30. https://doi.org/10.1186/1750-9378-8-30

Rohini Roy JCSNP (2012) BRCA1 and BRCA2: different roles in a common pathway of genome protection. Nat Rev Cancer 12:68. http://doi.org/10.1038/nrc3181

Rundle S, Bradbury A, Drew Y, Curtin NJ (2017) Targeting the ATR-CHK1 axis in cancer therapy. Cancers (Basel). https://doi.org/10.3390/cancers9050041

Saldivar JC, Cortez D, Cimprich KA (2017) The essential kinase ATR: ensuring faithful duplication of a challenging genome. Nat Rev Mol Cell Biol. https://doi.org/10.1038/nrm.2017.67

Satsuka A, Mehta K, Laimins L (2015) p38MAPK and MK2 pathways are important for the differentiation-dependent human papillomavirus life cycle. J Virol 89:1919–1924. https://doi.org/10.1128/JVI.02712-14

Schiffman M, Doorbar J, Wentzensen N et al (2016) Carcinogenic human papillomavirus infection. Nat Rev Dis Primers 2:16086. https://doi.org/10.1038/nrdp.2016.86

Shechter D, Costanzo V, Gautier J (2004) Regulation of DNA replication by ATR: signaling in response to DNA intermediates. DNA Repair (Amst) 3:901–908. https://doi.org/10.1016/j.dnarep.2004.03.020

Smits VAJ, Reaper PM, Jackson SP (2006) Rapid PIKK-dependent release of Chk1 from chromatin promotes the DNA-damage checkpoint response. Curr Biol 16:150–159. https://doi.org/10.1016/j.cub.2005.11.066

Spriggs CC, Laimins LA (2017) FANCD2 binds human papillomavirus genomes and associates with a distinct set of DNA repair proteins to regulate viral replication. mBio. https://doi.org/10.1128/mbio.02340-16

Straight SW, Herman B, McCance DJ (1995) The E5 oncoprotein of human papillomavirus type 16 inhibits the acidification of endosomes in human keratinocytes. J Virol

Uckelmann M, Sixma TK (2017) Histone ubiquitination in the DNA damage response. DNA Repair (Amst) 56:92–101. https://doi.org/10.1016/j.dnarep.2017.06.011

Valle GF, Banks L (1995) The human papillomavirus (HPV)-6 and HPV-16 E5 proteins co-operate with HPV-16 E7 in the transformation of primary rodent cells. J Gen Virol 76(Pt 5):1239–1245

Van Doorslaer K, Burk RD (2012) Association between hTERT activation by HPV E6 proteins and oncogenic risk. Virology 433:216–219. https://doi.org/10.1016/j.virol.2012.08.006

Vande Pol SB, Klingelhutz AJ (2013) Papillomavirus E6 oncoproteins. Virology 445:115–137. https://doi.org/10.1016/j.virol.2013.04.026

Wallace NA, Galloway DA (2014) Manipulation of cellular DNA damage repair machinery facilitates propagation of human papillomaviruses. Semin Cancer Biol 26:30–42. https://doi.org/10.1016/j.semcancer.2013.12.003

Yuan SS, Lee SY, Chen G et al (1999) BRCA2 is required for ionizing radiation-induced assembly of Rad51 complex in vivo. Cancer Res 59:3547–3551

Zhang B, Chen W, Roman A (2006) The E7 proteins of low- and high-risk human papillomaviruses share the ability to target the pRB family member p130 for degradation. Proc Natl Acad Sci USA 103:437–442. https://doi.org/10.1073/pnas.0510012103

Zhou Q, Zhu K, Cheng H (2013) Toll-like receptors in human papillomavirus infection. Arch Immunol Ther Exp 61:203–215. https://doi.org/10.1007/s00005-013-0220-7

Zimmermann H, Degenkolbe R, Bernard HU, O'Connor MJ (1999) The human papillomavirus type 16 E6 oncoprotein can down-regulate p53 activity by targeting the transcriptional coactivator CBP/p300. J Virol 73:6209–6219

Vaccination Strategies for the Control and Treatment of HPV Infection and HPV-Associated Cancer

Emily Farmer, Max A. Cheng, Chien-Fu Hung, and T.-C. Wu

1 Introduction

Human papillomavirus (HPV) is the most common sexually transmitted infection in the world (World Health Organization 2020; Brianti et al. 2017), affecting more than 600 million people worldwide (Gaspar et al. 2015). HPV can be transmitted through skin-to-skin contact or through contact between mucosal membranes (Brianti et al. 2017). HPV can cause a range of clinical diseases in the body, escalating in severity from benign warts to metastatic cancer. More than 200 types of HPV have been identified, which are broadly categorized into high-risk and low-risk types. Low-risk HPV types, such as HPV6 and 11, do not cause cancer. Instead, low-risk HPV types can generate genital warts around the anogenital region, known as condylomata acuminata, as well as benign tumors in the respiratory tract, known as recurrent respiratory papillomatosis or laryngeal papillomatosis. High-risk HPV types, including HPV16, 18, 31, 33, 35, 45, 51, 52, 56, 58,

E. Farmer · M. A. Cheng · C.-F. Hung · T.-C. Wu (✉)
Department of Pathology, The Johns Hopkins School of Medicine,
Cancer Research Building II, 1550 Orleans Street, Baltimore, MD 21287, USA
e-mail: wutc@jhmi.edu

C.-F. Hung · T.-C. Wu
Department of Oncology, The Johns Hopkins School of Medicine,
Cancer Research Building II, 1550 Orleans Street, Baltimore, MD 21287, USA

T.-C. Wu
Department of Obstetrics and Gynecology, The Johns Hopkins School of Medicine,
Cancer Research Building II, 1550 Orleans Street, Baltimore, MD 21287, USA

T.-C. Wu
Department of Pathology, Oncology, Obstetrics and Gynecology, and Molecular
Microbiology and Immunology, The Johns Hopkins Medical Institutions, Cancer Research
Building II, Room 309, 1550 Orleans Street, Baltimore, MD 21287, USA

© Springer Nature Switzerland AG 2021
T.-C. Wu et al. (eds.), *Viruses and Human Cancer*, Recent Results
in Cancer Research 217, https://doi.org/10.1007/978-3-030-57362-1_8

59, and 68, can cause cancer and are often necessary for oncogenic transformation. Virtually, all cases of cervical cancer, 95% of cases of anal cancer, 70% of cases of oropharyngeal cancer, 65% of cases of vaginal cancer, 50% of cases of vulvar cancer, and 35% of cases of penile cancer are caused by high-risk HPV types. Specifically, HPV16 and 18 are associated with the majority of these cancers, including over 70% of all cervical cancers (Roden and Stern 2018), ~90% of anogenital cancers, and up to 75% of oropharyngeal cancers (Elrefaey et al. 2014; Walboomers et al. 1999; National Cancer Institute 2020).

It is estimated that around 5% of all cancers worldwide are caused by HPV (de Martel et al. 2012, 2017). Of the aforementioned HPV-associated diseases, cervical cancer accounts for the largest number of HPV-associated cancer cases (de Martel et al. 2017). Cervical cancer is the fourth most common cancer in women worldwide, and the second most common cancer in women living in low- and middle-income countries (LMICs). In 2018 alone, cervical cancer was responsible for over ~311,000 deaths (World Health Organization 2019; World Cancer Research Fund 2018). The global prevalence of cervical cancer has decreased since the 1950s largely due to early detection, improved HPV testing, prophylactic vaccination, and wider treatment availability; however, more than 85% of cervical cancer-associated deaths occurred in LMICs where infrastructure and access to preventions and treatments may be limited (World Health Organization 2018, 2019, Vaccarella et al. 2013). Additionally, we have seen global rates of HPV-associated oropharyngeal cancer in men on the rise, especially in North America and Northern Europe (Gillison et al. 2015). Unlike for cervical cancer in women, no routine screening tests exist for oropharyngeal cancers, largely due to the inability to detect precancerous lesions as well as subclinical or early-stage cancer (Roden and Stern 2018; Gillison et al. 2015; American Cancer Society 2018). This information highlights the need for methods to control HPV, especially oncogenic types.

The identification of HPV as the etiological factor for HPV-associated diseases has afforded the opportunity to manage these cancers through vaccination (Yang et al. 2016). Our increased understanding of the molecular biology of HPV has powered the development of HPV-targeted vaccines. HPV is a small, non-enveloped, double-stranded DNA virus belonging to the *Papillomaviridae* family. The HPV genome is comprised of ~8000 base pairs, which encode for eight major proteins, six early (E) genes and two late (L) genes (Yang et al. 2016; Graham 2010). Early genes, E1, E2, E4, E5, E6, and E7, contribute to the regulatory function of the viral genome, including DNA replication and transcription (Graham 2010). Late genes, L1 and L2, are known as the major and minor capsid proteins, respectively. The late genes comprise the viral capsid, which is responsible for viral transmission, spread, and survival (Graham 2010). Upon infection, the HPV viral genome is integrated into the host genome where it carries out processes necessary for viral replication and transcription. Specifically, E1 is involved in viral DNA replication, while E2 is involved in RNA transcription. E4 is involved in regulating the cytoskeleton network of infected cells, cell cycle arrest, and virion assembly. E5 is considered an oncogenic protein and is responsible for cell growth and differentiation as well as immune modulation (Yang et al. 2016;

Graham 2010). Both E6 and E7 are oncoproteins expressed in transformed cells (Yang et al. 2016). E6 and E7 are responsible for the carcinogenesis of HPV-associated lesions and are necessary for the initiation and upkeep of HPV-associated malignancies. E6 inhibits apoptosis and differentiation through the degradation of the tumor suppressor gene p53 (Yim and Park 2005). E7 interacts with the Rb protein, a cell cycle regulator, rendering it inoperable. This interaction results in the unregulated proliferation of infected cells as well as the transformation into cancer (Yim and Park 2005). In most cases of HPV-associated cancer, the HPV viral DNA genome integrates into the host's genome. The integration process leads to the deletion of early genes E1, E2, E4, and E5, and late genes L1 and L2. E2 is a negative transcriptional regulator for E6 and E7. The deletion of E2 leads to the disruption of normal cell cycle regulation by interacting with p53 and Rb, respectively. This results in the progression of HPV-associated cervical cancer (Yang et al. 2016). Further, the deletion of L1 and L2 during the integration process are what render prophylactic vaccines ineffective against established HPV-associated diseases (Yang et al. 2016).

HPV types are tissue-trophic and infect keratinocytes, preferentially propagating in epithelial mucosa (Egawa et al. 2015). Viral expression is associated with the differentiation of keratinocytes and viral shedding from superficial epithelial layers, which express L1 and L2 viral capsid proteins (Williams et al. 2011; Roden and Wu 2006). HPV infections are restricted to the basal epithelial cells, which are often shielded from circulating immune cells during surveillance. Because HPV infection often does not generate a host immune response, HPV DNA goes undetected, enabling the virus to continue to amplify, eventually spreading to and infecting neighboring cells. HPV is also non-lytic and does not generate an inflammatory response, meaning that individuals who are infected may not know their disease status. Only after HPV-associated tumor cell has been sufficiently amplified to a level where it can be detected by immune surveillance cells, does it mount an active immune response. Unfortunately, this often occurs during the later stages of HPV transformation, sometimes years after the initial HPV infection (Williams et al. 2011). Prophylactic vaccines have traditionally be used to prevent disease prior to infection. Current prophylactic HPV vaccines are used to deliver HPV L1 and/or L2 capsid antigens, which self-assemble to form a virus-like particle (VLP). These vaccines stimulate the immune generation of neutralizing antibodies against VLPs, which can prevent the acquisition of real HPV infection in healthy individuals. Unfortunately, neutralizing antibodies against HPV are incapable of controlling or killing existing HPV-infected and/or transformed cells. Instead, HPV antigen-specific cytotoxic T cells (CD8+ T cells) and helper T cells (CD4+ T cells) are necessary for the targeted killing of infected and/or transformed cells (Yang et al. 2016). Unlike prophylactic HPV vaccines, therapeutic HPV vaccines rely on T cell-mediated immune responses to target and kill infected cells. This is facilitated by antigen-presented cells (APCs), such as dendritic cells (DCs). DCs present HPV antigen through major histocompatibility class I (MHC-I) and class II (MHC-II) molecules for recognition by HPV antigen-specific CD8+ and CD4+ T cells, respectively.

When devising a therapeutic vaccine, the target antigen of choice requires significant consideration. Because L1 and L2 are deleted during the integration process of HPV into the host genome, they are not suitable target antigens for the development of therapeutic HPV vaccines. However, HPV oncoproteins E6 and E7 present as ideal targets for the development of therapeutic HPV vaccines. E6 and E7 are only expressed in transformed cells and are necessary for initiating and maintaining HPV-associated malignancies (Yang et al. 2016). Therefore, therapeutic HPV vaccines targeting E6 and E7 are safe and can circumvent immune tolerance against self-antigens (Yang et al. 2016).

2 Current Preventive HPV Vaccines

In the last decade, a total of three prophylactic HPV vaccines have been developed for commercial use, Cervarix™ (from GlaxoSmithKline), Gardasil®, Gardasil9® (from Merck). These prophylactic vaccines have provided an opportunity to prevent the acquisition of HPV infection in unexposed, healthy individuals. All three vaccines use an L1 VLP vaccine platform, which constitutes the non-infectious papillomavirus particles without the viral genome (Yang et al. 2016). Cervarix™ is a bivalent vaccine containing HPV16 and HPV18 VLPs produced in insect cells (*Trichoplusia ni*) using a baculovirus expression vector system. Cervarix™ also incorporates Adjuvant System 04 (comprised of monophosphoryl lipid A and an aluminum hydroxide salt) (U.S. Food and Drug Administraiton 2018) to enhance the body's humoral immune responses after vaccination. Cervarix™ only protects against oncogenic HPV types HPV16 and HPV18; however, these HPV types are present in the majority of HPV-associated cancers (Elrefaey et al. 2014; National Cancer Institute 2019), and GlaxoSmithKline discontinued the marketing of Cervarix™ in the United States in 2016. Gardasil® is a recombinant quadrivalent vaccine prepared from HPV 6, 11, 16, and 18 VLPs. The L1 proteins are produced in yeast cells (*Saccharomyces cerevisiae*) and absorbed on an amorphous aluminum hydroxyphosphate sulfate adjuvant (U.S. Food and Drug Administration 2019). It provides protection against oncogenic types HPV16 and 18, as well as low-risk HPV types 6 and 11, which cause common genital warts.

Currently, around 15 oncogenic HPV types have been identified. While Gardasil® and Cervarix™ provide protection against the two most common oncogenic HPV types, HPV16 and HPV18, neither provide protection against any of the remaining ~ 13 oncogenic HPV types. To address this disparity in coverage, Gardasil®9 was developed to provide broader protection against more HPV types (Zhai and Tumban 2016; Manini and Montomoli 2018). Gardasil®9 is a recombinant nanovalent vaccine prepared from L1 VLPs of HPV6, 11, 16, 18, 31, 33, 45, 52, and 58. Gardasil®9 is produced using the same method as Gardasil® and contains an amorphous aluminum hydroxyphosphate sulfate adjuvant (U.S. Food and Drug Administration 2019) to enhance immunogenicity. Importantly, Gardasil®9 expanded existing vaccination coverage against HPV, protecting against the seven most common oncogenic HPV types (HPV16, 18, 31, 33, 45, 52, and 58) and two most common low-risk HPV types (HPV6 and 11) (Zhai and Tumban 2016; Immunization Action Coalition 2019). Gardasil®9 is currently the only HPV vaccine being distributed in the US and is licensed for females and males ages 9–45 (Immunization Action Coalition 2019). In fact, Cervarix™ and Gardasil® are no longer available for distribution or purchase in the US. Specifically, the sale of Cervarix™ was discontinued in the US due to low demand, while the sale of Gardasil® was discontinued in 2018, after the FDA-approved Gardasil®9 (Kaiser and Family Foundation 2018). However, Cervarix™ and Gardasil® are still widely used outside the use in both clinical practice and investigational trials.

In clinical testing, the efficacy of Gardasil® was assessed in 20,541 women and 4055 men ages 16–26. Vaccine efficacy, measured as protection from HPV types 6, 11, 16, and 18 after three doses, was evaluated in subjects who were HPV-naïve prior to the first vaccination dose. The results of this study showed that in both men and women aged 16–26 who were HPV-naïve, Gardasil® was effective at preventing the development of lesions caused by HPV6, 11, 16, and 18. Moreover, Gardasil® was shown to have: 98% efficacy against HPV16 and 18-associated cervical intraepithelial neoplasia grades 2 and 3 (CIN2/3) and adenomacarcinoma in situ (AIS); 100% efficacy against HPV16 and 18-associated vulvar intraepithelial neoplasia grades 2 and 3 (VIN2/3) and vaginal intraepithelial neoplasia grades 2–3 (VaIN2/3); 75% efficacy against HPV6, 11, 16, and 18-associated anal intraepithelial neoplasia grades 2–3 (AIN2/3); and 89% and 99% efficacy against HPV6 and 11-associated genital warts in males and females, respectively (Merck & Co. Inc. 2019).

Because Gardasil®9 was developed to protect against HPV strains not previously covered by the first generation of Gardasil®, a comparative clinical trial was led by Merck to empirically evaluate the efficacy of the two vaccines. The clinical study compared Gardasil® and Gardasil®9 in 14,204 women ages 16–26 worldwide. A total of 7099 women were randomized to receive Gardasil®9, while 7105 women were randomized to receive Gardasil®. Vaccine efficacy was evaluated in subjects who received three doses of vaccination and were HPV-naïve prior to the first vaccination dose. Compared to Gardasil®, Gardasil®9 demonstrated 97% clinical efficacy against HPV31, 33, 45, 52, and 58-associated CIN2/3, AIS, VIN2/3, and VaIN2/3, suggesting that Gardasil®9 provides protection against five more types of HPV (types 31, 33, 45, 52, and 58) than the first-generation vaccine Gardasil®. Moreover, since both vaccines are manufactured similarly and comprise four of the same HPV L1 VLPs the efficacy and effectiveness of Gardasil®9 against HPV6, 11, 16, and 18 were comparable to that of Gardasil® (Merck & Co. Inc. 2019). Notably, the efficacy of both Gardasil® and Gardasil®9 against oropharyngeal cancer was not tested in these trials, which could be attributed to the somewhat recent determination of the etiologic relationship between HPV and oropharyngeal cancer (Guo et al. 2016). An additional Phase III trial comparing Gardasil® to Gardasil®9 in 14,215 women confirmed the results of the first study, demonstrating that Gardasil®9 protected against 96.7% of CIN2/3, VIN2/3, and VaIN2/3 caused by HPV31, 33, 45, 52, and 58. Furthermore, the efficacies of Gardasil® and Gardasil®9 against HPV types 6, 11, 16, and 18 were also shown to be comparable. The investigators of this study also found that the geometric mean antibody titers (GMTs) one month after the third vaccine dose of Gardasil®9 were noninferior to Gardasil® for HPV6, 11, 16, and 18. Additionally, seroconversion for women in the Gardasil®9 group to all nine HPV types was >99% (Joura et al. 2015). Additional clinical trials have been developed to evaluate the efficacy of the described prophylactic HPV vaccines (Centers for Disease Control and Prevention 2015).

All three commercially available prophylactic HPV vaccines are administered intramuscularly in the arm muscle in a two-dose or three-dose regimen, spread out over the course of 6–12 months, depending on dose schedule (Centers for Disease Control and Prevention 2015; Merck & Co. Inc. 2017). However, in the last decade, recommended vaccination series have changed as a result of burgeoning data on dose recommendations as well as the development of Gardasil®9. Pre-adolescent girls (ages 9–15) now have the option to receive a two-dose HPV vaccination regimen at a 6-month or 12-month interval to protect against HPV. While this two-dose recommendation was first recommended by the World Health Organization (WHO) in 2015, a three-dose vaccine regimen is recommended for girls and women 15 years and older (Harper and DeMars 2017). The second scheduled vaccine dose is typically administered 1–2 months after the first dose, followed by the third dose 5–10 months later (Centers for Disease Control and Prevention 2015). Due to the novelty of Gardasil®9, there is little data to elucidate the ideal dose schedule for vaccination. In comparative studies, Gardasil®9 has demonstrated similar GMTs and seropositivity for anti-HPV6, 11, 16, and 18; however, the immunogenicity of two or three-dose HPV vaccine regimens are still being studied (Harper and DeMars 2017). To this end, a comparative phase III clinical trial is planned to determine the comparative immunogenicity of the two-dose to the three-dose vaccine schedule (Harper and DeMars 2017) (NCT02834637). Importantly, this study might not only uncover the optimal vaccination schedule for Gardasil®9, but it may also help inform future efforts in vaccine development to minimize necessary vaccine doses. Additional studies have studied the efficacy of a single dose of prophylactic HPV vaccine, demonstrating that it could provide similar protection when compared to two- to three-dose vaccination regimens (Safaeian et al. 2013; Kreimer et al. 2015). Several clinical trials investigating whether a single dose of HPV vaccine is efficacious in the prevention of HPV infection were recently completed or are ongoing (NCT03431246, NCT00635830, and NCT03675256, respectively).

3 Improving Preventive HPV Vaccine Development

The clinical efficacy of available prophylactic HPV vaccines represents a substantial improvement in the prevention of HPV and HPV-associated diseases for targeted-types, but problems in coverage and disease burden still remain. 85% of the burden of HPV and HPV-associated diseases occurs in LMICs, where infrastructure and access to HPV prevention, therapies, or treatments are often limited (World Health Organization 2018; Vaccarella et al. 2013). Furthermore, many persons living in LMICs face significant barriers to vaccination. The cost of the vaccine can pose a significant financial barrier at both the individual and institutional levels. Due to the cost associated with vaccination, many LMICs choose to fund vaccination for a single age group each year, rather than make the vaccine available to persons of all recommended age groups (Gallagher et al. 2018; Bruni et al. 2016). Likewise, the need for multiple doses (2–3 doses) of prophylactic HPV vaccines is a barrier for many, as persons may receive an incomplete vaccine schedule, which does not afford full immunologic protection. Another barrier to vaccination involves the physical access and availability of the vaccine. Many individuals in LMICs do not live in close proximity to a health clinic or a provider through which they can access and obtain the vaccine. Many LMICs also lack the infrastructure or capacity to store and distribute such vaccines (i.e., through cold chains) proving vaccine provision and dissemination a significant challenge.

Another challenge of prophylactic vaccines is limited cross-reactivity to multiple oncogenic HPV strains. Although individuals vaccinated with Gardasil®9 will be protected from seven of the most common oncogenic HPV types, including HPV16 and 18, Gardasil®9 still only covers fewer than half of the ~15 oncogenic HPV strains (Merck & Co. Inc. 2017), rendering vaccinated persons susceptible to subsequent infection. Because vaccination with the current prophylactic vaccine cannot protect close to 100% of all HPV infections, Pap smear screening or HPV testing for screening are still required. Although substantial progress has been made towards developing a more-protective HPV vaccine, continued efforts to improve prophylactic HPV vaccines are warranted. Broader, or even full-coverage, against oncogenic HPV types is a desirable attribute in future generations of prophylactic HPV vaccines that researchers should strive towards. Strategies to developed improved prophylactic HPV vaccines include, but are not limited to, the development of (1) L1-based capsomeres, (2) L2-based vaccines, and (3) chimeric L1–L2 based vaccines. Figure 1 summarizes the various strategies in the current, and the next generation of preventive HPV vaccines, including the newest multivalent VLP vaccine, Gardasil®9.

One potential method for creating new prophylactic HPV vaccines involves the development of L1 capsomere vaccines. Capsomeres are structural subunits, which self-assemble to form the virus capsid. HPV L1 VLPs are composed of 360 L1 monomers, which assemble into 72 pentavalent capsomeres (DiGiuseppe et al. 2017). L1 capsomeres can be purified from *Escherichia coli*, offering a cost-effective alternative to L1 VLP-based vaccines, which are produced in yeast cells (Schadlich et al. 2009). In preclinical models, L1 capsomere proteins produced in *E. coli*, attenuated measles virus, or recombinant *Salmonella enterica serotype Typhi* have successfully generated neutralizing antibodies against HPV-L1 and demonstrated immunogenicity comparable to that of VLPs (Barra et al. 2019). Thus, L1 capsomere vaccines may represent a potential lower-cost prophylactic HPV vaccine which provides comparable immunogenicity to existing VLP vaccines and potentially reduce the number of booster vaccinations currently required.

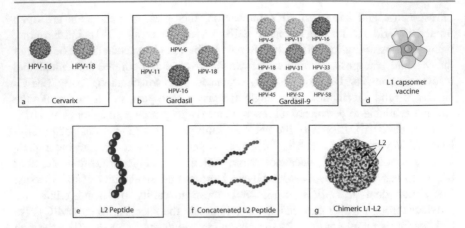

Fig. 1 Schematic diagram to depict the current and next generation of preventive HPV vaccines. **a** Cervarix™ composed of HPV16 and HPV18 VLPs. **b** Gardasil® composed of HPV6, HPV11, HPV16, and HPV18 VLPs. **c** Gardasil®9 composed of HPV6, HPV11, HPV16, HPV18, HPV31, HPV33, HP-45, HPV52, and HPV58 VLPs. **d** L1 capsomer vaccine. **e** L2 peptide vaccine. **f** Concatenated L2 peptide vaccine. **g** Chimeric L1-L2 VLP vaccine with L2 on the surface

Another potential approach to creating new prophylactic vaccines involves the use of the HPV minor capsid protein, L2-based vaccines, instead of L1 VLPs. The difficulties associated with manufacturing multivalent L1-VLP vaccines, such as Gardasil®9, limit the number of VLPs and hinder the vaccine's protective capacity. Unlike L1, L2 is highly conserved between different HPV types and contains type-common epitopes. Therefore, the L2 from one HPV strain could potentially induce broader protection against multiple HPV types through cross-neutralizing antibodies, even across species (Roden and Stern 2018; Schellenbacher et al. 2017; Gambhira et al. 2006; Kaliamurthi et al. 2019). Currently, more than three preventive L2 VLP-based prophylactic HPV vaccines are underway for evaluation in early phase clinical trials (for review see Schellenbacher et al. 2017). One downside to L2-based vaccines is low immunogenicity. Antibody levels induced by L2 peptides are significantly inferior to those induced by L1 VLP vaccines. Hence, the development of L2-based vaccines necessitates strategies to enhance immunogenicity and antibody response in vivo (Schellenbacher et al. 2017).

In general, L2-based vaccines are less immunogenic than L1-based vaccines (Roden and Stern 2018). Additionally, the current gold standard for the detection of L1-specific antibodies cannot reliably detect an L2-specific immune response in vivo. To address this challenge, Day et al. created a test to detect L2-directed neutralizing antibodies with significantly improved sensitivity (Day et al. 2012). With an improved method of detection now available, several strategies have been implemented to enhance the immunogenicity of L2, many of which have shown great potential. For example, in a preclinical study, mice vaccinated with a combination of L2-peptides derived from eight HPV types displayed on the surface of

PP& bacteriophage VLPs were protected from HPV pseudovirus challenge of all eight HPV types (Caldeira Jdo et al. 2010; Tumban et al. 2011). Other studies have also been used to enhance L2 immunogenicity, including the application of an *E. coli*-based concatenated multitype L2 fusion protein with multitype cross-neutralizing epitopes (Jagu et al. 2009), the oral administration of HPV16 L2 expressed on the surface of *Lactobacillus casei* (Yoon et al. 2012), and the administration of HPV16 L2 protein with bacterial thioredoxin, and T cell stimulator (Rubio et al. 2009). While L2-based vaccines have the potential to confer greater cross-reactivity, the immunogenicity of L2 proteins remain low. For this reason, more potent adjuvants and display methods continue to be explored. Currently, however, no L2-peptide-based prophylactic HPV vaccines have been approved for clinical trials (Kaliamurthi et al. 2019).

The potent immunogenicity of L1 vaccines and the broad cross-protection provided by L2 can also potentially be exploited through the combination in the form of chimeric L1/L2 VLPs. A single copy of the L2 protein is present in each L1 pentavalent capsomere, thus, each HPV virion contains 72 copies of the L2 protein (Kaliamurthi et al. 2019). L2 plays a critical role in the assembly of L1 into VLPs and has been demonstrated to facilitate the encapsulation of the viral genome (Kaliamurthi et al. 2019). Because L2 is less abundant than L1 and is predominantly found in the interior of the VLP, replacing some L1 immunodominant epitope regions of the VLP surface with a neutralizing epitope of L2 may generate stronger immunogenic, cross-protective immune responses against multiple HPV types. In L1-based vaccine models, the surface expression of the neutralizing epitope of L2 is necessary for the generation of L2 neutralizing antibodies. In a recent study, Kaliamurthi *et al.* constructed an L2-based chimeric HPV vaccine (SGD58) using two selected epitope sequences on the *N*-terminal region of the L2 sequence of HPV58 [the fourth most common high-risk HPV type in the world (Zhai and Tumban 2016)], two Toll-like receptors (TLR) adjuvants (Flagellin and RS09), and two T helper epitopes (PADRE and TpD) (Kaliamurthi et al. 2019). While this chimeric vaccine has not been tested in vivo, SGD58 demonstrated immunologic properties capable of producing both humoral and cellular immune responses against HPV through immunomics testing in vitro. The SGD58 vaccine also demonstrated cross-protection against 15 different high-risk HPV types (Kaliamurthi et al. 2019). Another candidate chimeric L1–L2 based vaccine uses the RG1 epitope, a single L2 epitope. RG1 can be incorporated into the capsid surface DE loop of HPV16 L1 or HPV18 L1 to create a chimeric L1–L2 based VLP vaccine. In preclinical studies, RG1-VLPs provided broad protection against heterologous high-risk HPV types (Schellenbacher et al. 2017; Boxus et al. 2016; Gambhira et al. 2007a, b). Specifically, chimeric RG1-VLP vaccines have demonstrated protection against challenge of high-risk HPV types 16, 18, 26, 32, 33, 34, 35, 39, 45, 51, 52, 53, 56, 58, 59, 66, 68, 73, and low-risk HPV types 6, 43, and 44 (Schellenbacher et al. 2013). Currently, chimeric RG1-VLPs are under cGMP production and are planned for testing in phase I clinical studies. RG1-VLPs offer a promising next-generation vaccine for wider protection against HPV (Schellenbacher et al. 2013, 2017). In short, the inclusion of the immunodominant neutralizing epitopes of

L2, such as RG1, into the L1 VLPs offer a promising prophylactic HPV vaccine capable of inducing broad-spectrum neutralizing antibodies against different HPV types.

4 Strategies for Therapeutic HPV Vaccine Development

4.1 Introduction to Therapeutic HPV Vaccines

While prophylactic HPV vaccines have been hugely successful in averting HPV infections, they are incapable of treating or eliminating existing HPV infections or HPV-associated lesions. Given that HPV is the most common sexually transmitted infection (STI), virtually anyone who is sexually active is susceptible to HPV exposure or infection during their lifetime (Yang et al. 2016; Centers for Disease Control and Prevention 2017). Because HPV infection is the known etiologic factor for HPV-associated diseases, including nearly all cases of cervical cancer, therapeutic HPV vaccines represent an ideal method for the eradication of HPV-infected cells and HPV-associated tumors. Most individuals who develop an HPV infection will clear the viral infection naturally through their immune system. However, individuals who are unable to clear the infection can develop persistent HPV infections, which may progress into precancerous lesions and eventually, invasive cancer. The progression of an HPV infection into invasive cancer can take years, and remain asymptomatic. Due to the latent nature of the HPV virus, regular screening is recommended to track the progression or regression of the disease. In a prolonged chronic infection, there is a considerable window for secondary preventive treatment for infections caught by cytologic screening and HPV DNA testing. Currently, the US Preventive Services Task Force (USPSTF), American Cancer Society (ACS), and the American College of Obstetricians and Gynecologists (ACOG) recommend cytologic screening (Pap smears) every three years in women aged 21–65 (U.S. Preventive Services Task Force 2012; Centers for Disease Control and Prevention 2012). In 2014, the FDA approved an HPV screening test for primary cervical cancer screening. Primary HPV testing, as well as HPV co-testing (Pap smear and HPV testing), have become widely accepted and utilized in clinical practice (Cooper and Saraiya 2017). While methods of early detection for HPV-associated diseases have seen significant improvement, these detection and screening strategies are still limited or impossible in some settings. Furthermore, some HPV-associated diseases, such as oropharyngeal cancer, do not have methods for routine screening (American Cancer Society 2018). Therefore, efficacious therapeutic HPV vaccines that can selectively target HPV-infected cells during HPV transformation and carcinogenesis represent an ideal strategy to treat HPV infection or HPV-associated diseases, as well as prevent the development of advanced cancer. If therapeutic vaccines are able to eradicate transformed cells before disease progression, the disease burden of HPV and HPV-associated malignancies worldwide may see a drastic decline.

Many different platforms have been used to develop therapeutic HPV vaccines, which have been tested at various phases in preclinical and clinical trials. Characteristics of an ideal therapeutic vaccine include (1) safety; (2) ability to mount a potent HPV antigen-specific T cell-mediated immune response; (3) tumor-targeting specificity; (4) lasting efficacy; (5) cost-effective; and (6) minimal dose requirements. The ability of a therapeutic vaccine to elicit antigen-specific T cell-mediated killing is vital to the vaccine's efficacy in recognizing and targeting transformed cells, and evading healthy cells. Reduced production and storage cost is also a highly desirable trait of any vaccine in order to increase access and availability of therapeutic vaccines for patients with persistent HPV infections or HPV-associated cancers worldwide. In addition, the need for fewer doses of the therapeutic HPV vaccine can potentially improve vaccination compliance and may help reinforce vaccination receipt in targeted populations. As described earlier, E6 and E7 are ideal targets for therapeutic HPV vaccines. To this end, several types of therapeutic HPV vaccines targeting E6 and/or E7 antigens have been developed and undergone preclinical and clinical studies, including live-vector-based, peptide-based, protein-based, dendritic cell-based, DNA-based, and combination vaccines. A graphic representation of how therapeutic HPV vaccines harness the host's immune system to fight HPV infection and associated disease is provided in Fig. 2 (adapted from Cheng et al. 2018).

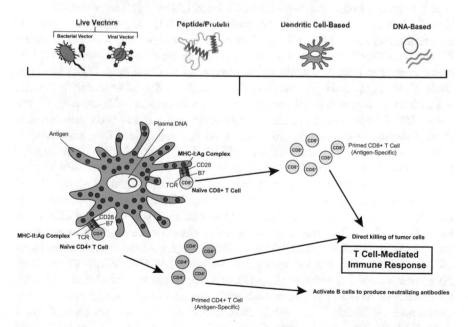

Fig. 2 Therapeutic HPV vaccination schematic. Therapeutic vaccines activate the adaptive immune system by targeting E6 and/or E7 antigen(s), producing a cell-mediated immune response for the control or treatment of HPV infection or HPV-associated disease. Therapeutic HPV vaccination methods include, live-vector-based vaccines (bacterial vector or viral vector); peptide- or protein-based vaccines; dendritic cell-based vaccines; and DNA-based vaccines

4.2 Live Vector-Based Therapeutic HPV Vaccines

Live vector vaccines utilize live bacterial or viral vectors as a vehicle to deliver recombinant antigens. These live vectors infect the host, replicate in the body, and spread the antigen in order to mount an immune response. Live vector-based therapeutic HPV vaccines can deliver E6 and/or E7 antigens to APCs to stimulate antigen presentation to immune cells through the MHC-I MHC-II pathways. Live vector-based vaccines are highly immunogenic and capable of mounting robust humoral and cell-mediated immune responses. One challenge to live vector-based vaccines is the potential safety risk they pose, particularly to individuals who are immunocompromised (Yang et al. 2016; da Silva et al. 2014).

4.2.1 Bacterial Vectors

Several bacterial vectors have been explored for the development of therapeutic HPV vaccines. Among these, *Listeria monocytogenes*, a facultative gram-positive intracellular bacterium, has garnered significant attention as a particularly promising vector for the delivery of HPV antigens (Guirnalda et al. 2012). HPV vaccine antigens can be expressed through fusion to the pore-forming toxin, listeriolysin O (LLO). Fused vaccine antigens are then processed and presented through both the MHC-I and MHC-II pathways. Live *L. monocytogenes*-based vaccines have been shown to induce both antigen-specific CD8+ and CD4+ T cell responses (Kim and Kim 2017; Cory and Chu 2014; Wallecha et al. 2013). In preclinical studies, *L. monocytogenes*-based HPV E7 vaccines have been shown to stimulate a potent E7-specific CD8+ T cell response and were able to slow tumor growth and reduce tumor burden in both transgenic and wild-type mice (Gunn et al. 2001; Lin et al. 2002; Verch et al. 2004; Hussain and Paterson 2004; Sewell et al. 2004; Souders et al. 2007). In response to the demonstrated immunogenicity of *L. monocytogenes*-based HPV E7 vaccines in preclinical studies, this research has been translated into clinical studies. For example, the vaccine ADXS11-001 is a live, attenuated *L. monocytogenes* bacterial vector in which HPV16 E7 protein is fused to a modified LLO molecule (Cory and Chu 2014; Miles et al. 2017). In preclinical studies, ADXS11-001 mounted strong humoral and E7 antigen-specific CD8+ T cell immune responses. In phase I/II clinical trials, ADXS11-001 has demonstrated efficacy in women with cervical cancer (Yang et al. 2016; Cory and Chu 2014; Miles et al. 2017). Currently, two phase II clinical trials are ongoing to assess the safety and efficacy of ADXS11-001 in patients with HPV-associated head and neck (NCT02002182) and cervical cancer, respectively (NCT01266460) (Yang et al. 2016) (for review see Miles et al. 2017). Another phase II clinical trial evaluating the efficacy of ADXS11-001 in patients with anorectal cancer was recently finished recruitment (NCT02399813). While clinical trial data has not yet been reported, ADXS11-001 has shown promising antitumor activity in patients with HPV-associated diseases in multiple studies and was well-tolerated in patients. These findings, along with the ongoing clinical trial findings may lay the groundwork for phase III clinical trials and the potential introduction of ADXS11-001 in the clinic (Guirnalda et al. 2012; Miles et al. 2017). In addition, other attenuated bacterial vectors can also be used to deliver antigens of interest to APCs, including *Salmonella* (Buttaro and Fruehauf 2010; Le Gouellec et al. 2012), *Shingella*, and *E. coli* (Yang et al. 2016; Buttaro and Fruehauf 2010; Hitzeroth 2018).

4.2.2 Viral Vectors

In addition to bacterial vectors, live viral vector-based HPV vaccines have been studied in preclinical and clinical studies due to their high immunogenicity. Viral vectors, including adenoviruses, adeno-associated viruses, alphaviruses, lentiviruses, and vaccinia viruses have been used to deliver HPV E6, E7, and E2 antigens (Yang et al. 2016; Kim and Kim 2017; Hung et al. 2008). Of these viral vectors, vaccinia virus has demonstrated the greatest immunogenic potential in clinical studies. The vaccinia virus is an enveloped, double-stranded DNA virus in the *Poxviridae* family. In particular, vaccinia virus represents a promising viral vector for vaccine delivery because it has a large genome and is highly infectious (Yang et al. 2016). In the past two decades, several vaccinia-based therapeutic HPV vaccines have been studied in clinical trials. The first vaccinia-based vaccine trial was a phase I/II trial using a recombinant vaccinia virus expressing HPV16 and 18 E6/E7 proteins (TA-HPV) in patients with advanced cervical cancer (Borysiewicz et al. 1996). Subsequent phase I/II clinical trials have evaluated the safety and immunogenicity of TA-HPV in patients with early-stage cervical cancer (Kaufmann et al. 2002), VIN (Davidson et al. 2003), and VaIN (Baldwin et al. 2003). Through various clinical studies, TA-HPV was found to be safe and successful in mounting a vaccinia-specific antibody, and HPV antigen-specific CD8+T cell-mediated response (Borysiewicz et al. 1996; Kaufmann et al. 2002; Davidson et al. 2003; Baldwin et al. 2003).

Another vaccinia-based therapeutic HPV vaccine MVA-E2, is a modified vaccinia Ankara virus (MVA) encoding bovine papillomavirus type 1 (BPV-1) E2 protein (Corona Gutierrez et al. 2004; Vici et al. 2016). In several early phase I/II clinical trials, MVA-E2 led to significant therapeutic effects in patients with CIN1/2/3 lesions, including complete regression of precancerous lesions and generation of robust HPV antigen-specific immune responses (Corona Gutierrez et al. 2004; Garcia-Hernandez et al. 2006)). More recently, MVA-E2 was tested in a phase III clinical study for the treatment of anogenital HPV-associated intraepithelial lesions (Rosales et al. 2014) in a total of 1176 female and 180 male patients. Patients received MVA-E2 vaccination at the lesion-site and were then monitored using colposcopy (females only) and cytology for lesion progression/regression. After MVA-E2 vaccination, ~90% of female study patients and all-male study patients had complete elimination of intraepithelial lesions. Furthermore, all patients in the study treated with MVA-E2 developed MVA-E2-specific antibodies, as well as cell-mediated cytotoxic activity, specifically against HPV-transformed cells. The MVA-E2 vaccine was found to eliminate CIN1/2/3 lesions as well as most other anogenital lesions. The results of this study demonstrate the therapeutic potential of the MVA-E2 vaccine for the treatment of HPV-associated anogenital intraepithelial lesions in both males and females (Rosales et al. 2014).

Another MVA-based therapeutic HPV vaccine, TG4001, has been tested in clinical studies. TG4001 is a suspension of MVATG8042 vector particles. MVATG8042 vector particles consist of an attenuated recombinant MVA, encoding modified HPV16 E6 and E7 proteins, and human interleukin-2 (IL-2) (Yang et al. 2016; ***Brun et al. 2011). In phase II clinical trial, the safety and efficacy of TG4001 were evaluated in 21 patients with HPV16-associated CIN2/3. All patients received three weekly subcutaneous injections of TG4001. Regression of CIN2/3 lesions and clearance of HPV infection were monitored using cytology, colposcopy, and DNA/mRNA detection. After six months of vaccination, nearly half the women in the study had a clinical response, showing improvement in their infection or lesion regression. Out of the ten responders in this study, HPV16 DNA clearance was observed in eight individuals, HPV16 mRNA clearance in seven individuals. Further, no recurrence of high-grade lesions was observed for 12 months after treatment (Yang et al. 2016; Kim and Kim 2017; Brun et al. 2011). Another ongoing phase Ib/II clinical trial is assessing the safety and efficacy of TG4001 in combination with avelumab in patients with HPV16-associated oropharyngeal squamous cell carcinoma of the head and neck (NCT03260023). While live vector-based therapeutic HPV vaccines have shown promising results in clinical studies, they pose a potential safety risk, especially for individuals who may be immunocompromised. Due to the possibility of inducing vector-specific neutralizing antibodies and/or having pre-existing vector-specific immunity, live vectors are not able to be repeatedly administered.

4.3 Peptide-Based Therapeutic HPV Vaccines

Short peptides derived from HPV antigens can be delivered to DCs for processing and presentation on MHC-I molecules in order to activate an antigen-specific immune response. Production of peptide vaccines involves the prior identification of specific CD8+ cytotoxic and CD4+ helper T cell epitopes of HPV antigens. Peptide-based vaccines offer several advantages over other therapeutic vaccine types; they are safe, stable, and easy to produce (Cheng et al. 2018). However, peptide-based vaccines also suffer from low immunogenicity and MHC restriction, which ultimately affects their strength and efficacy (Yang et al. 2016; Lin et al. 2010).

Peptide-based vaccines often require lipids or other adjuvants, such as chemokines, cytokines, or TLR ligands to enhance immunogenicity (Yang et al. 2016). Specific adjuvants that have been previously employed to enhance the immunogenicity of peptide-based therapeutic HPV vaccines include aluminum adjuvants (Khong and Overwijk 2016), immunoglobulin G fragment (Qin et al. 2005), streptavidin fused to the extracellular domain of murine 4-1BBL (Sharma et al. 2009), DC stimulatory cytokine bryostatin (Yan et al. 2010), and TLR agonists (Khong and Overwijk 2016; Daftarian et al. 2006; Wu et al. 2010; Zhang et al. 2010; Zwaveling et al. 2002). Another limitation of peptide-based vaccines is that they are MHC-specific, which can lead to challenges for large-scale vaccine production and treatment of HPV-associated diseases (Yang et al. 2016; Su et al. 2010). One potential strategy to circumvent such a challenge is to employ the use of overlapping long-peptide vaccines. Numerous HPV16 synthetic long-peptide vaccines (e.g., HPV16-SLP) have been studied in clinical trials to evaluate therapeutic effect against HPV-associated disease. A phase II study in patients with low-grade abnormalities of the cervix demonstrated that two doses of the HPV16-SLP vaccine were capable of eliciting a robust HPV16-specific T cell response lasting for more than one year (de Vos van Steenwijk et al. 2014). In a more recent study, Welters et al. explored whether HPV16-SLP vaccination could be combined with standard chemotherapy to enhance immunogenicity in patients with advanced cervical cancer (Yang et al. 2016; Welters et al. 2016). From this study, HPV16-SLP vaccination was found to enhance T cell response levels in patients, which remained unchanged even after six cycles of chemotherapy. While this vaccination regimen did not demonstrate significant tumor regression, the HPV16-SLP vaccine was observed to be well-tolerated with minimal adverse effects. As a result of this study, additional phase I and II clinical trials have been designed to evaluate the therapeutic potential of HPV16-SLP vaccines in advanced or recurrent cervical cancer (NCT02128126), as well as in other HPV-associated malignancies (NCT01923116) (Welters et al. 2016). Another method to enhance therapeutic HPV vaccine potency is the administration of therapeutic HPV vaccines in combination with immune checkpoint blockade, which has become popularized in recent years. Recently, a clinical study investigating the safety of an HPV peptide vaccine (ISA101) and nivolumab was completed in patients with HPV16+ incurable/recurrent solid tumors. In this study, patients received nivolumab, an anti-programmed cell death 1 (PD-1) antibody and ISA101 peptide vaccine, a therapeutic HPV16-SLP vaccine containing nine HPV16 E6 synthetic peptides, and four HPV16 E7 synthetic peptides in order to determine whether ISA101 amplified the efficacy of nivolumab in patients with incurable HPV16+ cancer (NCT02426892). The results of this study showed that, out of 24 patients with oropharyngeal cancer, eight patients responded, resulting in an overall response rate of 33% and median overall survival of 17.5 months. Based on the results of this trial, future studies investigating the efficacy of ISA101 and immune checkpoint blockade are warranted (Massarelli et al. 2019).

Several additional studies have been conducted to evaluate the immunogenicity of various peptide-based therapeutic HPV vaccines. In a phase I dose-escalation study, 31 patients with high-grade squamous intraepithelial lesions (HSIL) received PepCan, a therapeutic HPV vaccine containing four cGMP-manufactured synthetic peptides covering HPV16 E6 and Candin® as an adjuvant. No dose-limiting toxicities were observed in this study (Coleman et al. 2016). Varying histologic regression rates were observed and correlated with increasing vaccination dose (Coleman et al. 2016). Due to the demonstrated safety of PepCan, a phase II clinical trial to evaluate the efficacy and safety of PepCan in patients with HSIL is ongoing (NCT02481414). Similarly, another phase I/II clinical trial is underway investigating the safety and efficacy of PepCan in patients with head and neck cancer (NCT03821272). In a phase I dose-escalation study of GL-0810, an HPV peptide-based vaccine with Montanide^{TM} and granulocyte-macrophage colony-stimulating factor (GM-CSF) as adjuvants, patients with recurrent/metastatic squamous cell carcinoma of the head and neck (SCCHN) were shown to have an immune response of the vaccine, which was also well-tolerated (Zandberg et al. 2015). In addition, two more phase I clinical trials was designed to evaluate the safety and immunogenicity of PDS0101, a peptide-based therapeutic HPV vaccine. One of these trials was recently completed in women with high-risk HPV infection of CIN1 lesions (NCT02065973), while the other trial, a phase Ib/II trial in human leukocyte antigen (HLA)-A*02 positive patients with incurable HPV16-associated oropharyngeal, cervical, and anal cancer, is still ongoing (NCT02865135) (Yang et al. 2016).

4.4 Protein-Based Vaccines

Protein-based vaccines have also been developed as potential therapeutic vaccines for HPV-associated cancer. Not only are protein-based vaccines safer than live vector-based vaccines, but they also contain all HLA-binding epitopes, enabling protein-based vaccines to circumvent MHC restriction. However, one downside to protein-based vaccines is that they suffer from low immunogenicity. Additionally, most protein-based vaccines are presented through the MHC-II pathway resulting in the generation of antibody rather than antigen-specific cytotoxic T cell immune response (Su et al. 2010). Strategies to improve the immunogenicity and MHC-I processing and presentation of protein-based vaccines have been widely studied. To this end, numerous adjuvants, fusion proteins, and immunostimulating molecules have been tested, including the liposome-polycation-DNA (LPD) adjuvant (Cui and Huang 2005), saponin-based ISCOMATRIX (Frazer et al. 2004), and TLR agonist s (Kang et al. 2011).

One protein-based vaccine which has shown efficacy against HPV-associated diseases is TA-CIN, a fusion protein-based vaccine consisting of HPV16 E6, E7, and L2 (van der Burg et al. 2001). In several phase I/II clinical trials, TA-CIN has been shown to be safe and immunogenic against HPV infection. Additionally, in previous a phase II clinical trial, TA-CIN was administered in combination with imiquimod, a topical immunomodulator in patients with high-grade VIN. The combination treatment was well-tolerated in patients and these "responders" had increased levels of infiltrating CD4+ and CD8+ T cells both locally and in HPV-associated lesions (Daayana et al. 2010). Currently, a phase I clinical trial to determine the safety and feasibility of TA-CIN in patients with HPV16-associated cervical cancer is ongoing (NCT02405221). Another protein-based therapeutic HPV vaccine, which targets both HPV16 and 18 has undergone clinical studies (Van Damme et al. 2016). GTL001, is composed of recombinant HPV16 and 18 E7 proteins fused to catalytically inactive *Bordetella pertussis* CyaA expressed in *E. coli* (Yang et al. 2016; Van Damme et al. 2016). In a phase I clinical trial, GTL001 demonstrated tolerability and immunogenicity in women with HPV16 or 18 infections with normal cytology (Van Damme et al. 2016). A phase II clinical trial assessing the efficacy of GTL001 in women with HPV16 or 18 infections with normal cytology or atypical squamous cells of undetermined significance (ASC-US)/LSIL was recently completed; however, results have not yet been published (NCT01957878). Another phase II clinical trial to test the safety and efficacy of a different protein-based vaccine was recently completed. This phase IIa clinical trial was conducted to test the safety and efficacy of TVGV-1 vaccine construct, a fusion protein consisting of HPV16 E7 protein fused to pseudomonas aeruginosa exotoxin A (PE) and an endoplasmic reticulum retention signal (KDEL), to treat patients with cervical HSIL (NCT02576561). Data analysis from this trial is still ongoing.

4.5   Dendritic Cell-Based Vaccines

DCs are professional APCs that can induce an adaptive immune response by processing antigens to prime antigen-specific T cells. DCs play an important role in the regulation of the immune system and are commonly cited as the most efficient APCs (Yang et al. 2016; Lee et al. 2016). Autologous DCs can be pulsed *ex vivo* with peptides, proteins, or DNA-encoding antigens, and then reintroduced into the patient in order to elicit a cell-mediated immune response (Kim and Kim 2017). One benefit to DC-based vaccines is that DCs can serve as adjuvants to increase the strength of antigen-specific immunotherapies (Yang et al. 2016; Santin et al. 2005). Unfortunately, T cell-mediated apoptosis may limit the lifespan of DCs (Skeate et al. 2016); therefore, strategies to increase the immunogenicity and efficacy of DC-based vaccines, as well as prolong the survival of DCs are needed. Strategies have emerged, including the addition of adjuvants such as cholera toxin (Nurkkala et al. 2010) and TLR agonists (Chen et al. 2010). DCs have also been transfected with siRNAs, which target pro-apoptotic molecules to prevent the apoptosis of DCs (Kim et al. 2009; Peng et al. 2005; Ahn et al. 2015).

Several clinical trials have been developed to evaluate the therapeutic potential of DC-based therapeutic HPV vaccines. For example, in a phase I dose-escalation study to evaluate the safety, toxicity, and immunogenicity of DC-based vaccines in patients with stage IB or IIA cervical cancer, all patients developed CD4+ T cell and antibody responses to the DC-based vaccine. Furthermore, eight out of ten patients developed an E7-specific CD8+ T cell response. In their study, Santin et al. concluded that the HPV E7-loaded DC-based vaccine tested in this study was both safe and immunogenic (Santin et al. 2008). In another phase I clinical trial, the toxicity and immunogenicity of a DC-based vaccine was assessed in patients with HPV-associated, advanced, recurrent cervical cancer. The DC-based vaccine was shown to be well-tolerated by patients; however, it did not increase lymphocyte proliferation to a degree of statistical significance (Ramanathan et al. 2014). Nonetheless, there are several downsides to the use of DC-based vaccines. DC-based vaccines are technically difficult to manufacture and highly individualized, making them a poor vaccine choice for large-scale production. In addition, due to a lack of standard vaccine evaluation criteria and varying cell-culture techniques, vaccine quality can be inconsistent amongst batches, which is a particular challenge for large-scale production (Yang et al. 2016; Kim and Kim 2017; Skeate et al. 2016).

4.6 DNA-Based Vaccines

DNA-based vaccines have emerged as an attractive approach to therapeutic HPV vaccination as they offer several advantages over the other types of therapeutic vaccination. Specifically, DNA-based vaccines are safe, stable, and easy to produce on a large scale. DNA-based vaccination involves intramuscularly injecting a plasmid encoding the antigens of interest into the host's cells for uptake and processing by APCs. DNA-based vaccines are an especially attractive method for therapeutic vaccination because they are capable of mounting both cell-mediated and humoral immune responses against the encoded antigen (Kim and Kim 2017; Vici et al. 2016). While DNA vaccines have fewer safety risks than other types of therapeutic HPV vaccines, genomic integration remains a potential issue. However, no evidence of genomic integration has been observed (Yang et al. 2016; Wang et al. 2004). Naked DNA is easy to manufacture and can maintain antigen expression inside target cells for a longer duration than other therapeutic vaccine types. Unlike vector-based vaccines, DNA-based vaccines do not elicit neutralizing antibodies; therefore, they can be safely administered multiple times for booster vaccinations (Yang et al. 2016; Kim and Kim 2017; Vici et al. 2016). While DNA-based vaccines offer many advantages to therapeutic HPV vaccination, they have limited immunogenicity as they are unable to amplify and spread to surrounding cells. For this reason, DNA-based vaccines require strategies to improve antigen delivery to DCs in order to mount a potent immune response. In general, strategies to enhance therapeutic HPV DNA-based vaccine immunogenicity have focused on: (1) increasing the uptake of HPV antigen by DCs; (2) improving HPV antigen expression, processing, and presentation in DCs; and (3) enhancing DC and T cell interaction (for review see Cheng et al. 2018).

Several strategies have been employed to enhance the number of antigen-expressing/antigen-loaded DCs after DNA-based vaccination including, delivery by gene gun, microencapsulation, electroporation, laser, and microsphere-/nanoparticle-based delivery systems (Kim and Kim 2017). After DNA plasmids encoding HVP antigens are taken up by DCs, they must be expressed, processed, and presented. Enhancing antigen expression in transfected DCs, amplifying translation of HPV mRNA, increasing MHC-I and MHC-II expression in DCs, enhancing antigen processing through the MHC-II pathway, and increasing cross-presentation of HPV antigen through the MHC-I pathway are all methods that have been used to improve HPV antigen expression, processing, and presentation by DCs. Additionally, several methods to enhance DC and T cell function, survival, and interaction have also been employed to improve the therapeutic HPV DNA-based vaccine response. These methods include improving DC function and survival, boosting DC and T cell interactions, promoting T cell function and survival, as well as eliminating immunosuppressive regulatory T cells in the tumor microenvironment (TME) (for review see Cheng et al. 2018).

Therapeutic HPV DNA-based vaccines have undergone numerous clinical trials to evaluate safety and efficacy. In addition, DNA-based vaccines have been tested in conjunction with different adjuvants to improve vaccine potency. An ongoing phase I clinical study is evaluating the safety, tolerability, and feasibility of heterologous prime-boost regimen with therapeutic HPV DNA-based vaccine, pNGVL4a-Sig/E7(detox)/HSP70, and TA-HPV vaccine (NCT00788164). In this study, patients with HPV16-associated CIN3 will receive pNGVL4a-sig/E7(detox)/ HSP70 DNA vaccine and TA-HPV vaccine boost with or without imiquimod. Although this study is still ongoing, the preliminary results of this study suggest that this vaccine regimen can induce a localized immune response, which was largely responsible for the therapeutic effects observed in target lesions (Yang et al. 2016; Maldonado et al. 2014). Another phase I clinical trial, which assessed the efficacy and safety of various routes of administration of the pNGVL4a-CRT/E7 (detox) DNA vaccine in patients with HPV16-associated CIN2/3 was recently completed (NCT00988559). Patients were vaccinated with pNGVL4a-CRT/E7 (detox) either through intradermal, intramuscular, or intralesional injection. In this study, pNGVL4a-CRT/E7(detox) was shown to be well-tolerated by patients and elicited the most robust immune response when administered via intralesional injection (Alvarez et al. 2016). Many other clinical trials studying the efficacy of various therapeutic HPV DNA-based vaccines have been completed (NCT02139267) or are currently ongoing, including two early phase clinical studies evaluating the safety and efficacy of GX-188E DNA vaccine (Kim et al. 2014) (NCT02596243; NCT03444376) (for review see Yang et al. 2016). Additionally, more than ten clinical studies evaluating the safety and efficacy of HPV DNA vaccine VGX-3100 in patients with HPV-associated diseases have been completed (NCT03499795; NCT03180684; NCT03721978; NCT01188850; NCT03185013; NCT02163057; NCT00685412; NCT02172911; NCT03606213; NCT03439085) or are ongoing (NCT03603808). One such study, led by Trimble et al. in collaboration with Inovio Pharmaceuticals, found that half of the patients with HPV16/18 CIN2/3 who received VGX-3100 showed histopathological regression of their disease. This study is the first to demonstrate the therapeutic efficacy of VGX-3100 against HPV16/18-associated CIN2/3 as a potential alternative to surgical treatment (NCT01304524) (Trimble et al. 2015).

Numerous therapeutic HPV DNA-based vaccines have been developed or are under development for the control and treatment of HPV infections and HPV-associated diseases. As aforementioned, there are many benefits to the use of therapeutic HPV DNA-based vaccines and their adaptation in clinical studies has become favorable as novel methods to improve their immunogenicity have been generated and studied. Therapeutic HPV DNA-based vaccines also offer ample opportunity for the application of combinational treatment regimens to enhance vaccine potency. As described in Sect. 4.7, such as combinational regiments include prime-boost vaccination strategies, combination with conventional or standard-of-care cancer treatments such as chemotherapy and radiotherapy, as well as co-administration with different drugs (for review see Cheng et al. 2018). Thus, therapeutic HPV DNA-based vaccines are highly advantageous in the treatment of HPV infections and HPV-associated diseases, and continued efforts to develop more immunogenic DNA vaccines or DNA vaccine regimens are warranted for the amelioration of the global HPV-associated disease burden.

4.7 Combination Strategies

As briefly stated in Sect. 4.6, therapeutic HPV vaccines can be used in combination with other therapies to improve vaccine potency and host immune response. Combinatorial administration of therapeutic HPV vaccines along with established cancer treatment methods such as chemotherapy and/or radiation therapy may offer one method to improve the treatment of HPV-associated malignancies. Due to the growing popularity of HPV DNA-based vaccines, combination regimens of therapeutic HPV DNA-based vaccines have been explored in preclinical and clinical studies. The continued development of therapeutic HPV DNA-based vaccines largely relies on vaccine administration in combination with other treatment strategies to create synergistic effects capable of mounting potent CD8+ T cell-mediated immune responses (Cheng et al. 2018). For example, chemotherapy and radiotherapy are known to induce the apoptosis of tumor cells. In turn, this process releases HPV antigens into circulation where they can be more easily "seen" and up-taken by DCs traveling through the bloodstream. To this end, combination regimens of therapeutic HPV DNA-based vaccines in conjunction with chemotherapy or radiation therapy have been used to improve vaccine immunogenicity and demonstrated great potential. In preclinical studies, a therapeutic HPV DNA-based vaccine with calreticulin (CRT) fused to HPV16 E7 (CRT/E7) was administered in combination with chemotherapeutic agents cisplatin and bortezomib. This combinatorial approach was shown to generate a strong E7-specific CD8+ T cell responses in mice bearing an HPV16 E6/E7-expressing tumors (TC-1 tumor model) (Kim et al. 2014) compared to mice treated with CRT/E7 DNA vaccine only, or chemotherapy alone (Tseng et al. 2008a, b). Other chemotherapeutic agents have also been tested in preclinical studies, including 5,6-dimethylxanthenon-4-acetic acid (Peng et al. 2011), epigallocathechin-3-gallate (Kang et al. 2007), and 4′,5,7-trihydroxyflavone (Chuang et al. 2009) (for review see Cheng et al. 2018). Another method to enhance therapeutic HPV vaccine efficacy for the control and treatment of HPV-associated diseases is the combination of treatment with radiation therapy. In one such study, DNA vaccine with CRT linked to the modified form of HPV16 E7 antigen (CRT/E7(detox)) was administered in combination with radiotherapy in TC-1 tumor-bearing mice. Mice treated with CRT/E7(detox) and radiotherapy showed a significant increase in the number of E7-specific CD8+ T cells and had significantly better antitumor effects against E7-expressing tumors compared to mice treated with CRT/E7(detox) DNA vaccine, or radiotherapy alone (Tseng et al. 2009). The combination of chemotherapy and/or radiation therapy with therapeutic HPV vaccines has demonstrated promising antitumor results in preclinical studies and may offer significant benefits to patients with HPV-associated diseases.

Another example of a combinational strategy employed to enhance the potency of therapeutic HPV vaccines is prime-boost vaccination regimens. In previous studies, heterologous prime with DNA vaccine followed by boosting with recombinant protein has been shown to elicit both potent CD8+ and CD4+ T cell responses, as well as antibody responses. For example, in an ongoing clinical trial, a total of 12 patients with HPV16+ CIN2/3 were administered a heterologous DNA-prime-recombinant vaccinia vector-based boost vaccination regimen targeting HPV16 and HPV18 E6/E7 (NCT00788164). This prime-boost vaccination regimen was found to be safe and tolerable in patients with HPV16+ CIN2/3 lesions. The vaccination regimen included intramuscular prime vaccination with a DNA vaccine expressing HPV16 E7 (pNGVL4a-Sig/E7(detox)/HSP70), followed by a boost with TA-HPV, a recombinant vaccinia boost expressing HPV16 and HPV18 E6 and E7. This prime-boost vaccination strategy was also found to generate a more potent immune response than DNA vaccination alone (Cheng et al. 2018; Maldonado et al. 2014). Unfortunately, due to the risks associated with live vaccinia virus, TA-HPV may not be a suitable for administering prime-boost vaccination regimens in patients who are immune compromised (Peng et al. 2016). Additional studies have shown the therapeutic efficacy of prime-boost regimens using different recombinant proteins. For example, in a preclinical study, heterologous DNA-prime, TA-CIN protein boost was demonstrated to be safe and well-tolerated in both naïve and tumor-bearing mice (Peng et al. 2016). In this particular study, Peng et al. showed that heterologous DNA-prime with pNGVL4aCRTE6E7L2 DNA vaccine and TA-CIN protein boost regimen was capable of eliciting a more potent antigen-specific response than homologous DNA or protein vaccination alone. Specifically, the TA-CIN boosting regimen was able to generate an HPV16 L2, E6, and E7-specific immune response after heterologous DNA prime, representing the therapeutic efficacy of this regimen.

In addition to preclinical studies, several clinical studies have been conducted to evaluate the safety and efficacy of therapeutic HPV vaccines are prime-boost vaccination regimens in human patients. Currently, several ongoing clinical studies are testing the safety and tolerability of heterologous DNA prime vaccination with TA-CIN protein boost. Including a phase II clinical trial in patients with HPV16+ Atypical Squamous Cells of Undetermined Significance (ASC-US), which will evaluate the safety and tolerability of the prime-boost vaccination regimen with IM injection of pNGVL4a-Sig/E7(detox)/HSP70 DNA vaccine followed by IM TA-CIN protein boost (NCT03911076). In another phase I ongoing clinical study is assessing the safety and feasibility of IM administration of pNGVL4aCRTE6E7L2 DNA vaccine prime, TA-CIN protein boost in patients with persistent HPV16+ ASC-US/LSIL (NCT03913117). While these studies are ongoing, the potential therapeutic strategies of prime-boost vaccination regimens are promising and warrant further clinical investigation. Table 1 (adapted from Yang et al. 2017) lists several ongoing clinical trials evaluating the efficacy of various therapeutic HPV vaccines.

Table 1 Ongoing Therapeutic HPV vaccine clinical trials for HPV-associated diseases

Vaccine	Antigen(s)	Construct	Study sponsor	Trial design	Estimated date of trial completion	Clinical trials.gov identifier
Persistent HPV infection and Low-grade Squamous Intraepithelial Lesion (LSIL)						
Ad26-HPV16, Ad26-HPV18, and MVA-HPV16/18	HPV16/18 E6/E7	Adenovirus serotype 26 (Ad26)-human papillomavirus (HPV16 or HPV18) and Modified Vaccinia Ankara (MVA)-HPV16/18	Janssen Vaccines and prevention B.V.	Phase I/IIa in healthy female patients with persistent HPV16 or HPV18 infection of the cervix (66 estimated patients)	September 2020	NCT03610581
Atypical Squamous Cells of Undetermined Significance (ASC-US) or atypical squamous cells, cannot rule out high-grade SIL						
pNGVL4a-Sig/E7(detox)/ HSP70 + TA-CIN (PVX-2)	HPV16 E6/E7	Plasmid encoding mutated form of HPV16-E7 linked to Sig and HSP70 and HPV16 E6, E7, and L2 fusion protein	PapiVax Biotech Inc.	Phase II in female patients with confirmation of ASC-US, HSC-H or LSIL (122 estimated patients)	December 2021	NCT03911076
pNGVL4aCRTE6E7L2 + TA-CIN (PVX-6)	HPV16 E6/E7	Plasmid encoding Calreticulin, HPV16 E6, E7, and L2, and HPV16 E6, E7, and L2 fusion protein	PapiVax Biotech Inc.	Phase I in female patients with persistent ASC-US/LSIL (30 estimated patients)	December 2021	NCT03913117
Cervical Intraepithelial Neoplasia (CIN)/High-Grade Squamous Intraepithelial Lesion (HSIL)						
GX-188E	HPV16/18 E6/E7	Plasmid encoding fusion protein of HPV 16/18 E6/E7 linked to Flt3L and tpa	Genexine, Inc	Phase II in female patients with HPV16/18+ CIN2, CIN2/3, or CIN3 (134 enrolled patients)	August 2018	NCT02596243

(continued)

Table 1 (continued)

Vaccine	Antigen(s)	Construct	Study sponsor	Trial design	Estimated date of trial completion	Clinical trials.gov identifier
pNGVL4a-Sig/E7(detox)/HSP70 + TA-HPV	HPV16/18 E6/E7	Plasmid encoding mutated form of HPV16-E7 linked to Sig and HSP70 and vaccinia virus with HPV16/18 E6/E7	Sidney Kimmel Comprehensive Cancer Center	Phase I in female patients with HPV16+ CIN3 in combination with topical imiquimod (48 estimated patients)	June 2020	NCT00788164
TVGV-1 + GPI-0100	HPV16 E7	Fusion protein of HPV16 E7 and ER targeting sequence	THEVAX Genetics Vaccine Co.	Phase IIa in female patients with HPV16 or HPV16/18+ cervical HSIL (10 enrolled patients)	September 2018	NCT02576561
PepCan + Candin	HPV16 E6	HPV16 E6 peptides combined with Candida skin testing reagent candin.	University of Arkansas	Phase II in female patients with cervical HSIL (125 estimated patients)	August 2020	NCT02481414
Gardasil®9	hrHPV16, 18, 31, 33, 45, 52 and 58	Recombinant vaccine purified VLPs of the major capsid protein (L1) of HPV6, 11, 16, 18, 31, 33, 45, 52 and 58	AIDS Malignancy Consortium	Phase III trial in female patients with HIV-1 infection and co-infection with hrHPV16, 18, 31, 33, 35, 45, 52 or 58 (536 estimated patients)	October 2021	NCT03284866
Gardasil®9 + imiquimod	HPV6, 11, 16, 18, 31, 33, 45, 52 and 58	Gardasil®9: Recombinant vaccine purified VLPs of the major capsid protein (L1) of HPV6, 11, 16, 18, 31, 33, 45, 52 and 58 Imiquimod: TLR7/8 agonist	Yale University	Phase II trial in female patients with untreated HPV+ CIN2/3 (138 patients estimated)	April 2021	NCT02864147

(continued)

Table 1 (continued)

Vaccine	Antigen(s)	Construct	Study sponsor	Trial design	Estimated date of trial completion	Clinical trials.gov identifier
GX-188E + GX-I7 or imiquimod	HPV16/18 E6/E7	GX-188E: Plasmid encoding fusion protein of HPV 16/18 E6/E7 linked to Flt3L and tpa GX-17: recombinant human IL-7-hybrid Fc NT-I7 Imiquimod: TLR7/8 agonist	Seoul St. Mary's Hospital	Safety and efficacy trial in female patients with HPV16/18+ CIN3 (50 patients estimated)	October 2018	NCT03206138
Anal Intraepithelial Neoplasia (AIN)						
VGX-3100 + Electroporation	HPV16/18 E6/E7	Mixture of three plasmids encoding optimized consensus of E6 and E7 antigen of HPV 16 and 18	AIDS Malignancy Consortium	Phase II trial in male and female patients with HPV16/18+ anal HSIL (80 estimated patients)	September 2021	NCT03603808
HPV-associated incurable solid tumors						
ISA101 (SLP-HPV-01; HPV16-SLP) + nivolumab	HPV16 E6/E7	Combination of nine HPV-16 E6 and four HPV-16 E7 synthetic peptides with incomplete Freund's adjuvant	MD Anderson Cancer Center	Phase II trial in male and female patients with HPV16+ incurable solid tumors (OPSCC, cervical, vulvar, vaginal, anal, and penile cancer) as combination therapy with Nivolumab (34 enrolled patients)	December 2019	NCT02426892

(continued)

Table 1 (continued)

Vaccine	Antigen(s)	Construct	Study sponsor	Trial design	Estimated date of trial completion	Clinical trials.gov identifier
E7 TCR T cells (T cell-based vaccine)	HPV16 E7	Genetically engineered T cells with a T cell receptor (TCR) targeting HPV16 E7 (E7 TCR)	National Cancer Institute	Phase I/II trial in male and female patients with incurable metastatic or refractory/recurrent HPV16+ cervical, vulvar, vaginal, penile, anal, and oropharyngeal cancer) (180 estimated patients)	January 2026	NCT02858310
DPX-E7	HPV16 E7	Synthetic HPV16 E7 peptides packed into liposomes, freeze dried, then re-suspended in oil	Dana-Farber Cancer Institute	Phase Ib/II trial in male and female patients positive for HLA-A*02 with incurable HPV16+ head and neck, cervical or anal cancer	May 2023	NCT02865135
HARE-40	HPV16 E6/E7	Anti-CD40 IS-Ab ChiLob7/4	University of Southhampton	Phase I/II trial in male and female patients with advanced HPV+ head and neck, anogenital, penile, or cervical cancer	December 2020	NCT03418480
INO-3112 + durvalumab	HPV16/18 E6/E7	Mixture of three plasmids encoding optimized consensus of E6 and E7 antigen of HPV 16 and 18 and proprietary immune activator expressing IL-12	MD Anderson Cancer Center	Phase II trial in male and female patients with recurrent or metastatic HPV16/18+ cancers (anal, cervical, penial, vaginal, vulval) (77 estimated patients)	January 2020	NCT03439085

(continued)

Table 1 (continued)

Vaccine	Antigen(s)	Construct	Study sponsor	Trial design	Estimated date of trial completion	Clinical trials.gov identifier
Head and neck cancer						
ADXS11-001 (ADXS-HPV)	HPV16 E7	prfA-defective Listeria monocytogenes strain transformed with plasmid encoding HPV-16 E7 antigen fused to a fragment of nonhemolytic listeriolysin O (LLO)	Advaxis, Inc.	Phase II trial in male and female patients with HPV+ OPSCC (stage I-IV) prior to robot-assisted resection (30 estimated patients)	August 2021	NCT02002182
PepCan	HPV16 E6	HPV16 E6 peptides	University of Arkansas	Phase I/II trial in male and female patients with head and neck cancer (20 estimated participants)	December 2021	NCT03821272
MEDI0457 + durvalumab	HPV16/18 E6/E7	Mixture of three plasmids encoding optimized consensus of E6 and E7 antigen of HPV 16 and 18 and proprietary immune activator expressing IL-12	MedImmune LLC	Phase Ib/IIa trial in male and female patients with HPV-associated recurrent or metastatic head and neck cancer (50 estimated patients)	August 2020	NCT03162224
TG4001 + Avelumab	HPV16 E6/E7	Suspension of MVATG8042 vector particles, which consists of an attenuated recombinant MVA encoding modified HPV16 E6 and E7, and human interleukin-2 (IL-2)	Transgene	Phase Ib/II trial in male and female patients with HPV16+ recurrent or metastatic OPSCC of the head and neck (52 estimated patients)	December 2021	NCT03260023

(continued)

Table 1 (continued)

Vaccine	Antigen(s)	Construct	Study sponsor	Trial design	Estimated date of trial completion	Clinical trials.gov identifier
Cervical cancer						
ADXS11-001 (Lm-LLo-E7)	HPV16 E7	prfA-defective Listeria monocytogenes strain transformed with plasmid encoding HPV-16 E7 antigen fused to a fragment of nonhemolytic listeriolysin O (LLO)	Advaxis, Inc.	Phase II trial in female patients with persistent or recurrent squamous or non-squamous cell carcinoma of the cervix (67 estimated patients)	October 2018	NCT01266460
ISA101/ISA101b (SLP-HPV-01; HPV16-SLP)	HPV16 E6/E7	Combination of nine HPV-16 E6 and four HPV-16 E7 synthetic peptides with incomplete Freund's adjuvant	ISA Pharmaceuticals	Phase I/II trial in female patients with HPV-16+ advanced or recurrent cervical cancer (100 estimated patients)	April 2021	NCT02128126
TA-CIN	HPV16 E6/E7/L2	HPV16 E6, E7, L2 fusion protein	Sidney Kimmel Comprehensive Cancer Center	Phase I trial in female patients with a history of HPV16-associated cervical cancer (stage IB1-IV) (14 patients estimated)	November 2022	NCT02405221
GX-188E + pembrolizumab	HPV16/18 E6/E7	Plasmid encoding fusion protein of HPV 16/18 E6/E7 linked to Flt3L and tpa	Genexine, Inc	Phase Ib/II trial in female patients with advanced, non-resectable HPV+ cervical cancer (46 estimated patients)	June 2023	NCT03444376

(continued)

Table 1 (continued)

Vaccine	Antigen(s)	Construct	Study sponsor	Trial design	Estimated date of trial completion	Clinical trials.gov identifier
BVAC-C	HPV16/18 E6/E7	B cell and monocyte-based vaccine transfected with recombinant HPV E6/E7 gene and loaded with alpha-glactosyl ceramide	Celid Co., Ltd.	Phase I/II trial in females with multiple metastatic progressive or recurrent HPV16 or 18+ cervical cancer (30 estimated patients)	August 2020	NCT02866006
INO-3112 + durvalumab	HPV16/18 E6/E7	Mixture of three plasmids encoding optimized consensus of E6 and E7 antigen of HPV 16 and 18 and proprietary immune activator expressing IL-12	MD Anderson Cancer Center	Phase II trial in female patients with recurrent or metastatic HPV16/18+ cancers (77 estimated patients)	January 2020	NCT03439085

AIN Anal intraepithelial Neoplasia; *ASC-US* Atypical Squamous Cells of Undetermined Significance; *ASC-H* Atypical Squamous Cells, Cannot Rule out High-Grade SIL; *CIN* Cervical intraepithelial neoplasia; *ER* Endoplasmic reticulum; *HIV* Human immunodeficiency virus; *HPV* Human papillomavirus; *HSIL* High-grade squamous intraepithelial lesion; *LSIL* Low-grade squamous intraepithelial lesion; *OPSCC* oropharyngeal squamous cell carcinoma

5 Conclusion

While HPV infections and HPV-associated diseases remain highly prevalent across the globe, the development of effective, and accessible methods for the control of HPV infections and HPV-associated diseases continue to pose a significant challenge for researchers, healthcare providers, and public health workers. The current, commercially available prophylactic vaccines represent a significant triumph in HPV research and public health; however, there is still an urgent need to improve methods for the control of HPV-associated disease. Methods to improve HPV prevention and the control of HPV-associated diseases include: (1) develop broader vaccine coverage against oncogenic HPV types; (2) reduce vaccine-related costs; (3) reduce minimum dose requirements; (4) improve vaccine stability; (5) make prophylactic HPV vaccines more widely accessible; and (6) employ our current understanding of the immunology of HPV infections to create successful and efficacious therapeutic HPV vaccines. Numerous innovative strategies have been employed in order to develop therapeutic HPV vaccines, including many which have resulted in phase I, II, and III clinical trials. In order to best identify suitable therapeutic HPV vaccine candidates for the control of established HPV infections and HPV-associated lesions or disease, we must continue to devise and implement clinical studies, and improve upon existing therapeutic strategies that have demonstrated promising clinical translatability. Additionally, the control of advanced HPV-associated diseases will likely necessitate further investigation into the ideal combinatorial approaches using therapeutic HPV vaccines in conjunction with conventional cancer therapies such as chemotherapy and radiation therapy. Continued efforts to advance both prophylactic and therapeutic HPV vaccines will undoubtedly relieve the global burden of HPV and help reduce incidence rates of HPV infection in millions of people each year around the world.

Acknowledgements This review is not intended to be encyclopedic, and the authors apologize to those not cited. This work was funded by the National Institutes of Health Cervical Cancer SPORE (P50 CA098252).

References

Ahn YH, Hong SO, Kim JH, Noh KH, Song KH, Lee YH, Jeon JH, Kim DW, Seo JH, Kim TW (2015) The siRNA cocktail targeting interleukin 10 receptor and transforming growth factor-beta receptor on dendritic cells potentiates tumour antigen-specific CD8(+) T cell immunity. Clin Exp Immunol 181(1):164–178. Epub 2015/03/11. https://doi.org/10.1111/cei.12620. PubMed PMID: 25753156; PMCID: PMC4469167

Alvarez RD, Huh WK, Bae S, Lamb LS, Jr., Conner MG, Boyer J, Wang C, Hung CF, Sauter E, Paradis M, Adams EA, Hester S, Jackson BE, Wu TC, Trimble CL (2016) A pilot study of pNGVL4a-CRT/E7(detox) for the treatment of patients with HPV16+ cervical intraepithelial neoplasia 2/3 (CIN2/3). Gynecol Oncol 140(2):245–252. Epub 2015/12/01. https://doi.org/10.1016/j.ygyno.2015.11.026. PubMed PMID: 26616223; PMCID: PMC4724445

American Cancer Society (2018) Can oral cavity and oropharyngeal cancers be found early? Available from https://www.cancer.org/cancer/oral-cavity-and-oropharyngeal-cancer/detection-diagnosis-staging/detection.html. Updated 9 Mar 2018; Cited 2019

Baldwin PJ, van der Burg SH, Boswell CM, Offringa R, Hickling JK, Dobson J, Roberts JS, Latimer JA, Moseley RP, Coleman N, Stanley MA, Sterling JC (2003) Vaccinia-expressed human papillomavirus 16 and 18 e6 and e7 as a therapeutic vaccination for vulval and vaginal intraepithelial neoplasia. Clin Cancer Res 9(14):5205–5213 Epub 2003/11/14 PubMed PMID: 14614000

Barra F, Maggiore ULR, Bogani G, Ditto A, Signorelli M, Martinelli F, Chiappa V, Lorusso D, Raspagliesi F, Ferrero S (2019) New prophylactics human papilloma virus (HPV) vaccines against cervical cancer. J Obstet Gynaecol 39(1):1–10. Epub 2018/10/30. https://doi.org/10.1080/01443615.2018.1493441. PubMed PMID: 30370796

Borysiewicz LK, Fiander A, Nimako M, Man S, Wilkinson GW, Westmoreland D, Evans AS, Adams M, Stacey SN, Boursnell ME, Rutherford E, Hickling JK, Inglis SC (1996) A recombinant vaccinia virus encoding human papillomavirus types 16 and 18, E6 and E7 proteins as immunotherapy for cervical cancer. Lancet 347(9014):1523–1527 Epub 1996/06/01 PubMed PMID: 8684105

Boxus M, Fochesato M, Miseur A, Mertens E, Dendouga N, Brendle S, Balogh KK, Christensen ND, Giannini SL (2016) Broad cross-protection is induced in preclinical models by a human papillomavirus vaccine composed of L1/L2 chimeric virus-like particles J Virol 90 (14):6314–6325. Epub 2016/05/06. https://doi.org/10.1128/jvi.00449-16. PubMed PMID: 27147749; PMCID: PMC4936133

Brianti P, De Flammineis E, Mercuri SR (2017) Review of HPV-related diseases and cancers. New Microbiol 40(2):80–85 Epub 2017/04/04 PubMed PMID: 28368072

Brun JL, Dalstein V, Leveque J, Mathevet P, Raulic P, Baldauf JJ, Scholl S, Huynh B, Douvier S, Riethmuller D, Clavel C, Birembaut P, Calenda V, Baudin M, Bory JP (2011) Regression of high-grade cervical intraepithelial neoplasia with TG4001 targeted immunotherapy (2011) Am J Obstet Gynecol 204(2):169 e1–e8. Epub 2011/02/03. https://doi.org/10.1016/j.ajog.2010.09.020. PubMed PMID: 21284968

Bruni L, Diaz M, Barrionuevo-Rosas L, Herrero R, Bray F, Bosch FX, de Sanjose S, Castellsague X (2016) Global estimates of human papillomavirus vaccination coverage by region and income level: a pooled analysis. Lancet Glob Health 4(7):e453–e463. Epub 2016/06/25. https://doi.org/10.1016/s2214-109x(16)30099-7. PubMed PMID: 27340003

Buttaro C, Fruehauf JH (2010) Engineered E. coli as vehicles for targeted therapeutics. Curr Gene Ther 10(1):27–33. Epub 2010/02/17. PubMed PMID: 20156190

Caldeira Jdo C, Medford A, Kines RC, Lino CA, Schiller JT, Chackerian B, Peabody DS (2010) Immunogenic display of diverse peptides, including a broadly cross-type neutralizing human papillomavirus L2 epitope, on virus-like particles of the RNA bacteriophage PP7. Vaccine 28 (27):4384–4393. Epub 2010/05/04. https://doi.org/10.1016/j.vaccine.2010.04.049. PubMed PMID: 20434554; PMCID: PMC2881612

Centers for Disease Control and Prevention (2012) Cervical cancer screening guidelines for average-risk womena. Available from https://www.cdc.gov/cancer/cervical/pdf/guidelines.pdf. Cited 2019

Centers for Disease Control and Prevention (2015) Use of 9-valent human papillomavirus (HPV) vaccine: updated HPV vaccination recommendations of the advisory committee on immunization practices. Available from https://www.cdc.gov/mmwr/preview/mmwrhtml/mm6411a3.htm. Updated 27 Mar 2015; Cited 2019

Centers for Disease Control and Prevention (2017) Genital HPV infection—Fact sheet. Available from https://www.cdc.gov/std/hpv/stdfact-hpv.htm. Updated 16 Nov 2017; Cited 2019

Chabeda A, Yanez RJR, Lamprecht R, Meyers AE, Rybicki EP, Hitzeroth, II (2018) Therapeutic vaccines for high-risk HPV-associated diseases. Papillomavirus Res 5:46–58. Epub 2017/12/27. https://doi.org/10.1016/j.pvr.2017.12.006. PubMed PMID: 29277575; PMCID: PMC5887015

Chen XZ, Mao XH, Zhu KJ, Jin N, Ye J, Cen JP, Zhou Q, Cheng H (2010) Toll like receptor agonists augment HPV 11 E7-specific T cell responses by modulating monocyte-derived dendritic cells. Arch Dermatol Res 302(1):57–65. Epub 2009/07/07. https://doi.org/10.1007/s00403-009-0976-0. PubMed PMID: 19578865

Cheng MA, Farmer E, Huang C, Lin J, Hung CF, Wu TC (2018) Therapeutic DNA vaccines for human papillomavirus and associated diseases. Hum Gene Ther 29(9):971–976. Epub 2018/01/11. https://doi.org/10.1089/hum.2017.197. PubMed PMID: 29316817; PMCID: PMC6152857

Chuang CM, Monie A, Wu A, Hung CF (2009) Combination of apigenin treatment with therapeutic HPV DNA vaccination generates enhanced therapeutic antitumor effects. J Biomed Sci 16:49. Epub 2009/05/29. https://doi.org/10.1186/1423-0127-16-49. PubMed PMID: 19473507; PMCID: PMC2705346

Coleman HN, Greenfield WW, Stratton SL, Vaughn R, Kieber A, Moerman-Herzog AM, Spencer HJ, Hitt WC, Quick CM, Hutchins LF, Mackintosh SG, Edmondson RD, Erickson SW, Nakagawa M (2016) Human papillomavirus type 16 viral load is decreased following a therapeutic vaccination. Cancer Immunol Immunother 65(5):563–573. Epub 2016/03/17. https://doi.org/10.1007/s00262-016-1821-x. PubMed PMID: 26980480; PMCID: PMC4841729

Cooper CP, Saraiya M (2017) Primary HPV testing recommendations of US providers, 2015. Prev Med 105:372–377. Epub 2017/10/24. https://doi.org/10.1016/j.ypmed.2017.08.006. PubMed PMID: 29056319; PMCID: PMC5809311

Corona Gutierrez CM, Tinoco A, Navarro T, Contreras ML, Cortes RR, Calzado P, Reyes L, Posternak R, Morosoli G, Verde ML, Rosales R (2004) Therapeutic vaccination with MVA E2 can eliminate precancerous lesions (CIN 1, CIN 2, and CIN 3) associated with infection by oncogenic human papillomavirus. Hum Gene Ther 15(5):421–431. Epub 2004/05/18. https://doi.org/10.1089/10430340460745757. PubMed PMID: 15144573

Cory L, Chu C (2014) ADXS-HPV: a therapeutic Listeria vaccination targeting cervical cancers expressing the HPV E7 antigen. Hum Vaccin Immunother 10(11):3190–3195. Epub 2014/12/09. https://doi.org/10.4161/hv.34378. PubMed PMID: 25483687; PMCID: PMC4514130

Cui Z, Huang L (2005) Liposome-polycation-DNA (LPD) particle as a carrier and adjuvant for protein-based vaccines: therapeutic effect against cervical cancer. Cancer Immunol Immunother 54(12):1180–1190. Epub 2005/04/23. https://doi.org/10.1007/s00262-005-0685-2. PubMed PMID: 15846491

da Silva AJ, Zangirolami TC, Novo-Mansur MT, Giordano Rdc C, Martins EA (2014) Live bacterial vaccine vectors: an overview. Braz J Microbiol 45(4):1117–1129. Epub 2014/01/01. PubMed PMID: 25763014; PMCID: PMC4323283

Daayana S, Elkord E, Winters U, Pawlita M, Roden R, Stern PL, Kitchener HC (2010) Phase II trial of imiquimod and HPV therapeutic vaccination in patients with vulval intraepithelial neoplasia. Br J Cancer 102(7):1129–1136. Epub 2010/03/18. https://doi.org/10.1038/sj.bjc.6605611. PubMed PMID: 20234368; PMCID: PMC2853099

Daftarian P, Mansour M, Benoit AC, Pohajdak B, Hoskin DW, Brown RG, Kast WM (2006) Eradication of established HPV 16-expressing tumors by a single administration of a vaccine composed of a liposome-encapsulated CTL-T helper fusion peptide in a water-in-oil emulsion. Vaccine 24(24):5235–5244. Epub 2006/05/06. https://doi.org/10.1016/j.vaccine.2006.03.079. PubMed PMID: 16675074

Davidson EJ, Boswell CM, Sehr P, Pawlita M, Tomlinson AE, McVey RJ, Dobson J, Roberts JS, Hickling J, Kitchener HC, Stern PL (2003) Immunological and clinical responses in women with vulval intraepithelial neoplasia vaccinated with a vaccinia virus encoding human papillomavirus 16/18 oncoproteins. Cancer Res 63(18):6032–6041 Epub 2003/10/03 PubMed PMID: 14522932

Day PM, Pang YY, Kines RC, Thompson CD, Lowy DR, Schiller JT (2012) A human papillomavirus (HPV) in vitro neutralization assay that recapitulates the in vitro process of infection provides a sensitive measure of HPV L2 infection-inhibiting antibodies. Clin Vaccine

Immunol 19(7):1075–1082. Epub 2012/05/18. https://doi.org/10.1128/cvi.00139-12. PubMed PMID: 22593236; PMCID: PMC3393370

de Martel C, Ferlay J, Franceschi S, Vignat J, Bray F, Forman D, Plummer M (2012) Global burden of cancers attributable to infections in 2008: a review and synthetic analysis. Lancet Oncol 13(6):607–615. Epub 2012/05/12. https://doi.org/10.1016/s1470-2045(12)70137-7. PubMed PMID: 22575588

de Martel C, Plummer M, Vignat J, Franceschi S (2017) Worldwide burden of cancer attributable to HPV by site, country and HPV type. Int J Cancer 141(4):664–670. Epub 2017/04/04. https://doi.org/10.1002/ijc.30716. PubMed PMID: 28369882; PMCID: PMC5520228

de Vos van Steenwijk PJ, van Poelgeest MI, Ramwadhdoebe TH, Lowik MJ, Berends-van der Meer DM, van der Minne CE, Loof NM, Stynenbosch LF, Fathers LM, Valentijn AR, Oostendorp J, Osse EM, Fleuren GJ, Nooij L, Kagie MJ, Hellebrekers BW, Melief CJ, Welters MJ, van der Burg SH, Kenter GG (2014) The long-term immune response after HPV16 peptide vaccination in women with low-grade pre-malignant disorders of the uterine cervix: a placebo-controlled phase II study. Cancer Immunol Immunother 63(2):147–160. Epub 2013/11/16. https://doi.org/10.1007/s00262-013-1499-2. PubMed PMID: 24233343

DiGiuseppe S, Bienkowska-Haba M, Guion LGM, Keiffer TR, Sapp M (2017) Human papillomavirus major capsid protein L1 remains associated with the incoming viral genome throughout the entry process. J Virol. Epub 2017/06/02. https://doi.org/10.1128/jvi.00537-17. PubMed PMID: 28566382; PMCID: PMC5533910

Egawa N, Egawa K, Griffin H, Doorbar J (2015) Human papillomaviruses; epithelial tropisms, and the development of neoplasia. Viruses 7(7):3863–3890. Epub 2015/07/21. https://doi.org/10.3390/v7072802. PubMed PMID: 26193301; PMCID: PMC4517131

Elrefaey S, Massaro MA, Chiocca S, Chiesa F, Ansarin M (2014) HPV in oropharyngeal cancer: the basics to know in clinical practice. Acta Otorhinolaryngol Ital 34(5):299–309. Epub 2015/02/25. PubMed PMID: 25709145; PMCID: PMC4299160

Frazer IH, Quinn M, Nicklin JL, Tan J, Perrin LC, Ng P, O'Connor VM, White O, Wendt N, Martin J, Crowley JM, Edwards SJ, McKenzie AW, Mitchell SV, Maher DW, Pearse MJ, Basser RL (2004) Phase 1 study of HPV16-specific immunotherapy with E6E7 fusion protein and ISCOMATRIX adjuvant in women with cervical intraepithelial neoplasia. Vaccine 23 (2):172–181. Epub 2004/11/09. https://doi.org/10.1016/j.vaccine.2004.05.013. PubMed PMID: 15531034

Gallagher KE, LaMontagne DS, Watson-Jones D (2018) Status of HPV vaccine introduction and barriers to country uptake. Vaccine 36(32 Pt A):4761–4767. Epub 2018/03/28. https://doi.org/10.1016/j.vaccine.2018.02.003. PubMed PMID: 29580641

Gambhira R, Gravitt PE, Bossis I, Stern PL, Viscidi RP, Roden RB (2006) Vaccination of healthy volunteers with human papillomavirus type 16 L2E7E6 fusion protein induces serum antibody that neutralizes across papillomavirus species. Cancer Res 66(23):11120–11124. Epub 2006/12/06. https://doi.org/10.1158/0008-5472.can-06-2560. PubMed PMID: 17145854

Gambhira R, Jagu S, Karanam B, Gravitt PE, Culp TD, Christensen ND, Roden RB (2007a) Protection of rabbits against challenge with rabbit papillomaviruses by immunization with the N terminus of human papillomavirus type 16 minor capsid antigen L2. J Virol 81(21):11585–11592. Epub 2007/08/24. https://doi.org/10.1128/jvi.01577-07. PubMed PMID: 17715230; PMCID: PMC2168774

Gambhira R, Karanam B, Jagu S, Roberts JN, Buck CB, Bossis I, Alphs H, Culp T, Christensen ND, Roden RB (2007b) A protective and broadly cross-neutralizing epitope of human papillomavirus L2. J Virol 81(24):13927–13931. Epub 2007/10/12. https://doi.org/10.1128/jvi.00936-07. PubMed PMID: 17928339; PMCID: PMC2168823

Garcia-Hernandez E, Gonzalez-Sanchez JL, Andrade-Manzano A, Contreras ML, Padilla S, Guzman CC, Jimenez R, Reyes L, Morosoli G, Verde ML, Rosales R (2006) Regression of papilloma high-grade lesions (CIN 2 and CIN 3) is stimulated by therapeutic vaccination with MVA E2 recombinant vaccine. Cancer Gene Ther 13(6):592–597. Epub 2006/02/04. https://doi.org/10.1038/sj.cgt.7700937. PubMed PMID: 16456551

Gaspar J, Quintana SM, Reis RK, Gir E (2015) Sociodemographic and clinical factors of women with HPV and their association with HIV. Rev Lat Am Enfermagem 23(1):74–81. Epub 2015/03/26. https://doi.org/10.1590/0104-1169.3364. PubMed PMID: 25806634; PMCID: PMC4376034

Gillison ML, Chaturvedi AK, Anderson WF, Fakhry C (2015) Epidemiology of human papillomavirus-positive head and neck squamous cell carcinoma. J Clin Onco l33(29):3235–3242. Epub 2015/09/10. https://doi.org/10.1200/jco.2015.61.6995. PubMed PMID: 26351338; PMCID: PMC4979086

Graham SV (2010) Human papillomavirus: gene expression, regulation and prospects for novel diagnostic methods and antiviral therapies. Future Microbiol 5(10):1493–1506. Epub 2010/11/16. https://doi.org/10.2217/fmb.10.107. PubMed PMID: 21073310; PMCID: PMC3527891

Guirnalda P, Wood L, Paterson Y (2012) Listeria monocytogenes and its products as agents for cancer immunotherapy. Adv Immunol 113:81–118. Epub 2012/01/17. https://doi.org/10.1016/b978-0-12-394590-7.00004-x. PubMed PMID: 22244580

Gunn GR, Zubair A, Peters C, Pan ZK, Wu TC, Paterson Y (2001) Two Listeria monocytogenes vaccine vectors that express different molecular forms of human papilloma virus-16 (HPV-16) E7 induce qualitatively different T cell immunity that correlates with their ability to induce regression of established tumors immortalized by HPV-16. J Immunol 167(11):6471–6479 Epub 2001/11/21 PubMed PMID: 11714814

Guo T, Eisele DW, Fakhry C (2016) The potential impact of prophylactic human papillomavirus vaccination on oropharyngeal cancer. Cancer 122(15):2313–2323. Epub 2016/05/07. https://doi.org/10.1002/cncr.29992. PubMed PMID: 27152637; PMCID: PMC4956510

Harper DM, DeMars LR (2017) HPV vaccines—A review of the first decade. Gynecol Oncol 146 (1):196–204. Epub 2017/04/27. https://doi.org/10.1016/j.ygyno.2017.04.004. PubMed PMID: 28442134

Hung CF, Ma B, Monic A, Tsen SW, Wu TC (2008) Therapeutic human papillomavirus vaccines: current clinical trials and future directions. Expert Opin Biol Ther 8(4):421–439. Epub 2008/03/21. https://doi.org/10.1517/14712598.8.4.421. PubMed PMID: 18352847; PMCID: PMC3074340

Hussain SF, Paterson Y (2004) CD4+CD25+ regulatory T cells that secrete TGFbeta and IL-10 are preferentially induced by a vaccine vector. J Immunother 27(5):339–346 Epub 2004/08/18 PubMed PMID: 15314542

Immunization Action Coalition (2019) Human papillomavirus (HPV). Available from http://www.immunize.org/askexperts/experts_hpv.asp. Updated 4 Mar 2019; Cited 2019

Jagu S, Karanam B, Gambhira R, Chivukula SV, Chaganti RJ, Lowy DR, Schiller JT, Roden RB (2009) Concatenated multitype L2 fusion proteins as candidate prophylactic pan-human papillomavirus vaccines. J Natl Cancer Inst 101(11):782–792. Epub 2009/05/28. https://doi.org/10.1093/jnci/djp106. PubMed PMID: 19470949; PMCID: PMC2689872

Joura EA, Giuliano AR, Iversen OE, Bouchard C, Mao C, Mehlsen J, Moreira ED, Jr., Ngan Y, Petersen LK, Lazcano-Ponce E, Pitisuttithum P, Restrepo JA, Stuart G, Woelber L, Yang YC, Cuzick J, Garland SM, Huh W, Kjaer SK, Bautista OM, Chan IS, Chen J, Gesser R, Moeller E, Ritter M, Vuocolo S, Luxembourg A (2015) Broad spectrum HPVVS. A 9-valent HPV vaccine against infection and intraepithelial neoplasia in women. N Engl J Med372(8):711–723. Epub 2015/02/19. https://doi.org/10.1056/nejmoa1405044. PubMed PMID: 25693011

Kaiser KJ, Family Foundation (2018) The HPV vaccine: access and use in the U.S. Available from https://www.kff.org/womens-health-policy/fact-sheet/the-hpv-vaccine-access-and-use-in-the-u-s/. Updated 9 Oct 2018; Cited 2019

Kaliamurthi S, Selvaraj G, Chinnasamy S, Wang Q, Nangraj AS, Cho WC, Gu K, Wei DQ (2019) Exploring the papillomaviral proteome to identify potential candidates for a chimeric vaccine against cervix papilloma using immunomics and computational structural vaccinology. Viruses 11(1). Epub 2019/01/18. https://doi.org/10.3390/v11010063. PubMed PMID: 30650527; PMCID: PMC6357041

Kang TH, Lee JH, Song CK, Han HD, Shin BC, Pai SI, Hung CF, Trimble C, Lim JS, Kim TW, Wu TC (2007) Epigallocatechin-3-gallate enhances CD8+ T cell-mediated antitumor immunity induced by DNA vaccination. Cancer Res 67(2):802–811. Epub 2007/01/20. https://doi.org/10.1158/0008-5472.can-06-2638. PubMed PMID: 17234792; PMCID: PMC3181129

Kang TH, Monie A, Wu LS, Pang X, Hung CF, Wu TC (2011) Enhancement of protein vaccine potency by in vivo electroporation mediated intramuscular injection. Vaccine 29(5):1082–1089. Epub 2010/12/07. https://doi.org/10.1016/j.vaccine.2010.11.063. PubMed PMID: 21130752; PMCID: PMC3026065

Kaufmann AM, Stern PL, Rankin EM, Sommer H, Nuessler V, Schneider A, Adams M, Onon TS, Bauknecht T, Wagner U, Kroon K, Hickling J, Boswell CM, Stacey SN, Kitchener HC, Gillard J, Wanders J, Roberts JS, Zwierzina H (2002) Safety and immunogenicity of TA-HPV, a recombinant vaccinia virus expressing modified human papillomavirus (HPV)-16 and HPV-18 E6 and E7 genes, in women with progressive cervical cancer. Clin Cancer Res 8 (12):3676–3685 Epub 2002/12/11 PubMed PMID: 12473576

Khong H, Overwijk WW (2016) Adjuvants for peptide-based cancer vaccines. J Immunother Cancer 4:56. Epub 2016/09/24. https://doi.org/10.1186/s40425-016-0160-y. PubMed PMID: 27660710; PMCID: PMC5028954

Kim HJ, Kim HJ (2017) Current status and future prospects for human papillomavirus vaccines. Arch Pharm Res 40(9):1050–1063. Epub 2017/09/07. https://doi.org/10.1007/s12272-017-0952-8. PubMed PMID: 28875439

Kim JH, Kang TH, Noh KH, Bae HC, Kim SH, Yoo YD, Seong SY, Kim TW (2009) Enhancement of dendritic cell-based vaccine potency by anti-apoptotic siRNAs targeting key pro-apoptotic proteins in cytotoxic CD8(+) T cell-mediated cell death. Immunol Lett 122 (1):58–67. Epub 2009/01/13. https://doi.org/10.1016/j.imlet.2008.12.006. PubMed PMID: 19135479

Kim TJ, Jin HT, Hur SY, Yang HG, Seo YB, Hong SR, Lee CW, Kim S, Woo JW, Park KS, Hwang YY, Park J, Lee IH, Lim KT, Lee KH, Jeong MS, Surh CD, Suh YS, Park JS, Sung YC (2014) Clearance of persistent HPV infection and cervical lesion by therapeutic DNA vaccine in CIN3 patients. Nat Commun 5:5317. Epub 2014/10/31. https://doi.org/10.1038/ncomms6317. PubMed PMID: 25354725; PMCID: PMC4220493

Kreimer AR, Struyf F, Del Rosario-Raymundo MR, Hildesheim A, Skinner SR, Wacholder S, Garland SM, Herrero R, David MP, Wheeler CM, Costa Rica Vaccine Trial Study Group A, Gonzalez P, Jimenez S, Lowy DR, Pinto LA, Porras C, Rodriguez AC, Safaeian M, Schiffman M, Schiller JT, Schussler J, Sherman ME, Authors PSG, Bosch FX, Castellsague X, Chatterjee A, Chow SN, Descamps D, Diaz-Mitoma F, Dubin G, Germar MJ, Harper DM, Lewis DJ, Limson G, Naud P, Peters K, Poppe WA, Ramjattan B, Romanowski B, Salmeron J, Schwarz TF, Teixeira JC, Tjalma WA, Collaborators HPPIC-PI, Group GSKVCSS (2015) Efficacy of fewer than three doses of an HPV-16/18 AS04-adjuvanted vaccine: combined analysis of data from the costa rica vaccine and PATRICIA trials. Lancet Oncol. 16(7):775–786. Epub 2015/06/14. https://doi.org/10.1016/s1470-2045(15)00047-9. PubMed PMID: 26071347; PMCID: PMC4498478

Le Gouellec A, Chauchet X, Polack B, Buffat L, Toussaint B (2012) Bacterial vectors for active immunotherapy reach clinical and industrial stages. Hum Vaccin Immunother 8(10):1454–1458. Epub 2012/08/17. https://doi.org/10.4161/hv.21429. PubMed PMID: 22894945; PMCID: PMC3660766

Lee SJ, Yang A, Wu TC, Hung CF (2016) Immunotherapy for human papillomavirus-associated disease and cervical cancer: review of clinical and translational research. J Gynecol Oncol 27 (5):e51. Epub 2016/06/23. https://doi.org/10.3802/jgo.2016.27.e51. PubMed PMID: 27329199; PMCID: PMC4944018

Lin CW, Lee JY, Tsao YP, Shen CP, Lai HC, Chen SL (2002) Oral vaccination with recombinant Listeria monocytogenes expressing human papillomavirus type 16 E7 can cause tumor growth in mice to regress. Int J Cancer 102(6):629–637. Epub 2002/11/26. https://doi.org/10.1002/ijc.10759. PubMed PMID: 12448006

Lin K, Doolan K, Hung CF, Wu TC (2010) Perspectives for preventive and therapeutic HPV vaccines. J Formos Med Assoc 109(1):4–24. Epub 2010/02/04. PubMed PMID: 20123582; PMCID: PMC2908016

Maldonado L, Teague JE, Morrow MP, Jotova I, Wu TC, Wang C, Desmarais C, Boyer JD, Tycko B, Robins HS, Clark RA, Trimble CL (2014) Intramuscular therapeutic vaccination targeting HPV16 induces T cell responses that localize in mucosal lesions. Sci Transl Med 6 (221):221ra13. Epub 2014/01/31. https://doi.org/10.1126/scitranslmed.3007323. PubMed PMID: 24477000; PMCID: PMC4086631

Manini I, Montomoli E (2018) Epidemiology and prevention of Human Papillomavirus. Ann Ig30 (4 Suppl 1):28–32. Epub 2018/08/01. https://doi.org/10.7416/ai.2018.2231. PubMed PMID: 30062377

Massarelli E, William W, Johnson F, Kies M, Ferrarotto R, Guo M, Feng L, Lee JJ, Tran H, Kim YU, Haymaker C, Bernatchez C, Curran M, Zecchini Barrese T, Rodriguez Canales J, Wistuba I, Li L, Wang J, van der Burg SH, Melief CJ, Glisson B (2019) Combining immune checkpoint blockade and tumor-specific vaccine for patients with incurable human papillomavirus 16-related cancer: a phase 2 clinical trial. JAMA Oncol 5(1):67–73. Epub 2018/09/30. https://doi.org/10.1001/jamaoncol.2018.4051. PubMed PMID: 30267032

Merck & Co. Inc. (2017) Gardasil 9 human papillomavirus 9-valent vaccine, recombinant. Available from https://www.gardasil9.com/. Updated 2017; Cited 2019

Merck & Co. Inc. (2019) Efficacy of GARDASIL 9. Available from https://www.merckvaccines.com/Products/Gardasil9/efficacy#add5Types. Updated Feb 2019; Cited 2019

Miles BA, Monk BJ, Safran HP (2017) Mechanistic insights into ADXS11-001 human papillomavirus-associated cancer immunotherapy. Gynecol Oncol Res Pract 4:9. Epub 2017/06/08. https://doi.org/10.1186/s40661-017-0046-9. PubMed PMID: 28588899; PMCID: PMC5455112

National Cancer Institute (2020) HPV and cancer. Available from: https://www.cancer.gov/about-cancer/causes-prevention/risk/infectious-agents/hpv-and-cancer. Updated 1 Mar 2019; Cited 2019

Nurkkala M, Wassen L, Nordstrom I, Gustavsson I, Slavica L, Josefsson A, Eriksson K (2010) Conjugation of HPV16 E7 to cholera toxin enhances the HPV-specific T-cell recall responses to pulsed dendritic cells in vitro in women with cervical dysplasia. Vaccine 28(36):5828–5836. Epub 2010/07/06. https://doi.org/10.1016/j.vaccine.2010.06.068. PubMed PMID: 20600477

Peng S, Kim TW, Lee JH, Yang M, He L, Hung CF, Wu TC (2005) Vaccination with dendritic cells transfected with BAK and BAX siRNA enhances antigen-specific immune responses by prolonging dendritic cell life. Hum Gene Ther 16(5):584–593. Epub 2005/05/27. https://doi.org/10.1089/hum.2005.16.584. PubMed PMID: 15916483; PMCID: PMC3181105

Peng S, Monie A, Pang X, Hung CF, Wu TC (2011) Vascular disrupting agent DMXAA enhances the antitumor effects generated by therapeutic HPV DNA vaccines. J Biomed Sci 18:21. Epub 2011/03/10. https://doi.org/10.1186/1423-0127-18-21. PubMed PMID: 21385449; PMCID: PMC3062584

Peng S, Qiu J, Yang A, Yang B, Jeang J, Wang JW, Chang YN, Brayton C, Roden RBS, Hung CF, Wu TC (2016) Optimization of heterologous DNA-prime, protein boost regimens and site of vaccination to enhance therapeutic immunity against human papillomavirus-associated disease. Cell Biosci 6:16. Epub 2016/02/27. https://doi.org/10.1186/s13578-016-0080-z. PubMed PMID: 26918115; PMCID: PMC4766698

Qin Y, Wang XH, Cui HL, Cheung YK, Hu MH, Zhu SG, Xie Y (2005) Human papillomavirus type 16 E7 peptide(38–61) linked with an immunoglobulin G fragment provides protective immunity in mice. Gynecol Oncol 96(2):475-83. Epub 2005/01/22. https://doi.org/10.1016/j.ygyno.2004.10.028. PubMed PMID: 15661238

Ramanathan P, Ganeshrajah S, Raghanvan RK, Singh SS, Thangarajan R (2014) Development and clinical evaluation of dendritic cell vaccines for HPV related cervical cancer—A feasibility study. Asian Pac J Cancer Prev 15(14):5909–5916 Epub 2014/08/02 PubMed PMID: 25081721

Roden RBS, Stern PL (2018) Opportunities and challenges for human papillomavirus vaccination in cancer. Nat Rev Cancer. 18(4):240–254. Epub 2018/03/03. https://doi.org/10.1038/nrc.2018.13. PubMed PMID: 29497146

Roden R, Wu TC (2006) How will HPV vaccines affect cervical cancer? Nat Rev Cancer 6 (10):753–763. Epub 2006/09/23. https://doi.org/10.1038/nrc1973. PubMed PMID: 16990853; PMCID: PMC3181152

Rosales R, Lopez-Contreras M, Rosales C, Magallanes-Molina JR, Gonzalez-Vergara R, Arroyo-Cazarez JM, Ricardez-Arenas A, Del Follo-Valencia A, Padilla-Arriaga S, Guerrero MV, Pirez MA, Arellano-Fiore C, Villarreal F (2014) Regression of human papillomavirus intraepithelial lesions is induced by MVA E2 therapeutic vaccine. Hum Gene Ther 25 (12):1035–1049. Epub 2014/10/03. https://doi.org/10.1089/hum.2014.024. PubMed PMID: 25275724; PMCID: PMC4270165

Rubio I, Bolchi A, Moretto N, Canali E, Gissmann L, Tommasino M, Muller M, Ottonello S (2009) Potent anti-HPV immune responses induced by tandem repeats of the HPV16 L2 (20–38) peptide displayed on bacterial thioredoxin. Vaccine 27(13):1949–1956. Epub 2009/04/17. https://doi.org/10.1016/j.vaccine.2009.01.102. PubMed PMID: 19368776

Safaeian M, Porras C, Pan Y, Kreimer A, Schiller JT, Gonzalez P, Lowy DR, Wacholder S, Schiffman M, Rodriguez AC, Herrero R, Kemp T, Shelton G, Quint W, van Doorn LJ, Hildesheim A, Pinto LA, Group CVT (2013) Durable antibody responses following one dose of the bivalent human papillomavirus L1 virus-like particle vaccine in the Costa Rica Vaccine Trial. Cancer Prev Res (Phila). 6(11):1242–1250. Epub 2013/11/06. https://doi.org/10.1158/1940-6207.capr-13-0203. PubMed PMID: 24189371

Santin AD, Bellone S, Roman JJ, Burnett A, Cannon MJ, Pecorelli S (2005) Therapeutic vaccines for cervical cancer: dendritic cell-based immunotherapy. Curr Pharm Des 11(27):3485–3500 Epub 2005/10/27 PubMed PMID: 16248803

Santin AD, Bellone S, Palmieri M, Zanolini A, Ravaggi A, Siegel ER, Roman JJ, Pecorelli S, Cannon MJ (2008) Human papillomavirus type 16 and 18 E7-pulsed dendritic cell vaccination of stage IB or IIA cervical cancer patients: a phase I escalating-dose trial. J Virol 82(4):1968–1979. Epub 2007/12/07. https://doi.org/10.1128/jvi.02343-07. PubMed PMID: 18057249; PMCID: PMC2258728

Schadlich L, Senger T, Gerlach B, Mucke N, Klein C, Bravo IG, Muller M, Gissmann L (2009) Analysis of modified human papillomavirus type 16 L1 capsomeres: the ability to assemble into larger particles correlates with higher immunogenicity. J Virol 83(15):7690–7705. Epub 2009/05/22. https://doi.org/10.1128/jvi.02588-08. PubMed PMID: 19457985; PMCID: PMC2708645

Schellenbacher C, Kwak K, Fink D, Shafti-Keramat S, Huber B, Jindra C, Faust H, Dillner J, Roden RBS, Kirnbauer R (2013) Efficacy of RG1-VLP vaccination against infections with genital and cutaneous human papillomaviruses. J Invest Dermatol133(12):2706–2713. Epub 2013/06/12. https://doi.org/10.1038/jid.2013.253. PubMed PMID: 23752042; PMCID: PMC3826974

Schellenbacher C, Roden RBS, Kirnbauer R (2017) Developments in L2-based human papillomavirus (HPV) vaccines. Virus Res 231:166–175. Epub 2016/11/28. https://doi.org/10.1016/j.virusres.2016.11.020. PubMed PMID: 27889616; PMCID: PMC5549463

Sewell DA, Shahabi V, Gunn GR, 3rd, Pan ZK, Dominiecki ME, Paterson Y (2004) Recombinant Listeria vaccines containing PEST sequences are potent immune adjuvants for the tumor-associated antigen human papillomavirus-16 E7. Cancer Res 64(24):8821–8825. Epub 2004/12/18. https://doi.org/10.1158/0008-5472.can-04-1958. PubMed PMID: 15604239

Sharma RK, Elpek KG, Yolcu ES, Schabowsky RH, Zhao H, Bandura-Morgan L, Shirwan H (2009) Costimulation as a platform for the development of vaccines: a peptide-based vaccine containing a novel form of 4-1BB ligand eradicates established tumors. Cancer Res 69 (10):4319–4326. Epub 2009/05/14. https://doi.org/10.1158/0008-5472.can-08-3141. PubMed PMID: 19435920; PMCID: PMC2755220

Skeate JG, Woodham AW, Einstein MH, Da Silva DM, Kast WM (2016) Current therapeutic vaccination and immunotherapy strategies for HPV-related diseases. Hum Vaccin Immunother 12(6):1418–1429. Epub 2016/02/03. https://doi.org/10.1080/21645515.2015.1136039. PubMed PMID: 26835746; PMCID: PMC4964648

Souders NC, Sewell DA, Pan ZK, Hussain SF, Rodriguez A, Wallecha A, Paterson Y (2007) Listeria-based vaccines can overcome tolerance by expanding low avidity CD8+ T cells capable of eradicating a solid tumor in a transgenic mouse model of cancer. Cancer Immun 7:2. Epub 2007/02/07. PubMed PMID: 17279610; PMCID: PMC3077294

Su JH, Wu A, Scotney E, Ma B, Monie A, Hung CF, Wu TC (2010) Immunotherapy for cervical cancer: Research status and clinical potential. BioDrugs 24(2):109–129. Epub 2010/03/05. https://doi.org/10.2165/11532810-000000000-00000. PubMed PMID: 20199126; PMCID: PMC2913436

Trimble CL, Morrow MP, Kraynyak KA, Shen X, Dallas M, Yan J, Edwards L, Parker RL, Denny L, Giffear M, Brown AS, Marcozzi-Pierce K, Shah D, Slager AM, Sylvester AJ, Khan A, Broderick KE, Juba RJ, Herring TA, Boyer J, Lee J, Sardesai NY, Weiner DB, Bagarazzi ML (2015) Safety, efficacy, and immunogenicity of VGX-3100, a therapeutic synthetic DNA vaccine targeting human papillomavirus 16 and 18 E6 and E7 proteins for cervical intraepithelial neoplasia 2/3: a randomised, double-blind, placebo-controlled phase 2b trial. Lancet 386(10008):2078–2088. Epub 2015/09/21. https://doi.org/10.1016/s0140-6736 (15)00239-1. PubMed PMID: 26386540; PMCID: PMC4888059

Tseng CW, Hung CF, Alvarez RD, Trimble C, Huh WK, Kim D, Chuang CM, Lin CT, Tsai YC, He L, Monie A, Wu TC (2008a) Pretreatment with cisplatin enhances E7-specific CD8+ T-cell-mediated antitumor immunity induced by DNA vaccination. Clin Cancer Res 14 (10):3185–3192. Epub 2008/05/17. https://doi.org/10.1158/1078-0432.ccr-08-0037. PubMed PMID: 18483387; PMCID: PMC3066100

Tseng CW, Monie A, Wu CY, Huang B, Wang MC, Hung CF, Wu TC (2008b) Treatment with proteasome inhibitor bortezomib enhances antigen-specific CD8+ T-cell-mediated antitumor immunity induced by DNA vaccination. J Mol Med (Berl). 86(8):899–908. Epub 2008/06/11. https://doi.org/10.1007/s00109-008-0370-y. PubMed PMID: 18542898; PMCID: PMC2535907

Tseng CW, Trimble C, Zeng Q, Monie A, Alvarez RD, Huh WK, Hoory T, Wang MC, Hung CF, Wu TC (2009) Low-dose radiation enhances therapeutic HPV DNA vaccination in tumor-bearing hosts. Cancer Immunol Immunother 58(5):737–748. Epub 2008/09/26. https://doi.org/10.1007/s00262-008-0596-0. PubMed PMID: 18815785; PMCID: PMC2647576

Tumban E, Peabody J, Peabody DS, Chackerian B (2011) A pan-HPV vaccine based on bacteriophage PP7 VLPs displaying broadly cross-neutralizing epitopes from the HPV minor capsid protein, L2. PLoS One 6(8):e23310. Epub 2011/08/23. https://doi.org/10.1371/journal.pone.0023310. PubMed PMID: 21858066; PMCID: PMC3157372

U.S. Food and Drug Administration (2018) Cervarix. Available from https://www.fda.gov/biologicsbloodvaccines/vaccines/approvedproducts/ucm186957.htm. Updated 26 Feb 26 2018; Cited 2019

U.S. Food and Drug Administration (2019) Package insert—Gardasil 9. Available from https://www.fda.gov/downloads/biologicsbloodvaccines/vaccines/approvedproducts/ucm111263.pdf

U.S. Preventive Services Task Force (2012) Cervical cancer: screening. Available from https://www.uspreventiveservicestaskforce.org/Page/Document/UpdateSummaryFinal/cervical-cancer-screening. Updated Mar 2012; Cited 2019

Vaccarella S, Lortet-Tieulent J, Plummer M, Franceschi S, Bray F (2013) Worldwide trends in cervical cancer incidence: impact of screening against changes in disease risk factors. Eur J Cancer 49(15):3262–3273. Epub 2013/06/12. https://doi.org/10.1016/j.ejca.2013.04.024. PubMed PMID: 23751569

Van Damme P, Bouillette-Marussig M, Hens A, De Coster I, Depuydt C, Goubier A, Van Tendeloo V, Cools N, Goossens H, Hercend T, Timmerman B, Bissery MC (2016) GTL001, A therapeutic vaccine for women infected with human papillomavirus 16 or 18 and normal cervical cytology: results of a phase I clinical trial. Clin Cancer Res 22(13):3238–3248. Epub 2016/06/03. https://doi.org/10.1158/1078-0432.ccr-16-0085. PubMed PMID: 27252412

van der Burg SH, Kwappenberg KM, O'Neill T, Brandt RM, Melief CJ, Hickling JK, Offringa R (2001) Pre-clinical safety and efficacy of TA-CIN, a recombinant HPV16 L2E6E7 fusion protein vaccine, in homologous and heterologous prime-boost regimens. Vaccine 19(27):3652–3660 Epub 2001/06/08 PubMed PMID: 11395199

Verch T, Pan ZK, Paterson Y (2004) Listeria monocytogenes-based antibiotic resistance gene-free antigen delivery system applicable to other bacterial vectors and DNA vaccines. Infect Immun 72(11):6418–6425. Epub 2004/10/27. https://doi.org/10.1128/iai.72.11.6418-6425.2004. PubMed PMID: 15501772; PMCID: PMC523039

Vici P, Pizzuti L, Mariani L, Zampa G, Santini D, Di Lauro L, Gamucci T, Natoli C, Marchetti P, Barba M, Maugeri-Sacca M, Sergi D, Tomao F, Vizza E, Di Filippo S, Paolini F, Curzio G, Corrado G, Michelotti A, Sanguineti G, Giordano A, De Maria R, Venuti A (2016) Targeting immune response with therapeutic vaccines in premalignant lesions and cervical cancer: hope or reality from clinical studies. Expert Rev Vaccines 15(10):1327–1336. Epub 2016/04/12. https://doi.org/10.1080/14760584.2016.1176533. PubMed PMID: 27063030; PMCID: PMC5152541

Walboomers JM, Jacobs MV, Manos MM, Bosch FX, Kummer JA, Shah KV, Snijders PJ, Peto J, Meijer CJ, Munoz N (1999) Human papillomavirus is a necessary cause of invasive cervical cancer worldwide. J Pathol 189(1):12–19. Epub 1999/08/19. https://doi.org/10.1002/(sici)1096-9896(199909)189:1%3c12::aid-path431%3e3.0.co;2-f. PubMed PMID: 10451482

Wallecha A, Wood L, Pan ZK, Maciag PC, Shahabi V, Paterson Y (2013) Listeria monocytogenes-derived listeriolysin O has pathogen-associated molecular pattern-like properties independent of its hemolytic ability. Clin Vaccine Immunol 20(1):77–784. Epub 2012/11/09. https://doi.org/10.1128/cvi.00488-12. PubMed PMID: 23136118; PMCID: PMC3535771

Wang Z, Troilo PJ, Wang X, Griffiths TG, Pacchione SJ, Barnum AB, Harper LB, Pauley CJ, Niu Z, Denisova L, Follmer TT, Rizzuto G, Ciliberto G, Fattori E, Monica NL, Manam S, Ledwith BJ (2004) Detection of integration of plasmid DNA into host genomic DNA following intramuscular injection and electroporation. Gene Ther 11(8):711–721. Epub 2004/01/16. https://doi.org/10.1038/sj.gt.3302213. PubMed PMID: 14724672

Welters MJ, van der Sluis TC, van Meir H, Loof NM, van Ham VJ, van Duikeren S, Santegoets SJ, Arens R, de Kam ML, Cohen AF, van Poelgeest MI, Kenter GG, Kroep JR, Burggraaf J, Melief CJ, van der Burg SH (2016) Vaccination during myeloid cell depletion by cancer chemotherapy fosters robust T cell responses. Sci Transl Med 8(334):334ra52. Epub 2016/04/15. https://doi.org/10.1126/scitranslmed.aad8307. PubMed PMID: 27075626

Williams VM, Filippova M, Soto U, Duerksen-Hughes PJ (2011) HPV-DNA integration and carcinogenesis: putative roles for inflammation and oxidative stress. Future Virol 6(1):45–57. Epub 2011/02/15. https://doi.org/10.2217/fvl.10.73. PubMed PMID: 21318095; PMCID: PMC3037184

World Cancer Research Fund (2018) Worldwide cancer data. Available from: https://www.wcrf.org/dietandcancer/cancer-trends/worldwide-cancer-data. Cited 2019

World Health Organization (2018) Human papillomavirus (HPV). Available from: https://www.who.int/immunization/diseases/hpv/en/. Cited 2019

World Health Organization (2020) Human papillomavirus (HPV) and cervical cancer. Available from https://www.who.int/immunization/diseases/hpv/en/. Updated 24 Jan 2019; Cited 2019

Wu CY, Monie A, Pang X, Hung CF, Wu TC (2010) Improving therapeutic HPV peptide-based vaccine potency by enhancing CD4+ T help and dendritic cell activation. J Biomed Sci 17:88. Epub 2010/11/26. https://doi.org/10.1186/1423-0127-17-88. PubMed PMID: 21092195; PMCID: PMC3000388

Yan W, Chen WC, Liu Z, Huang L (2010) Bryostatin-I: a dendritic cell stimulator for chemokines induction and a promising adjuvant for a peptide based cancer vaccine. Cytokine 52(3):238–244. Epub 2010/09/28. https://doi.org/10.1016/j.cyto.2010.08.010. PubMed PMID: 20869878

Yang A, Farmer E, Wu TC, Hung CF (2016) Perspectives for therapeutic HPV vaccine development. J Biomed Sci 23(1):75. Epub 2016/11/05. https://doi.org/10.1186/s12929-016-0293-9. PubMed PMID: 27809842; PMCID: PMC5096309

Yang A, Farmer E, Lin J, Wu TC, Hung CF (2017) The current state of therapeutic and T cell-based vaccines against human papillomaviruses. Virus Res 231:148–615. Epub 2016/12/10. https://doi.org/10.1016/j.virusres.2016.12.002. PubMed PMID: 27932207; PMCID: PMC5325765

Yim EK, Park JS (2005) The role of HPV E6 and E7 oncoproteins in HPV-associated cervical carcinogenesis. Cancer Res Treat 37(6):319–324. Epub 2005/12/01. https://doi.org/10.4143/crt.2005.37.6.319. PubMed PMID: 19956366; PMCID: PMC2785934

Yoon SW, Lee TY, Kim SJ, Lee IH, Sung MH, Park JS, Poo H (2012) Oral administration of HPV-16 L2 displayed on Lactobacillus casei induces systematic and mucosal cross-neutralizing effects in Balb/c mice. Vaccine 30(22):3286–3294. Epub 2012/03/20. https://doi.org/10.1016/j.vaccine.2012.03.009. PubMed PMID: 22426329

Zandberg DP, Rollins S, Goloubeva O, Morales RE, Tan M, Taylor R, Wolf JS, Schumaker LM, Cullen KJ, Zimrin A, Ord R, Lubek JE, Suntharalingam M, Papadimitriou JC, Mann D, Strome SE, Edelman MJ (2015) A phase I dose escalation trial of MAGE-A3- and HPV16-specific peptide immunomodulatory vaccines in patients with recurrent/metastatic (RM) squamous cell carcinoma of the head and neck (SCCHN) Cancer Immunol Immunother 64(3):367–379. Epub 2014/12/30. https://doi.org/10.1007/s00262-014-1640-x. PubMed PMID: 25537079; PMCID: PMC4381442

Zhai L, Tumban E (2016) Gardasil-9: a global survey of projected efficacy. Antiviral Res.130:101–109. Epub 2016/04/05. https://doi.org/10.1016/j.antiviral.2016.03.016. PubMed PMID: 27040313

Zhang YQ, Tsai YC, Monie A, Hung CF, Wu TC (2010) Carrageenan as an adjuvant to enhance peptide-based vaccine potency. Vaccine 28(32):5212–5219. Epub 2010/06/15. https://doi.org/10.1016/j.vaccine.2010.05.068. PubMed PMID: 20541583; PMCID: PMC2908183

Zwaveling S, Ferreira Mota SC, Nouta J, Johnson M, Lipford GB, Offringa R, van der Burg SH, Melief CJ (2002) Established human papillomavirus type 16-expressing tumors are effectively eradicated following vaccination with long peptides. J Immunol. 169(1):350–358 Epub 2002/06/22 PubMed PMID: 12077264

Epstein–Barr Virus-Associated Post-transplant Lymphoproliferative Disease

Richard F. Ambinder

1 Introduction

Epstein–Barr virus (EBV)-associated post-transplant lymphoproliferative disease (PTLD) is recognized in solid organ and bone marrow or hematopoietic cell transplant (hereafter referred to as BMT) recipients. The occurrence of these tumors is important not only for the life-threatening problems they lead to but also for the resultant limitations in approaches to therapeutic immunosuppression in the context of transplant. This chapter begins with a brief overview of EBV and cancer and EBV biology so as to put EBV-PTLD into context. This is followed by discussion of EBV DNA monitoring in blood, approaches to treatment and approaches to prevention.

2 EBV, Cancer, and Aspects of EBV Biology

When Burkitt called attention to a tumor common in young children in Africa, the distribution of the tumor in equatorial Africa led to interest in the possibility that the tumor might be virus associated (reviewed in (Balfour et al. 2019)). While efforts to culture virus from tumor specimens were unsuccessful, tumor cells did grow in culture, and examination of resultant cell lines by electron microscopy showed the presence of virions with the morphology of herpesviruses. With time, it was appreciated that the distribution of EBV was not the distribution of Burkitt lymphoma in equatorial Africa, but that EBV was ubiquitous infecting more than 90% of adults. Further, it was appreciated that the virus is associated with many different

R. F. Ambinder (✉)
Department of Oncology, Johns Hopkins School of Medicine, Baltimore, MD, US
e-mail: rambind1@jhmi.edu

© Springer Nature Switzerland AG 2021
T.-C. Wu et al. (eds.), *Viruses and Human Cancer*, Recent Results
in Cancer Research 217, https://doi.org/10.1007/978-3-030-57362-1_9

cancers including lymphoproliferative diseases, carcinomas, and a mesenchymal tumor (see Table 1). Among these many tumors, some are very consistently associated with the virus (African BL), nasopharyngeal carcinoma, nasal NK/T cell lymphoma, post-transplant lymphoproliferative disease), whereas others are variably associated with the virus (diffuse large B cell lymphoma, classic Hodgkin lymphoma and gastric carcinoma) (Ambinder and Cesarman 2007). Although the list of EBV-associated cancers is long, it is important to appreciate that the virus infects most adults worldwide and is only rare associated with cancer (Young et al. 2016).

Viral genomes packaged in virions are linear double-stranded DNA molecules (Young et al. 2016). The viral genome is approximately 171 kb and codes for more than 80 open-reading frames as well as many non-coding RNAs. B lymphocyte infection involves an interaction between gp350/220, a viral envelope glycoprotein, and CD21, a component of the complement receptor. The virion attaches and is endocytosed. The nucleocapsid is released into the cytoplasm. The viral genome is transported to the nucleus, where the ends of linear viral genome fuse to form a closed circle.

There is an initial amplification of episomes (Hammerschmidt and Sugden 2013). Thereafter, the virus will remain latent (not producing virions), and some of these latently infected cells will proliferate. The numbers of viral episomes per cell remain approximately stable over many cell divisions. Replication of the viral genome in these latently infected cells relies on a single viral protein, EBNA1, that activates the viral origin of replication and allows cellular enzymes including the cellular DNA polymerase to replicate the episome. In other cells, productive infection will ensue with activation of lytic viral gene expression and the production of linear viral genomes in a process that involves many viral replication enzymes including a viral DNA polymerase. Most of the open-reading frames encoded by the virus are expressed in the lytic program.

When resting B cells are infected in vitro, some of the B cells become immortalized growing indefinitely in tissue culture as an EBV lymphoblastoid cell line (LCL) (Thorley-Lawson 2001). In immunodeficient mice, LCL will grow as EBV-driven human B cell tumors. There are nine viral proteins that are expressed: six nuclear antigens and three latency membrane proteins. In genetic experiments, no single viral protein is sufficient to generate an LCL. Rather coordinated expression of five viral proteins is required. These are Epstein–Barr nuclear antigen 1 (EBNA1), EBNA2, EBNA3A, EBNA3C, and latency membrane protein 1 (LMP1). EBNA1 is a sequence-specific DNA-binding protein required for the maintenance of the viral episome. EBNA2 is a transcriptional transactivator that leads to expression of a variety of cellular and viral genes. LMP1 is a member of the tumor necrosis factor receptor (TNFR) superfamily. It is an integral membrane protein and is constitutively activated leading in turn to the activation of NF-kB. In addition to proteins, the virus expresses many non-coding RNAs. These include two small polymerase III transcripts, EBER 1 and 2. Although the functions of these transcripts remain uncertain, their abundance in latency, as high as 10^7

Table 1 EBV-associated cancers

Cancer	Tissue	EBV association	Comments
BL	B lymphocytes	Nearly 100% in equatorial Africa, ~20% elsewhere	The tumor is much more common in equatorial Africa, and it appears to be associated with malaria
Hodgkin lymphoma	B lineage cells	20–75% in the general population	The EBV association is much greater in Africa, Latin America, and parts of Asia
PTLD	B lymphocytes	See text	
AIDS primary central nervous system lymphomas	B lymphocytes	Nearly 100%	Typically associated with profound immunocompromise, end-stage HIV disease
Other AIDS non-Hodgkin lymphomas	B cells	Varies by histology but approximately 40% overall	
HIV Hodgkin lymphoma	B lineage cells	Approximately 90%	
AIDS primary effusion lymphoma	B cells	Approximately 80%	All are coinfected with KSHV/HHV8
Lymphoma in inherited immune deficiency disorders	B cells	Nearly 100%	
Nasal NK/T cell lymphoma	NK/T lymphocytes	Nearly 100%	The incidence of the tumor varies with geography, much more common in parts of Asia
Nasopharyngeal carcinoma	Epithelium	Nearly 100%	Much more common in certain populations, particularly southern Chinese
Gastric carcinoma	Epithelium	Approximately 10%	The incidence of the tumor varies substantially with geography, but the EBV association is relatively constant
Leiomyosarcoma	Mesenchymal	Nearly 100%	Occurs only in profoundly immunocompromised populations (HIV, post-transplant)

copies/cell by some estimates has made them important in the detection of latent infection by in situ hybridization (Wu et al. 1990; Ambinder and Mann 1994).

EBV is transmitted person-to-person through saliva and infects B lymphocytes in the oral mucosa (Cohen 2015). Primary infection in childhood is usually asymptomatic. In adolescence and adulthood, primary infection is sometimes associated with the syndrome of infectious mononucleosis (Balfour and Verghese 2013). Some of the latently infected B lymphocytes proliferate, leading to the spread of infected B cells throughout the B cell compartment. Immune response follows and ultimately prevents uncontrolled EBV-infected B cell proliferation (Thorley-Lawson et al. 2013). Cytotoxic T-lymphocyte responses to viral antigens limit the production of virions and virus-driven B cell proliferation. In healthy individuals, EBV genomes are harbored in a tiny percentage of resting B cells (Yang et al. 2000; Thorley-Lawson et al. 2013).

These viral proteins modulate a multitude of complex cellular pathways involving notch and NF-κB among others. They drive cell proliferation and confer resistance to apoptosis.

The B cells that harbor EBV after the cellular immune response do not express the viral proteins that drive the proliferation of LCLs. They are resting memory B cells that are able to evade immune surveillance in part because they do not express the antigens commonly targeted by cytotoxic T cells (Hadinoto et al. 2008). Periodic activation of viral lytic gene expression is presumed to lead to intermittent virion production and viral shedding that occurs throughout the lifetime of the host.

The humoral response to acute infection is the development of IgM and later IgG responses to virus capsid antigen (VCA). There are also IgG responses to early antigens and the EBNAs. However, the IgM response to VCA and the early antigens wane, the IgG response to VCA and EBNA. GP350/220 neutralizing antibodies may hinder the spread of infection resulting from virion production and inhibit superinfection with other strains of virus, but humoral immunity is thought to have minimal effect on latently infected cells insofar as the target antigens are not expressed on the surface of latently infected cells. The T cell response to viral antigens in healthy volunteers is very strong targeting with latency antigens EBNA3A, EBNA3B, and EBNA3C being targeted as well as lytic antigens. EBNA1 evades CD8 cytotoxic T cells because the protein contains a cis-acting repetitive glycine-alanine sequence which inhibits MHC class I presentation in part by inhibition of translation of the protein or antigen processing (Levitskaya et al. 1995; Apcher et al. 2010).

3 PTLD

PTLD refers to lymphomas that occur following organ or hematopoietic cell transplantation (Dierickx and Habermann 2018). Many of these lymphomas are EBV-associated, but the association is not a requirement for the diagnosis of PTLD. The histologic appearance of PTLD can be quite variable. The World Health

Organization recognizes non-destructive PTLD, polymorphic PTLD, monomorphic PTLD, and Hodgkin lymphoma-like PTLD (Swerdlow et al. 2016). Note that some lymphoma types are not considered post-transplant lymphoma even if they occur in a patient with a history of transplantation. For example, mantle cell lymphoma or follicular lymphoma when they occur in a patient after transplantation are not classified as PTLD. Among PTLD types, the non-destructive and polymorphic cases are generally not associated with chromosomal translocations and other cytogenetic abnormalities. Often there is a broad pattern of viral latency gene expression that resembles the patterns in EBV immortalized LCLs. This expression pattern is variously referred to as "latency 3" or the "growth program" and includes expression of EBNA2, 3A, 3B, and 3C (Rowe et al. 1992, Thorley-Lawson and Gross 2004). In tumors that express the full spectrum of latency genes, these viral genes appear to be the major force driving proliferation (Vereide and Sugden 2011). The EBNA3 proteins are especially good targets for EBV-specific cytotoxic T cells and tumors that express these proteins are almost never seen except in profoundly immunocompromised patients.

4   Risk Factors

PTLD can occur at any time after transplant but usually occurs within the first year (Luskin et al. 2015). The earlier a lymphoma occurs, the more likely it is to be EBV-associated. In one series, PTLD in the first year was 84% EBV-associated, whereas PTLD occurring later was only 57% EBV-associated (Luskin et al. 2015). Among the factors associated with increased risk for PTLD in transplant, recipients are EBV-seronegativity (Caillard et al. 2006; Kasiske et al. 2011; Sampaio et al. 2012). Primary EBV infection in the post-transplant period is a major risk factor. Many reports have identified age as a risk factor for PTLD but insofar as the youngest patients are most likely to be seronegative, age, and seronegativity which are confounded and whether age is an independent risk factor which is not clear. Pharmacologic immunosuppression compromises the ability of EBV-naïve patients to establish an immune response.

PTLD in solid organ transplant recipients and BMT recipients differs in several regards. First in solid organ transplant recipients, most PTLD develops in recipient B cells (Swerdlow et al. 2017). In allogeneic BMT recipients, most PTLD develops in donor B cells. In solid organ transplant recipients, the risk of PTLD remains substantial for many years, whereas in BMT patients, the risk is almost exclusively within the first six months (Al Hamed et al. 2019). The explanation for this difference may relate to the duration of immunosuppression. Some solid organ recipients often continue on immunosuppression indefinitely.

Among solid organ transplant recipients, lung, heart, heart/lung, and gut, transplants are associated with the highest incidences of PTLD among organ transplant recipients (Jagadeesh et al. 2019). With the use of calcineurin inhibitors, there was initially an increase in the incidence of PTLD. When drug level monitoring became available, average calcineurin doses fell, and there was a concomitant decline in the incidence of PTLD. Many reports indicate that T cell depletion, particularly with monoclonal antibody therapy, targeting CD3 is associated with increased risk.

Among BMT recipients, risk factors include age of the patient, the use of reduced intensity conditioning, degree of HLA mismatch, seropositivity mismatch between donor and recipient, use of umbilical cord blood, and particular preparative regimens or graft vs host disease prevention strategies (Al Hamed et al. 2019). Antithymocyte globulin in particular is associated with a high rate or PTLD (Brunstein et al. 2006). A variety of T cell depletion strategies have been associated with PTLD—but approaches that deplete B and T cells such as elutriation are associated with lower risk of PTLD than approaches that selectively deplete T cells (Landgren et al. 2009). With the use of post-transplant cyclophosphamide, PTLD is vanishingly rare—even with elderly recipients, reduced intensity conditioning, haploidentical or mismatched unrelated donors, or the use of ATG as part of the preparative regimen (Bolanos-Meade et al. 2012; Kanakry et al. 2013; Kasamon et al. 2017; Imus et al. 2019). The explanation for the absence of PTLD with post-transplant cyclophosphamide could relate to improved T cell reconstitution or to the impact of cyclophosphamide on B cells.

5 Virus Monitoring

Monitoring EBV DNA in whole blood, mononuclear cells, or plasma has been used to guide immunosuppression or pre-emptive interventions (Table 2) (Ru et al. 2018). It should be noted that although publications commonly refer to "viral load" and "viremia," what is being measured are viral genomes. Viral genomes detected in plasma may be from virions, i.e., infectious virus particles or may be DNA released from cells that is not packaged in infectious particles. Latently infected tumor cells that undergo apoptosis release tumor cell DNA but not virion DNA. The distinction is important because immuno-suppressed patients may have high levels of virions in plasma but not have tumor, whereas patients with EBV-associated tumors may have high levels of viral DNA but no virions. When plasma DNA is compared to whole blood cell DNA, plasma DNA has better sensitivity and specificity for PTLD (Tsai et al. 2008). In a recent series from Johns Hopkins, EBV in plasma DNA was more useful than in PBMC DNA for diagnosing EBV-associated disease including malignancies—but even so the majority of patients with markedly elevated plasma EBV DNA had no underlying EBV-associated disease (Kanakry et al. 2016). Measurement of intracellular viral DNA has a major shortcoming. Treatment with rituximab may clear the blood of measurable B lymphocytes and measurable intracellular EBV without any impact on tumor (Yang et al. 2000).

Table 2 Measuring EBV DNA in blood

Specimen type	Derivation of the EBV DNA being measured	Comment
Plasma	Virions, DNA released from latently infected cells including tumor cells that do not circulate	Viral DNA will often be detected in plasma even when there are no B lymphocytes in the circulation such as in patients treated with rituximab
Lymphocytes	Latently infected lymphocytes, lytically infected lymphocytes	Viral DNA will sometimes not be detected even when there is tumor progression
Whole blood	All of the above	More sensitive than either plasma or lymphocyte measurement, but less specific for EBV-associated disease

6 Treatment

The details of treatment are dictated in part by the kind of transplant (hematopoietic versus solid organ; if solid organ, the organ(s) transplanted), the location and extent of the disease, the subtype of the disease, and patient comorbidities. General approaches include withdrawal or reduction of immunosuppression, surgical excision or radiation, cytotoxic chemotherapy, treatment with monoclonal antibodies, particularly those targeting CD20 or other cellular antigens, antiviral agents, and EBV-specific T cells or other adoptive cellular therapies. The details of treatment are beyond the scope of this chapter. However, there are some general principles to be noted.

Reduction or withdrawal of immunosuppression is often a component of treatment. In the setting of renal transplant where the transplanted organ can be sacrificed and the patient's life still saved, this was a mainstay of treatment in the past. With a variety of alternative treatments, there is an reluctance to risk sacrificing the transplanted organ, and complete withdrawal of immunosuppression is less commonly used. Rituximab, a monoclonal antibody targeting CD20-positive B cells, is commonly used (Burns et al. 2020). Recently other monoclonal antibodies or antibody conjugates have been used including brentuximab vedotin to target CD30 or CD38 sometimes expressed in PTLD (Pearse et al. 2019; Chaulagain et al. 2020). These antibodies may be used alone or in combination with cytotoxic chemotherapy such as cyclophosphamide, doxorubicin, vincristine, and prednisone (Caillard and Green 2019).

With regard to reduction or withdrawal of immunosuppression, it should be noted that there is considerable uncertainty. For example, steroids are commonly part of immunosuppression and are often tapered or stopped in response to the development of PTLD. However, steroids are part of many front line lymphoma treatment regimens. When withdrawal or reduction of immunosuppression fails, it is common to begin treatment with lymphoma regimens that include prednisone

(Burns et al. 2020). Similarly, sirolimus and everolimus are widely used to prevent graft rejection or graft vs host disease and may be stopped or dose-reduced in response to the development of PTLD. However, these mTOR inhibitors have activity against some lymphomas, and there is evidence that this class of drugs may be useful in the treatment of Kaposi sarcoma, another opportunistic neoplasia seen in the post-transplant setting (Stallone et al. 2005; Krown et al. 2012). Changing immunosuppressive regimens to include an mTOR inhibitor have been associated with tumor regression in case reports (Cullis et al. 2006; Nanmoku et al. 2019). Thus, whether stopping or dose-reducing either steroids or mTOR inhibitors is helpful in the treatment of PTLD is not known.

There are also EBV-targeted therapies. These include autologous EBV-specific cytotoxic T cells (Kim et al. 2017), "off the shelf" third party partially HLA-matched cytotoxic T cells (Prockop et al. 2019), and chemotherapies meant to induce EBV lytic infection so as to sensitize EBV harboring cells to killing by ganciclovir (Hui et al. 2016). Although there have been successes reported with each of these approaches, none are standardly accepted as front line therapies.

Outcomes differ among solid organ transplant recipients as a function of transplant type (Jagadeesh et al. 2019). Cardiac and lung transplant recipients with PTLD have the worst five year overall survival.

7 Prevention

There are many approaches to prevention. Avoiding certain immunosuppressive approaches or regimens that are particularly likely to lead to PTLD is standard. Thus, CD3 targeted antibody therapies are now rarely used for the prevention of graft rejection. Similarly, aggressive T cell depletion to prevent graft vs host disease in hematopoietic transplantation is less commonly used than in the past. As noted above, the incidence of PTLD in hematopoietic cell transplantation using post-transplant cyclophosphamide is vanishingly low.

Because EBV-seronegative transplant recipients are at the highest risk for PTLD, there has been interest in vaccination of seronegative recipients prior to transplantation. There are several vaccine efforts that are ongoing. These have recently been reviewed (Balfour et al. 2019). Several different strategies have been investigated that might be relevant for preventing EBV-PTLD. One involves the EBV gp350 antigen. Antibodies to this antigen are virus neutralizing and might prevent or reduce the severity of primary infection. An alternative strategy is to immunize with CD8 T cell epitopes from the immunodominant EBNA proteins in hopes of generating cellular immunity that might protect against infection. None of these approaches is very close to approval but both are likely to be especially relevant to seronegative potential solid organ transplant recipients awaiting transplantation.

References

Al Hamed R, Bazarbachi AH, Mohty M (2019) Epstein-barr virus-related post-transplant lymphoproliferative disease (EBV-PTLD) in the setting of allogeneic stem cell transplantation: a comprehensive review from pathogenesis to forthcoming treatment modalities. Bone Marrow Transplant 1–15

Ambinder RF, Cesarman E (2007) Clinical and pathological aspects of EBV and KSHV infection. In: Arvin A, Campadelli-Fiume G, Moore PS (eds.) Human herpesviruses biology, therapy, and immunoprophylaxis, pp. 885–914. New York, Cambridge University Press

Ambinder RF, Mann RB (1994) Detection and characterization of epstein-barr virus in clinical specimens. Am J Pathol 145(2):239–252

Apcher S, Daskalogianni C, Manoury B, Fåhraeus R (2010) Epstein barr virus-encoded EBNA1 interference with MHC class I antigen presentation reveals a close correlation between mRNA translation initiation and antigen presentation. PLoS Pathogens 6(10)

Balfour HH Jr, Verghese P (2013) Primary epstein-barr virus infection: impact of age at acquisition, coinfection, and viral load. J Infect Dis 207(12):1787–1789

Balfour HH, Schmeling DO, Grimm-Geris JM (2019) The promise of a prophylactic epstein–barr virus vaccine. Pediatric Res 1–9

Bolanos-Meade J, Fuchs EJ, Luznik L, Lanzkron SM, Gamper CJ, Jones RJ, Brodsky RA (2012) HLA-haploidentical bone marrow transplantation with posttransplant cyclophosphamide expands the donor pool for patients with sickle cell disease. Blood 120(22):4285–4291

Brunstein CG, Weisdorf DJ, DeFor T, Barker JN, Tolar J, van Burik J-AH, Wagner JE (2006) Marked increased risk of epstein-barr virus-related complications with the addition of antithymocyte globulin to a nonmyeloablative conditioning prior to unrelated umbilical cord blood transplantation. Blood 108(8):2874–2880

Burns DM, Clesham K, Hodgson YA, Fredrick L, Haughton J, Lannon M, Hussein H, Shin J-S, Hollows RJ, Robinson L (2020) Real-world outcomes with rituximab-based therapy for posttransplant lymphoproliferative disease arising after solid organ transplant. Transplantation

Caillard S, Green M (2019) Prevention and treatment of EBV-related complications. Infectious diseases in solid-organ transplant recipients, pp. 81–91. Springer

Caillard S, Lelong C, Pessione F, Moulin B, Group FPW (2006) Post-transplant lymphoproliferative disorders occurring after renal transplantation in adults: report of 230 cases from the French registry. Am J Transplant 6(11):2735–2742

Chaulagain CP, Diacovo JM, Elson L, Comenzo RL, Samaras C, Anwer F, Khouri J, Landau H, Valent J (2020) Daratumumab-based regimen in treating clonal plasma cell neoplasms in solid organ transplant recipients. Clin Lymphoma Myeloma Leuk

Cohen JI (2015) Epstein-barr virus vaccines. Clin Transl Immunology 4(1):e32

Cullis B, D'Souza R, McCullagh P, Harries S, Nicholls A, Lee R, Bingham C (2006) Sirolimus-induced remission of posttransplantation lymphoproliferative disorder. Am J Kidney Dis 47(5):e67–e72

Dierickx D, Habermann TM (2018) Post-transplantation lymphoproliferative disorders in adults. N Engl J Med 378(6):549–562

Hadinoto V, Shapiro M, Greenough TC, Sullivan JL, Luzuriaga K, Thorley-Lawson DA (2008) On the dynamics of acute EBV infection and the pathogenesis of infectious mononucleosis. Blood 111(3):1420–1427

Hammerschmidt W, Sugden B (2013) Replication of epstein-barr viral DNA. Cold Spring Harb Perspect Biol 5(1):a013029

Hui KF, Cheung AKL, Choi CK, Yeung PL, Middeldorp JM, Lung ML, Tsao SW, Chiang AKS (2016) Inhibition of class I histone deacetylases by romidepsin potently induces E pstein-B arr virus lytic cycle and mediates enhanced cell death with ganciclovir. Int J Cancer 138(1): 125–136

Imus PH, Tsai HL, Luznik L, Fuchs EJ, Huff CA, Gladstone DE, Lowery P, Ambinder RF, Borrello IM, Swinnen LJ, Wagner-Johnston N, Gocke CB, Ali SA, Bolanos-Meade FJ, Varadhan R, Jones RJ (2019) Haploidentical transplantation using posttransplant cyclophosphamide as GVHD prophylaxis in patients over age 70. Blood Adv 3(17):2608–2616

Jagadeesh D, Tsai D, Wei W, Wagner-Johnston N, Xie E, Berg S, Smith S, Koff J, Barot S, Hwang D (2019) Post-transplant lymphoproliferative disorder after solid organ transplant: survival and prognostication among 570 patients treated in the modern era. Hematol Oncol 37:159–160

Kanakry JA, Hegde AM, Durand CM, Massie AB, Greer AE, Ambinder RF, Valsamakis A (2016) The clinical significance of EBV DNA in the plasma and peripheral blood mononuclear cells of patients with or without EBV diseases. Blood 127(16):2007–2017

Kanakry JA, Kasamon YL, Bolanos-Meade J, Borrello IM, Brodsky RA, Fuchs EJ, Ghosh N, Gladstone DE, Gocke CD, Huff CA, Kanakry CG, Luznik L, Matsui W, Mogri HJ, Swinnen LJ, Symons HJ, Jones RJ, Ambinder RF (2013) Absence of post-transplantation lymphoproliferative disorder after allogeneic blood or marrow transplantation using post-transplantation cyclophosphamide as graft-versus-host disease prophylaxis. Biol Blood Marrow Transplant 19(10):1514–1517

Kasamon YL, Ambinder RF, Fuchs EJ, Zahurak M, Rosner GL, Bolaños-Meade J, Levis MJ, Gladstone DE, Huff CA, Swinnen LJ (2017) Prospective study of nonmyeloablative, HLA-mismatched unrelated BMT with high-dose posttransplantation cyclophosphamide. Blood Adv 1(4):288–292

Kasiske BL, Kukla A, Thomas D, Ives JW, Snyder JJ, Qiu Y, Peng Y, Dharnidharka VR, Israni AK (2011) Lymphoproliferative disorders after adult kidney transplant: epidemiology and comparison of registry report with claims-based diagnoses. Am J Kidney Dis 58(6):971–980

Kim WS, Ardeshna KM, Lin Y, Oki Y, Ruan J, Jacobsen ED, Yoon DH, Suh C, Suarez F, Porcu P (2017) Autologous EBV-specific T cells (CMD-003): early results from a multicenter, multinational phase 2 trial for treatment of EBV-associated NK/T-cell lymphoma. Blood 130 (Supplement 1):4073

Krown SE, M.D., Roy D, Ph.D., Lee JY, Ph.D., Dezube BJ, M.D., Reid EG, M.D., Venkataramanan R, Ph.D., Han K, Ph.D., Cesarman E, M.D., Ph.D., and Dittmer DP, Ph.D (2012) Rapamycin with antiretroviral therapy in AIDS-associated kaposi sarcoma: an AIDS malignancy consortium study. J Acquir Immune Defic Syndr 59(5): 447–454. https://doi.org/: 10.1097/QAI.0b013e31823e7884. PMCID: PMC3302934. NIHMSID: NIHMS340254. PMID: 22067664

Landgren O, Gilbert ES, Rizzo JD, Socié G, Banks PM, Sobocinski KA, Horowitz MM, Jaffe ES, Kingma DW, Travis LB (2009) Risk factors for lymphoproliferative disorders after allogeneic hematopoietic cell transplantation. Blood, J Am Soc Hematol 113(20):4992–5001

Levitskaya J, Coram M, Levitsky V, Imreh S, Steigerwald-Mullen PM, Klein G, Kurilla MG, Masucci MG (1995) Inhibition of antigen processing by the internal repeat region of the epstein-barr virus nuclear antigen-1. Nature 375(6533):685–688

Luskin MR, Heil DS, Tan KS, Choi S, Stadtmauer EA, Schuster SJ, Porter DL, Vonderheide RH, Bagg A, Heitjan DF (2015) The impact of EBV status on characteristics and outcomes of posttransplantation lymphoproliferative disorder. Am J Transplant 15(10):2665–2673

Nanmoku K, Shinzato T, Kubo T, Shimizu T, Yagisawa T (2019) Remission of epstein-barr virus-positive post-transplant lymphoproliferative disorder by conversion to everolimus in a kidney transplant recipient. Transpl Infect Dis 21(4):e13116

Pearse W, Pro B, Gordon LI, Karmali R, Winter JN, Ma S, Behdad A, Klein A, Petrich AM, Jovanovic B (2019) A phase I/II trial of brentuximab vedotin (BV) plus rituximab (R) as frontline therapy for patients with immunosuppression-associated CD30 + and/or EBV + lymphomas. American Society of Hematology. Washington, DC

Prockop SE, Suser S, Doubrovina ES, Castro-Malaspina H, Papadopoulos EB, Sauter CS, Young JW, Szenes V, Slocum A, Baroudy K (2019) Efficacy of donor and 'third party'(tabelecleucel) EBV-specific T cells for treatment of central nervous system (CNS) EBV-PTLD. Biol Blood Marrow Transplant 25(3):S72

Rowe M, Lear A, Croom-Carter D, Davies A, Rickinson A (1992) Three pathways of epstein-barr virus gene activation from EBNA1-positive latency in B lymphocytes. J Virol 66(1):122–131

Ru Y, Chen J, Wu D (2018) Epstein-barr virus post-transplant lymphoproliferative disease (PTLD) after hematopoietic stem cell transplantation. Eur J Haematol 101(3):283–290

Sampaio MS, Cho YW, Shah T, Bunnapradist S, Hutchinson IV (2012) Impact of epstein-barr virus donor and recipient serostatus on the incidence of post-transplant lymphoproliferative disorder in kidney transplant recipients. Nephrol Dial Transplant 27(7):2971–2979

Stallone G, Schena A, Infante B, Di Paolo S, Loverre A, Maggio G, Ranieri E, Gesualdo L, Schena FP, Grandaliano G (2005) Sirolimus for Kaposi's sarcoma in renal-transplant recipients. N Engl J Med 352(13):1317–1323

Swerdlow S, Campo E, Harris NL, Jaffe E, Pileri S, Stein H, Thiele J, Arber D, Hasserjian R, Le Beau M (2017) WHO classification of tumours of haematopoietic and lymphoid tissues (Revised 4th edition). IARC Lyon 421

Swerdlow SH, Campo E, Pileri SA, Harris NL, Stein H, Siebert R, Advani R, Ghielmini M, Salles GA, Zelenetz AD, Jaffe ES (2016) The 2016 revision of the World Health Organization classification of lymphoid neoplasms. Blood 127(20):2375–2390

Thorley-Lawson DA (2001) Epstein-Barr virus: exploiting the immune system. Nat Rev Immunol 1(1):75–82

Thorley-Lawson DA, Gross A (2004) Persistence of the Epstein-Barr virus and the origins of associated lymphomas. N Engl J Med 350(13):1328–1337

Thorley-Lawson DA, Hawkins JB, Tracy SI, Shapiro M (2013) The pathogenesis of epstein-barr virus persistent infection. Curr Opin Virol 3(3):227–232

Tsai D, Douglas L, Andreadis C, Vogl D, Arnoldi S, Kotloff R, Svoboda J, Bloom R, Olthoff K, Brozena S (2008) EBV PCR in the diagnosis and monitoring of posttransplant lymphoproliferative disorder: results of a two-arm prospective trial. Am J Transplant 8(5):1016–1024

Vereide DT, Sugden B (2011) Lymphomas differ in their dependence on epstein-barr virus. Blood, J Am Soc Hematol 117(6):1977–1985

Wu TC, Mann RB, Charache P, Hayward SD, Staal S, Lambe BC, Ambinder RF (1990) Detection of EBV gene expression in reed-sternberg cells of Hodgkin's disease. Int J Cancer 46(5):801–804

Yang J, Tao Q, Flinn IW, Murray PG, Post LE, Ma H, Piantadosi S, Caligiuri MA, Ambinder RF (2000) Characterization of epstein-barr virus-infected B cells in patients with posttransplantation lymphoproliferative disease: disappearance after rituximab therapy does not predict clinical response. Blood 96(13):4055–4063

Young LS, Yap LF, Murray PG (2016) Epstein-barr virus: more than 50 years old and still providing surprises. Nat Rev Cancer 16(12):789–802

HTLV-1 Replication and Adult T Cell Leukemia Development

Chou-Zen Giam

1 Introduction

Human T-cell leukemia virus type 1 (HTLV-1) is a complex human delta retrovirus that infects an estimated 10–20 million people worldwide (Gessain and Cassar 2012). HTLV-1 infection is mostly asymptomatic, and HTLV-1 viral RNAs or proteins are virtually undetectable in the blood of infected persons. Most infected cells harbor a single copy of the proviral DNA, and the proviral DNA load (PVL) is used to quantify the extent of HTLV-1 infection. Via mechanisms not fully understood, 3–5% of HTLV-1-infected individuals develop an intractable leukemia/lymphoma of CD4+ T cells known as adult T cell leukemia/lymphoma (ATLL, and abbreviated as ATL hereafter) after a latency period of 3–6 decades (Taylor and Matsuoka 2005; Matsuoka and Jeang 2007). HTLV-1 also causes several inflammatory and immune-mediated diseases, most notably HTLV-1-associated myelopathy (HAM)/tropical spastic paraparesis (TSP), and to a lesser extent, HTLV-1 uveitis, infective dermatitis, myositis, arthritis, and more recently, bronchiecstasis in a large percentage of adults of the indigenous people in Central Australia (Einsiedel et al. 2016; Gruber 2018). Determinants for ATL and HAM/TSP development include routes of infection (breastfeeding versus blood transfusion), HLA subtypes, and proviral DNA loads.

The leukemic cells of ATL are monoclonal and in most cases harbor a single copy of HTLV-1 proviral DNA at random chromosomal integration sites. Two viral regulatory proteins, Tax and HBZ, encoded by the sense and "antisense" viral transcripts, respectively, are thought to drive ATL oncogenesis (Matsuoka and

C.-Z. Giam (✉)
Department of Microbiology and Immunology, Uniformed Services University of the Health Sciences, 4301 Jones Bridge Rd., Bethesda, MD 20814, USA
e-mail: chou-zen.giam@usuhs.edu

© Springer Nature Switzerland AG 2021
T.-C. Wu et al. (eds.), *Viruses and Human Cancer*, Recent Results in Cancer Research 217, https://doi.org/10.1007/978-3-030-57362-1_10

Jeang 2007; Matsuoka and Green 2009). Tax is a potent activator of HTLV-1 viral transcription. It also exerts pleiotropic effect on cell signaling, activating IKK-NF-κB, JNK, mTOR, and other signaling pathways. Importantly, Tax is a strong clastogen: Its expression induces DNA double-strand breaks and represses DNA damage repair (Majone et al. 1993; Giam and Semmes 2016; Marriott and Semmes 2005). HBZ, in contrast, antagonizes many activities of Tax and promotes cell proliferation (Ma et al. 2016). While Tax plays an important role in leukemia development, its expression is frequently lost from ATL cells (in more than 50% of ATL cases). In contrast, the expression of HBZ is ubiquitous (Kataoka et al. 2015). The loss of Tax expression from ATL cells suggests that the oncogenic effects of Tax are likely exerted early during ATL development, with HBZ playing a role in promoting ATL proliferation (Satou et al. 2006, 2008, 2011; Arnold et al. 2008).

ATL is exceptional among hematological malignancies for its extensive genomic instability, a feature that has been fully borne out by an integrated whole-genome sequencing (WGS), transcriptomic, and targeted resequencing analysis of ATL, which has identified on the average 59.5 structural variations/ATL sample and 7.9 point mutations/10^6 bases of ATL genome, almost 2–3 times of those of multiple myeloma (21 chromosomal rearrangements/sample and 2.9 point mutations/10^6 bases) (Chapman et al. 2011). The WGS analysis has revealed frequent gain-of-function mutations in PLCG1 (phospholipase Cγ1), PKCB (protein kinase Cβ), CARD11, DLG1, VAV1, CD28, IRF4, STAT3, Notch1, etc., in ATL genomes (Kataoka et al. 2015). Many of the alterations converge on genes involved in the $^T/_B$ cell receptor-NF-κB and the CD28 co-stimulatory signaling pathways, and genes responsible for T cell trafficking and immune surveillance. Interestingly, many of the ATL significant genes show remarkable functional overlap with the Tax interactome (Kataoka et al. 2015).

Although the molecular basis for the genomic instability of ATL is not fully understood, earlier studies have implicated a causal link to Tax, which represses DNA damage repair (DDR), induces DNA double-strand breaks (DSBs), and disrupts mitotic processes (Majone et al. 1993; Marriott and Semmes 2005; Majone and Jeang 2000; Semmes et al. 1996). Given its mutagenic/clastogenic effects, Tax likely acts as an initiator of leukemia development while HBZ serves as a promoter of ATL cell proliferation. It should also be noted that the full picture of the roles of Tax and HBZ in leukemogenesis defies simple caricature. By virtue of its ability to activate multiple signaling pathways, especially IKK-NF-κB, Tax can also promote proliferation, survival, apoptosis and senescence. Likewise, HBZ has been reported to induce genomic instability recently. As might be expected, the activities of these two viral regulatory proteins require interaction with a large assembly of cellular factors as elaborated below.

2 HTLV-1 Infection and Replication

2.1 Epidemiology of HTLV-1 Infection

HTLV-1 is endemic in specific regions of the world, including Kyushu and Okinawa in Japan, the Caribbean region, parts of South America, sub-Saharan Africa, Middle East, Papua New Guinea and Australia. The virus is transmitted mainly by breastfeeding, and to a lesser extent, via sexual intercourse and exposure to cell-containing infected blood components through transfusion or needle sharing. In agreement with the epidemiological findings, HTLV-1 infection in cell culture is strictly dependent on cell-to-cell contact. Cell-free HTLV-1 infection is virtually undetectable, and the difference in efficiency between cell-mediated and cell-free HTLV-1 infection is in the order of 10^5 to 1 (Mazurov et al. 2010).

2.2 Diagnosis of HTLV-1 Infection

Clinically, HTLV-1 infection is diagnosed by using ELISA or chemiluminescence microparticle immunoassay (CMIA) to detect the presence of serum/plasma antibodies against inactivated viral lysates or $gp21^{TM}$ and $gp46^{SU}$ proteins. ELISA- or CMIA-positive blood samples are further tested by western blotting and/or PCR to confirm and differentiate between HTLV-1 and HTLV-2 infections. Since HTLV-1 mRNA or protein is not detectable in the blood, real-time PCR of the peripheral blood mononuclear cells (PBMCs) is used to quantify proviral DNA loads (PVLs) in infected persons. Because most infected cells in PBMCs harbor one copy of the proviral DNA, PVL serves as an indicator of the abundance of infected cells in the PBMCs of an infected person and can range from 0.01 to 50% or higher. A higher PVL reflects the mitotic expansion of infected cells and is associated with development of HTLV-1-related diseases including ATL and HAM/TSP.

2.3 HTLV-1 Tropism

HTLV-1 uses a ubiquitous cell surface molecule, glucose transporter 1 (GLUT1), as the receptor for virus entry (Manel et al. 2003). In addition, neuropilin 1 (NRP1), the co-receptor of vascular endothelial growth factor (VEGF) receptors, functions as a co-receptor for HTLV-1 (Lambert et al. 2009). NRP1's relation to GLUT1 appears to be similar to that of HIV co-receptors: CXCR4 and CCR5 to HIV receptor CD4. NRP1 forms a complex with GLUT1 and co-localizes with HTLV-1 Env on the cell surface, and its over-expression in deficient cells augments HTLV-1 infection. Heparan sulfate proteoglycans also contribute to viral infection, likely by serving as a viral attachment factor that facilitates $NRP1\text{-}GLUT1\text{-}gp46^{SU}$ interaction (Lambert et al. 2009; Jones et al. 2005).

In vivo, HTLV-1 is found primarily in CD4+ T cells. It also infects CD8+ T cells, monocytes/macrophages, and dendritic cells (DCs). While infection of T cells by HTLV-1 requires cell-to-cell contact, it has been shown that DCs exposed to cell-free HTLV-1 particles not only become productively infected themselves (*cis*-infection), but also rapidly transmit the virus to CD4+ T cells (*trans*-infection) (Jones et al. 2008; Jain et al. 2009; De Castro-Amarante et al. 2015) (reviewed in Gross and Thoma-Kress 2016). This may explain the in vivo tropism of HTLV-1. As GLUT1 is broadly expressed, HTLV-1 infection in cell culture can occur in a wide variety of cells including T and B lymphocytes, monocytes, endothelial cells, and fibroblasts.

Cell-to-cell transmission of HTLV-1 occurs through the "virological synapse" formed in part through LFA1 and ICAM1 (Barnard et al. 2005; Nejmeddine et al. 2005; Igakura et al. 2003). Tax expression and ICAM1 engagement cause the microtubule polarization associated with the virological synapse. Tax is also localized to the region of the cell-cell contact formed between an HTLV-1-producing cell and its target cell, and at the vicinity of the microtubule-organizing center associated with the cis-Golgi (Nejmeddine et al. 2005, 2009). Interestingly, HTLV-1 viral particles were found to be stored as carbohydrate-rich, biofilm-like extracellular structures that rapidly attached to target cells for virus transmission (Pais-Correia et al. 2010).

2.4 HTLV-1 Genome Organization, Gene Expression, and Regulation

The genomic organization of HTLV-1 is shown in Fig. 1. In addition to *Gag*, *Pol*, and *Env*, HTLV-1 encodes six viral accessory proteins, Tax, Rex, $p12^I$, $p13^{II}$, $p30^{II}$, and HTLV-1 basic zipper protein (HBZ) from partially overlapping open reading frames (ORFs) in both directions of the viral genome. For a more recent review on $p12^I$, $p13^{II}$, $p30^{II}$, the readers are referred to this article (Edwards et al. 2011). $P12^I$, an ER and Golgi membrane-associated protein, plays a role in enhancing T-cell activation and signaling. It also counteracts innate and adaptive immune responses by binding MHC class I heavy chain, targeting it for degradation (Pise-Masison et al. 2014); and down-regulates ICAM-1 and ICAM-2, thereby mitigating autologous natural killer cell cytotoxicity for the infected CD4+ T cells (Banerjee et al. 2007). Interestingly, the proteolytic cleavage product of $p12^I$, p8, localizes to the cell surface where it increases the formation of cellular conduits and facilitates viral transmission (Van Prooyen et al. 2010; Galli et al. 2019). The $p13^{II}$ is an inner mitochondria membrane protein with anti-proliferation activity (Hiraragi et al. 2005). It also becomes ubiquitinated in the presence of Tax and translocates to the nucleus where it disrupts Tax-CBP/p300 interaction and inhibits viral and cellular transcription (Andresen et al. 2011). Its role in HTLV-1 infection and replication cycle is not fully defined. The anti-proliferation activity of $p13^{II}$ appears to be related to its interaction with farnesyl pyrophosphate synthetase (Lefebvre et al. 2002) and increased sensitivity to Ca++-mediated stimulation and enhanced

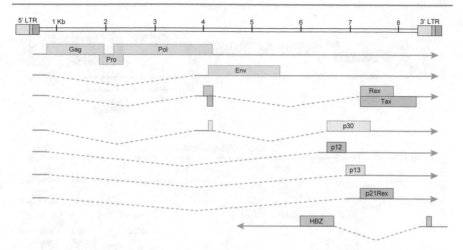

Fig. 1 Genomic organization of HTLV-1 proviral DNA and viral mRNA trasnscripts. The open reading frames for Gag, Pro, Pol, Env, Tax, Rex, p12I, p30II, p13II, and HBZ genes are as indicated (*Courtesy of* Dr. Patrick Green)

C2-ceramide-induced apoptosis of p13[II]-expressing T lymphocytes (Silic-Benussi et al. 2004). P30[II] is a nuclear protein that functions as a post-transcriptional modulator of viral replication. Data suggest that p30[II] retains the doubly spliced Tax/Rex mRNA in the nucleus and thereby down-modulates viral gene expression and particle production by reducing the levels of Tax and Rex (Edwards et al. 2011; Nicot et al. 2004). Rex regulates the transport of unspliced and singly spliced mRNA, while Tax, the viral transcriptional activator, is thought to be one of the major oncoproteins of HTLV-1. Tax expression is sufficient to effect cellular transformation in rodent fibroblasts and in primary human lymphocytes, and expression of Tax through a variety of promoters induces neoplasia in transgenic mice (discussed below). In promoting viral replication, HTLV-1 Tax interacts with key regulators of cell signaling pathways, and these interactions ultimately perturb several basic cellular processes, many or all of which contribute to leukemia development. The major antisense HBZ mRNA (spliced HBZ, sHBZ) spans the open sequence between the *env* and the *tax/rex* ORFs. A minor unspliced HBZ transcript (usHBZ) is also made. Both sHBZ and usHBZ mRNAs encode basic domain–leucine zipper proteins with minor differences in their respective NH$_2$-termini, and both forms of HBZ have been shown to negatively regulate Tax trans-activation (Yoshida et al. 2008) (see Sect. 4.1). Importantly, the spliced HBZ protein and RNA are expressed in all ATL cells and can stimulate cell proliferation (Matsuoka and Green 2009), and CRISP-Cas9-mediated deletion of HBZ gene causes ATL cells to cease proliferation (Nakagawa et al. 2018). A schematic summarizing the roles of Tax, Rex, and HBZ in viral replication is shown in Fig. 2.

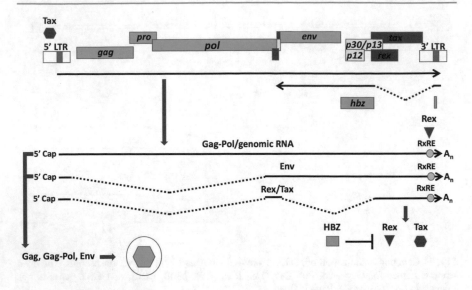

Fig. 2 The balance between Tax and HBZ expression regulates the outcome of HTLV-1 infection. Proviral integration sites and host transcription factors availability determine the levels of Tax and HBZ expression in infected cells. Most HTLV-1-infected cells express Tax and Rex strongly. Through the trans-activation and RNA export activities of Tax and Rex, HTLV-1 structural proteins are abundantly expressed and viral particles are produced. In these cells, Tax not only activates viral transcription, but also constitutively activates NF-κB, triggering cellular senescence or apoptosis. In a small fraction of HTLV-1-infected cells, Tax expression is low, intermittent, or silenced. When the levels of Tax/Rex expression are low, LTR trans-activation and senescence induction by Tax, and viral mRNA nuclear export by Rex are inhibited by HBZ. As such, no viral structural proteins are expressed and these cells undergo mitotic expansion possibly driven by both Tax and HBZ

2.5 Models for Studying HTLV-1 Infection

2.5.1 HTLV-1 Infection in Cell Culture

HTLV-1 infection in cell culture is achieved by co-cultivation of target cells including naïve peripheral blood mononuclear cells (PBMCs) and cells of established cell lines together with HTLV-1-producing cells that have been mitotically inactivated by γ-irradiation or DNA damaging agents such as mitomycin C (Anderson et al. 2004; Derse et al. 2004). MT-2, a human T-cell line chronically produces HTLV-1 particles and HEK293 cells transfected with an infectious molecular clone of HTLV-1 have been routinely used as sources of HTLV-1 (Anderson et al. 2004; Miyoshi et al. 1981). Cell-free infection using VSV G-pseudotyped viral particles (Derse et al. 2001) or using DCs as an intermediary to transmit viral particles has also been reported (Jones et al. 2008). As the efficiency of HTLV-1 infection is poor and the detection of newly infected cells is technically challenging, studies of the outcome of HTLV-1 infection often involve either reporter cell lines or long-term culture of infected PBMCs with immortalization/transformation of virus-infected primary

CD4+ T cells as the experimental end-point (Anderson et al. 2004). The early events of viral infection, especially in primary T cells, have not been examined in depth. It has been shown that ATL and HTLV-1-infected cells overexpress CADM1 (cell adhesion molecule 1)/TSLC1 (tumor suppressor in lung cancer 1) (Sasaki et al. 2005), and CD4+ CADM1+ cells carry 65% of proviral copies in the peripheral blood of infected individuals (Manivannan et al. 2016). This cell surface marker may afford a means to enrich HTLV-1-infected primary T cells for an in-depth analysis of the early events of viral replication.

2.5.2 HTLV-1 Infection in Humanized Immune-Deficient Mice

Non-obese diabetic/severe combined immunodeficiency (NOD/SCID) mice and NOD/SCID/IL2Rγ-null (NSG) mice engrafted with human peripheral or bone marrow hematopoietic stem cells have been used in the study of HTLV-1 infection. Tezuka et al. have established a system where NSG mice engrafted by intra-bone marrow injection of human CD133 + hematopoietic stem cells were infected by HTLV-1 via the injection of γ-irradiated MT-2 cells (Tezuka et al. 2014). Four to five months after infection, two of the eight humanized mice developed ATL-like leukemic conditions including lymphocytosis of CD4+ T cells in the periphery, initial proliferation of both CD4+ CD25- and CD4+ CD25+ T cells, followed by the appearance of atypical lymphocytes with lobulated nuclei resembling the ATL-specific flower cells and dramatic expansion of specific and dominant CD25 ι CD4+ T cell clones. This system recapitulated many of the pathological features of HTLV-1 infection and ATL development and has been used recently to study the roles of p12^{II} and p8 in HTLV-1 replication and pathogenesis (Galli et al. 2019) and the nature of the HTLV-1 provirus (Katsuya et al. 2019). A similar system generated by injecting human umbilical cord stem cells into the livers of neonatal NSG mice has been used to demonstrate that the envelope glycoproteins of HTLV-1 and HTLV-2 were responsible for the lymphoproliferation of CD4+ and CD8+ T cells, respectively (Huey et al. 2018).

2.5.3 HTLV-1 Infection in Rabbits

Of the animal models of HTLV-1 infection, the rabbit model is perhaps the most amenable to experimentation, although none of the human diseases associated with HTLV-1 infection could be recapitulated with the model (Panfil et al. 2013; Dodon et al. 2012). With the rabbit model, Li et al. have shown that the Tax/Rex mRNA levels in PBMCs peak one week after inoculation of rabbits with γ-irradiated HTLV-1-producing cells (Li et al. 2009). They then rapidly declined to low levels when the infection progressed beyond the second week. As might be expected, Gag-Pol mRNA expression is coincident with, but at levels that are approximately ¼ that of the levels of Tax/Rex mRNA, and became mostly undetectable 2–3 weeks after infection. By contrast, HBZ mRNA levels were low initially, but slowly increased and stabilized at 4 weeks and beyond. This is followed by a rise in proviral DNA loads that mostly peaked at 6 weeks after inoculation. These results suggest that immediately after HTLV-1 infection, there is a strong selective pressure to silence viral gene expression, and only infected cells expressing Tax/Rex and HBZ at low and steady levels, respectively, persist. Whether the rapid decline in Tax/Rex and Gag-Pol expression soon after infection is due solely to the elimination of infected cells by cytotoxic T lymphocytes (CTLs) is unclear. As detailed below, the senescence/cell cycle arrest triggered by Tax may play a role in selecting for infected cells that express viral genes minimally or intermittently.

2.6 Clonality of HTLV-1-Infected T Cells

2.6.1 HTLV-1 Clonality and Clone Abundance in Natural Infection

Longitudinal studies of newly infected seroconverters and asymptomatic carriers (ACs) have indicated that the clonality of HTLV-1-infected T cells is heterogeneous and unstable in initial infection (Tanaka et al. 2005). Analyses of pre-diagnostic PBMCs (3–8 years prior to ATL onset) showed that prior to the onset of ATL, there was a significant rise in PVLs. In one ATL case for which both leukemic and pre-diagnostic samples existed, pre-leukemic cells harboring the same integrated provirus as the leukemic cells could be detected as early as 2, 5, and 8 years prior to ATL diagnosis, supporting the notion that clonal expansion, selection, and evolution drive ATL development (Okayama et al. 2004). A US National Cancer Institute study of HTLV-1 infection in Jamaica followed a group of three children who acquired HTLV-1 perinatally (Umeki et al. 2009). The study spanned a period of more than a decade and indicated that HTLV-1 PVLs were variable (10^2–10^3 copies/10^5 PBMCs) in ACs. Some of the infected cell clones persisted for years. However, proviral integration patterns continued to evolve over time, indicating de novo viral infection in ACs. Importantly, two unique cell clones in one subject underwent significant expansion a decade or longer after the initial infection, causing PVLs to increase more than 40 fold (from 3×10^3–1.3×10^4 copies/10^5 PBMCs). While the clonal expansion did not result in HAM/TSP or ATL, lymphadenopathy, seborrheic dermatitis, and hyperreflexia were observed in the subject (Umeki et al. 2009), supporting the notion that a higher PVL correlates with a greater risk of disease development.

HTLV-1 clones that persist in chronically infected persons show little detectable Tax mRNA or protein expression in vivo and ex vivo, and silencing of Tax expression due to 5′ LTR methylation, 5′ LTR deletion, or nonsense mutations within its coding sequence is correlated with the expansion of malignant T cells (Takeda et al. 2004; Taniguchi et al. 2005). The 3′ region of the viral genome and the 3′ LTR, however, remain largely intact and unmethylated, thus favoring HBZ expression (Taniguchi et al. 2005). In vitro culture of infected but Tax-negative T cells from donors induces Tax expression, but clones that remain Tax-negative in culture are more abundant (Niederer and Bangham 2014).

High-throughput DNA sequencing has been used to characterize the chromosomal integration sites of HTLV-1 proviral DNA and the clone abundance of each integrated provirus in ACs, HAM/TSP, and ATL patients (reviewed in Niederer and Bangham 2014). These studies have revealed in ACs and HAM/TSP patients a large number of distinct proviral integration sites ($>10^4$) in each host and a large majority of infected cells harbor a single integrated provirus (Cook et al. 2012). This contrasts with ATL (91%) where often a predominant and malignant T cell clone containing a single provirus is detected (Cook et al. 2014). Proviral clones detected by ligation-mediated polymerase chain reaction (LM-PCR in ACs) ranged in clone abundance (defined as the number of cells in a given proviral clone per 10^4 PBMCs) from <0.1 to ~100; less abundant clones are undetectable however due to the technical limitation of LM-PCR. More recently, a study of 98 naturally infected individuals (24 ACs, 29 HAM/TSPs, 45 ATLs) using a viral DNA capture sequencing approach has shown that the median clone abundance of a given provirus was lower in ACs (slightly lower than 1) with narrow distribution, higher for HAM/TSP patients (around 1) also with narrow distribution, but significantly greater for ATL patients and with broad distribution as might be expected (Katsuya et al. 2019). Defective proviruses were seen not only in ATL patients where deletions of 5' LTR occurred in 26.7% of proviruses, but also across other clinical conditions of infection albeit to lower extents. The HTLV-1 provirus integrates frequently in intergenic regions and introns, rarely within exons, and with no preference in orientation (Katsuya et al. 2019).

2.7 Clonal Abundance of HTLV-1-Infected Cells in Cell Culture and Humanized Mouse Models

Cells infected by HTLV-1 in cell culture (using transfected 293 cells as the source of the virus and Jurkat cells as the target) showed an average clone abundance of less than 1 (Katsuya et al. 2019), likely because many target cells were infected by more than one provirus, a frequent occurence in cell culture infections. It appears clonal expansion only occurred in a small number of infected cells despite the absence of any immune control. Likewise, the clonal abundance of the majority of HTLV-1-infected T cells in the NSG-humanized mouse model (Tezuka et al. 2014) hovered around slightly less than 1. Here, again, only a small number of infected cells underwent clonal expansion (Tezuka et al. 2014; Katsuya et al. 2019). These results are reminiscent of the infection outcomes of reporter cells in culture where 98–99% of infected cells became senescent or cell cycle arrested, with a small fraction (1–2%) that express Tax and Rex at lower levels underwent clonal expansion. Whether the expanded clones in the study by Katsuya et al. (Katsuya et al. 2019) express Tax and other viral antigens at low levels or intermittently was not determined. Whether some of the clones with an abundance of 1 or <1 are arrested in senescence or G1 is also unknown. Notably, the frequencies of defective proviruses (especially with deletion in the 5' LTR and other 5' region of the proviral DNA) in infections in cell culture and humanized mice were low initially, and increased over time, suggesting a positive selection for a loss of Tax and viral gene expression during persistent infection.

3 HTLV-1 Tax

3.1 Tax Is a Potent Activator of HTLV-1 Transcription/Replication

Tax specifically and potently activates viral transcription via three Tax-responsive 21-bp repeat enhancer elements, each consisting of a cAMP response element (CRE, which consists of a palindromic consensus sequence: TGACGTCA that often contains significant degeneracy in the 3' CA nucleotides) flanked by 5' G-rich or 3' C-rich sequences, referred to as the viral CRE (vCRE) or Tax response element (TxRE). The CRE core of the TxRE binds the homodimer or heterodimer of a basic domain–leucine zipper (bZip) transcription factor, CREB (cAMP response element binding protein), and its close family member, ATF1 (activating transcription factor 1). Tax by itself does not bind TxRE. Rather, Tax, CREB/ATF-1, and TxRE form a stable ternary nucleoprotein complex in which Tax interacts with the basic domain of CREB/ATF-1 (Zhao and Giam 1992) and via this interaction contacts the flanking G-/C-rich sequences in the minor groove (Paca Uccaralertkun et al. 1994; Lenzmeier et al. 1999). In this ternary complex, Tax further recruits transcriptional co-activators/histone acetyl trans-ferases, CREB-binding protein (CBP)/p300 (Kwok et al. 1996), and p300-CBP associated factor (P/CAF) (Harrod et al. 1998), and transducers of regulated CREB (TORCs) (Siu et al. 2006) to establish a nucleosome-free region for potent HTLV-1 viral mRNA transcription (reviewed in (Nyborg et al. 2009)). This mode of action explains the exquisite DNA sequence specificity and great potency of Tax-mediated LTR trans-activation. Interestingly, whereas CRE-mediated cellular transcriptional activation requires the phosphorylation of Ser-133 of CREB by kinases such as protein kinase A to promote CBP/p300 recruitment (Chrivia et al. 1993), CBP/p300 recruitment and trans-activation of TxRE by Tax can occur independently of cel-lular signal transduction. Thus, Tax serves as an adaptor that confers both sequence specificity and signal independence to viral gene expression.

3.2 Tax Hijacks Ubiquitin E3 Ligase RING Finger Protein 8 (RNF8) for Canonical IKK-NF-κB Activation

In addition to activating viral transcription, Tax also exerts pleiotropic influence on cell signaling. It activates both canonical and noncanonical NF-κB pathways; the transcriptional activities of AP1, serum response factor (SRF), and nuclear factor of activated T cells (NFAT); and the kinase activity of IKK, JNK, and the mammalian target of rapamycin (mTOR), etc. (Matsuoka and Jeang 2007). Tax has been shown to interact directly or indirectly with many components of the IKK-NF-κB pathway (Harhaj and Giam 2018; Ho et al. 2015; Rauch and Ratner 2011; Shibata et al. 2017), including NF-κB, NEMO (NF-κB essential modulator): the regulatory subunit of IKK, protein phosphatase 2A (PP2A), TAX1BP1, etc. In fact, NEMO was first isolated as a gene whose loss prevented Tax-induced transformation/foci formation of Rat-1 fibroblasts (Yamaoka et al. 1998).

For a long time, the interactome of Tax appears extremely complex and defies a simple explanation. Ho et al. have recently used in vitro reconstitution to demonstrate that Tax hijacks a E3 ligase, RING finger protein 8 (RNF8), to assemble K63-linked polyubiquitin chains for canonical IKK and NF-κB activation (Ho et al. 2015). In the presence of Tax, RNF8 and E2 conjugating enzymes, Ubc13/Uev1a or Ubc13/Uev2, become greatly stimulated in vitro and in vivo, and assemble long unanchored K63-linked polyubiquitin chains as cytosolic signaling scaffolds for the activation of TAK1, IKK, and other downstream kinases such as JNK and mTOR (Ho et al. 2015) (Fig. 3). This mechanism explains the pleiotropic effect of Tax on multiple cell signaling pathways mentioned above (Summarized in Fig. 4). The multitude of cellular Tax-binding partners described in the literature could very well be associated with Tax via the K63-linked, M1-linked, and hybrid polyubiquitin chains (detailed below) whose formation is greatly stimulated by Tax. It is expected that many aspects of viral replication will be impacted upon by this mechanism as well.

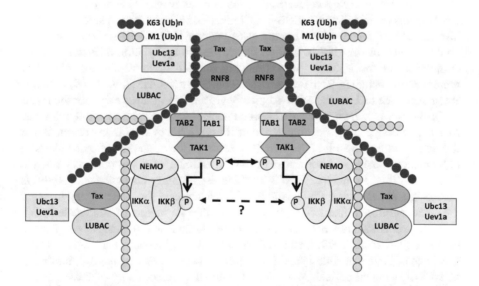

Fig. 3 Tax hijacks RNF8 and LUBAC to activate TAK1 and IKK, and multiple downstream signaling pathways. Tax interacts with and stimulates RNF8 and LUBAC to assemble hybrid K63-linked and M1-linked (linear) polyubiquitin chains as signaling scaffolds for recruiting TAK1 and IKK. TAK1 undergoes autophosphorylation and auto-activation, and phosphorylates and activates mitogen-activated kinase kinases (MKKs) and IKK, and downstream p38 mitogen-activated protein kinase, c-Jun kinase (JNK), canonical NF-κB, and mammalian target of rapamycin (mTOR) pathways. LUBAC can also be recruited to K63-linked polyubiquitin chains directly via the ubiquitin-binding domain of its subunit HOIP

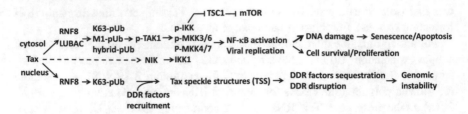

Fig. 4 Pleiotropic effect of Tax explained. Aberrant activation of RNF8 and LUBAC by Tax leads to the assembly of K63-pUb and K63-M1-hybrid pUb chains in the cytosol, and K63-pUb chains in the nucleus. The K63-M1-hybrid pUb chains trigger a cascade of kinase activation and phosphorylation, culminating in constitutive IKK/NF-κB activation, increased HTLV-1 replication, and senescence/apoptosis or cell survival/proliferation (upper pathway, Zhi 2020). The nuclear K63-pUb chains and Tax form microscopically visible Tax speckle structures that sequester DDR factors such as BRCA1, DNA-PK, and MDC1, disrupting DDR and inhibit DNA damage repair (lower pathway, Zhi 2020)

3.3 Aberrant Activation of RNF8 by Tax

During interphase, RNF8 is localized in the nucleus where it becomes recruited to DNA double-strand breaks (DSBs) to initiate K63 polyubiquitylation of linker histone H1 that signals the recruitment of another K63 E3 ligase, RNF168. The chromatin-bound RNF168 in turn propagates the ubiquitylation of histones H2A, culminating in the assembly of a signaling scaffold upon which multiple DNA repair factors including RNF168 itself, RAP80, 53BP1, RAD18, RNF169, BRCA1, etc., congregate to form the microscopically visible DNA damage foci for the repair of DSBs. Tax is known to shuttle between nuclear and cytoplasmic compartments. In Tax-expressing cells, the abundance of RNF8 in the cytosol is increased. While the activation of cytosolic RNF8 by Tax leads to the assembly of cytosolic K63 polyubiquitin chains and activation of TAK1, IKK-NF-κB, and other kinases, the impact of Tax on nuclear RNF8 remains to be fully characterized. RNF8 is responsible for initiating the assembly of DNA damage foci. Interestingly, recent data have indicated that RNF8 is critical for the formation of the nuclear Tax speckle structures known to sequester DDR factors and repress DNA damage response (Zhi et al. 2020; Giam and Semmes 2016; Haoudi et al. 2003). Other functions of RNF8 include cytokinesis, protection of telomere integrity, inhibition of NOTCH1 signaling, etc. While Tax is known to affect many of these cellular processes, whether such effects of Tax are mediated through RNF8 and K63-linked polyubiquitin chains remains unknown.

3.4 Tax Promotes the Assembly of K63/M1-Linked Hybrid Polyubiquitin Chains

TAK1 and IKK (IKK1-IKK2 or IKKα-IKKβ) holoenzyme complexes contain regulatory subunits: TAK1 binding protein 2/3 (TAB2/3) and NEMO, respectively, that facilitate their recruitment to polyubiquitin chains. TAB2/3 preferentially binds K63-linked polyubiquitin chains through a conserved zinc finger domain. In contrast, NEMO through its UBAN (UBD in ABIN proteins and NEMO) domain binds linear polyubiquitin chains with 100-fold higher affinity than K63 chains. Shibata et al. (Shibata et al. 2017) have recently demonstrated that Tax also interacts with and recruits the linear ubiquitin assembly complex (LUBAC) E3 ligase complex and together with a K63-specific E3 ligase, i.e., RNF8, generates K63/M1-linked hybrid polyubiquitin chains. It is important to note that K63/M1-linked hybrid polyubiquitin chains are generated during IL-1 and other immune signaling pathways as a mechanism to colocalize TAK1 and NEMO-containing IKK complexes (Emmerich et al. 2013; Emmerich et al. 2016). Upon recruitment of TAK1 and IKK to the hybrid polyubiquitin chains, TAK1 becomes activated by autophosphorylation, and then phosphorylates and activates IKK. While it has been proposed that the K63/M1 hybrid polyubiquitin chains bind to NEMO and promote IKK oligomerization, autophosphorylation, and activation (Shibata et al. 2017), prevailing evidence supports a key role of TAK1 in IKK phosphorylation and activation by Tax, in keeping with immune activation of the IKK-NF-κB pathway. It should also be noted that the assembly of K63-linked polyubiquitin chains is a prerequisite for the recruitment of LUBAC, and the LUBAC catalytic subunit, heme-oxidized IRP2 ubiquitin ligase 1 interacting protein (HOIP), specifically interacts with K63-linked polyubiquitin chains (Emmerich et al. 2013; Haas et al. 2009) (see Fig. 3).

In addition to NEMO, several M1-linear polyubiquitin chain-binding proteins have been identified including ABIN 1–3 (A20-binding inhibitor of NF-κB 1–3), the selective autophagy receptor: Optineurin, HOIL-1L chain, and A20 (Nakazawa et al. 2016). Several of them are involved in down-regulating IKK-NF-κB, but it is unclear whether any of them plays a role in Tax-induced IKK activation.

It should be pointed out that a recent report suggests that Tax itself may be a ubiquitin E3 ligase that together with a group of E2 enzymes including UbcH7, UbcH5b, UbcH5c, and UbcH2 assembles mixed-linkage polyubiquitin chains for IKK activation in a Ubc13- and TAK1-independent manner (Wang et al. 2016). This conclusion contrasts with published results from multiple laboratories (Ho et al. 2012, 2015; Wu and Sun 2007; Shembade et al. 2007; Shibata et al. 2011; Hayakawa 2012) and goes against the established mechanism of receptor-mediated IKK activation. Importantly, no RING finger or HECT domain commonly seen in E3 ligases has been found in Tax.

Finally, to maintain persistent NF-κB activation, Tax not only causes constitutive IKK activation and IκBα degradation, but also utilizes multiple mechanisms to counteract negative feedback regulators of IKK-NF-κB. These mechanisms include inhibition of the protein phosphatase, PP2A, and A20 and have been summarized in a recent review (Harhaj and Giam 2018).

3.5 Tax-Mediated Activation of NIK and IKK1

Tax activates both the canonical and the noncanonical NF-κB pathways (Senftleben et al. 2001; Xiao et al. 2001). Activation of the noncanonical NF-κB2 pathway in lymphoid organs by cytokines requires the NF-κB-inducing kinase, NIK, which activates downstream NEMO-free IKK1 (i.e., IKKα) to induce phosphorylation, ubiquitination, and processing of p100 (Tao and Ghosh 2012). The proteolytic conversion of p100 to p52 de-represses p52/RelB and up-regulates specific NF-κB target genes (Sun 2012). Under physiological conditions, newly synthesized NIK is targeted by TRAF3, which recruits NIK to the TRAF2:TRAF3:cIAP1:cIAP2 (TRAF-cIAP) E3 ligase complex for continuous K48 ubiquitination and proteasomal degradation. Upon engagement of specific receptors by ligands such as lymphotoxin β and B cell activating factor (BAFF), the TRAF-cIAP E3 ligase complex becomes recruited to the receptor. The dissociation of TRAF-cIAP complex stabilizes NIK. Furthermore, upon TRAF-cIAP complex binding to the receptor, the K63 ubiquitin ligase activity of TRAF2 is activated, leading to K63 polyubiquitylation and activation of cIAP1/2, which targets TRAF3 for K48 polyubiquitylation and degradation, again leading to NIK stabilization and

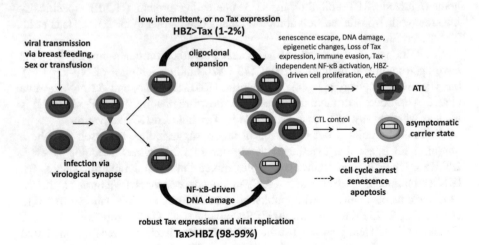

Fig. 5 A model for HTLV-1 infection and leukemogenesis. HTLV-1 is transmitted by cell-to-cell contact. The expression levels of Tax and HBZ modulate the outcomes of infection. Robust viral replication stimulated by Tax is accompanied by cellular senescence. In infected cells where Tax expression is low, intermittent or silenced, HBZ inhibits Tax and Rex and viral replication and facilitates oligoclonal expansion of T cells that are "latently" infected. Cytotoxic T lymphocyte (CTL) killing controls virus replication, resulting in an asymptomatic carrier state. HTLV-1-infected cells develop Tax/NF-κB-induced DNA damage, senescence/apoptosis and genomic instability. Genetic and epigenetic changes selected early during infection help sustain Tax expression and NF-κB activation, and set the stage for the development of Tax-independent NF-κB activation later. Loss of Tax expression is positively selected because Tax is a major CTL target and a potent clastogen that induces DNA damage and cellular senescence. The mitogenic activities of HBZ and HBZ mRNA maintain the ATL tumor phenotype and sustain its proliferation

activation (Sun 2012). Interestingly, NIK stability is also closely linked to the canonical NF-κB pathway. NEMO knockout or loss-of-function NEMO mutations lead to NIK stabilization and basal noncanonical NF-κB activation. Tax activates noncanonical NF-κB by promoting RelB and NF-κB2 p100 expression and by activating IKKα to signal proteolytic maturation of p100 into NF-κB2 p52. While it is known that noncanonical NF-κB activation by Tax is NIK-dependent, the exact mechanism remains to be elucidated.

3.6 The Biological Consequences of Tax-Mediated NF-κB Activation

3.6.1 Cell Transformation

The mitogenic effect of Tax (Kelly et al. 1992) correlates with its ability to activate multiple signaling pathways including IKK/NF-κB, JNK, mTOR, and cell cycle progression. Early studies have indicated that Tax could directly transform rat fibroblasts in vitro (Tanaka et al. 1990; Yamaoka et al. 1992; Pozzatti et al. 1990). Primary human T lymphocytes could be immortalized/transformed to grow continuously by a Herpesvirus saimiri vector carrying the *tax* gene (Grassmann et al. 1989). The transforming activity of Tax is NF-κB-dependent (Yamaoka et al. 1996; Matsumoto et al. 1997). Indeed, a flat revertant of Tax-transformed rat fibroblast cell line R5 lacked NEMO and was instrumental for the complementation cloning of the NEMO gene (Yamaoka et al. 1998). Tax activates telomerase (hTERT) expression via the NF-κB pathway (Sinha-Datta et al. 2004). In *tax*-transgenic and humanized mouse models, T cell immortalization and tumor development occur after *tax* transgene expression and viral infection (Tezuka et al. 2014; Hasegawa et al. 2006; Nerenberg et al. 1987; Grossman et al. 1995). However, with only a few exceptions (Yamaoka et al. 1992), constitutive expression of Tax in cultured cell lines is difficult to achieve, suggesting that in order for Tax to drive cell immortalization/transformation and oncogenesis, specific cellular alterations are needed or alternative viral factor(s) (such as HBZ) is (are) involved.

3.6.2 Functional Inactivation of P53

ATL and HTLV-1-transformed cell lines often retain the wild-type TP53 gene; however, p53 is functionally impaired (Cereseto et al. 1996). This functional impairment is Tax-induced and NF-κB-dependent (Pise-Masison et al. 2000) and has been attributed to the formation of a p65/RelA-p53 complex that is functionally inactive in promoting the transcription of p53-regulated genes including MDM2 (Pise-Masison et al. 2000; Jeong et al. 2004). It should be noted that although Tax impairs p53 functionally, TP53 mutations develop at a higher frequency in the aggressive acute and lymphomatous subtypes of ATL where Tax expression is often lost.

3.6.3 Apoptosis Induction and Inhibition

Tax is known to cause or prevent apoptosis in an NF-κB-dependent manner in a variety of lymphoid and non-lymphoid cell lines and experimental settings (Chlichlia et al. 1995; Hall et al. 1998; Haoudi and Semmes 2003; Rivera-Walsh et al. 2001; Yamada et al. 1994; Tsukahara et al. 1999; Takahashi et al. 2013; El Sabban et al. 2000; De La et al. 2003). Intermittent *tax* expression in MT-1, an ATL cell line, has been shown recently to induce anti-apoptotic proteins such as CFLAR, GADD45B, TRAF1, and TNFAIP3 that increase cell survival (Mahgoub et al. 2018). These seemingly conflicting activities of Tax are likely due to the different level and duration of NF-κB activation by Tax and the cellular backgrounds in which Tax is expressed.

3.6.4 Senescence Induction

Contrary to the notion that HTLV-1 infection leads to cell proliferation, most HTLV-1-infected lymphoid or non-lymphoid cells in culture cease proliferation immediately after infection (Liu et al. 2008; Philip et al. 2014; Zahoor et al. 2014). Similarly, CD34+ hematopoietic progenitor cells infected by HTLV-1 have also been shown to undergo G1 arrest (Tripp et al. 2005). The infected cells express high levels of cyclin-dependent kinase inhibitors: $p21^{CIP1/WAF1}$ (p21) and $p27^{KIP1}$ (p27), develop DNA damage and mitotic abnormalities accompanied by cytokinesis failure, and arrest in senescence (Kuo and Giam 2006; Yang et al. 2011). This is due to Tax-mediated hyperactivation of NF-κB (Zhi et al. 2011). Indeed, when NF-κB is blocked by ΔN-IκBα, a degradation-resistant mutant of IκBα that constitutively inhibits NF-κB, Tax-induced senescence is prevented, and cells expressing ΔN-IκBα continue to proliferate after HTLV-1 infection (Philip et al. 2014; Zahoor et al. 2014). As such, cell lines chronically infected by HTLV-1 and express all viral proteins and mRNAs can be readily established upon NF-κB inhibition (Philip et al. 2014; Zahoor et al. 2014).

The increase in p21 and p27 levels during Tax-induced senescence is caused by p53-independent transcriptional up-regulation and stabilization of p21 mRNA (Cereseto et al. 1996; Kuo and Giam 2006; Zhi et al. 2011; De La et al. 2000), and stabilization of p27 protein as a result of Skp2 degradation associated with the prematurely activated anaphase promoting complex (Kuo and Giam 2006; Zhang et al. 2009). The physiological significance of HTLV-1/Tax-induced senescence is unclear at present. Senescent cells are known to develop a "senescence-associated secretory phenotype" (SASP) that turns them into proinflammatory cells that secrete inflammatory cytokines, chemokines, proteases, etc., that attract and activate innate immune cells. Whether senescence and SASP contribute to the spread of HTLV-1 remains to be seen.

3.7 Tax Expression and Outcomes of HTLV-1 Infection

The senescence and apoptosis responses induced by Tax are causally linked to NF-κB activation. Notably, Tax-induced DNA damage has also been associated with NF-κB activation (Baydoun et al. 2012). Both senescence and apoptosis may represent cellular responses to NF-κB-hyperactivation-associated DNA damage that develops during productive HTLV-1 infection. In this sense, HTLV-1-infected cells are not only eliminated by CTL killing, the cellular mechanism (DNA damage response?) that guards against NF-κB hyperactivation may also restrict cells that are productively infected by HTLV-1 (Philip et al. 2014). In the latter scenario, only infected cells with minimal to no viral (sense) gene expression are able to persist and expand. Indeed, a large majority of HTLV-1-infected cells in cell culture became senescent or cell-cycle arrested. Only a small fraction (1–2%) managed to continue to proliferate. The latter expressed low levels of Tax, Rex, and HBZ, but not viral structural proteins, and the activities of both Tax and Rex were inhibited by HBZ (Philip et al. 2014). Whether this outcome occurs in human infection is unknown. Finally, a recent study has indicated that Tax expression in MT-1, an ATL cell line, occurs transiently and sporadically. The short burst of Tax expression induces anti-apoptotic factors that facilitate MT-1 cell survival (Mahgoub et al. 2018). Together, these results suggest that only HTLV-1-infected cells with no, low, or intermittent Tax expression are able to persist in infected individuals. The expansion of infected cells in vivo likely depends on a variety of factors: (1) low or transitory expression of Tax; (2) the anti-apoptotic activities induced by bouts of Tax expression; (3) the growth-promoting activity of HBZ; (4) down-regulation of NF-κB; (5) somatic mutations that mitigate or overcome senescence; or (6) a combination of some or all of the above. As MT-1, an ATL cell line, expresses Tax intermittently, this pattern of Tax expression is clearly associated with leukemia development. Other ATL cells that have lost Tax expression completely or those that express Tax and other viral proteins at low levels may have evolved different strategies to grow and expand. The outcomes of HTLV-1 infection are summarized in Fig. 5.

4 HTLV-1 HBZ

4.1 HTLV-1 HBZ: Gene Expression and Viral Persistence

HBZ gene is located at the 3' end of the viral genome. Its mRNA is synthesized from the minus or antisense strand of the viral genome and is of the opposite polarity of the major HTLV-1 viral transcript. The transcription of HBZ is regulated by three SP1 binding sites in a TATA-less promoter located in the 3' LTR of the proviral DNA (Yoshida et al. 2008). The major HBZ mRNA is spliced once and encodes a protein of 206 a. a. residues with a molecular size of 25 kDa. The spliced HBZ (sHBZ) transcript is predominant in ATL cells. It and its protein product are commonly referred to as HBZ. The minor unspliced HBZ mRNA transcript encodes a protein that is slightly larger than HBZ in size and contains a stretch of seven additional NH_2-terminal amino acid residues: MVNFVSV has a much shorter half-life compared to HBZ and much lower in abundance. Tax has been reported to up-regulate HBZ expression, and this up-regulation is influenced by the proviral integration sites (Landry et al. 2009). HBZ is ubiquitously expressed in ATL cells. It antagonizes several of the activities of Tax including LTR trans-activation and NF-κB activation (Zhao et al. 2009) and plays a critical role in the persistence of HTLV-1 infection. Importantly, both HBZ protein and RNA have been shown to promote T cell proliferation (Satou et al. 2006).

For details of the domain organization and activities of HBZ, readers are referred to several excellent reviews (Matsuoka and Green 2009; Barbeau et al. 2013; Mesnard et al. 2006; Matsuoka and Mesnard 2020; Gazon et al. 2017). HBZ is a basic domain–leucine zipper protein. Via its leucine zipper domain, HBZ interacts with and modulates the DNA binding or transcriptional activities of CREB-2, JunB, and c-Jun (AP-1) (Basbous et al. 2003). It binds the KIX domain of CBP/p300 and inhibits its HAT activity, thereby blocking Tax-driven viral mRNA expression (Lemasson et al. 2007; Clerc et al. 2008). HBZ also down-regulates the nuclear export of full-length and singly spliced HTLV-1 mRNAs by Rex to inhibit the production Gag, Gag-Pol, and Env proteins (Philip et al. 2014). HBZ has also been reported to dampen NF-κB activity by preventing NF-κB binding to DNA and induce p65/RelA degradation, mitigating the senescence response triggered by Tax-mediated NF-κB hyperactivation. These activities of HBZ promote latent HTLV-1 infection and facilitate the persistence of infected cells (Fig. 2).

4.2 HBZ Promotes T Cell Proliferation

Soon after HBZ was found to be ubiquitously expressed in ATL cells, both HBZ mRNA and protein were shown to stimulate T cell proliferation. Importantly, CD4+ T lymphocyte-specific expression of HBZ transgene in mice induces T-cell lymphoma and systemic inflammation (Satou et al. 2011). HBZ interacts with JunD and activates JunD-mediated transcription (Gazon et al. 2012). By forming a ternary complex with JunD and Sp1, HBZ activates its own expression and the expression of hTERT (Kuhlmann et al. 2007). The transcriptional activity of JunD is repressed by a tumor suppressor known as menin, which binds JunD directly. Interestingly, HBZ interacts with the JunD-menin complex to form a ternary complex that recruits the p300 co-activator/histone acetyl transferase to promote hTERT expression (Kuhlmann et al. 2007). Most recently, HBZ has been shown to assemble onto a transcriptional super-enhancer to transactivate the expression of a basic leucine zipper ATF/AP1-like transcription factor, BATF3, which then forms a complex with IRF4 to promote ATL cell proliferation (Nakagawa et al. 2018). Indeed, CRISPR-Cas9-medIted ablation of HBZ inhibited ATL cell growth (Nakagawa et al. 2018). These results clearly demonstrate the growth-promoting and oncogenic properties of HBZ and explain why HBZ is persistently expressed in ATL cells. Whether the same mechanism underlies the pleiotropic effect of HBZ remains to be seen.

It has been reported that HBZ protein can induce apoptosis, but HBZ mRNA prevents it, in part via up-regulation of the survivin gene (Kawatsuki et al. 2016).

HBZ protein has also been reported to target the Rb/E2F1 complex to activate the transcription of genes under E2F1 control that are critical for DNA replication and cell cycle progression (Kawatsuki et al. 2016). The activation of E2F-regulated genes by HBZ induces both T cell proliferation and apoptosis. HBZ transgene also stimulates the expression of Foxp3 in mouse CD4+ T cells. This may explain in part why some ATL cells are Foxp3+ (Yamamoto-Taguchi et al. 2013). Other activities of HBZ include the induction of a co-inhibitory immune receptor molecule, TIGIT (T cell immunoglobulin and ITIM domain), to facilitate IL-10 production, possibly to suppress anti-viral immune responses (Yasuma et al. 2016). HBZ also down-regulates DICER expression by diverting JunD from DICER promoter and thereby impairs the expression of some miRNAs (Gazon et al. 2016). Finally, the expression of HBZ in CD4+ T-cells correlates with the expression of OncomiRs, which have been associated with a wide range of oncogenic activities (Vernin et al. 2014). How HBZ mRNA promotes cell proliferation also awaits further investigations.

5 HTLV-1 and ATL

ATL develops mostly in individuals that become HTLV-1-infected during early childhood. High PVL is a risk factor for progression to ATL. The clinical manifestations of ATL include leukemia, lymphadenopathy, hepatosplenomegaly, hypercalcemia, and leukemia infiltration of the skin, central nervous system, and gastrointestinal tract. ATLs are clinically classified as acute, lymphomatous, chronic, and smoldering types based on criteria defined by Shimoyama (1991), including abnormal T lymphocytes level, site of infiltration, lactate dehydrogenase value, and hypercalcemia. Smoldering ATL likely represents the early stage of the disease that often can progress to acute ATL (Tsukasaki et al. 2009). The prognosis for ATL is poor. A recent Japanese study of 1594 ATL patients indicated that the median survival of acute, lymphoma, chronic, and smoldering ATL subtypes was 8.3, 10.6, 31.5, and 55.0 months, respectively (Katsuya et al. 2015).

5.1 Tax and HBZ in Cellular Transformation and Tumorigenesis

As described above (Sect. 3.6.1), Tax expression is sufficient to induce transformation in rodent fibroblasts and in primary human lymphocytes, and the transforming activity of Tax is associated with NF-κB activation and activation of other signaling pathways (Yamaoka et al. 1996; Matsumoto et al. 1997; Yoshita et al. 2012). Expression of Tax via a variety of enhancer/promoter cassettes including HTLV-1 LTR (Nerenberg et al. 1987), granzyme B promoter (Grossman et al. 1995), Lck proximal promoter (Hasegawa et al. 2006), and Lck distal promoter

(Ohsugi et al. 2007) induces neoplasia and/or inflammatory diseases in transgenic mice. The tumors induced, respectively, are neurofibroma/mesenchymal tumors (Nerenberg et al. 1987; Hinrichs et al. 1987), large granular lymphocytic leukemia (Grossman et al. 1995), lymphoma of immature T cells (Hasegawa et al. 2006), and diffuse large T-cell leukemia/lymphoma with leukemia cells of mature CD4+, CD8 +, and CD4+CD8+ types (Ohsugi et al. 2007). A significant fraction of the LTR-Tax founder mice developed thymic atrophy and growth retardation and died prematurely (Nerenberg et al. 1987). Whether the latter phenotype is related to the senescence/apoptosis effects of Tax is unknown.

Transgenic expression of HBZ in CD4+ T cells using a murine CD4-specific promoter/enhancer induced T-cell lymphomas and systemic inflammation in mice (Satou et al. 2011). Tax-HBZ double transgenic mice that expressed both proteins in CD4+ cells had similar phenotypes as HBZ transgenic, with enhanced proliferation of memory T cells and Foxp3+ Treg cells. Curiously, in this model, mice expressing the Tax transgene alone did not develop diseases (Zhao et al. 2014). In aggregate, these results support the notion that both Tax and HBZ are oncogenic through their abilities to promote the activation of IKK/NF-κB and other signaling pathways, destabilize the genome, and stimulate cell proliferation.

5.2 Genomic Instability in ATL

5.2.1 Overview

The ubiquitous expression of HBZ in ATL and its roles in viral persistence, growth stimulation, and tumor development have led some to propose it as the sole HTLV-1viral oncogene. A critical review of the current literature suggests that the full picture is far more complex, as is often the case with most human cancers. It has been estimated that at least five genetic alterations are involved in driving ATL development during its long clinical course (Watanabe 2017). Several striking features of ATL revealed by the comprehensive whole-genome sequence analysis (WGS) (Kataoka et al. 2015) are directly linked to Tax, including (1) the extensive instability of ATL genomes; (2) the constitutive activation of NF-κB in ATL; and (3) the functional similarity/overlap between the pathways/molecules that Tax perturb and the genetic alterations that occur in ATL, with both converging on signaling molecules in the $^T/_B$ CR-NF-κB and the CD-28 co-stimulatory pathways (Kataoka et al. 2015). Of note, each ATL genome on the average contains 59.5 structural alterations that involve the breaking and joining of DNA, almost three times that of the multiple myeloma genome (21 structural variations/genome). This is likely associated with the clastogenic effect of Tax.

5.2.2 Recurrent Mutations in ATL

The major recurrent mutations detected in ATL (Kataoka et al. 2015, 2018; Kogure and Kataoka 2017) are listed in Table 1. They can be classified as (1) gain-of-function (GOF) activating mutations that cause the activation of $^T/_B$ CR signaling and constitutive NF-κB activation (CD237, VAV1, PLCG1, PKCB,

Table 1 Frequently mutated genes in ATL

Biological functions	T/B-CR signaling NF-κB	RAS-RAF-ERK	Tumor suppressors	JAK/STAT and transcription factors	Apoptosis	Immune surveillance
Gain of function mutations	CD237 VAV1 PLCG1 PRKCB CARD11 CTLA4-CD28 ICOS-CD28	RHOA		IRF4 NOTCH1 FBXW7 STAT3		
Loss-of-function mutations	TNFAIP3/A20 NFKBIA/IκBα TRAF3 CBLB		TP53 CDKN2	GATA3 EP300	FAS WWOX	HLA-B B2M

The nature of the mutations in the ATL significant genes and the frequencies of their occurrence in ATL can be found in Kataoka et al. (2015).

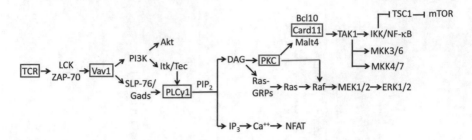

Fig. 6 Genes encoding mediators of the T cell receptor (TCR) signaling pathway are frequently mutationally activated in ATL. ATL genomes harbor frequent gain-of-function mutations in genes boxed in red. Other frequently mutated significant ATL genes are listed in Table 1. Details on TCR signaling can be found in these references (Katzav 2015; Charvet et al. 2006; Joshi and Koretzky 2013; Chen and Flies 2013). Vav1: a guanine nucleotide exchange factor that activates PI3 kinase and PLCγ1. PLCγ1: phospholipase G γ1. It converts PIP_2 into diacylglycerol (DG), which activates PKC and Ras-MEK-ERK; and inositide triphosphate (IP_3), which signals Ca^{++} mobilization and NF-AT activation. PKCβ: protein kinase C β, which activates Ras-MEK-ERK and IKK/cNF-κB pathways

CARD11, NIK, and CTLA4-CD28 and ICOS-CD28 fusions) (Fig. 6); loss-of-function (LOF) mutations that inactivate negative feedback regulators of NF-κB (TNFAIP3/A20, NFKBIA/IκBα, TRAF3, and CBLB); (2) loss-of-function mutations that inactivate tumor suppressor genes including those that safeguard genomic integrity (TP53) or regulate cell cycle control (p16INKa/CDKN2A): apoptosis mediator: FAS and WWOX; (3) GOF mutations including gene amplification, base substitutions, and indels that lead to over-expression/stabilization/activation of transcription factors IRF4, STAT3, and NOTCH1 that are involved in cell growth and proliferation or loss-of-function mutations in a E3 ligase (FBXW7) that targets NOTCH1 degradation; (4) not surprisingly, EP300 (p300), the gene that encodes the transcriptional co-activator of Tax-CREB-and Tax-NF-κB-mediated transcription and HLA-B and B2M, two genes that encode class 1 MHC components, are inactivated, likely selected to reduce viral replication and evade immune detection of HTLV-1 infection and/or ATL.

5.3 Loss of Tax Expression and Acquisition of Tax-Independent NF-κB Activation in ATL Cells

As HTLV-1 causes ATL in vivo and transforms T cell in culture, it was initially thought that HTLV-1 promotes T cell proliferation (Matsuoka and Jeang 2007). Because HTLV-1 Tax activates viral transcription and potently stimulates IKK-NF-κB and other signaling pathways, it has been proposed that these activities, especially IKK-NF-κB activation, drive HTLV-1-infected T cells to proliferate and expand, ultimately resulting in ATL. However, Tax expression is silenced in >50% of ATL cells via mechanisms including nonsense or frameshift mutations

(10%), 5′ LTR deletions (20–30%) or hypermethylation of CpG sites in 5′ LTR (10–20%). In these ATL cells, constitutive NF-κB activation is Tax-independent and driven by gain-of-function (GOF) and loss-of-function (LOF) somatic mutations in the$^T/_B$ cell receptor-NF-κB signaling and the CD28 co-stimulatory pathways that are selected during the long disease course of ATL (see Sect. 5.2.2 above for details.) Interestingly, PKCB and CARD11 mutations frequently occurred together and acted synergistically to stimulate NF-κB.

5.4 ATL Cells Evolve Adaptive Changes to Accommodate Constitutive NF-κB Activation

As NF-κB hyperactivation by Tax induces DNA damage and senescence or apoptosis, and NF-κB is constitutively activated in ATL, it is expected that ATL cells must have evolved from infected cells that have developed genetic/epigenetic changes that facilitate adaptation to constitutively activated NF-κB during viral infection. Indeed, while most HTLV-1-negative T cell lines became growth-arrested/senescent immediately upon *tax* transduction, Tax-negative ATL cell lines are resistant to Tax senescence and continue to proliferate upon re-introduction of *tax* and restoration of *tax* expression (Shudofsky and Giam 2019). Whether this resistance phenotype is due to HBZ or other adaptive cellular alterations is of significant interest. Based on present data, senescence mitigation likely involves (1) down-regulation of NF-κB; (2) down-regulation of mediators of DNA damage response; (3) down-regulation/functional inactivation of tumor suppressors that mediate G1 arrest/senescence; and (iv) up-regulation of drivers of cell proliferation. In this vein, clonal expansion of HTLV-1-infected cells readily occurred after NF-κB inhibition by a super-repressor, ΔN-IκBα (Zhi et al. 2011); and upon shutdown of NF-κB activities cell lines that are chronically infected by HTLV-1 can be readily established (Philip et al. 2014; Zahoor et al. 2014). Similarly, heterologous expression of the Kaposi's sarcoma herpesvirus (KSHV) viral cyclin (vCyclin), which gives rise to a Cdk6/vCyclin complex that resists p21/p27 inhibition and targets p27 for degradation, prevents Tax-induced senescence and promotes proliferation of Tax-expressing cells (Zhi et al. 2014). Some of the recurrent mutations detected in ATL genomes that cause CDKN2 and EP300 inactivation, and gain-of-function mutations that activate transcription factors IRF4, NOTCH1, and STAT3 may have effects similar to KSHV vCyclin in antagonizing the senescence effect of constitutively active NF-κB.

5.5 Epigenetic Changes in ATL Cells

Gene expression profile of ATL cells is markedly different from that of normal CD4 + T cells. This is due not only to the genetic mutations that accumulate in ATL described above, but also to the extensive epigenetic changes caused by DNA methylation and histone modification (Watanabe 2017; Kogure and Kataoka 2017).

5.5.1 Aberrant DNA Methylation

Widespread CpG island hypermethylation has been detected in one-third of ATL cases. Genes encoding MHC class I molecules and Cys2-His2 (C2H2) zinc finger proteins are often hypermethylated and silenced in ATL. Loss of MHC class I expression is likely selected to facilitate immune evasion of infected and ATL cells. Many of the C2H2 zinc-finger proteins contain a KRAB (Krupple-associated box) domain thought to interact with a cofactor known as KAP1/TRIM28 to silence transcription of both endogenous and exogenous retroviruses. How hypermethylation of the C2H2 genes impacts HTLV-1 gene expression and ATL development remains to be determined.

5.5.2 Aberrant Histone Modifications

A recent epigenome and transcriptome analysis of primary ATL cells by Toshiki Watanabe and co-workers has indicated that polycomb repressive complex (PRC) 2 components including the histone H3K27 methylase and EZH2 are highly expressed in ATL (Fujikawa et al. 2016). This results in increased trimethylation of H3K27 (H3K27m3) and reprogramming of over half of the genes in ATL, with progressive down-regulation of gene expression as the severity of the disease increases from indolent to aggressive ATL. Indeed, simultaneous silencing of EZH1 and EZH2 reduces H3K27m3 levels and induces ATL cell growth arrest (Fujikawa et al. 2016).

6 Concluding Remarks

HTLV-1 is unique among retroviruses in that it has evolved a bimodal replicative strategy similar to γ-herpesviruses such as EBV and KSHV. In the "lytic replication" mode, HTLV-1 uses Tax and Rex as early proteins to drive produtive viral replication, resulting in cytopathic effects including cell cycle arrest, senescence or apotosis in infected cellsand viral spread (Fig. 5). In the "latency" mode where Tax and Rex expressions are low or silenced, the HBZ protein and RNA antagonize Tax and Rex functions and facilitate the mitotic expansion of infected cells to produce a reservoir of latently infected cells for viral persistence and future transmission. The fate of the infected cells is determined by the chromosomal environment of the integrated proviral DNA, the physiological state of the infected cells, and the host immune response.

What can the replicative strategy of HTLV-1 and the activities of Tax and HBZ inform on carcinogenesis in general and ATL development in particular? The path of chemical carcinogenesis has been conceptually divided into four stages: tumor initiation, tumor promotion, malignant conversion, and tumor progression. Carcinogenic changes are initiated by genetic and epigenetic changes that occur in cells. The selective clonal expansion of initiated cells then provides a large pool of pre-cancerous cells that have the potential for further genetic evolution, eventually leading to malignant conversion and tumor progression. As a potent mutagen and a powerful mitogen, respectively, Tax and HBZ can be thought of as initiator and promoter of ATL development. It should be noted however that Tax and HBZ are evolved to regulate viral replication, spread and transmission, not leukemia development, and as discussed in this chapter, many of their activities do not fall neatly into the simple classification of the conventional carcinogenesis model.

The mutagenic/clastogenic property of Tax and its pleiotropic effect on cell signaling are connected to the hijacking of RNF8 and constitutive activation of the TAK1-IKK-NF-κB pathway. The fact that many genetic alterations in ATL cells reside in signaling molecules in the $^{T/}_B$ CR-NF-κB and the CD-28 co-stimulatory pathways that lead to constitutive NF-κB activation strongly suggests a link between Tax and the development of these mutations. It is conceivable that senescence induction by Tax during HTLV-1 infection selects for cells with genetic/epigenetic changes that facilitate adaptation to Tax and constitutively active NF-κB, setting the stage for the establishment of somatic mutations that drive Tax-independent NF-κB activation. How Tax/constitutive NF-κB activation induces senescence is unknown and how ATL cells maintain chronic NF-κB activation without triggering a senescence response also remains to be elucidated.

It should be pointed out that constitutive IKK/NF-κB activation is a common feature of HTLV-unrelated $^{T/}_B$ cell malignancies including Sézary syndrome (SS), mycosis fungoides (MF), diffuse large B cell lymphoma (DLBCL), multiple myeloma (MM), chronic lymphocytic leukemia (CLL), and lymphoproliferative disorders associated with Epstein–Barr virus and Kaposi sarcoma herpes virus (KSHV) infections (Chapman et al. 2011; Ungewickell et al. 2015; Wang et al. 2015; Pasqualucci et al. 2011). Many of these leukemias/lymphomas contain similar GOF and LOF mutations in the same genes (e.g., CARD11, IRF4, NOTCH1, p53, CDKN2A, etc.) Thus, a clear understanding of how HTLV-1 interacts with these genetic changes will impact our understanding of how B- and T-cell leukemia/lymphoma develop and uncover treatment targets and therapeutic strategies hitherto unrecognized.

Disclosure of Conflicts of Interest The author declares no conflicts of interest.
Disclaimer Statement The opinions and assertions expressed herein are those of the author's and do not necessarily reflect the official policy or position of the Uniformed Services University or the Department of Defense.

References

Anderson MD, Ye J, Xie L, Green PL (2004) Transformation studies with a human T-cell leukemia virus type 1 molecular clone. J Virol Methods 116(2):195–202

Andresen V, Pise-Masison CA, Sinha-Datta U, Bellon M, Valeri V, Washington Parks R et al (2011) Suppression of HTLV-1 replication by Tax-mediated rerouting of the P13 viral protein to nuclear speckles. Blood. 118(6):1549–1559. https://doi.org/10.1182/blood-2010-06-293340. Pubmed Pmid: 21677314; Pubmed Central Pmcid: Pmcpmc3156045

Arnold J, Zimmerman B, Li M, Lairmore MD, Green PL (2008) Human T-cell leukemia virus type-1 antisense-encoded gene, HBZ, promotes T-lymphocyte proliferation. Blood 112 (9):3788–3797

Banerjee P, Feuer G, Barker E (2007) Human T-cell leukemia virus type 1 (HTLV-1) P12I down-modulates ICAM-1 and -2 and reduces adherence of natural killer cells, thereby protecting HTLV-1-infected primary Cd4+ T cells from autologous natural killer cell-mediated cytotoxicity despite the reduction of major histocompatibility complex class I molecules on infected cells. J Virol 81(18):9707–9717

Barbeau B, Peloponese JM, Mesnard JM (2013) Functional comparison of antisense proteins of HTLV-1 and HTLV-2 in viral pathogenesis. Front Microbiol 4:226. https://doi.org/10.3389/fmicb.2013.00226 [Doi]

Barnard AL, Igakura T, Tanaka Y, Taylor GP, Bangham CR (2005) Engagement of specific T-cell surface molecules regulates cytoskeletal polarization in HTLV-1-infected lymphocytes. Blood 106(3):988–995

Basbous J, Arpin C, Gaudray G, Piechaczyk M, Devaux C, Mesnard JM (2003) The HBZ factor of human T-cell leukemia virus type I dimerizes with transcription factors JunB and c-Jun and modulates their transcriptional activity. J Biol Chem 278(44):43620–43627

Baydoun HH, Bai XT, Shelton S, Nicot C (2012) HTLV-I Tax increases genetic instability by inducing dna double strand breaks during dna replication and switching repair to Nhej. Plos One 7(8):E42226. Epub 2012/08/24. https://doi.org/10.1371/journal.pone.0042226. Pubmed Pmid: 22916124; Pubmed Central Pmcid: Pmcpmc3423393

Cereseto A, Diella F, Mulloy JC, Cara A, Michieli P, Grassmann R et al (1996) P53 functional impairment and high P21waf1/CIP1 expression in human T-cell lymphotropic/leukemia virus type I-transformed T cells. Blood 88(5):1551–1560

Chapman MA, Lawrence MS, Keats JJ, Cibulskis K, Sougnez C, Schinzel AC et al (2011) Initial genome sequencing and analysis of multiple myeloma. Nature 471(7339):467–472. Doi: Nature09837 [Pii];https://doi.org/10.1038/nature09837 [Doi]

Charvet C, Canonigo Aj, Becart S, Maurer U, Miletic AV, Swat W (2006) VAV1 Promotes T cell cycle progression by linking TCR/CD28 costimulation to FOXO1 and P27KIP1 expression. J Immunol 177(8):5024–50231. Pubmed Pmid: 17015685

Chen L, Flies DB (2013) Molecular mechanisms of T cell co-stimulation and co-inhibition. Nat Rev Immunol 13(4):227–242. https://doi.org/10.1038/nri3405. Pubmed Pmid: 23470321; Pubmed Central Pmcid: Pmcpmc3786574

Chlichlia K, Moldenhauer G, Daniel PT, Busslinger M, Gazzolo L, Schirrmacher V et al (1995) Immediate effects of reversible HTLV-1 Tax function: T-cell activation and apoptosis. Oncogene 10:269–277

Chrivia JC, Kwok RP, Lamb N, Hagiwara M, Montminy MR, Goodman RH (1993) Phosphorylated creb binds specifically to the nuclear protein CBP. Nature 365:855–859

Clerc I, Polakowski N, Ndre-Arpin C, Cook P, Barbeau B, Mesnard JM (2008) An interaction between the HTLV-1 bZip factor (Hbz) and the KIX domain of P300/CBP contributes to the downregulation of Tax-dependent viral transcription by HBZ J Biol Chem 283:23903–23913

Cook LB, Rowan AG, Melamed A, Taylor GP, Bangham CR (2012) HTLV-1-infected T cells contain a single integrated provirus in natural infection. Blood 120(17):3488–3490. Doi: Blood-2012-07-445593 [Pii];https://doi.org/10.1182/blood-2012-07-445593 [Doi]

Cook LB, Melamed A, Niederer H, Valganon M, Laydon D, Foroni L et al (2014) The role of HTLV-1 clonality, proviral structure, and genomic integration site in adult T-cell leukemia/lymphoma. Blood 123(25):3925–3931. https://doi.org/10.1182/blood-2014-02-553602. Pubmed Pmid: 24735963; Pubmed Central Pmcid: Pmcpmc4064332

De Castro-Amarante MF, Pise-Masison CA, Mckinnon K, Washington Parks R, Galli V, Omsland M et al (2015) Human T cell leukemia virus type 1 infection of the three monocyte subsets contributes to viral burden in humans. J Virol 90(5):2195–2207. https://doi.org/10.1128/jvi.02735-15. Pubmed Pmid: 26608313; Pubmed Central Pmcid: Pmcpmc4810698

De La Fc, Santiago F, Chong Sy, Deng L, Mayhood T, Fu P et al (2000) Overexpression of P21 (WAF1) in human T-cell lymphotropic virus type 1-infected cells and its association with cyclin A/CDK2. J Virol 74(16):7270–7283

De La FC, Wang L, Wang D, Deng L, Wu K, Li H et al (2003) Paradoxical effects of a stress signal on pro- and anti-apoptotic machinery in HTLV-1 Tax expressing cells. Molcell Biochem 245(1–2):99–113

Derse D, Hill SA, Lloyd PA, Chung H, Morse BA (2001) Examining human T-lymphotropic virus type 1 infection and replication by cell-free infection with recombinant virus vectors 7. J Virol 75(18):8461–8468

Derse D, Heidecker G, Mitchell M, Hill S, Lloyd P, Princler G (2004) Infectious transmission and replication of human T-cell leukemia virus type 1. Front Biosci 9:2495–2499

Dodon MD, Villaudy J, Gazzolo L, Haines R, Lairmore M (2012) What we are learning on HTLV-1 pathogenesis from animal models. Front Microbiol. 3:320. https://doi.org/10.3389/fmicb.2012.00320. Pubmed Pmid: 22969759; Pubmed Central Pmcid: Pmcpmc3431546

Edwards D, Fenizia C, Gold H, De Castro-Amarante MF, Buchmann C, Pise-Masison CA et al (2011) ORF-I and ORF-II-encoded proteins in HTLV-1 infection and persistence. Viruses 3 (6):861–885. https://doi.org/10.3390/v3060861. Pubmed Pmid: 21994758; Pubmed Central Pmcid: Pmcpmc3185781

Einsiedel LJ, Pham H, Woodman RJ, Pepperill C, Taylor KA (2016) The prevalence and clinical associations of HTLV-1 infection in a remote indigenous community. Med J Aust 205(7):305–309. Epub 2016/09/30. https://doi.org/10.5694/mja16.00285. Pubmed Pmid: 27681971

El Sabban ME, Nasr R, Dbaibo G, Hermine O, Abboushi N, Quignon F et al (2000) Arsenic-interferon-alpha-triggered apoptosis in HTLV-I transformed cells is associated with Tax down-regulation and reversal of NF-KappaB activation. Blood 96(8):2849–2855

Emmerich CH, Ordureau A, Strickson S, Arthur JS, Pedrioli PG, Komander D et al (2013) Activation of the canonical IKK complex by K63/M1-linked hybrid ubiquitin chains. Proc Natl Acad Sci U S A 110(38):15247–15252. https://doi.org/10.1073/pnas.1314715110. Pubmed Pmid: 23986494; Pubmed Central Pmcid: Pmc3780889

Emmerich CH, Bakshi S, Kelsall IR, Ortiz-Guerrero J, Shpiro N, Cohen P (2016) Lys63/Met1-hybrid ubiquitin chains are commonly formed during the activation of innate immune signalling. Biochem Biophys Res Commun. 474(3):452–461. https://doi.org/10.1016/j.bbrc.2016.04.141. Pubmed Pmid: 27133719; Pubmed Central Pmcid: Pmcpmc4880150

Fujikawa D, Nakagawa S, Hori M, Kurokawa N, Soejima A, Nakano K et al (2016) Polycomb-dependent epigenetic landscape in adult T-cell leukemia. Blood 127(14):1790–1802. Epub 2016/01/17. https://doi.org/10.1182/blood-2015-08-662593. Pubmed Pmid: 26773042

Galli V, Nixon CC, Strbo N, Artesi M, De Castro-Amarante MF, Mckinnon K et al (2019) Essential role of human T cell leukemia virus type 1 ORF-I in lethal proliferation of Cd4(+) cells in humanized mice. J Virol 93(19). Epub 2019/07/19. https://doi.org/10.1128/jvi.00565-19. Pubmed Pmid: 31315992; Pubmed Central Pmcid: Pmcpmc6744231

Gazon H, Lemasson I, Polakowski N, Cesaire R, Matsuoka M, Barbeau B et al (2012) Human T-cell leukemia virus type 1 (HTLV-1) Bzip factor requires cellular transcription factor jund to upregulate HTLV-1 antisense transcription from the 3′ long terminal repeat. J Virol 86 (17):9070–9078. Doi:Jvi.00661-12 [Pii]; https://doi.org/10.1128/jvi.00661-12 [Doi]

Gazon H, Belrose G, Terol M, Meniane JC, Mesnard JM, Cesaire R et al (2016) Impaired expression of Dicer and some micrornas in HBZ expressing cells from acute adult T-cell leukemia patients. Oncotarget 7(21):30258–30275. Epub 2016/02/06. https://doi.org/10.18632/oncotarget.7162. Pubmed Pmid: 26849145; Pubmed Central Pmcid: Pmcpmc5058679

Gazon H, Barbeau B, Mesnard JM, Peloponese JM Jr (2017) Hijacking of the AP-1 signaling pathway during development of ATl. Front Microbiol 8:2686. Epub 2018/01/31. https://doi.org/10.3389/fmicb.2017.02686. Pubmed Pmid: 29379481; Pubmed Central Pmcid: Pmcpmc5775265

Gessain A, Cassar O (2012) Epidemiological aspects and world distribution of HTLV-1 infection. Front Microb 3:388. https://doi.org/10.3389/fmicb.2012.00388 [Doi]

Giam CZ, Semmes OJ (2016) HTLV-1 Infection and adult T-cell leukemia/lymphoma—a tale of two proteins: Tax and HBZ. Viruses 8(6). Epub 2016/06/21. https://doi.org/10.3390/v8060161. Pubmed Pmid: 27322308; Pubmed Central Pmcid: Pmcpmc4926181

Grassmann R, Dengler C, Muller Fleckenstein I, Fleckenstein B, Mcguire K, Dokhelar MC et al (1989) Transformation to continuous growth of primary human T lymphocytes by human T-cell leukemia virus type I X-region genes transduced by a herpesvirus saimiri vector. Proc Natl Acad Sci U S A86:3351–3355

Gross C, Thoma-Kress AK (2016) Molecular mechanisms of HTLV-1 cell-to-cell transmission. Viruses 8(3). https://doi.org/10.3390/v8030074. Pubmed Pmid: 27005656; Pubmed Central Pmcid: Pmcpmc4810264

Grossman WJ, Kimata JT, Wong FH, Zutter M, Ley TJ, Ratner L (1995) Development of leukemia in mice transgenic for the Tax gene of human T-cell leukemia virus type I. Proc Natl Acad Sci U S A92:1057–1061

Gruber K (2018) Australia tackles HTLV-1. Lancet Infect Dis 18(10):1073–1074. Epub 2018/10/12. https://doi.org/10.1016/s1473-3099(18)30561-9. Pubmed Pmid: 30303105

Haas TL, Emmerich CH, Gerlach B, Schmukle AC, Cordier SM, Rieser E et al (2009) Recruitment of the linear ubiquitin chain assembly complex stabilizes the TNF-R1 signaling complex and is required for TNF-mediated gene induction. Mol Cell 36(5):831–844. Epub 2009/12/17. Doi: S1097-2765(09)00778-3 [Pii] https://doi.org/10.1016/j.molcel.2009.10.013. Pubmed Pmid: 20005846

Hall AP, Irvine J, Blyth K, Cameron ER, Onions DE, Campbell ME (1998) Tumours derived from HTLV-I Tax transgenic mice are characterized by enhanced levels of apoptosis and oncogene expression. J Pathol 186(2):209–214

Haoudi A, Semmes OJ (2003) The HTLV-1 Tax oncoprotein attenuates DNA damage induced G1 arrest and enhances apoptosis in P53 null cells 6. Virology 305(2):229–239

Haoudi A, Daniels RC, Wong E, Kupfer G, Semmes OJ (2003) Human T-cell leukemia virus-I Tax oncoprotein functionally targets a subnuclear complex involved in cellular DNA damage-response. J Biol Chem 278(39):37736–37744

Harhaj EW, Giam CZ (2018) NF-kappaB signaling mechanisms in HTLV-1-induced adult T-cell leukemia/lymphoma. Febs J 285(18):3324–3336. Epub 2018/05/04. https://doi.org/10.1111/febs.14492. Pubmed Pmid: 29722927

Harrod R, Tang Y, Nicot C, Lu HS, Vassilev A, Nakatani Y et al (1998) An exposed kid-like domain in human t-cell lymphotropic virus type 1 tax is responsible for the recruitment of coactivators cbp/p300. Mol Cell Biol 18(9):5052–5061

Hasegawa H, Sawa H, Lewis MJ, Orba Y, Sheehy N, Yamamoto Y et al (2006) Thymus-derived leukemia-lymphoma in mice transgenic for the Tax gene of human T-lymphotropic virus type I. Nat Med 12(4):466–472

Hasegawa H, Sawa H, Lewis MJ, Orba Y, Sheehy N, Yamamoto Y et al (2006) Thymus-derived leukemia-lymphoma in mice transgenic for the Tax gene of human T-lymphotropic virus type I. Nat Med 12(4):466–472. Epub 2006/03/22. https://doi.org/10.1038/nm1389. Pubmed Pmid: 16550188

Hayakawa M (2012) Role of K63-linked polyubiquitination in NF-kappaB signalling: which ligase catalyzes and what molecule is targeted? J Biochem 151(2):115–118. Doi:Mvr139 [Pii]; https://doi.org/10.1093/jb/mvr139 [Doi]

Hinrichs SH, Nerenberg M, Reynolds RK, Khoury G, Jay G (1987) A transgenic mouse model for human neurofibromatosis. Science 237:1340–1343

Hiraragi H, Michael B, Nair A, Silic-Benussi M, Ciminale V, Lairmore M (2005) Human T-lymphotropic virus type 1 mitochondrion-localizing protein P13II sensitizes jurkat T cells to RAS-mediated apoptosis. Jvirol. 79(15):9449–9457

Ho YK, Zhi H, Debiaso D, Philip S, Shih HM, Giam CZ (2012) HTLV-1 Tax-induced rapid senescence is driven by the transcriptional activity of NF-kappaB and depends on chronically activated ikkalpha and P65/Rela. J Virol 86(17):9474–9483. Doi:Jvi.00158-12 [Pii];https://doi.org/10.1128/jvi.00158-12 [Doi]

Ho YK, Zhi H, Bowlin T, Dorjbal B, Philip S, Zahoor MA et al (2015) HTLV-1 Tax stimulates ubiquitin E3 ligase, ring finger protein 8, to assemble lysine 63-linked polyubiquitin chains for

Tak1 and IKK activation. Plos Pathog 11(8):E1005102. https://doi.org/10.1371/journal.ppat. 1005102. Pubmed Pmid: 26285145; Pubmed Central Pmcid: Pmc4540474

Huey DD, Bolon B, La Perle KMD, Kannian P, Jacobson S, Ratner L et al (2018) Role of wild-type and recombinant human T-cell leukemia viruses in lymphoproliferative disease in humanized NSG mice. Comp Med 68(1):4–14. Epub 2018/02/21. Pubmed Pmid: 29460716; Pubmed Central Pmcid: Pmcpmc5824134

Igakura T, Stinchcombe JC, Goon PK, Taylor GP, Weber JN, Griffiths GM et al (2003) Spread of HTLV-I between lymphocytes by virus-induced polarization of the cytoskeleton. Science 299 (5613):1713–176

Jain P, Manuel Sl, Khan ZK, Ahuja J, Quann K, Wigdahl B (2009) DC-sign mediates cell-free infection and transmission of human T-cell lymphotropic virus type 1 by dendritic cells. J Virol 2009;83(21):10908–109021. https://doi.org/10.1128/jvi.01054-09. Pubmed Pmid: 19692463; Pubmed Central Pmcid: Pmcpmc2772783

Jeong SJ, Radonovich M, Brady JN, Pise-Masison CA (2004) HTLV-I Tax induces a novel interaction between P65/Rela and P53 that results in inhibition of P53 transcriptional activity. Blood 104(5):1490–1497

Jones KS, Petrow-Sadowski C, Bertolette DC, Huang Y, Ruscetti FW (2005) Heparan sulfate proteoglycans mediate attachment and entry of human T-cell leukemia virus type 1 virions into Cd4+ T cells. J Virol 79(20):12692–12702

Jones KS, Petrow-Sadowski C, Huang YK, Bertolette DC, Ruscetti FW (2008) Cell-free HTLV-1 infects dendritic cells leading to transmission and transformation of Cd4(+) T cells. Nat Med 14 (4):429–436

Joshi RP, Koretzky GA (2013) Diacylglycerol kinases: regulated controllers of T cell activation, function, and development. Int J Mol Sci 14(4):6649–73. https://doi.org/10.3390/Ijms14046649. Pubmed Pmid: 23531532; Pubmed Central Pmcid: Pmcpmc3645659

Kataoka K, Nagata Y, Kitanaka A, Shiraishi Y, Shimamura T, Yasunaga J et al (2015) Integrated molecular analysis of adult T cell leukemia/lymphoma. Nat Genet 47(11):1304–1315. https://doi.org/10.1038/Ng.3415. Pubmed Pmid: 26437031

Kataoka K, Iwanaga M, Yasunaga JI, Nagata Y, Kitanaka A, Kameda T et al (2018) Prognostic relevance of integrated genetic profiling in adult T-cell leukemia/lymphoma. Blood 131 (2):215–225. Epub 2017/11/01. https://doi.org/10.1182/blood-2017-01-761874. Pubmed Pmid: 29084771

Katsuya H, Ishitsuka K, Utsunomiya A, Hanada S, Eto T, Moriuchi Y et al (2015) Treatment and survival among 1594 patients with ATL. Blood 126(24):2570–2577. Epub 2015/09/13. https://doi.org/10.1182/blood-2015-03-632489. Pubmed Pmid: 26361794

Katsuya H, Islam S, Tan BJY, Ito J, Miyazato P, Matsuo M et al (2019) The nature of the HTLV-1 provirus in naturally infected individuals analyzed by the viral DNA-capture-SEQ approach. Cell Rep 29(3):724–35 E4. Epub 2019/10/17. https://doi.org/10.1016/j.celrep.2019.09.016. Pubmed Pmid: 31618639

Katzav S (2015) VAV1: a Dr. Jekyll and Mr. Hyde protein–good for the hematopoietic system, bad for cancer. Oncotarget 6(30):28731–28742. https://doi.org/10.18632/oncotarget.5086. Pubmed Pmid: 26353933; Pubmed Central Pmcid: Pmcpmc4745688

Kawatsuki A, Yasunaga Ji, Mitobe Y, Green Pl, Matsuoka M (2016) HTLV-1 bZip factor protein targets the RB/E2F-1 pathway to promote proliferation and apoptosis of primary CD4 T cells. Oncogene. https://doi.org/10.1038/onc.2015.510. Pubmed Pmid: 26804169

Kelly K, Davis P, Mitsuya H, Irving S, Wright J, Grassmann R et al (1992) A high proportion of early response genes are constitutively activated in T cells by HTLV-I. Oncogene 7:1463–1470

Kogure Y, Kataoka K (2017) Genetic alterations in adult T-Cell leukemia/lymphoma. Cancer Sci 108(9):1719–1725. Epub 2017/06/20. https://doi.org/10.1111/cas.13303. Pubmed Pmid: 28627735; Pubmed Central Pmcid: Pmcpmc5581529

Kuhlmann AS, Villaudy J, Gazzolo L, Castellazzi M, Mesnard JM, Duc DM (2007) HTLV-1 HBZ cooperates with jund to enhance transcription of the human telomerase reverse transcriptase gene (Htert). Retrovirology 4:92

Kwok RP, Laurance ME, Lundblad JR, Goldman PS, Shih H, Connor LM et al (1996) Control of camp-regulated enhancers by the viral transactivator tax through creb and the co-activator cbp. Nature 380(6575):642–646

Lambert S, Bouttier M, Vassy R, Seigneuret M, Petrow-Sadowski C, Janvier S et al (2009) HTLV-1 uses HSPG and neuropilin-1 for entry by molecular mimicry of VEGF165. Blood 113 (21):5176–5185

Landry S, Halin M, Vargas A, Lemasson I, Mesnard JM, Barbeau B (2009) Upregulation of human T-cell leukemia virus type 1 antisense transcription by the viral Tax protein. J Virol 83 (4):2048–2054. Doi:Jvi.01264-08 [Pii]; https://doi.org/10.1128/jvi.01264-08 [Doi]

Lefebvre L, Vanderplasschen A, Ciminale V, Heremans H, Dangoisse O, Jauniaux JC et al (2002) Oncoviral bovine leukemia virus G4 and human T-cell leukemia virus type 1 P13(II) accessory proteins interact with Farnesyl pyrophosphate synthetase. J Virol 76(3):1400–1414. Pubmed Pmid: 11773414; Pubmed Central Pmcid: Pmcpmc135811

Lemasson I, Lewis MR, Polakowski N, Hivin P, Cavanagh MH, Thebault S et al (2007) Human T-cell leukemia virus type 1 (HTLV-1) bZip Protein interacts with the cellular transcription factor creb to inhibit HTLV-1 transcription. J Virol 81(4):1543–1453

Lenzmeier BA, Giebler HA, Nyborg JK (1998) Human T-cell leukemia virus type I Tax requires direct access to DNA for recruitment of CREB binding protein to the viral promoter. Mol Cell Biol 18:721–731

Li M, Kesic M, Yin H, Yu L, Green PL (2009) Kinetic analysis of human T-cell leukemia virus type 1 gene expression in cell culture and infected animals. J Virol 83(8):3788–3797. Jvi.02315-08 [Pii];https://doi.org/10.1128/jvi.02315-08 [Doi]

Liu M, Yang L, Zhang L, Liu B, Merling R, Xia Z et al (2008) Human T-Cell leukemia virus type 1 infection leads to arrest in the G1 phase of the cell cycle. J Virol 82(17):8442–8455

Ma G, Yasunaga J, Matsuoka M (2016) Multifaceted functions and roles of HBZ in HTLV-1 pathogenesis. Retrovirology 13:16. Epub 2016/03/17. https://doi.org/10.1186/s12977-016-0249-x. Pubmed Pmid: 26979059; Pubmed Central Pmcid: Pmcpmc4793531

Mahgoub M, Yasunaga JI, Iwami S, Nakaoka S, Koizumi Y, Shimura K et al (2018) Sporadic on/off switching of HTLV-1 Tax expression is crucial to maintain the whole population of virus-induced leukemic cells. Proc Natl Acad Sci U S A 115(6):E1269–E1278. Epub 2018/01/24. https://doi.org/10.1073/pnas.1715724115. Pubmed Pmid: 29358408; Pubmed Central Pmcid: Pmcpmc5819419

Majone F, Jeang KT (2000) Clastogenic effect of the human T-cell leukemia virus type I Tax oncoprotein correlates with unstabilized DNA breaks. J Biol Chem 275(42):32906–32910

Majone F, Semmes OJ, Jeang KT (1993) Induction of micronuclei by HTLV-I Tax: a cellular assay for function. Virology 193:456–459

Manel N, Kim FJ, Kinet S, Taylor N, Sitbon M, Battini JL (2003) The ubiquitous glucose transporter GLUT-1 is a receptor for HTLV. Cell 115(4):449–459

Manivannan K, Rowan AG, Tanaka Y, Taylor GP, Bangham CR (2016) CADM1/TSLC1 identifies HTLV-1-infected cells and determines their susceptibility to CTL-mediated lysis. PLoS Pathog 12(4):E1005560. Epub 2016/04/23. https://doi.org/10.1371/journal.ppat.1005560 . Pubmed Pmid: 27105228; Pubmed Central Pmcid: Pmcpmc4841533

Marriott SJ, Semmes OJ (2005) Impact of HTLV-I Tax on cell cycle progression and the cellular DNA damage repair response. Oncogene 24(39):5986–5995

Matsumoto K, Shibata H, Fujisawa JI, Inoue H, Hakura A, Tsukahara T et al (1997) Human T-cell leukemia virus type 1 Tax protein transforms RAT fibroblasts via two distinct pathways 12. J Virol 71(6):4445–4451

Matsuoka M, Green PL (2009) The HBZ gene, a key player in HTLV-1 pathogenesis. Retrovirology 6:71. https://doi.org/10.1186/1742-4690-6-71. Pubmed Pmid: 19650892; Pubmed Central Pmcid: Pmc2731725

Matsuoka M, Jeang KT (2007) Human T-cell leukaemia virus type 1 (HTLV-1) infectivity and cellular transformation. Natrevcancer 7(4):270–280

Matsuoka M, Mesnard JM (2020) HTLV-1 bZip factor: the key viral gene for pathogenesis. Retrovirology 17(1):2. Epub 2020/01/10. https://doi.org/10.1186/s12977-020-0511-0. Pubmed Pmid: 31915026; Pubmed Central Pmcid: Pmcpmc6950816

Mazurov D, Ilinskaya A, Heidecker G, Lloyd P, Derse D (2010) Quantitative comparison of HTLV-1 and HIV-1 cell-to-cell infection with new replication dependent vectors. PLoS Pathog 6(2):E1000788. https://doi.org/10.1371/journal.ppat.1000788 [Doi]

Mesnard JM, Barbeau B, Devaux C (2006) HBZ, a new important player in the mystery of adult T-cell leukemia. Blood. 108(13):3979–3982. https://doi.org/10.1182/blood-2006-03-007732. Pubmed Pmid: 16917009

Miyoshi I, Kubonishi I, Yoshimoto S, Shiraishi Y (1981) A T-cell line derived from normal human cord leukocytes by co-culturing with human leukemic T-cells. GANN 72(6):978–981

Nakagawa M, Shaffer AL 3rd, Ceribelli M, Zhang M, Wright G, Huang DW et al (2018) Targeting the HTLV-I-regulated BATF3/IRF4 transcriptional network in adult T cell leukemia/lymphoma. Cancer Cell 34(2):286–297 E10. Epub 2018/07/31. https://doi.org/10.1016/j.ccell.2018.06.014. Pubmed Pmid: 30057145

Nakazawa S, Oikawa D, Ishii R, Ayaki T, Takahashi H, Takeda H et al (2016) Linear ubiquitination is involved in the pathogenesis of optineurin-associated amyotrophic lateral sclerosis. Nat Commun 7:12547. https://doi.org/10.1038/Ncomms12547. Pubmed Pmid: 27552911; Pubmed Central Pmcid: Pmcpmc4999505

Nejmeddine M, Barnard AL, Tanaka Y, Taylor GP, Bangham CR (2005) Human T-lymphotropic virus, type 1, Tax protein triggers microtubule reorientation in the virological synapse. J Biol Chem 280(33):29653–29660

Nejmeddine M, Negi VS, Mukherjee S, Tanaka Y, Orth K, Taylor GP et al (2009) HTLV-1-Tax and ICAM-1 act on T-cell signal pathways to polarize the microtubule-organizing center at the virological synapse. Blood 114(5):1016–1025. https://doi.org/10.1182/blood-2008-03-136770. Pubmed Pmid: 19494354

Nerenberg M, Hinrichs SH, Reynolds RK, Khoury G, Jay G (1987) The TAT gene of human T-lymphotropic virus type 1 induces mesenchymal tumors in transgenic mice. Science 237:1324–1329

Nicot C, Dundr M, Johnson JM, Fullen JR, Alonzo N, Fukumoto R et al (2004) HTLV-1-encoded P30II is a post-transcriptional negative regulator of viral replication. Nat Med 10(2):197–201

Niederer HA, Bangham CR (2014) Integration site and clonal expansion in human chronic retroviral infection and gene therapy. Viruses 6(11):4140–4164. https://doi.org/10.3390/V6114140. Pubmed Pmid: 25365582; Pubmed Central Pmcid: Pmcpmc4246213

Nyborg JK, Egan D, Sharma N (2009) The HTLV-1 Tax protein: revealing mechanisms of transcriptional activation through histone acetylation and nucleosome disassembly. Bio Chim Biophys Acta 1799(3–4):266–274

Ohsugi T, Kumasaka T, Okada S, Urano T (2007) The Tax protein of HTLV-1 promotes oncogenesis in not only immature T cells but also mature T cells. Nat Med 13(5):527–528. Epub 2007/05/05. https://doi.org/10.1038/nm0507-527. Pubmed Pmid: 17479090

Okayama A, Stuver S, Matsuoka M, Ishizaki J, Tanaka G, Kubuki Y et al (2004) Role of HTLV-1 proviral DNA load and clonality in the development of adult T-cell leukemia/lymphoma in asymptomatic carriers. Intjcancer. 110(4):621–625

Pais-Correia AM, Sachse M, Guadagnini S, Robbiati V, Lasserre R, Gessain A et al (2010) Biofilm-like extracellular viral assemblies mediate HTLV-1 cell-to-cell transmission at virological synapses. Nat Med 16(1):83–89

Panfil AR, Al-Saleem JJ, Green PL (2013) Animal models utilized in HTLV-1 research. Virology (Auckl) 4:49–59. https://doi.org/10.4137/Vrt.S12140. Pubmed Pmid: 25512694; Pubmed Central Pmcid: Pmcpmc4222344

Paca Uccaralertkun S, Zhao LJ, Adya N, Cross JV, Cullen BR, Boros IM et al (1994) In vitro selection of DNA elements highly responsive to the human T-cell lymphotropic virus type I transcriptional activator, Tax. Mol Cell Biol 14:456–462

Pasqualucci L, Trifonov V, Fabbri G, Ma J, Rossi D, Chiarenza A (2011) Analysis of the coding genome of diffuse large B-cell lymphoma. Nat Gen 43(9):830–837. https://doi.org/10.1038/Ng. 892. Pubmed Pmid: 21804550; Pubmed Central Pmcid: Pmcpmc3297422

Philip S, Zahoor MA, Zhi H, Ho YK, Giam CZ (2014) Regulation of human t-lymphotropic virus type I latency and reactivation by HBZ and Rex. PLoS Pathog 10(4):E1004040. https://doi.org/ 10.1371/journal.ppat.1004040. Pubmed Pmid: 24699669; Pubmed Central Pmcid: Pmc3974842

Pise-Masison CA, Mahieux R, Jiang H, Ashcroft M, Radonovich M, Duvall J et al (2000) Inactivation of P53 by human T-cell lymphotropic virus type 1 Tax requires activation of the NF-kappaB pathway and is dependent on P53 Phosphorylation. Molcell Biol. 20(10):3377–3386

Pise-Masison CA, De Castro-Amarante MF, Enose-Akahata Y, Buchmann RC, Fenizia C, Washington Parks R et al (2014) Co-dependence of HTLV-1 P12 and P8 functions in virus persistence. PLoS Pathog 10(11):E1004454. https://doi.org/10.1371/journal.ppat.1004454. Pubmed Pmid: 25375128; Pubmed Central Pmcid: Pmcpmc4223054

Pozzatti R, Vogel J, Jay G (1990) The human T-lymphotropic virus type I Tax gene can cooperate with the RAS Oncogene to induce neoplastic transformation of cells. Molcell Biol 10:413–417

Rauch DA, Ratner L (2011) Targeting HTLV-1 activation of NF-kappaB in mouse models and Atll patients. Viruses 3(6):886–900. https://doi.org/10.3390/v3060886. Pubmed Pmid: 21994759; Pubmed Central Pmcid: Pmc3185776

Rivera-Walsh I, Waterfield M, Xiao G, Fong A, Sun SC (2001) NF-KappaB signaling pathway governs trail gene expression and human T-cell leukemia virus-I Tax-induced T-cell death. J Biol Chem 276(44):40385–40388

Sasaki H, Nishikata I, Shiraga T, Akamatsu E, Fukami T, Hidaka T et al (2005) Overexpression of a cell adhesion molecule, TSLC1, as a possible molecular marker for acute-type adult T-cell leukemia. Blood 105(3):1204–1213

Satou Y, Yasunaga J, Yoshida M, Matsuoka M (2006) HTLV-I basic leucine zipper factor gene mRNA supports proliferation of adult T cell leukemia cells. Procnatlacadsciusa 103(3):720–725

Satou Y, Yasunaga J, Yoshida M, Matsuoka M (2008) HTLV-1 bZip factor (HBZ) gene has a growth-promoting effect on adult T-cell leukemia cells. Rinsho Ketsueki 49(11):1525–1529

Satou Y, Yasunaga J, Zhao T, Yoshida M, Miyazato P, Takai K et al (2011) HTLV-1 bZip factor induces T-cell lymphoma and systemic inflammation in vivo. PLoS Pathog 7(2):E1001274. https://doi.org/10.1371/journal.ppat.1001274 [Doi]

Semmes OJ, Majone F, Cantemir C, Turchetto L, Hjelle B, Jeang KT (1996) HTLV-I and HTLV-II Tax: differences in induction of micronuclei in cells and transcriptional activation of viral LTRs. Virology 217(1):373–379

Senftleben U, Cao Y, Xiao G, Greten FR, Krahn G, Bonizzi G et al (2001) Activation by Ikkalpha of a second, evolutionary conserved, NF-KappaB signaling pathway. Science 293(5534):1495–1499

Shembade N, Harhaj NS, Yamamoto M, Akira S, Harhaj EW (2007) The human T-cell leukemia virus type 1 Tax oncoprotein requires the ubiquitin-conjugating enzyme Ubc13 for NF-kappaB activation. J Virol 81(24):13735–13742

Shibata Y, Tanaka Y, Gohda J, Inoue J (2011) Activation of the IkappaB kinase complex by HTLV-1 Tax requires cytosolic factors involved in Tax-induced polyubiquitination. J Bio Chem 150(6):679–686. Doi:Mvr106 [Pii]; https://doi.org/10.1093/jb/mvr106 [Doi]

Shibata Y, Tokunaga F, Goto E, Komatsu G, Gohda J, Saeki Y et al (2017) HTLV-1 Tax induces formation of the active macromolecular IKK complex by generating Lys63- and Met1-linked hybrid polyubiquitin chains. Plos Pathog 13(1):E1006162. https://doi.org/10.1371/Journal. Ppat.1006162. Pubmed Pmid: 28103322; Pubmed Central Pmcid: Pmcpmc5283754

Shimoyama M (1991) Diagnostic criteria and classification of clinical subtypes of adult T-cell leukaemia-lymphoma. A report from the lymphoma study group (1984–87). Br J Haematol 79 (3):428–437

Shudofsky AMD, Giam CZ (2019) Cells of adult t-cell leukemia evade HTLV-1 Tax/NF-KappaB hyperactivation-induced senescence. Blood Adv. 3(4):564–569. Epub 2019/02/23. https://doi. org/10.1182/bloodadvances.2018029322. Pubmed Pmid: 30787019; Pubmed Central Pmcid: Pmcpmc6391679

Silic-Benussi M, Cavallari I, Zorzan T, Rossi E, Hiraragi H, Rosato A et al (2004) Suppression of tumor growth and cell proliferation by P13II, a mitochondrial protein of human T cell leukemia virus type 1. Proc Natl Acad Sci USA 101(17):6629–6634

Sinha-Datta U, Horikawa I, Michishita E, Datta A, Sigler-Nicot JC, Brown M et al (2004) Transcriptional activation of htert through the NF-kappaB pathway in HTLV-I-transformed cells. Blood 104(8):2523–2531

Siu YT, Chin KT, Siu KL, Yee Wai CE, Jeang KT, Jin DY (2006) Torc1 and Torc2 coactivators are required for Tax activation of the human T-Cell leukemia virus type 1 long terminal repeats. J Virol 80(14):7052–7059

Sun SC (2012) The noncanonical NF-kappaB pathway. Immunol Rev 246(1):125–1240. https:// doi.org/10.1111/J.1600-065x.2011.01088.X. Pubmed Pmid: 22435551; Pubmed Central Pmcid: Pmc3313452

Takahashi M, Higuchi M, Makokha GN, Matsuki H, Yoshita M, Tanaka Y et al (2013) HTLV-1 Tax oncoprotein stimulates ROS production and apoptosis in T cells by interacting with USP10. Blood 122(5):715–725. https://doi.org/10.1182/blood-2013-03-493718. Pubmed Pmid: 23775713

Takeda S, Maeda M, Morikawa S, Taniguchi Y, Yasunaga J, Nosaka K et al (2004) Genetic and epigenetic inactivation of Tax gene in adult T-cell leukemia cells. Int J Cancer 109(4):559–567

Tanaka A, Takahashi C, Yamaoka S, Nosaka T, Maki M, Hatanaka M (1990) Oncogenic transformation by the Tax gene of human T-cell leukemia virus type I in vitro. Proc Natl Acad Sci U S A 87:1071–1075

Tanaka G, Okayama A, Watanabe T, Aizawa S, Stuver S, Mueller N et al (2005) The clonal expansion of human T lymphotropic virus type 1-infected T cells: a comparison between seroconverters and long-term carriers. J Infect Dis 191(7):1140–1147

Taniguchi Y, Nosaka K, Yasunaga J, Maeda M, Mueller N, Okayama A et al (2005) Silencing of human T-cell leukemia virus type I gene transcription by epigenetic mechanisms. Retrovirology 2:64

Tao Z, Ghosh G (2012) Understanding NIK regulation from its structure. Structure 20(10):1615–1617. https://doi.org/10.1016/J.Str.2012.09.012. Pubmed Pmid: 23063006

Taylor GP, Matsuoka M (2005) Natural History of adult T-cell leukemia/lymphoma and approaches to therapy. Oncogene 24(39):6047–6057

Tezuka K, Xun R, Tei M, Ueno T, Tanaka M, Takenouchi N et al (2014) An animal model of adult T-cell leukemia: humanized mice with HTLV-1-specific immunity. Blood 123(3):346–355. Blood-2013-06-508861 [Pii]; https://doi.org/10.1182/blood-2013-06-508861 [Doi]

Tripp A, Banerjee P, Sieburg M, Planelles V, Li F, Feuer G (2005) Induction of cell cycle arrest by human T-Cell lymphotropic virus type 1 Tax in hematopoietic progenitor (Cd34+) cells: modulation of P21CIP1/WAF1 and P27KIP1 expression. Jvirol. 79(22):14069–14078

Tsukahara T, Kannagi M, Ohashi T, Kato H, Arai M, Nunez G et al (1999) Induction of BCL-X(L) expression by human T-cell leukemia virus type 1 Tax through NF-KappaB in apoptosis-resistant T-cell transfectants with Tax. J Virol 73(10):7981–7987. Pubmed Pmid: 10482545; Pubmed Central Pmcid: Pmcpmc112812

Tsukasaki K, Hermine O, Bazarbachi A, Ratner L, Ramos JC, Harrington W Jr et al (2009) Definition, prognostic factors, treatment, and response criteria of adult T-cell leukemia-lymphoma: a proposal from an international consensus meeting. J Clin Oncol 27 (3):453–459. Doi:Jco.2008.18.2428 [Pii]; https://doi.org/10.1200/jco.2008.18.2428 [Doi]

Umeki K, Hisada M, Maloney EM, Hanchard B, Okayama A (2009) Proviral loads and clonal expansion of HTLV-1-infected cells following vertical transmission: a 10-year follow-up of children in jamaica. Inter Virol 52(3):115–122. Doi:000219384 [Pii]; https://doi.org/10.1159/000219384 [Doi]

Ungewickell A, Bhaduri A, Rios E, Reuter J, Lee CS, Mah A et al (2015) Genomic analysis of mycosis fungoides and sezary syndrome identifies recurrent alterations in TNFR2. Nat Gen 47 (9):1056–60. https://doi.org/10.1038/ng.3370. Pubmed Pmid: 26258847

Van Prooyen N, Andresen V, Gold H, Bialuk I, Pise-Masison C, Franchini G (2010) Hijacking the T-cell communication network by the human T-cell leukemia/lymphoma virus type 1 (HTLV-1) P12 and P8 proteins. Mol Aspects Med 31(5):333–343. Epub 2010/08/03. https://doi.org/10.1016/j.mam.2010.07.001. Pubmed Pmid: 20673780; Pubmed Central Pmcid: Pmcpmc2967610

Vernin C, Thenoz M, Pinatel C, Gessain A, Gout O, Delfau-Larue MH et al (2014) HTLV-1 bZip factor hbz promotes cell proliferation and genetic instability by activating oncomirs. Cancer Res 74(21):6082–6093. https://doi.org/10.1158/0008-5472.Can-13-3564. Pubmed Pmid: 25205102

Wang L, Ni X, Covington KR, Yang By, Shiu J, Zhang X (2015) Genomic profiling of sezary syndrome identifies alterations of key T cell signaling and differentiation genes. Nat Gen 47 (12):1426–1434. https://doi.org/10.1038/ng.3444. Pubmed Pmid: 26551670; Pubmed Central Pmcid: Pmcpmc4829974

Wang C, Long W, Peng C, Hu L, Zhang Q, Wu A, et al (2016) HTLV-1 Tax functions as a ubiquitin E3 ligase for direct IKK activation via synthesis of mixed-linkage polyubiquitin chains. Plos Pathog 12(4):E1005584. https://doi.org/10.1371/Journal.Ppat.1005584. Pubmed Pmid: 27082114; Pubmed Central Pmcid: Pmcpmc4833305

Watanabe T (2017) Adult T-Cell leukemia: molecular basis for clonal expansion and transformation of HTLV-1-infected T cells. Blood 129(9):1071–1081. Epub 2017/01/25. https://doi.org/10.1182/blood-2016-09-692574. Pubmed Pmid: 28115366; Pubmed Central Pmcid: Pmcpmc5374731

Wu X, Sun SC (2007) Retroviral oncoprotein Tax deregulates NF-kappaB by activating Tak1 and mediating the physical association of Tak1-IKK. Embo Rep 8(5):510–515

Xiao G, Cvijic ME, Fong A, Harhaj EW, Uhlik MT, Waterfield M et al (2001) Retroviral oncoprotein Tax induces processing of NF-kappaB2/P100 in T Cells: evidence for the involvement of Ikkalpha. Embo J 20(23):6805–6815

Yamada T, Yamaoka S, Goto T, Nakai M, Tsujimoto Y, Hatanaka M (1994) The human T-cell leukemia virus type I Tax protein induces apoptosis which is blocked by the BCL-2 protein. Jvirol. 68:3374–3379

Yamamoto-Taguchi N, Satou Y, Miyazato P, Ohshima K, Nakagawa M, Katagiri K (2013) HTLV-1 bZip factor induces inflammation through labile Foxp3 expression. PLoS Pathog 9(9): E1003630. https://doi.org/10.1371/Journal.Ppat.1003630. Pubmed Pmid: 24068936; Pubmed Central Pmcid: Pmc3777874

Yamaoka S, Tobe T, Hatanaka M (1992) Tax protein of human T-cell leukemia virus type I is required for maintenance of the transformed phenotype. Oncogene 7:433–437

Yamaoka S, Inoue H, Sakurai M, Sugiyama T, Hazama M, Yamada T et al (1996) Constitutive activation of NF-kappaB is essential for transformation of RAT fibroblasts by the human T-cell leukemia virus type I Tax protein. Embo J 15(4):873–887

Yamaoka S, Courtois G, Bessia C, Whiteside ST, Weil R, Agou F et al (1998) Complementation cloning of nemo, a component of the IkappaB kinase complex essential for NF-kappaB activation. Cell 93(7):1231–1240

Yang L, Kotomura N, Ho YK, Zhi H, Bixler S, Schell MJ et al (2011) Complex cell cycle abnormalities caused by human T-lymphotropic virus type 1 Tax. J Virol. 85(6):3001–3009. Doi:Jvi.00086-10 [Pii]; https://doi.org/10.1128/jvi.00086-10 [Doi]

Yasuma K, Yasunaga J, Takemoto K, Sugata K, Mitobe Y, Takenouchi N et al (2016) HTLV-1 bZip factor impairs anti-viral immunity by inducing co-inhibitory molecule, T cell immunoglobulin and itim domain (Tigit). PLoS Pathog. 12(1):E1005372. https://doi.org/10.1371/Journal.Ppat.1005372. Pubmed Pmid: 26735971; Pubmed Central Pmcid: Pmcpmc4703212

Yl Kuo, Cz Giam (2006) Activation Of the anaphase promoting complex by HTLV-1 Tax leads to senescence. EMBO J 25(8):1741–1752

Yoshida M, Satou Y, Yasunaga J, Fujisawa J, Matsuoka M (2008) Transcriptional control of spliced and unspliced human T-cell leukemia virus type 1 bZip factor (HBZ) gene. Jvirol. 82 (19):9359–9368

Yoshita M, Higuchi M, Takahashi M, Oie M, Tanaka Y, Fujii M (2012) Activation of MTOR by human T-cell leukemia virus type 1 Tax is important for the transformation of mouse T cells to interleukin-2-independent growth. Cancer Sci 103(2):369–374. https://doi.org/10.1111/J.1349-7006.2011.02123.X Pubmed Pmid: 22010857

Zahoor MA, Philip S, Zhi H, Giam CZ (2014) NF-KappaB inhibition facilitates the establishment of cell lines that chronically produce human T-lymphotropic virus type 1 viral particles. J Virol 88(6):3496–3504. Doi:Jvi.02961-13 [Pii]; https://doi.org/10.1128/jvi.02961-13 [Doi]

Zhang L, Zhi H, Liu M, Kuo YL, Giam CZ (2009) Induction of P21(CIP1/WAF1) expression by human T-lymphotropic virus type 1 Tax requires transcriptional activation and mRNA stabilization. Retrovirology 6:35

Zhao T, Yasunaga J, Satou Y, Nakao M, Takahashi M, Fujii M et al (2009) Human T-cell leukemia virus type 1 bZip factor selectively suppresses the classical pathway of NF-KappaB. Blood 113(12):2755–2764

Zhao LJ, Giam CZ (1992) Human T-cell lymphotropic virus type I (HTLV-I) transcriptional activator, Tax, enhances CREB binding to HTLV-I 21-base-pair repeats by protein-protein interaction. Proc Natl Acad Sci USA 89:7070–7074

Zhao T, Satou Y, Matsuoka M (2014) Development of T cell lymphoma in HTLV-1 bZip factor and Tax double transgenic mice. Arch Virol 159(7):1849–1856. Epub 2014/05/14. https://doi.org/10.1007/s00705-014-2099-y. Pubmed Pmid: 24818712

Zhi H, Yang L, Kuo Yl, Ho YK, Shih HM, Giam CZ (2011) NF-KappaB hyper-activation by HTLV-1 Tax induces cellular senescence, but can be alleviated by the viral anti sense protein HBZ. PLoS Pathog 7(4):E1002025. https://doi.org/10.1371/journal.ppat.1002025 [Doi]

Zhi H, Zahoor MA, Shudofsky AM, Giam CZ (2014) KSHV vcyclin counters the senescence/G1 arrest response triggered by NF-KappaB hyperactivation. Oncogene. Doi:Onc2013567 [Pii]; https://doi.org/10.1038/onc.2013.567 [Doi]

Zhi H, Guo X, Ho YK, Pasupala N, Engstrom HAA, Semmes OJ et al (2020) RNF8 Dysregulation and Down-regulation During HTLV-1 Infection Promote Genomic Instability in Adult T-Cell Leukemia. PLoS Pathog 16(5):e1008618. Epub 2020/05/27. https://doi.org/10.1371/journal.ppat.1008618. PubMed PMID: 32453758

Novel Functions and Virus–Host Interactions Implicated in Pathogenesis and Replication of Human Herpesvirus 8

Young Bong Choi, Emily Cousins, and John Nicholas

1 Introduction

Human herpesvirus 8 (HHV-8) is classified as a γ2-herpesvirus and is related to Epstein–Barr virus (EBV), a γ1-herpesvirus. One important aspect of the γ-herpesviruses is their association with neoplasia, either naturally or in animal model systems. HHV-8 is associated with B-cell-derived primary effusion lymphoma (PEL) and multicentric Castleman's disease (MCD), endothelial-derived Kaposi's sarcoma (KS), and KSHV inflammatory cytokine syndrome (KICS) (Arvanitakis et al. 1996; Carbone et al. 2000; Chang and Moore 1996; Gaidano et al. 1997; Goncalves et al. 2017a). EBV is also associated with a number of B-cell malignancies, such as Burkitt's lymphoma, Hodgkin's lymphoma, and posttransplant lymphoproliferative disease, in addition to epithelial nasopharyngeal and gastric carcinomas, T-cell lymphoma, and muscle tumors (Kawa 2000; Okano 2000; Young and Murray 2003). Despite the similarities between these viruses and their associated malignancies, the particular protein functions and activities involved in key aspects of virus biology and neoplastic transformation appear to be quite distinct. Indeed, HHV-8 specifies a number of proteins for which counterparts had not previously been identified in EBV, other herpesviruses, or even viruses in general, and these proteins are believed to play vital functions in virus biology and to be involved centrally in viral pathogenesis.

One such gene is viral interleukin-6 (vIL-6), which was immediately upon its discovery implicated as a candidate contributor to HHV-8 pathogenesis (Chang and Moore 1996; Neipel et al. 1997a; Nicholas et al. 1997). Previous reports had

Y. B. Choi (✉) · E. Cousins · J. Nicholas
Sidney Kimmel Comprehensive Cancer Center at Johns Hopkins,
Department of Oncology, Johns Hopkins University School of Medicine,
1650 Orleans Street, Baltimore, MD 21287, USA
e-mail: ychoi15@jhmi.edu

© Springer Nature Switzerland AG 2021
T.-C. Wu et al. (eds.), *Viruses and Human Cancer*, Recent Results
in Cancer Research 217, https://doi.org/10.1007/978-3-030-57362-1_11

indicated that IL-6 was produced by and supported the growth of KS cells, promoted inflammation and angiogenesis typical of KS, served as an important B-cell growth factor, and was found at elevated levels in the sera of MCD (Burger et al. 1994; Ishiyama et al. 1994; Miles et al. 1990; Roth 1991; Yoshizaki et al. 1989; Polizzotto et al. 2012). Similarly, the discovery of viral chemokines, vCCLs 1-3, and the demonstration of their pro-angiogenic activities in experimental systems suggested that these proteins may also contribute to disease, in addition to their suspected roles in immune evasion during HHV-8 productive replication (Boshoff et al. 1997; Stine et al. 2000). The chemokine receptor homolog, vGPCR, was found to induce angiogenic cellular cytokines of the type produced in and suspected to promote the growth of KS lesions (Cannon et al. 2003; Pati et al. 2001; Schwarz and Murphy 2001). The constitutively active membrane receptors encoded by HHV-8 open reading frames (ORFs) K1 and K15 could function similarly (Brinkmann et al. 2007; Samaniego et al. 2001; Wang et al. 2006). vGPCR and K1 also acted as oncogenes, promoting cell transformation and inducing tumorigenesis in animal models (Bais et al. 1998; Lee et al. 1998a; Yang et al. 2000). However, like the v-cytokines, vGPCR and K1 are expressed predominantly or exclusively during productive, lytic replication; therefore, any contributions to malignant pathogenesis are likely to be mediated through paracrine signaling. There is ample evidence that cytokine-mediated paracrine signal transduction plays a role in KS, and B-cell growth can also be influenced via this route, as discussed below. Apart from the likely involvement of these viral proteins in HHV-8-associated pathogenesis, the roles of some of these "unique" viral products in virus biology are only beginning to be appreciated. For example, the pro-survival signaling induced by vCCLs and vGPCR and the anti-apoptotic activities of vIRF-1 have been demonstrated to enhance productive replication. Therefore, functions that serve normal virus biology, such as inhibiting infection-induced apoptosis, may have the "side effect" of promoting virus-associated neoplasia. This concept is familiar to viral oncologists, but the precise mechanisms deployed by HHV-8 are novel. Classical oncogene and tumor suppressor activities are mediated in an autocrine fashion, and viral genes expressed during latency are potential contributors to malignant disease. Chief among these for HHV-8 is the latency-associated nuclear antigen (LANA) which specifies essential replication and genome segregation activities in dividing cells and impacts several host pathways to promote cell survival and proliferation (Verma et al. 2007). These activities of LANA have obvious connections to processes involved in malignant transformation. Likewise, the viral homolog of cellular FLICE-inhibitory protein, vFLIP, is both latently expressed and crucially important for maintaining cell viability. vFLIP acts predominantly via induction of NF-κB activity and associated anti-apoptotic mechanisms rather than via inhibition of receptor-mediated caspase activation (Chugh et al. 2005; Guasparri et al. 2004). Latency genes v-cyclin (ORF72) and microRNAs (miRNAs) have also been implicated in viral pathogenesis (Gottwein et al. 2011; Liang et al. 2011a; Qin et al. 2017; Verschuren et al. 2004a). In addition to these latency products, vIL-6, vIRF-1, vIRF-3, K1, and K15, while expressed maximally during productive replication, have also been detected in latently infected cells (of some types) and may contribute in a direct, autocrine fashion to viral neoplasia. The interplay between lytic and latent activities is likely to be important for KS, in which cytokine dysregulation is believed to drive the disease, and this interplay may also be significant in PEL and MCD. These issues and details of the molecular biology of virus–host interactions involving these various HHV-8-encoded factors are the topic of this review.

2 HHV-8 Latency Products and Autocrine Activities

2.1 Latency-Associated Nuclear Antigen (LANA)

LANA is specified by ORF73 of HHV-8, and homologs are encoded by other γ2-herpesviruses. The basic functions of these proteins are to serve as a latency origin-binding protein and to tether viral genomes to host chromosomes for appropriate segregation to daughter cell nuclei during cell division (Barbera et al. 2006; Verma et al. 2007). These activities are equivalent to those of EBNA1 of γ1 subfamily Epstein–Barr virus (Lindner and Sugden 2007). However, LANA has further activities that are likely to play roles in viral pathogenesis in addition to contributing to the maintenance of HHV-8 latency.

One such property reported for LANA is its association with and inhibition of the cell cycle checkpoint protein and tumor suppressor p53 (Friborg et al. 1999). However, while the presence of wild-type p53 in most PEL cell lines suggests that inactivation of p53 could be biologically relevant, the susceptibility of PEL cells to p53 activation indicates that LANA is not fully able to inhibit the tumor suppressor (Chen et al. 2010; Petre et al. 2007). LANA also interacts with retinoblastoma protein (Rb) to mediate activation of E2F-responsive targets and can transform rat embryo fibroblasts in combination with transduced H-Ras (Radkov et al. 2000). Additionally, LANA was found to suppress cyclin-dependent kinase inhibitor p16INK4a-mediated cell cycle arrest and to induce E2F-mediated S-phase entry in lymphoid cells (An et al. 2005). However, as in the case of p53, the actual relevance of this experimental finding has been questioned because Rb function appears to be fully intact in PEL cells (Platt et al. 2002). Moreover, LANA was recently reported to interfere directly with the spindle assembly checkpoint by inducing the degradation of Bub1, which is a histone H2A kinase that mediates recruitment of a guardian of centromeric cohesin, Shugoshin-1, to kinetochores (Sun et al. 2014; Lang et al. 2018). Thus, LANA-induced chromosome instability and aneuploidy likely contribute to HHV-8-associated tumorigenesis.

LANA also interacts with glycogen synthase kinase 3 beta (GSK3β), a kinase that targets various proteins involved in cell cycle regulation. GSK3β targets include the pro-mitogenic transcriptional regulator β-catenin and proto-oncoprotein c-Myc; phosphorylation of these proteins by GSK3β promotes their proteolytic degradation (Karim et al. 2004; Sears et al. 2000). β-catenin, in combination with the transcription factor TCF, induces expression of various genes, including *c-myc*, *c-jun*, and *cyclin D1*; these genes are involved in cell cycle promotion and are dysregulated in oncogenesis. LANA binding of GSK3β leads to nuclear sequestration and inactivation of the kinase, removing its negative regulation of β-catenin and promoting cell proliferation (Fujimuro and Hayward 2003; Fujimuro et al. 2003; Liu et al. 2007a). c-Myc residue T-58, the target of GSK3β phosphorylation, was found to be hypophosphorylated in PEL cells, and this under-phosphorylation and consequent stabilization of c-Myc were dependent on LANA expression (Bubman et al. 2007). In addition, LANA interacts directly with c-Myc and induces ERK-mediated activation of c-Myc via phosphorylation of residue S-62 (Liu et al. 2007b). LANA binding of c-Myc and activation of ERK activity occur independently of LANA interaction with GSK3β. Thus, through these various interactions, LANA can activate proliferative pathways of likely significance to both HHV-8 latency and pathogenesis.

LANA has been reported to induce and interact with angiogenin (ANG), a mediator of angiogenesis, which itself upregulates LANA expression and appears to play a role in the establishment of latency and promotion of cell viability (Paudel et al. 2012; Sadagopan et al. 2011). Furthermore, the interaction of LANA and ANG with annexin A2 has been identified in both HHV-8 latently infected telomerase immortalized endothelial (TIME) cells and BCBL-1 PEL cells. Based on the results from confocal microscopy analyses, it appears that these proteins colocalize and can form complexes together in addition to establishing separate ANG-LANA and ANG-annexin A2 interactions (Paudel et al. 2012). Annexin A2 is involved in the regulation of cell proliferation, apoptosis, and cytoskeletal reorganization, among other activities (Shim et al. 2007; Thomas and Augustin 2009). Suppression of annexin A2 or ANG expression in PEL cells was found to increase cell death, and depletion of annexin A2 led to decreased expression of ANG and LANA (Paudel et al. 2012; Sadagopan et al. 2011). Thus, there appears to be an integrated and functionally important relationship between LANA, ANG, and annexin A2 that promotes the viability of latently infected cells. Furthermore, the increased level of ANG in HHV-8-infected cells may contribute to KS pathogenesis via induction of endothelial cell activation, migration, and angiogenesis (Sadagopan et al. 2009). While the mechanisms involved in LANA, ANG, and annexin A2 mutual regulation and functional interactions remain to be elucidated, ANG interaction with and destabilization of p53 may be significant with respect to pro-survival effects of LANA mediated via ANG (Sadagopan et al. 2012). Furthermore, LANA has been shown to upregulate the expression of epidermal growth factor like domain 7 (EGFL7) by sequestering the transcriptional repressor Daxx away from the EGFL7 promoter, thereby promoting angiogenesis in both a paracrine and autocrine manner (Thakker et al. 2018). In addition to the direct LANA-cellular protein interactions outlined above, LANA can also mediate transcriptional regulation of cellular genes via more general mechanisms. One such mechanism involves regulation of transcriptionally suppressive DNA methylation. LANA interacts with DNA methyltransferase DNMT3a, leading to its recruitment to and methylation of LANA-targeted promoters (Shamay et al. 2006). LANA also associates with histone methyltransferase SUV39H1 and transcriptional histone deacetylase-associated corepressors mSin3, SAP30, and CIR; these interactions implicate additional mechanisms of direct, promoter-specific repression by LANA (Krithivas et al. 2000; Sakakibara et al. 2004). These mechanisms are believed to be important for the suppression of both viral lytic and cellular gene programs for the general purpose of promoting viral latency and long-term cell viability (Verma et al. 2007). LANA-targeted epigenetic repression of specific cellular genes that are silenced in various cancers could contribute to HHV-8-associated malignancies in addition to viral latency (Shamay et al. 2006; Ziech et al. 2010). For example, LANA together with DNMT3a strongly represses the expression of heart-cadherin (H-cadherin, also termed CDH13), a tumor suppressor that is frequently silenced in many different cancers (Berx and van Roy 2009), by inducing hypermethylation of the CDH13 promoter in endothelial and PEL cells (Shamay et al. 2006; Journo et al. 2018).

Intriguingly, in addition to its role in latency, LANA has been reported to play a role, in the cytoplasm, during lytic replication, where it appears to antagonize cGAS and the Rad50-Mre11-CARD9 complex innate immune DNA sensors (that activate IRF3 and NF-κB signaling, respectively), thus leading to the promotion of virus replication (Mariggio et al. 2017; Garrigues et al. 2017; Zhang et al. 2016a). These novel findings suggest that LANA is indispensable for both latency and lytic replication.

2.2 Viral FLICE-Inhibitory Protein (vFLIP)

HHV-8-encoded Fas-associated death domain-like IL-1β-converting enzyme (FLICE) inhibitory protein (vFLIP) is specified by ORF K13, and the protein is often referred to simply as K13. vFLIP/K13 is related structurally to death effector domain (DED)-containing and death receptor-interacting vFLIPs of other viruses, such as molluscum contagiosum virus MC159L, rhesus monkey rhadinovirus RRV-vFLIP, and equine herpesvirus2 E8; these vFLIPs are protective against Fas/CD95- and TNF receptor-induced apoptosis (Bertin et al. 1997; Hu et al. 1997; Thome et al. 1997; Ritthipichai et al. 2012). HHV-8 vFLIP/K13 also mediates the protection of mouse lymphoma and rat pheochromocytoma cell lines from Fas- and TNFα-induced apoptosis (Belanger et al. 2001; Djerbi et al. 1999). However, the unique ability of HHV-8 vFLIP/K13 to induce NF-κB signaling and its inability to effectively suppress Fas-induced apoptosis suggest that vFLIP/K13 functions primarily through activation of NF-κB rather than via death receptor/caspase inhibition (Chugh et al. 2005; Chaudhary et al. 1999; Matta and Chaudhary 2004). vFLIP/K13 activates the canonical and non-canonical NF-κB pathways by interacting directly with the inhibitory κ-kinase (IKK) complexes (IKKα: IKKβ ± IKKγ/Nemo) to stimulate kinase activity, leading to disruption of IκB interaction with p50/p65(RelA)-subunit NF-κB and to protease-mediated release of RelB/p52 (active form) from RelB/p100 (Field et al. 2003; Liu et al. 2002; Matta et al. 2007). Thus, vFLIP/K13 is able to activate NF-κB independently of upstream receptor-associated mechanisms involving signaling adaptors and kinases such as TNF receptor-associated factors (TRAFs) and receptor-interacting serine/threonine protein kinases (RIPKs); in doing so, vFLIP/K13 can avoid activation of c-Jun N-terminal kinase (JNK) stress signaling (Matta et al. 2007). While it has been reported that the vFLIP/K13 interaction with TRAF2 is required for vFLIP/K13 binding to IKKγ in BC-3 PEL cells (Guasparri et al. 2006), TRAF2-dependent interaction between vFLIP/K13 and IKKγ and activation of JNK/AP1 signaling by vFLIP/K13 were not evident in subsequent studies (Matta et al. 2007).

NF-κB activation by vFLIP/K13 is significant in that NF-κB is a suppressor of lytic reactivation in latently infected PEL cells, and vFLIP/K13 and NF-κB promote survival of these cells (Guasparri et al. 2004; Brown et al. 2003; Godfrey et al. 2005; Grossmann and Ganem 2008; Keller et al. 2000; Zhao et al. 2007). These effects have clear implications for the maintenance of long-term latency and for the potential contribution of vFLIP/K13 to HHV-8 malignancies. In addition to these biological effects, NF-κB signaling also induces pro-inflammatory and angiogenic cytokines, such as IL-6 and IL-8. These cytokines are produced by KS lesions and are predicted to promote KS pathogenesis and to mediate vFLIP/K13-induced cellular proliferation and transformation in experimental systems (An et al. 2003; Grossmann et al. 2006; Sun et al. 2006; Ballon et al. 2015). Additionally, vFLIP/K13-activated NF-κB activation downregulates the expression of glucose transporters GLUT1 and GLUT3 to suppress aerobic glycolysis and promote cell viability (Zhu et al. 2016). Reduced levels of the glucose transporters or aerobic glycolysis are often detected in HHV-8-infected cells in KS lesions and in several HHV-8-infected PEL cell lines (Zhu et al. 2016), suggesting the ability of HHV-8 to regulate a key metabolic pathway to adapt to stress conditions such as nutrient starvation. Furthermore, vFLIP/K13 inhibits the expression of C-X-C chemokine receptor type 4 (CXCR4) by induction of microRNA (miR)-146a via NF-κB activation (Punj et al. 2010). As CXCR4 plays an important role in developing vascular endothelial cells (Tachibana et al. 1998), downregulation of CXCR4 by vFLIP/K13 may contribute to KS development by promoting premature release of HHV-8-infected endothelial progenitor cells into the circulation. In addition to NF-κB activation, vFLIP/K13 was shown to inhibit autophagy and associated cell death induced by starvation or rapamycin in an HHV-8-infected PEL cell line (Lee et al. 2009a). Together, these findings suggest that vFLIP/K13 contributes to both viral latency maintenance and to PEL and KS pathogenesis through the constitutive activation of pro-survival signals, protection of infected cells from stress-induced cell death, and induction of antilytic NF-κB signaling.

Intriguingly, vFLIP/K13 protein levels remain very low during active infection, and it is often difficult to efficiently express the protein in recombinant vectors. This is probably due to its suboptimal codon usage which consequently causes mRNA instability and inefficient translation (Bellare et al. 2015). The inefficient codon usage may have evolved to prevent hyperactivation of NF-κB, which would lead to dysregulated expression of inflammatory cytokines, a strong immune targeting of infected cells, and a thorough blockade of lytic reactivation. vFLIP/K13 usurps host cell chaperone and signaling proteins to enhance its protein stability. Specifically, vFLIP/K13 interacts with HSP90, and HSP90 inhibitors induce vFLIP/K13 degradation and inhibit vFLIP/K13-induced NF-κB, leading to apoptosis and tumor growth inhibition in a mouse model (Gopalakrishnan et al. 2013; Nayar et al. 2013). Furthermore, a recent report demonstrated that vFLIP/K13 localizes to peroxisomes where it is stabilized by the MAVS/TRAF signaling complex via K63-linked polyubiquitination, and MAVS-mediated vFLIP/K13 stabilization is important for protecting virus-infected cells from autophagy-associated cell death (Choi et al. 2018). These studies suggest that targeted destabilization of vFLIP/K13 via inhibition of its polyubiquitination through its MAVS/TRAF interaction may be an effective treatment strategy for HHV-8-associated tumors.

2.3   Kaposins

The K12 transcription unit comprises the K12 ORF and two sets of GC-rich repeat units (DR1, DR2) positioned upstream of K12 (Sadler et al. 1999). Three proteins, kaposins A, B, and C, are produced from this locus by virtue of alternative transcriptional and translational initiation. Kaposin A, corresponding to the K12 translation product, is initiated from a conventional AUG codon in a transcript originating proximal to K12. Kaposins B and C initiate from CUG codons in different reading frames in transcripts containing the upstream repeat elements. Kaposin B is translated from DR1 and DR2 in "frame 1," while frame 2-translated kaposin C contains DR1/2-translated sequences fused to K12. A larger, spliced transcript initiating 5 kb upstream of K12 has also been identified, and this sequence has the potential to encode non-AUG-initiated protein(s) with novel N-terminal sequences derived from codons upstream of DR2 (Li et al. 2002a; Pearce et al. 2005). K12-locus transcripts are found in high abundance in latently infected cells but are induced during lytic replication (Sadler et al. 1999; Li et al. 2002a; Staskus et al. 1997; Sturzl et al. 1997; Zhong et al. 1996). The relative expression of kaposins A, B, and C in different cell types and tissues varies. For example, kaposins A and C are predominant in primary PEL cells, whereas kaposin B is most abundant in the BCBL-1 PEL cell line (Li et al. 2002a). While there have been no functional studies of kaposin C, activities of kaposins A and B have been reported.

As the direct product of ORF K12, kaposin A was the first identified and studied protein from this locus. In transfection experiments, the 6 kDa protein was found to transform immortalized rat-3 and NIH3T3 cells, forming cell colony foci in culture and tumors in athymic mice (Muralidhar et al. 1998). Transformation was dependent on cytohesin-1, a guanine nucleotide exchange factor, which binds to kaposin A (Kliche et al. 2001). This interaction promotes membrane recruitment and activity of cytohesin-1, which acts on membrane-associated target GTPases such as ARF1 (Kliche et al. 2001). Increased activities of kinases, such as cdc2, PKC, ERK, and CAM kinase II, have been detected in kaposin A-transduced cells, but the underlying mechanisms have not been established (Muralidhar et al. 2000). Studies using gene arrays and signaling assays have implicated activation of MEK/ERK, PI3K/AKT, and STAT3 pathways by kaposin A (Chen et al. 2009a). Initial immunofluorescence studies indicated possible Golgi localization of kaposin A (Muralidhar et al. 2000). However, subsequent confocal fluorescence microscopy, cell fractionation, and flow cytometry analyses detected mostly perinuclear kaposin A with some plasma membrane localization also detectable (Tomkowicz et al. 2002, 2005). A LXXLL motif resembling ligand-interacting regions of nuclear receptors was required for immortalized cell transformation, and mutation of this motif led to greatly diminished nuclear association and newly acquired cytoplasmic localization of kaposin A (Tomkowicz et al. 2005). However, both wild-type and motif-mutated kaposin A were equally capable of activating an AP1 reporter, indicating that transformation occurs via a mechanism distinct from AP1 activation. Kaposin A was also found to interact with a variant of GTP-binding protein septin-4, a protein which localizes to mitochondria and promotes apoptosis (Lin et al. 2007; Mandel-Gutfreund et al. 2011). Co-expression of kaposin A with the septin-4 variant led to suppression of septin-4 variant-induced apoptosis in trans-fected cells (Lin et al. 2007). Therefore, inhibition of septin-4 function may be one mechanism by which kaposin A can influence malignant pathogenesis by pro-moting cell survival. Normally, this activity would be expected to serve virus latency. This mechanism would be biologically significant if septin-4 was expressed and functional in HHV-8 latently infected cell types.

Kaposin B, translated from the DR repeats and K12, interacts via DR2-encoded sequences with the stress-responsive kinase MK2 and enhances its activity (McCormick and Ganem 2005). Kaposin B binds the "C-lobe" region of MK2, a region also targeted by p38 kinase. Kaposin B binding of the C-lobe, like its phosphorylation by p38, prevents inhibitory intramolecular association of C-lobe and C-tail sequences of MK2, which results in activation of the kinase. A single DR1 together with a single DR2 repeat, but neither element alone, is sufficient to mediate MK2 activation (McCormick and Ganem 2006). MK2 activity leads to stabilization of high turnover mRNAs containing AU-rich elements (AREs), and many of these mRNAs specify cytokines, such as pro-inflammatory and angiogenic IL-6. In addition to MK2 activation, kaposin B activates the small GTPase RhoA in endothelial cells to induce actin stress fiber formation, increased cell motility, and angiogenesis (Corcoran et al. 2015). Thus, kaposin B has the potential to influence KS pathogenesis. Kaposin B is predicted to function in the maintenance of latently infected cell populations and/or expansion of latent cell pools through pro-survival and mitogenic activities of induced cellular proteins. Importantly, kaposin B sta-bilizes the mRNA encoding PROX1, the "master regulator" of lymphatic endothelial cell differentiation. PROX1 is targeted by ARE-binding protein HuR, and kaposin B-activated p38 kinase promotes nucleo-cytoplasmic export of HuR (Yoo et al. 2010). Reprogramming of blood endothelial cells to cells expressing lymphatic markers is induced by HHV-8 infection and is believed to be a key process in KS development (Pyakurel et al. 2006; Wang et al. 2004). Stabilization of PROX1 mRNA by kaposin B is likely to represent an important mechanism by which blood-to-lymphatic endothelial reprogramming is induced by HHV-8. Therefore, in addition to kaposin A, kaposin B is likely to contribute significantly to HHV-8-associated disease.

2.4 Viral Cyclin (v-Cyclin)

Viral cyclin (v-cyclin) shares 54% homology and 32% identity with the sequences of human cyclin D2 (Chang et al. 1996; Li et al. 1997). The oncogenic potential of v-cyclin was first shown in experiments in which its expression in p53-ablated mice was able to induce tumors (Verschuren et al. 2002, 2004b). Furthermore, a more recent study using an HHV-8 mutant virus defective in the v-cyclin gene demonstrated that v-cyclin promotes proliferation of HHV-8-infected rat primary cells by regulating cell cycle progression and the G1/S transition in the context of contact-inhibited cell culture conditions; the deletion of v-cyclin also decreased tumor incidence in athymic nude mice injected with HHV-8-transformed cells (Jones et al. 2014). The oncogenic function of v-cyclin is believed to be mediated primarily by its activation of cyclin-dependent kinases (CDKs) including CDK6. The kinases in turn phosphorylate and inhibit CDK inhibitors such as p21/Cip and p27/Kip by inducing their cytoplasmic sequestration or degradation (Jones et al. 2014; Mann et al. 1999; Ellis et al. 1999; Sarek et al. 2006; Jarviluoma et al. 2006).

2.5 Viral Interleukin-6 (vIL-6) in PEL

Viral IL-6 (vIL-6) shares approximately 25% sequence identity with its cellular counterpart, human IL-6 (hIL-6); this viral cytokine was independently discovered by multiple groups (Neipel et al. 1997a; Nicholas et al. 1997; Moore et al. 1996). Although sequence identity between human and viral homologs is low, the cytokines adopt equivalent 4-α-helical bundle structures and have similar receptor interactions and signaling activities (Boulanger et al. 2003; Chow et al. 2001; Heinrich et al. 2003; Kishimoto et al. 1995). Signaling by hIL-6 requires interaction with gp130 signal transducer and gp80 receptor subunits, which leads to Janus kinase (JAK) activation and phosphorylation, dimerization, and nuclear translocation of STATs 1 and 3 (Heinrich et al. 1998). Several groups have shown that vIL-6 utilizes the same signaling components employed by human IL-6 but that it does not require gp80 for active complex formation and can signal through tetrameric $(gp130_2:vIL-6_2)$ or hexameric $(gp130_2:gp80_2:vIL-6_2)$ complexes (Chow et al. 2001; Aoki et al. 2001; Chen and Nicholas 2006; Boulanger et al. 2004). Ultimately, both cytokines share functional characteristics, such as the ability to sustain the growth of IL-6-dependent cell lines (Nicholas et al. 1997; Burger et al. 1998). vIL-6, however, is distinct in its ability to signal not only through gp130 complexes located on the plasma membrane but also intracellularly within the endoplasmic reticulum (ER), via gp80-devoid tetrameric complexes; vIL-6, unlike hIL-6, is secreted inefficiently and localizes in large part to the ER. These unique properties of the viral homolog are likely to be involved in the maintenance of viral latency and important for HHV-8 pathogenicity.

Several studies have shown that vIL-6 is critical for the growth of PEL cells. The viral cytokine, in addition to IL-10, was detected in PEL cell culture media and both were found to support PEL cell proliferation (Aoki and Tosato 1999; Jones et al.

1999). The detection of vIL-6 in these cultures was initially assumed to be the result of spontaneous lytic reactivation, either full or abortive, in a small proportion of cells because vIL-6 expression is induced during productive replication. However, vIL-6 is now known to be expressed at low levels in latent PEL cells (Chandriani and Ganem 2010; Chen et al. 2009b). Depletion of vIL-6 in these cells induces apoptosis and slows the rate of cell growth (Chen et al. 2009b). Similar growth effects were observed with intracellularly delivered single-chain antibody and peptide-conjugated phosphorodiamidate morpholino oligomers (PPMO) directed to vIL-6 and its transcript (Kovaleva et al. 2006; Zhang et al. 2008). Fully ER-retained vIL-6 (cloned to include an ER-targeting KDEL motif) is capable of rescuing the growth effects mediated by vIL-6 depletion (Chen et al. 2009b). These data indicate a prominent role of vIL-6 intracellular, autocrine signaling in support of the growth and survival of latently infected PEL cells.

The mechanisms through which vIL-6 acts in the ER to promote growth and viability of PEL cells are not entirely clear, but evidence suggests that a novel interaction with the ER membrane protein vitamin K epoxide reductase complex subunit 1 variant 2 (VKORC1v2) is critical. VKORC1v2 was identified as a novel-binding partner of vIL-6 and was found to be required for PEL cell survival (Chen et al. 2012). Depletion of VKORC1v2 yielded similar growth effects to those observed in vIL-6-depleted cell cultures. Furthermore, a small peptide inhibitor capable of disrupting the VKORC1v2:vIL-6 interaction recapitulated growth and apoptosis effects observed upon vIL-6 or VKORC1v2 depletion, confirming the biological relevance of the vIL-6:VKORC1v2 interaction (Chen et al. 2012). Subsequent studies revealed some aspects of the molecular mechanisms underlying the activities of this viral cytokine through VKORC1v2. First, vIL-6 together with VKORC1v2 promoted ER-associated degradation (ERAD) of pro-apoptotic protease cathepsin D through association with the ERAD machinery (Chen and Nicholas 2015). Second, vIL-6 and VKORC1v2 interacted with calnexin cycle components including UDP-glucose glycoprotein glucosyltransferase 1 (UGGT1) and glucosidase II (GlucII) to promote proper folding of nascent protein (Chen et al. 2017). Lastly, vIL-6 increased the expression of insulin-like growth factor 2 receptor (IGF2R) by inhibiting VKORC1v2-mediated ER-associated degradation of IGF2R (Li et al. 2018). IGF2R is a multifunctional protein related to lysosomal biogenesis and the regulation of growth and development (El-Shewy and Luttrell 2009), and its depletion in PEL cells resulted in cell death and reduced virus replication (Li et al. 2018). Therefore, vIL-6/VKORC1v2-associated activities are likely to contribute to virus biology and pathogenesis via the promotion of both latently infected cell viability and productive replication.

Increased levels of phosphorylated (active) STAT3 have been detected in several PEL cell lines (Aoki et al. 2003). STAT3 is activated upon vIL-6 signaling through gp130 complexes. Depletion of STAT3 in PEL cells leads to an increase in apoptosis and a decrease in the levels of survivin, which has been demonstrated to be critical for PEL cell viability (Aoki et al. 2003). Survivin is a member of the inhibitors of apoptosis (IAP) family of proteins and has been shown to inhibit apoptosis in several cancer cell lines (Ambrosini et al. 1997). These results are significant because they link anti-apoptotic activities of survivin to STAT3 signaling and potentially to vIL-6. It is noteworthy that gp130 depletion in PEL cells leads to diminished growth and increased apoptosis in culture (Cousins and Nicholas 2013). In addition to vIL-6:gp130 signaling, STAT3 can also be activated by VEGF (Bartoli et al. 2000). Importantly, vIL-6 has been found to induce VEGF in experimentally transduced cell lines and to play a significant role in PEL growth and dissemination in a xenograft model (Aoki and Tosato 1999; Aoki et al. 1999). Therefore, vIL-6 is involved in a complex set of activities in PEL cells and is not only capable of initiating pro-growth and survival signaling through the ER compartment (Fig. 1) but may also contribute to PEL pathogenesis through activation of STAT3/survivin and VEGF signaling. In addition to latency, vIL-6/gp130/STAT3 signaling was shown to be important for HHV-8 productive replication in PEL and endothelial cells (Cousins et al. 2014).

2.6   Viral Interferon Regulatory Factor-3 (vIRF-3) in PEL

HHV-8 specifies four viral interferon (IFN) regulatory factor homologs, vIRFs 1-4 (Cunningham et al. 2003; Lee et al. 2009b), which serve to counter the effects of cellular IRFs and to inhibit innate responses of the cell to virus infection and productive replication (see below). While all of the vIRFs are expressed during lytic replication, consistent with their presumed primary functions in evasion of antiviral host cell defenses, vIRF-3 is expressed as a *bona fide* latent product in PEL cells (Jenner et al. 2001; Paulose-Murphy et al. 2001; Rivas et al. 2001). As such, vIRF-3 has been referred to by some investigators as latency-associated nuclear antigen-2 (LANA2), despite its partial localization to the cytoplasm and the absence of demonstrable latent expression in any other cell type examined (Munoz-Fontela et al. 2005). Nonetheless, in the context of PEL/B-cells, vIRF-3 has the potential to impact cellular pathways that may be of biological relevance to viral latency and virus-associated pathogenesis. In common with other vIRFs, vIRF-3 can inhibit cellular IRF function. vIRF-3 does so by interfering with the transcriptional activities of IRFs 3, 5, and 7 in addition to inhibiting PKR, a pro-apoptotic kinase that is activated by IFN and dsRNA (Esteban et al. 2003; Joo et al. 2007; Lubyova and Pitha 2000; Wies et al. 2009). Importantly, vIRF-3 has also been found to interact directly with tumor suppressors including p53 and the pocket proteins pRb, p107, and p130 to inhibit their anti-tumor function and to induce c-Myc-directed transcription by recruiting the F-box of Skp2 protein, a key component of the Skp, cullin, F-box (SCF) ubiquitin ligase complex, to c-Myc-regulated promoters and stabilizing c-Myc (Rivas et al. 2001; Lubyova et al. 2007; Baresova et al. 2012, 2014; Marcos-Villar et al. 2014; Laura et al. 2015). Recently, vIRF-3 has been found to bind directly to ubiquitin-specific protease 7 (USP7, also called HAUSP), a hydrolase that deubiquitinates and stabilizes target proteins such as MDM2 and p53 (Li et al. 2002b, 2004). This interaction of vIRF-3 has been demonstrated to promote latently infected PEL cell viability and also to inhibit productive

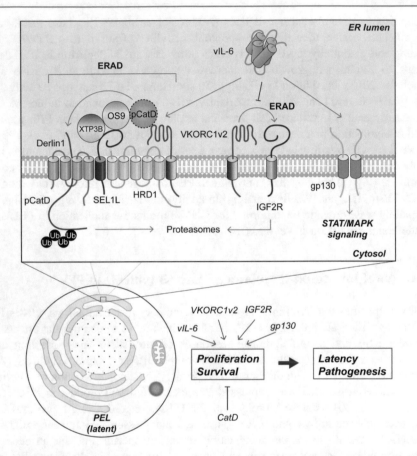

Fig. 1 ER-localized interactions and activities of vIL-6 in latently infected PEL cells. In the context of PEL latency, vIL-6 is expressed at low but functional levels and is largely sequestered in the ER compartment. Latently expressed vIL-6 supports PEL cell growth and viability. In vIL-6-depleted PEL cells, these activities can be complemented by ER-restricted (KDEL-tagged) transduced vIL-6, demonstrating sufficiency of ER-localized vIL-6 activity. The vIL-6 signal transducer, gp130, and a novel splice-variant protein, vitamin K epoxide reductase complex subunit 1 variant 2 (VKORC1v2), each binds vIL-6 within the ER, and depletion of each inhibits PEL cell growth and viability. Available evidence indicates that vIL-6 activity via each of these ER receptors occurs independently. While gp130-mediated activation of STAT and MAPK signaling has been detected in PEL cells, gp130 signaling (in contrast to effects on growth and survival) is not affected by VKORC1v2 depletion or by peptide-mediated disruption of the vIL-6: VKORC1v2 interaction. vIL6 regulates VKORC1v2-mediated ER-associated protein degradation (ERAD) either to promote the degradation of a precursor form (pCatD) of cathepsin D, a pro-apoptotic protease, or to inhibit the degradation of insulin-like growth factor 2 receptor (IGF2R), which is involved in promoting PEL cell viability. ERAD components: ER lectins (XTP3B and OS9) and ER translocon proteins (HRD1, Derlin, and Sel1L)

replication (Xiang et al. 2018). vIRF-3 can also interact with and inhibit FOXO3a, a transcription factor targeting pro-apoptotic genes (Munoz-Fontela et al. 2007). These activities indicate that vIRF-3 plays a significant role in promoting cell survival and proliferation in the context of PEL and potentially in general B-cell latency. Indeed, vIRF-3 has been demonstrated to be critically important for PEL cell viability in culture, as its depletion triggers apoptosis (Wies et al. 2008). Pro-survival effects of vIRF-3 may in part be the result of its inhibition of PML-mediated repression of survivin (Marcos-Villar et al. 2009). Furthermore, vIRF-3 has been reported to repress CIITA transcription factor-directed expression of IFN-γ and class II major histocompatibility complex in PEL cells (Schmidt et al. 2011). This immune evasion activity of vIRF-3 is likely to be vital for the long-term survival of these cells in vivo. Therefore, vIRF-3 activities are likely to be important for latency persistence in at least some cell types (where vIRF-3 is latently expressed) and are probably significant contributors to PEL malignancy.

Interestingly, a recent study showed that vIRF-3 is also required for KS pathogenesis. vIRF-3 is readily detected in approximately 42% of KS lesions from various organs, and it serves as a pro-angiogenic factor that promotes sprouting formation of lymphatic endothelial cells (Lee et al. 2018). This lymphangiogenesis was proposed to be mediated by vIRF-3 inhibition of phosphorylation and nucleo-cytoplasmic shuttling of histone deacetylase 5 (HDAC5), ultimately resulting in the retention of HDAC5 in the nucleus and epigenetic repression of antiangiogenic gene expression (Lee et al. 2018).

2.7 MicroRNAs (miRNAs)

MicroRNAs (miRNAs) are $\sim$ 22 nucleotide (nt) non-coding RNAs that regulate the expression of mRNAs via cleavage or by inhibiting translation. miRNAs are encoded as primary miRNAs (pri-miRNAs), synthesized by RNA polymerase II, and processed to pre-miRNAs (stem–loop structures) by the RNase III domain of DROSHA prior to nuclear export by exportin 5 (Lee et al. 2003a; Lund et al. 2004). In the cytoplasm, pre-miRNAs are cleaved into 21–24 nt double-stranded RNAs by an RNase III domain of DICER, and one strand of the miRNA duplex is then incorporated into RISC (RNA-induced silencing complex) (Lee et al. 2003a; Bartel 2004). The incorporated miRNA guides the loaded RISC to the mRNA target (Schwarz et al. 2003). Generally, the mRNA target is degraded if the miRNA is perfectly complementary to the targeted sequence. Alternatively, binding of a miRNA lacking perfect complementarity inhibits translation of the mRNA (Zeng et al. 2003).

While miRNAs have been detected in all metazoans, virus-encoded miRNAs were discovered relatively recently (Pfeffer et al. 2004). To date, 12 pre-miRNAs (termed miR-K1 to miR-K12) and 25 mature miRNAs from the HHV-8 genome have been identified. A total of ten of the 12 pre-miRNAs are located between latently expressed ORFs 71 and K12 in the HHV-8 genome; miR-K10 is located within ORF K12, and miR-K12 is within the 3′ UTR of K12 (Pfeffer et al. 2005;

Samols et al. 2005). All 12 of the HHV-8 miRNAs are orientated "in sense" with
ORFs 71 to K12 and are expressed during latency (Cai et al. 2005). Most of the
miRNAs can be detected during latent infection (Cai et al. 2005; Umbach and
Cullen 2010). Bioinformatic approaches have utilized miRNA seed sequences (nt
2-7/8 of the miRNA) to search for miRNA gene targets (Gottwein and Cullen 2010;
Lu et al. 2010b; Nachmani et al. 2009; Qin et al. 2010a), but these approaches yield
large numbers of potential candidates. Additionally, targets with less than perfect
seed sequence complementarity may be overlooked, and experimental validation of
identified candidates is necessary to assess authenticity. Functional approaches
involving transduction of recombinant viruses containing single or a combination of
HHV-8 miRNAs and monitoring potential changes in mRNA levels via microarray
have also been utilized (Ziegelbauer et al. 2009). In addition, methods based on
immunoprecipitation of RISC and/or associated argonaute proteins and subsequent
high-throughput sequencing of co-precipitated miRNAs have been employed
(Gottwein et al. 2011; Dolken et al. 2010; Haecker et al. 2012; Gay et al. 2018).
These methods have identified multiple mRNA targets within the host cell. From
these data, the virally encoded miRNAs have been deduced to (1) promote
angiogenesis and cell migration, (2) modulate programmed cell death and cell
cycle, (3) alter a variety of cell signaling pathways that are involved in the regulation of cell growth, differentiation, innate immunity and inflammation, and
(4) maintain latency. These targets and pathways are summarized in Table 1 and are
discussed further below.

HHV-8 employs many strategies to remain undetected by the host immune
response, and viral miRNAs are believed to play a vital role. HHV-8 miRNAs play
significant roles in cell growth, angiogenesis, and cell migration, which are considered important factors in KS pathogenesis. Transforming growth factor beta
(TGF-β) signaling can be downregulated via the direct targeting of SMAD5 by
miR-K11(Liu et al. 2012); downregulation of TGF-β signaling induces cell proliferation. It has been noted that HHV-8+ B-cell lines have decreased levels of
miR-155, the cellular homolog of miR-K11 that represses SMAD2 protein
expression and thereby modulates the cellular response to TGF-β (Louafi et al.
2010); miR-K11 may compensate for the limited levels of miR-155 in these cells
(Skalsky et al. 2007). Inhibition of miR-K11 was found to derepress TGF-β signaling in HHV-8+ B-cells (Liu et al. 2012). TGF-β signaling can be modulated by
thrombospondin 1 (THBS1), a target of HHV-8 miRNAs miR-K1, miR-K3,
miR-K6, and miR-K11 (Samols et al. 2007). THBS1 is an antiangiogenic factor,
and its downregulation leads to repressed TGF-β signaling (Samols et al. 2005). In
addition, miR-K3 represses the expression of GPCR kinase 2 (GRK2) to induce the
expression of the chemokine receptor CXCR2, thereby promoting migration and
invasion of endothelial cells (Hu et al. 2015; Li et al. 2016a). miR-K2 and miR-K5
are known to downregulate protein levels of the tumor suppressor protein tropomyosin 1 (TPM1) and thereby prevent anoikis, programmed cell death that occurs
after detachment of endothelial cells, and enhance tube formation of endothelial
cells (Kieffer-Kwon et al. 2015). Thus, multiple viral microRNAs are capable of
altering growth signaling and increasing angiogenesis to support the establishment

Table 1 HHV-8-encoded microRNAs and their activities

	Viral microRNA	Target	Activity/function	References
Angiogenesis	miR-K1, 3, 6, 11	THBS1	Cell adhesion, migration, and angiogenesis	Samols et al. (2007)
	miR-K3	GRK2	Migration, invasion, and angiogenesis	Hu et al. (2015), Li et al. (2016a)
	miR-K2, 5	TPM1	Enhanced tube formation of endothelial cells	Kieffer-Kwon et al. (2015)
	miR-K6	BCR	Increases Rac-1-mediated angiogenesis	Ramalingam et al. (2015)
	miR-K6-3p	SH3BGR	STAT3 activation, cell migration and angiogenesis	Li et al. (2016b)
	miR-K6-5p	CD82	Invasion and angiogenesis	Li et al. (2017)
Apoptosis	miR-K1, 3, 4-3p	CASP3	Inhibition of apoptosis	Dolken et al. (2010)
	miR-K5, 9, 10	BCLAF1	Inhibition of apoptosis	Ziegelbauer et al. (2009)
	miR-K9	GADD45B	Inhibition of apoptosis	Liu et al. (2017)
	miR-K10a	TWEAKR	Reduced caspase activation	Abend et al. (2010)
	miR-K10a/b	TGF-β receptor II	Inhibition of apoptosis	Lei et al. (2012)
Cell cycle	miR-K1	p21	Inhibition of cell cycle arrest	Gottwein and Cullen (2010)
	miR-K9	GADD45B	Inhibition of cell cycle arrest	Liu et al. (2017)
Cell growth Differentiation	miR-K1	IκBα	Activation of NF-κB/IL-6/STAT3 signaling	Chen et al. (2016)
	miR-K6, 8, 10, 11, 12	Not known	Activation of the STAT3 pathway to upregulate ASS1 expression	Li et al. (2019)
	miR-K6, 11	MAF	Endothelial cell reprogramming	Hansen et al. (2010)
	miR-K11	SMAD5	Suppression of TGF-β signaling	Liu et al. (2012)
		BACH-1	Regulation of B-cell development Induction of heme oxygenase-1 (HO-1)	Skalsky et al. (2007), Gottwein et al. (2007), Botto et al. (2015)
		C/EBPβ	Expansion of infected B-cells	Boss et al. (2011)
Innate immunity Inflammation	miR-K3, 7	C/EBPβ	Induction of IL-6 and IL-10 expression	Qin et al. (2010b)
	miR-K5 miR-K12-7	MYD88	Inhibition of IL-1 signaling	Abend et al. (2012)

(continued)

Table 1 (continued)

	Viral microRNA	Target	Activity/function	References
	miR-K7	MICB	Evasion from NK cell killing	Nachmani et al. (2009)
	miR-K9	IRAK1	Inhibition of the TLR/IL-1R signaling	Abend et al. (2012)
	miR-K11	IKKε	Attenuation of IFN signaling	Liang et al. (2011b)
Latency maintenance	miR-K1	IκBα	Activation of NF-κB, Inhibition of lytic replication	Lei et al. (2010)
	miR-K3	NF-I/B	RTA repression	Lu et al. (2010b)
		GRK2	Activation of CXCR2/AKT	Li et al. (2016a)
	miR-K4-5p	RBL2	Epigenetic reprogramming via DNMT1/3	Lu et al. (2010a)
	miR-K5, 9, 10a/b	BCLAF1	Inhibition of lytic replication	Ziegelbauer et al. (2009)
	miR-K7-5p, 9	RTA	Inhibition of lytic replication	Bellare and Ganem (2009), Lin et al. (2011)
	miR-K11	MYB	RTA repression	Plaisance-Bonstaff et al. (2014)

of HHV-8-associated neoplasia. Additionally, viral miRNAs can alter the transcriptomes of endothelial cells and assist in cellular reprogramming. Both blood-vessel endothelial cell (BEC) and lymphatic endothelial cell (LEC) expression markers are found in KS tissue, and the tissue can be reprogrammed toward the LEC or BEC fate under the appropriate stimuli. Specifically, miR-K6 and miR-K11 target the cellular transcription factor musculoaponeurotic fibrosarcoma oncogene homolog (MAF) (Hansen et al. 2010), which is found in LECs but not in BECs. Silencing of MAF by the microRNAs increases expression of BEC marker genes within the KS tissue. Furthermore, miR-K6 is known to target other factors, including breakpoint cluster region protein (BCR), SH3BGR, and CD82 to promote angiogenesis and cell migration (Ramalingam et al. 2015; Li et al. 2016b; Li et al. 2017).

HHV-8 miRNAs are also involved in counteracting pro-apoptotic pathways induced by the host cell upon viral infection. Caspase 3 is a target of multiple viral miRNAs (miR-K1, miR-K3, and miR-K4-3p), and its inhibition desensitizes HHV-8-infected cells to caspase-induced apoptosis (Dolken et al. 2010). Similarly, miR-K10a suppresses TWEAK receptor (TWEAKR) expression, which limits

TWEAK-induced caspase activation and apoptosis (Abend et al. 2010). TWEAKR inactivation also reduces production of pro-inflammatory cytokines IL-8/CXCL8 and MCP-1/CCL2 (Abend et al. 2010). miR-K10a also inhibits TGF-β-induced apoptosis by targeting TGF-β-type II receptor (Lei et al. 2012). BCLAF1, a transcriptional repressor, can be targeted by several of the HHV-8 miRNAs (miR-K5, miR-K9, and miR-K10), leading to decreased etoposide-induced apoptosis in PEL cells (Ziegelbauer et al. 2009). miR-K9 is known also to target growth arrest DNA damage-inducible gene 45 beta (GADD45B) to protect cells from cell cycle arrest and apoptosis induced by DNA damage (Liu et al. 2017). Finally, miR-K11, a homolog of cellular miR-155, alters expression of BACH1, which leads to a variety of phenotypic changes including upregulation of HMOX1 (increased cell survival), upregulation of xCT (increased permissiveness to infection in macrophages and endothelial cells), and protection against apoptosis mediated by reactive nitrogen species (Qin et al. 2010a; Skalsky et al. 2007; Gottwein et al. 2007).

In addition, HHV-8 miRNAs play a role in modulating innate immune responses and inflammation. miR-K7 targets MICB, a viral infection-induced cell surface marker that functions to induce natural killer (NK) cell recognition and killing via engagement with the NK-expressed NKG2D receptor (Nachmani et al. 2009; Glas et al. 2000). MICB targeting by virus-encoded miRNAs is conserved in human cytomegalovirus and EBV infection (Nachmani et al. 2009; Stern-Ginossar et al. 2007, 2008). miR-K3 and miR-K7 reduce the expression of the C/EBPβ p20 (LIP) isoform, which functions as a negative transcriptional regulator, to increase the expression of IL-6 and IL-10 in macrophages (Qin et al. 2010b). These anti-inflammatory cytokines are more frequently found in patients with KS, PEL, and MCD (Jones et al. 1999; Aoki et al. 2000; Oksenhendler et al. 2000; Drexler et al. 1999) and inhibit innate immune responses (Cirone et al. 2008). Conversely, IL-1 pro-inflammatory cytokine receptor signaling is inhibited by targeting of adaptor proteins MYD88 and IRAK1 by miR-K5 and miR-K9, respectively (Abend et al. 2012). Thus, regulation of anti- and pro-inflammatory cytokines by HHV-8 miRNAs may favor HHV-8 immune escape. Furthermore, miR-K11 reduces IFN expression by targeting I-κB kinase epsilon (IKKε) and inhibiting IKKε-mediated IRF3/IRF7 phosphorylation (Liang et al. 2011b). miR-K11 targeting of IKKε is also known to contribute to maintenance of HHV-8 latency.

In addition to immune evasion, HHV-8 miRNAs are involved in development or proliferation of HHV-8-infected cells, related directly to the oncogenic potential of HHV-8. For example, miR-K11, an ortholog of miR-155 that is involved in B-cell development (Li et al. 2013), downregulates the expression of C/EBPβ, a transcriptional repressor of IL-6 that is coupled to B-cell lymphoproliferative diseases (Alonzi et al. 1997), and induces splenic B-cell expansion in a mouse model (Gottwein et al. 2007; Boss et al. 2011). In addition, miR-K11 represses the expression of BACH1, a transcriptional repressor of heme oxygenase-1 (HO-1), and induces sustained expression of HO-1 during latency, thereby promoting cell proliferation (Botto et al. 2015). miR-K1 is known to target the NF-κB inhibitor IκBα to induce NF-κB-dependent expression of IL-6, in turn leading to STAT3 activation (Chen et al. 2016). Furthermore, a recent study demonstrated that numerous HHV-8 miRNAs including miR-K6 to -K12 upregulate the expression of argininosuccinate synthase 1 (ASS1), a key enzyme in the citrulline–nitric oxide (NO) cycle; the NO product then activates STAT3, but the direct targets of the miRNAs have not been identified (Li et al. 2019). The study further demonstrated that miRNA-induced ASS1/STAT3 activation contributes to cell proliferation and transformation.

Several HHV-8 miRNAs target RTA, thereby functioning to maintain viral latency and inhibit viral replication. Several research groups have identified miR-K3, miR-K5, miR-K9, and miR-K11 as directly or indirectly targeting RTA expression (Lu et al. 2010a; Bellare and Ganem 2009; Lin et al. 2011; Plaisance-Bonstaff et al. 2014; Lu et al. 2010b). miR-K3 also is involved in the inhibition of virus replication by activating the CXCR2/AKT signaling axis via GRK2 suppression (Li et al. 2016a). Viral replication is also limited through the inhibition of the NF-κB inhibitor IκBα by miR-K1; upregulation of NF-κB signaling abrogates lytic replication of HHV-8 (Lei et al. 2010). By reducing virion production, the virus is able to limit the host immune response. RBL2 (retinoblastoma-like protein 2) has been identified as a target of miR-K4-5p (Lu et al. 2010a). RBL2 is an inhibitor of specific DNA methyltransferases (DNMT3a and 3b). Epigenetic changes have been observed following inhibition of RBL2, and the consequences of these alterations may contribute to latency maintenance (Lu et al. 2010a).

3 Novel Virus–Host Interactions via Lytic Activities

3.1 Viral Interleukin-6 (vIL-6)

In contrast to its direct autocrine role in PEL pathogenesis, vIL-6 is believed to contribute to KS and MCD predominantly via paracrine signaling. Newly infected cells and those undergoing lytic replication express vIL-6 as an early gene product, and vIL-6 is rapidly induced following RTA expression in these cells (Sun et al. 1999). The majority of HHV-8-infected cells in the KS lesion remains latently infected, but small subsets of cells are lytically active. This minority of cells expresses lytic proteins, including vIL-6, vGPCR, and K1; these proteins ultimately enhance the expression of cellular inflammatory and angiogenic cytokines (Mesri et al. 2010). For example, vIL-6 can induce expression of VEGF (Aoki and Tosato 1999), considered to be a key contributor to KS development. VEGF, cytokines such as IL-6 and IL-10, and bFGF are secreted from lytically active cells and can promote proliferation and/or viability of nearby latently infected and uninfected cells in a paracrine fashion. The secreted proteins modulate survival of HHV-8-infected cells, angiogenesis (predominantly through VEGF), and recruitment of uninfected cells to the lesion. Inflammatory cytokines (including IL-6) and angiogenic factors have been proposed to play a role in the initial development of KS (Ensoli and Sturzl 1998). Additionally, vIL-6 modulates innate immune responses by regulating the expression of chemokines. For example, vIL-6 enhances basal and IL-1β-induced CCL2 expression and inhibits IL-1β-induced CXCL8 expression to block the infiltration of neutrophils during acute infection of B-cells (Fielding et al. 2005).

In MCD patients, levels of IL-6 are substantially higher in affected lymph nodes when compared to unaffected nodes of the same patient (Yoshizaki et al. 1989). Additionally, disease severity correlates with IL-6 levels, and when affected lymph nodes are resected, IL-6 levels decrease (Yoshizaki et al. 1989). Similarly, cell lines derived from HIV-positive KS patients express IL-6, and KS tissue produces increased amounts of IL-6 compared to normal tissue (Miles et al. 1990). IL-6 antisense oligonucleotides were found to suppress the growth of KS cells from HIV-positive patients as well as the production of IL-6, and addition of exogenous, recombinant IL-6 was able to restore growth and proliferation of these cells (Miles et al. 1990). An additional study observed substantially elevated levels of IL-6 produced by malignant plasma cells from MCD patients though not from other B-cell tumors (Burger et al. 1994). At the time of these discoveries, the mechanism of disease-associated IL-6 dysregulation was not understood. More recent reports have demonstrated that vIL-6 can induce the expression of IL-6 and VEGF in some cell types (Aoki et al. 1999; Mori et al. 2000). It is likely that vIL-6 also contributes directly to MCD. Relative to MCD, but pathologically distinct, is KSHV inflammatory cytokine syndrome (KICS) (Polizzotto et al. 2012; Goncalves et al. 2017b). Patients with KICS exhibit elevated HHV-8 viral loads and cytokines including vIL-6, IL-6, and IL-10 in peripheral blood (Polizzotto et al. 2012). Thus, it is likely that vIL-6 may also contribute to KICS pathogenesis by inducing cellular cytokines and promoting lytic replication, although the role of vIL-6 in KICS development remains to be determined.

3.2 Viral CC-Chemokine Ligands (vCCLs)

The three HHV-8 chemokines, vCCL-1, vCCL-2, and vCCL-3 (previously/alternatively referred to vMIP-I, vMIP-II, and vMIP-III, respectively), are encoded by ORFs K6, K4, and K4.1, respectively (Nicholas et al. 1997; Moore et al. 1996; Neipel et al. 1997b; Russo et al. 1996). All are expressed early during the lytic cycle (Jenner et al. 2001; Paulose-Murphy et al. 2001). vCCL-1 and vCCL-2 are most closely related structurally to cellular chemokines CCL3 and CCL4, while vCCL-3 shares significant primary sequence similarity with a number of CC-chemokines. However, the properties of the v-chemokines are distinct from those of their cellular counterparts. With respect to receptor usage, vCCL-1 is an agonist for CCR8; vCCL-2 signals through CCR3, CCR8, and ACKR3 (formerly known as CXCR7); and vCCL-3 functionally targets CCR4 and XCR1 (Boshoff et al. 1997; Stine et al. 2000; Dairaghi et al. 1999; Luttichau 2008; Luttichau et al. 2007; Nakano et al. 2003; Szpakowska et al. 2016). In addition, vCCL-2 binds several CCR- and CXCR-type chemokine receptors and CX_3CR1 as a neutral (non-signaling) ligand and effectively inhibits cellular chemokine activity through these receptors (Dairaghi et al. 1999; Chen et al. 1998; Crump et al. 2001; Kledal et al. 1997; Luttichau et al. 2001). The nature of the v-chemokine-targeted receptors suggests that the v-chemokines may mediate immune evasion via Th2 polarization and blocking of leukocyte trafficking, as has been demonstrated for vCCL-2 in in vivo experiments (Chen et al. 1998; Weber et al. 2001).

Apart from these immune evasion properties, each of the HHV-8 v-chemokines has been demonstrated to promote angiogenesis, in part via induction of VEGF (Boshoff et al. 1997; Stine et al. 2000; Liu et al. 2001). Like vIL-6, HHV-8 vCCLs have the potential to promote KS pathogenesis via paracrine signaling and may also play roles in PEL, where VEGF has been implicated as an important factor based on data from murine studies (Aoki and Tosato 1999; Ensoli and Sturzl 1998; Haddad et al. 2008). Additional contributions of vCCL-1 and vCCL-2 to pathogenesis may include pro-survival signaling via CCR8, demonstrated in uninfected and HHV-8-infected endothelial cells (Choi and Nicholas 2008). vCCL-1 and vCCL-2 were also found to promote survival of PEL and murine cell lines (Liu et al. 2001; Louahed et al. 2003). Unlike most (non-secreted) viral proteins, the v-chemokines have the potential to function in a paracrine manner. Therefore, these chemokines may promote cell survival of latently infected and uninfected cells surrounding those supporting lytic replication, thus contributing to viral pathogenesis. Nonetheless, an important aspect of the pro-survival activities of vCCL-1 and vCCL-2 is the positive contribution to productive replication via autocrine signaling. The endogenously produced v-chemokines inhibit lytic cycle-induced apoptosis and increase virus yields in HHV-8-infected endothelial cultures (Choi and Nicholas 2008). This activity involves CCR8 signaling-dependent suppression of lytic cycle stress-induced pro-apoptotic protein Bim, a powerful inhibitor of productive viral replication.

3.3 Viral G Protein-Coupled Receptor (vGPCR)

While not unique to HHV-8, the vGPCR encoded by this virus is structurally and functionally diverged from other γ2-herpesvirus vGPCRs and is strongly implicated as a paracrine contributor to KS development (Cannon 2007; Nicholas 2005; Rosenkilde et al. 2001; Verzijl et al. 2004). HHV-8 vGPCR is unusual in its promiscuous functional association with three classes of Gα proteins (i, q, and 13) in addition to its direct association with and activation of the signaling protein SHP2 (Couty et al. 2001; Liu et al. 2004; Philpott et al. 2011; Shepard et al. 2001). Initial reports that vGPCR can function as a classical "autocrine" oncogene in in vitro and in vivo experimental systems employing vGPCR-transduced cell lines implied that vGPCR may be expressed as a latent protein (enabling it to contribute directly to HHV-8 oncogenesis) (Bais et al. 1998). However, no evidence of vGPCR latent expression has been forthcoming. Nonetheless, vGPCR can induce KS-like tumors in receptor-transduced mice, and this phenotype is supported despite only a minority of cells expressing vGPCR (Yang et al. 2000; Montaner et al. 2003). This result can be explained by angiogenic, mitogenic, and inflammatory cellular cytokine induction via vGPCR signaling, leading to local endothelial activation, proliferation, and tumorigenesis (Montaner et al. 2004). As stated above, cytokine dysregulation is considered to be the principle driver of KS disease. vGPCR is known to induce the expression of key factors, such as VEGF, bFGF, CXCL8, IL-6, and CCL2 (also termed MCP-1) that are found in KS lesions and believed to promote and be required for KS development and progression, via the activation of multiple signaling pathways to induce activation of transcription factors including AP-1, NF-κB, CREB, NFAT, and C/EBPβ (Cannon et al. 2003; Pati et al. 2001; Schwarz and Murphy 2001; Bais et al. 1998; Choi and Nicholas 2010; Cannon and Cesarman 2004). Thus, vGPCR produced in a small minority of spontaneously reactivating endothelial cells may induce levels of cellular cytokines sufficient to promote KS. It should be noted that in animal models, vGPCR-expressing cells are able to cooperate with cells expressing latency genes v-cyclin and/or vFLIP to increase the frequency of KS achieved with inoculated vGPCR$^+$ cells alone (Montaner et al. 2003). This supports the notion that both autocrine latent and paracrine lytic activities can function together in HHV-8-associated neoplasia.

In addition, a genetic study using a vGPCR-deleted HHV-8 genome and shRNA-mediated depletion demonstrated that vGPCR supports virus-productive replication via MAPK pathway activation in infected cells (Sandford et al. 2009). The pro-replication activity of vGPCR is mediated specifically via the Gα$_q$-ERK pathway, indicating direct autocrine effects of vGPCR signaling in the support of lytic replication. Consistent with this, MAPK signaling was identified as important for murine γ-herpesvirus 68 (MHV-68) vGPCR-mediated pro-replication activity (Lee et al. 2003b).

3.4 Viral Interferon Regulatory Factors (vIRFs)

HHV-8 vIRFs 1-4 are specified by the genomic region encompassing ORFs K9 to K11; an equivalent locus (encoding 8 vIRFs) has been identified only in the closely related rhesus rhadinovirus (Moore et al. 1996; Cunningham et al. 2003; Alexander et al. 2000; Searles et al. 1999). All four of the vIRFs are expressed during lytic replication, but vIRF-1, vIRF-2, and vIRF-3 are also expressed, to varying degrees, during PEL latency (Cunningham et al. 2003; Rivas et al. 2001; Pozharskaya et al. 2004; Burysek and Pitha 2001). Additionally, vIRF-1 transcripts have been detected in KS cells using reverse transcription polymerase chain reaction techniques (Dittmer 2003). As outlined above (Sect. 2.6), vIRF-3 is required for PEL cell viability (Wies et al. 2008). In addition, vIRF-1 depletion influences PEL cell growth in culture (Xiang et al. 2018; Hwang and Choi 2016).

The vIRFs appear to function primarily to evade host cell defenses against de novo infection and virus-productive replication, which trigger cellular IRF and IFN signaling cascades leading to cell cycle arrest and pro-apoptotic signaling (Lee et al. 2009b; Offermann 2007). The vIRFs counter these cellular signals in several ways. IRF5 and IRF7 functional dimerization and promoter association are antagonized by direct binding of vIRF-3 to these cellular factors, and vIRF-3 also inhibits IRF3 activity (Joo et al. 2007; Wies et al. 2009). vIRF-1 mediates transcriptional repression of IRF-targeted genes by blocking IRF-directed promoter recruitment of p300/CBP via competitive binding to the transcriptional coactivators (Burysek et al. 1999a; Li et al. 2000; Lin et al. 2001; Seo et al. 2000; Zimring et al. 1998). Transcriptional repression of IRF1-, IRF3-, and ISGF3-targeted genes is mediated by vIRF-2, in part by activation of caspase 3-mediated destabilization of IRF3 (Areste et al. 2009; Burysek et al. 1999b; Fuld et al. 2006). vIRF-2 has recently been shown also to inhibit FOXO3A-mediated apoptosis via activation of the PI3K/AKT pathway (Kim et al. 2016). In addition, vIRF-2 and vIRF-3 directly target and/or inhibit dsRNA-activated PKR kinase activity; this suppresses PKR promotion of protein translation and inhibits IFN signaling (Esteban et al. 2003; Burysek and Pitha 2001). Interestingly, vIRF-3 associates with and stabilizes the pro-angiogenic transcription factor HIF-1α (Shin et al. 2008), and this could serve not only to promote endothelial cell survival but may also contribute to KS and PEL pathogenesis via induction of cytokines such as VEGF. vIRF-4 binds to CSL, the target of Notch, but the significance of this interaction is unclear (Heinzelmann et al. 2010).

While each of the cellular IRFs contains a well-conserved DNA-binding domain (DBD) of about 120 amino acids containing a unique tryptophan pentad repeat, HHV-8 vIRFs share limited homology with the DBD of cellular IRFs and lack some of the conserved and functionally important tryptophan residues (Takaoka et al. 2008; Jacobs and Damania 2011). Thus, it was predicted that this incomplete conservation may render vIRFs incapable of binding to cellular or viral DNA. However, some studies have demonstrated that vIRF-1 and vIRF-2 have the ability to bind to DNA (Park et al. 2007). In vitro pull-down assay of the 24-mer random oligonucleotide pool with purified recombinant vIRF-1 protein revealed that vIRF-1 binds directly to DNA elements with the consensus sequence, 5'-GCGTCNNGACGC-3', a similar sequence of which is found in the K3-viral dihydrofolate reductase-vIL-6 promoter region in the HHV-8 genome. Furthermore, the crystal structure of the vIRF-1 DBD in complex with DNA was solved (Hew et al. 2013). The structural study also found, however, that the full-length vIRF-1 does not bind to DNA, indicating that vIRF-1 may require activation to release the DBD from inhibitory intramolecular interaction. Therefore, further studies are required to assess how vIRF-1 is activated for DNA binding and what cellular or viral DNA elements are the in vivo targets of vIRF-1. Moreover, a recent genome-wide study using the ChIP-seq method demonstrated that vIRF-2 binds to the promoters of cellular genes including *PIK3C3*, *HMGCR* (encoding HMG-coenzyme A), and *HMGCL* (encoding HMG-CoA lyase) and regulates their expression (Hu et al. 2016). However, the functional significance of vIRF-2 as a transcription factor in infected cells is unclear. Structural and mutational studies of vIRF-1 and vIRF-2 demonstrated that the arginine residues in the putative DNA-binding helix of the DBD region are essential for binding to DNA and/or gene expression (Hew et al. 2013; Hu et al. 2016). Similarly, it has also been reported that vIRF-4 binds to its own promoter and those of vIRF-1 and PAN in the

HHV-8 genome via the arginine residues in its putative DNA-binding helix to enhance the expression of the lytic genes in collaboration with RTA (Xi et al. 2012).

vIRFs 1, 3, and 4 have been shown to inhibit p53 activity via (1) direct binding to the tumor suppressor (vIRF-1 and vIRF-3), (2) interaction with p53-phosphorylating and p53-activating ATM kinase (vIRF-1), or (3) stabilization of MDM2 (vIRF-4), which promotes ubiquitination and proteasomal degradation of p53 (Lee et al. 2009c; Nakamura et al. 2001; Seo et al. 2001; Shin et al. 2006). Inhibitory interactions of vIRF-4 with deubiquitinase USP7 have been reported to be involved in p53 destabilization (Lee et al. 2011). In addition, vIRF-1 has been reported to interact directly with USP7, and this can affect decreased levels of p53 in transfected cells and contributes to enhanced viability of PEL cells (Chavoshi et al. 2016; Xiang et al. 2018). Therefore, p53 represents a major vIRF target that is likely to be regulated by the vIRFs to promote cell survival and productive replication.

vIRF-1 associates with and inhibits the activities of other cellular proteins involved in innate cellular responses to infection and the promotion of apoptosis. These targets include retinoic acid/IFN-induced protein GRIM19 and TGFβ receptor-activated transcription factors SMAD3 (tumor suppressor) and SMAD4 (co-SMAD) (Angell et al. 2000; Ma et al. 2007; Seo et al. 2002, 2005). vIRF-1 also binds directly to members of the so-called BH3-only protein (BOP) family (Choi et al. 2012; Choi and Nicholas 2010). BOPs are Bcl-2-related proteins that function to promote apoptosis either via inhibitory interactions with pro-survival members of the Bcl-2 family or by direct activation of apoptotic executioners Bax and Bak (Kuwana et al. 2005; Willis and Adams 2005). A region of vIRF-1 comprising residues 170-184 (BOP-binding domain, BBD) interacts with BOP BH3 domains, required for pro-apoptotic activities of these proteins, thereby mediating direct inhibition, demonstrated for tBid, through interference with BH3-mediated interactions and, for Bim, inactivation also through nuclear sequestration (Choi et al. 2012; Choi and Nicholas 2010). vIRF-1 BBD, a predicted amphipathic α-helix, resembles the Bid BH3-inhibitory BH3-B domain and represents only the second example of a BH3-B-type BH3-inhibitory domain, and therefore, a novel viral mechanism of apoptotic inhibition. BBD-mediated interactions with BOPs are functionally important, as indicated by the following findings: (1) BBD-mutated vIRF-1 is less active than wild-type vIRF-1 in promoting productive replication and inhibiting apoptosis in lytically infected endothelial cells; (2) vIRF-1: BOP-disrupting BBD peptide causes significant inhibition of virus production in these cells; and (3) depletion of vIRF-1-targeted BOPs Bim or Bid leads to substantial increases in replicative titers (Choi et al. 2012; Choi and Nicholas 2010). Intriguingly, vIRF-1 localizes, in part, to mitochondria by targeting to the detergent-resistant microdomains on the outer mitochondrial membranes via its N-terminal proline-rich domain upon virus replication (Hwang and Choi 2016). This mitochondrial targeting of vIRF-1 is functionally important for downregulation of IFN expression and apoptosis induced by MAVS and for activation of NIX-mediated mitophagy (Vo et al. 2019), thereby promoting virus replication. The proposed functions of mitochondria-targeted vIRF-1 are depicted in Fig. 2. In addition, vIRF-1 has been found to inhibit IFN-β expression induced by the activation of the cGAS-STING DNA-sensing pathway (Ma et al. 2015). This study demonstrated that vIRF-1 interacts with STING and hinders TBK1-mediated phosphorylation of STING, possibly on the ER membranes. Taken together, vIRF-1 appears to downregulate antiviral innate immune and apoptotic responses in multiple cellular compartments including the nucleus, mitochondria, and ER.

In summary, a wealth of published data indicates that the HHV-8 vIRFs represent an effective panel of anti-apoptotic proteins that promote productive replication via interactions with an important group of cellular proteins involved in host cell defenses against infection (summarized in Fig. 3).

3.5 Viral BCL-2 (vBcl-2)

In common with other γ-herpesviruses, HHV-8 specifies a homolog of cellular BCL-2 proteins, vBcl-2, encoded by ORF16. While the overall amino acid sequence identity between vBcl-2 and cellular BCL-2 family members is low (15–20%), vBcl-2 contains significantly conserved sequences in both the BH1 and BH2 domains, and its structure within the BH3 and BH4 domains in solution is homologous to other BCL-2 family members (Cheng et al. 1997; Sarid et al. 1997; Huang et al. 2002). Interestingly, a profiling study of binding to the BH3 domains of BH3-only proteins suggested that vBcl-2 more closely resembles MCL-1 than BCL-2, interacting selectively with the BH3 domains of BIM, BID, NOXA, BIK, PUMA, and BMF (Flanagan and Letai 2008). vBcl-2 was also shown to interact with Aven, which interferes with the ability of pro-apoptotic protein APAF-1 to

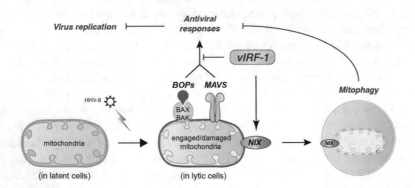

Fig. 2 Summary of the proposed roles of mitochondria-targeted vIRF-1 in the regulation of host antiviral responses. Upon virus lytic reactivation, mitochondria-mediated antiviral responses including BH3-only proteins (BOPs such as Bim and Bid)-activated apoptosis and MAVS-mediated innate immune signaling are inhibited by vIRF-1. In addition, vIRF-1 activates NIX-mediated mitophagy to remove the mitochondria involved in antiviral responses or damaged during virus replication

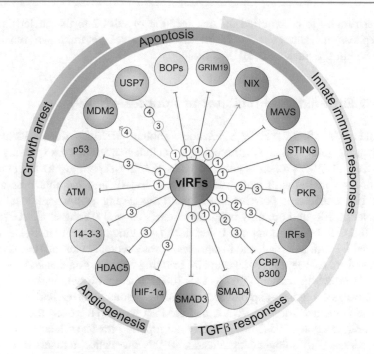

Fig. 3 Summary of vIRF interactions with cellular proteins. The particular vIRFs interacting with each target are indicated by the numbering within the open red (inhibitory), blue (stimulatory), and gray (unknown function) circles, and the effects of each interaction are indicated by red line (inhibitory) and blue arrow (stimulatory); the gray line indicates vIRF–USP7 interactions only, as the associated functions are uncertain. The activities of vIRF interactions are grouped into five general and overlapping biological categories, as indicated by the shells and color coding. The vIRF:protein interactions and their significance are discussed fully in the text

self-associate and activate caspase-9 (Chau et al. 2000). These structural features contribute to the anti-apoptotic activity of vBcl-2. Furthermore, like BCL-2, vBcl-2 inhibits Beclin-1-mediated autophagy by binding directly to Beclin-1 via the BH3 binding groove of vBcl-2 (Pattingre et al. 2005; Liang et al. 2006). Nonetheless, it is not clear that these anti-apoptotic and anti-autophagic activities of vBcl-2 are required for virus replication and production. Recent studies using recombinant HHV-8 viruses showed that rather than vBcl-2 biological activity through apoptosis and autophagy regulation, the pro-replication activity of vBcl-2 is dependent on a glutamate residue (E14), which is not involved in the anti-apoptotic or anti-autophagic activities of vBcl-2 (Liang et al. 2015; Gelgor et al. 2015). Furthermore, a recent report demonstrated that vBcl-2 promotes virus assembly and replication by interacting with the ORF55-encoded tegument protein via vBcl-2 residues 11-20, encompassing E14 (Liang et al. 2018). These results suggest that vBcl-2 can promote virus replication independently of its anti-apoptotic and anti-autophagic activities. Intriguingly, the N-terminal 17 amino acids of vBcl-2 have been found to be required for translocation of vBcl-2 to the nucleus and for virus replication (Gallo et al. 2017), implicating a novel, yet unknown, function of vBcl-2 within the nucleus in HHV-8 biology.

3.6 K7-Encoded Viral Inhibitor of Apoptosis (vIAP)

In addition to vBcl-2, HHV-8 encodes another BH-like domain-containing anti-apoptotic protein that is a homolog of cellular survivin ΔEx3 (Wang et al. 2002). This HHV-8 protein is referred to as K7 (corresponding to the encoding ORF) or viral inhibitor of apoptosis (vIAP). K7/vIAP is a membrane-associated protein that contains a putative mitochondrial-targeting signal and localizes to mitochondria, ER, and possibly other membranes as well (Wang et al. 2002; Feng et al. 2002). It has been reported that K7/vIAP binds to cellular Bcl-2 via its C-terminal BH2-like domain and to activated (proteolytically cleaved) caspase 3 via its baculovirus IAP repeat (BIR) domain, bridging the two proteins and inhibiting caspase 3 proteolytic activity (Wang et al. 2002). Interaction with and inhibition of terminal caspase 3 in the apoptotic cascade are analogous to the activities of cellular IAPs, which include survivin, XIAP, and cIAPs 1 and 2. However, the interaction between K7/vIAP and Bcl-2 is a property not reported for its cellular counterparts. The functional and biological significance of this interaction remains to be determined. Nonetheless, K7/vIAP is able to inhibit pro-apoptotic signaling in transfected cells treated with agents such as Fas antibody and TNF-α, indicating its potential to act as a promoter of lytic replication via its pro-survival activity during lytic cycle-induced stress (Wang et al. 2002). K7/vIAP also interacts with calcium-modulating cyclophilin ligand (CAML), which regulates intracellular calcium ion concentrations (Feng et al. 2002; Bram and Crabtree 1994). The functional significance of this interaction is evident from the ability of wild type but not CAML binding-refractory K7/vIAP to inhibit chemically induced mitochondrial depolarization (i.e., apoptotic triggering) in transfected cells (Feng et al. 2002). Thus, in addition to its inhibitory binding to caspase 3, K7/vIAP appears to mediate apoptotic inhibition via its CAML interaction. K7/vIAP also interacts with the cellular protein protein-linking integrin-associated protein and cytoskeleton-1, also called ubiquilin (PLIC1), which associates with ubiquitin-conjugated proteins to inhibit their proteasomal degradation (Feng et al. 2004; Kleijnen et al. 2000; Wang et al. 2012). K7/vIAP appears to antagonize PLIC1 activity, thereby destabilizing ubiquitinated proteins, as demonstrated for p53 and NF-κB-inhibitory IκB (Feng et al. 2004). Together, the inhibitory interactions between K7/vIAP and cellular proteins PLIC1, caspase 3/Bcl-2, and CAML may promote cell survival during lytic replication. However, whether the K7/vIAP activities contribute to virus production is not clear because deletion of the K7/vIAP gene in the HHV-8 genome (BAC16) had little or no effect on viral lytic replication and virus production in infected cells (Liang et al. 2015, 2013).

4 Terminal Membrane Proteins

The human γ-herpesviruses EBV and HHV-8 and simian γ-herpesviruses including herpesvirus saimiri (HVS), herpesvirus ateles (HVA), and rhesus monkey rhadinovirus (RRV) all contain genes, located adjacent to the terminal-repeat region of their genomes, encoding membrane-bound proteins interacting with a variety of cellular signaling molecules. Due to their localization in the viral genome, the proteins are often called "terminal membrane proteins" (Brinkmann and Schulz 2006).

4.1 K1/Variable ITAM-Containing Protein (VIP)

The K1 ORF of HHV-8 is located at the left end of the genome and is collinear with other γ-herpesvirus genes encoding signaling membrane proteins. These include saimiri transformation-associated protein (STP) of HVS, latency membrane protein-1 (LMP-1) of EBV, and the K1-homologous R1 receptor of RRV (Albrecht et al. 1992; Damania et al. 1999; Kaye et al. 1993; Lagunoff and Ganem 1997; Murthy et al. 1989). The K1 protein is a type I transmembrane signaling protein containing a functional immunoreceptor tyrosine-based activation motif (ITAM) in its cytoplasmic C-tail (Lagunoff et al. 2001; Lee et al. 1998b). Sequencing of K1 in different HHV-8 isolates identified an unusual degree of amino acid sequence variability in the extracellular regions of the encoded proteins (Zong et al. 1999; Nicholas et al. 1998), hence the naming of the K1 protein as variable ITAM-containing protein (VIP). While the functional significance of this variability has not been established, the K1 locus has served as a basis of epidemiological studies of HHV-8 strain distribution and infectivity (Hayward and Zong 2007; Mbulaiteye et al. 2006; Whitby et al. 2004). Based initially on the genomic position of K1 and subsequently on the constitutive signaling and transforming properties of K1/VIP, the protein was implicated as a potential contributor to HHV-8 pathogenesis. K1/VIP, like RRV R1, was able to substitute functionally for the positionally equivalent ORF1/STP of HVS in in vivo tumorigenesis assays and to promote cell growth and transformation in isolation (Lee et al. 1998a; Prakash et al. 2002). K1/VIP activation of the AKT pathway and consequent activation of mTOR (associated with cell growth) and the inactivation of pro-apoptotic GSK3, BAD, and forkhead transcription factors have been implicated in these activities (Wang et al. 2006; Tomlinson and Damania 2004). However, as K1 appears to be expressed primarily or exclusively during lytic replication (Jenner et al. 2001; Paulose-Murphy et al. 2001; Lagunoff and Ganem 1997; Nakamura et al. 2003), its potential role in KS, PEL, and MCD may be restricted to paracrine effects of K1/VIP-induced cellular cytokines (see below) rather than direct effects suggested by initial functional analyses. While immunodetection of K1/VIP in KS and MCD tissues has been reported, this has not been associated with latently infected cells (Wang et al. 2006; Lee et al. 2003c). It should be noted, however, that in situ detection of K1 transcripts in some KS cells lacking detectable lytic marker (major capsid protein) mRNA expression suggests the possibility that K1/VIP may be expressed in at least some latently infected KS cells (Wang et al. 2006).

K1/VIP recruits and activates Src family kinases, PI3K, and PLCγ to mediate signal transduction via several pathways by ligand-independent, constitutive signaling (Samaniego et al. 2001; Lagunoff et al. 2001; Lee et al. 1998b; Tomlinson and Damania 2004; Lee et al. 2005). It has been suggested that K1/VIP may contribute to HHV-8-associated disease, especially KS, by induction of cellular cytokines. K1/VIP-induced cytokines include pro-inflammatory IL-1β, IL-6, and GM-CSF and angiogenic factors VEGF, CXCL8, and bFGF (Samaniego et al. 2001; Wang et al. 2006; Prakash et al. 2002; Lee et al. 2005). Contributions of the HHV-8 receptor to pathogenesis via cellular cytokine induction could theoretically occur during lytic replication or during latency. In KS, PEL, and MCD, small proportions of cells support lytic reactivation, enabling lytically expressed proteins, like K1, to exert paracrine influence on surrounding latently infected and uninfected cells (Aoki et al. 2003; Aoki and Tosato 2003; Ensoli et al. 2001).

In addition to the paracrine function of K1/VIP, recent genetic studies using the recombinant HHV-8 genome mutants defective in the K1/VIP gene suggest that autocrine function may contribute to HHV-8 pathogenesis via promoting cell survival under metabolic stress conditions. For example, deletion of K1/VIP rendered infected cells sensitive to nutrient deprivation (Anders et al. 2016). This study further revealed that K1/VIP interacts with the γ1-subunit of AMPK in subcellular membrane compartments at perinuclear regions and that this interaction is important for the ability of K1/VIP to enhance cell survival. Another genetic study demonstrated that cells infected with K1/VIP-null HHV-8 displayed reduced AKT activation and produced lower viral titers than cells infected with wild-type virus (Zhang et al. 2016b), indicating a positive role of K1/VIP in virus replication. Taken together, K1/VIP may contribute to HHV-8 pathogenesis in both paracrine and autocrine fashions.

4.2 K15-Encoded Membrane Protein

The K15-encoded protein in its full-length form is a twelve-transmembrane domain-containing signaling receptor. Like K1, K15 may play a role in pathogenesis via cytokine dysregulation, and it could conceivably contribute to malignant disease through pro-survival signaling during latency. Transcripts from the K15 locus are expressed predominantly during the lytic cycle, but some K15 products have been detected in resting (latent) PEL cultures (Choi et al. 2000; Glenn et al. 1999; Sharp et al. 2002). The issue is complex because the K15 primary transcript contains eight exons and can be differentially spliced; the resulting mRNAs and encoded proteins may be expressed differently based on cell type and whether the virus is in the latent or productive phase. All forms of K15 contain C-terminal protein sequences with functional signaling motifs (see below), but the protein isoforms differ in their complement of transmembrane domains. K15 transcripts and a 23-kDa protein isoform have been detected during latency in PEL cells, but K15 mRNA levels are induced considerably upon lytic reactivation (Choi et al. 2000; Glenn et al. 1999; Sharp et al. 2002; Tsai et al. 2009).

Full-length K15 protein has been detected in HHV-8 bacmid-containing HEK293 cells, though only after lytic induction with butyrate treatment (Brinkmann et al. 2007). However, the full-length protein has not been observed in cells naturally infected with HHV-8. The ability of immediate early, lytic trigger protein RTA to activate transcription from the K15 promoter is consistent with predominant lytic expression of K15 (Wong and Damania 2006). Nonetheless, uncertainty remains regarding the expression characteristics of K15 transcripts and proteins and whether K15 receptor signaling could contribute to latency and HHV-8 neoplasia in an autocrine manner.

Signaling motifs in the cytoplasmic tail of all K15 isoforms and in both M and P allelic types include two SH2- and single SH3- and TRAF-binding sites (Brinkmann et al. 2003; Choi et al. 2000; Glenn et al. 1999; Poole et al. 1999). SH2 binding-mediated interactions with Src family kinases occur via the Y_{481}EEV motif, which is the primary site of K15 phosphorylation (Brinkmann et al. 2003; Choi et al. 2000). This, together with the SH3-binding sequence (PPLP), leads to inhibition of B-cell receptor (BCR) signaling. PPLP–motif interactions with intersectin 2 (endocytic adaptor protein) and with Src kinases (such as Lyn and Hck) are important for this activity (Lim et al. 2007; Pietrek et al. 2010). BCR inhibition by the K15 receptor parallels that of the collinearly encoded LMP-2 of EBV, and each is likely to promote latency by inhibiting lytic cycle reactivation promoted by BCR signaling. The Y_{481}EEV motif has been implicated in the activation of NF-κB and mitogen-activated protein kinases (MAPKs) ERK and JNK, which occurs after Y_{481} phosphorylation (Brinkmann and Schulz 2006; Choi et al. 2000; Pietrek et al. 2010). Interaction of the K15 receptor with TRAFs 1, 2, and 3 is likely to contribute to NF-κB and MAPK signaling (Brinkmann and Schulz 2006; Glenn et al. 1999). The second tyrosine-containing motif, Y_{432}ASI, is not detectably phosphorylated, and its significance is uncertain. However, its interaction with apoptotic regulatory protein HAX-1 and the ER and mitochondrial colocalization of HAX-1 with K15 suggests that the viral receptor may function to promote cell survival via this motif (Sharp et al. 2002).

Examination of the downstream effects of K15 signaling has provided insight into possible functions of the receptor in HHV-8 biology and its potential contributions to viral pathogenesis. In addition to suspected anti-apoptotic activity via interaction with HAX-1, K15 can induce the expression of several anti-apoptotic genes, including A20, Bcl-2A1, Birc2, and Birc3 (Brinkmann et al. 2007). Induction of these genes may help promote cell survival during the lytic cycle, further enhancing productive replication. If K15 is expressed during latency, its pro-survival signaling could contribute to prolonged latent cell viability and maintenance of latency pools in vivo and to viral pathogenesis. On the other hand, the observed induction of cellular cytokines in K15-transduced cells suggests a mechanism by which K15 can affect surrounding cells (latently infected and uninfected) by paracrine signaling from lytically infected cells. K15-expressing latently infected cells could exert similar effects on the microenvironment. Cytokines induced by K15 receptor signaling include IL-6, CCL2, CXCL3, and CXCL8; each of these possesses angiogenic activity and has been implicated in KS pathogenesis (Brinkmann et al. 2007; Ensoli and Sturzl 1998; Caselli et al. 2007). It is intriguing that K15 also induces expression of genes representing downstream targets of angiogenic VEGF signaling. This would clearly implicate K15 as an autocrine contributor to pathogenesis should the receptor be expressed during latency, but such activity of K15 could also contribute to productive replication. Angiogenic targets of K15 include *Dscr-1* and *Cox-2* (Brinkmann et al. 2007). It is notable that Cox-2 has been reported to be induced during de novo and subsequent latent infection of endothelial cells by HHV-8, and Cox-2 is important for the production of several inflammatory and angiogenic factors (Sharma-Walia et al. 2010). K15 promotes cell motility by inducing the expression of miR-21 and miR-31 via its SH2-binding motif (Tsai et al. 2009). Additionally, recent reports indicate that K15 interacts with the class II PI3K-C2α in the perinuclear region of HHV-8-infected endothelial cells and activates its downstream targets PLCγ1 and ERK1/2, thereby promoting lytic replication and virus production (Abere et al. 2018). As PI3K-C2α is localized mainly in the trans-Golgi network (TGN) (Domin et al. 2000), K15 is likely to also localize to the TGN. In fact, small isoforms (23 to 24 kDa) of K15 are detected predominantly in the TGN of latent PEL cells (Smith et al. 2017). Furthermore, levels of the small isoforms diminish during lytic replication in PEL cells, while the full length (45 kDa) K15 accumulates and is dispersed to peripheral areas during lytic replication, reflecting the alternative roles of K15 in the latent and lytic cycles. In short, pro-survival and paracrine-mediated pro-angiogenic roles of K15 in HHV-8 lytic replication and pathogenesis seem likely, and there is potential for autocrine activity via pro-survival signaling contributing to latency and neoplasia.

5 Summary

The discovery and study of HHV-8 have provided the opportunity to identify unique virus-specified activities encoded by proteins either not previously known among viruses or those not previously investigated or characterized in depth in other viral systems. HHV-8 has also provided a model for the identification and characterization of viral miRNAs, a new area of research that has yielded unique and important insights into viral manipulation of host cell processes as part of normal virus biology and potentially in viral pathogenesis. The properties of the characterized protein and miRNA players in these processes have been described in detail, and several key points emerge. First, the notion that only autocrine, latent viral activities are relevant to virus-associated neoplasia needs revision, certainly in the case of KS and possibly for PEL and MCD. Paracrine factors (viral and/or cellular) produced during lytic replication can contribute to pro-proliferative, pro-survival, and other functions of pathogenic relevance. The latent and lytic viral proteins implicated in HHV-8 pathogenesis and their potential autocrine and paracrine contributions to disease are summarized in Fig. 4. Secondly, "lytic" and "latent" classifications for viral products are not as distinct as once thought. For example, vIL-6, vIRF-1, vIRF-2, vIRF-3, K1, and K15 are clearly expressed

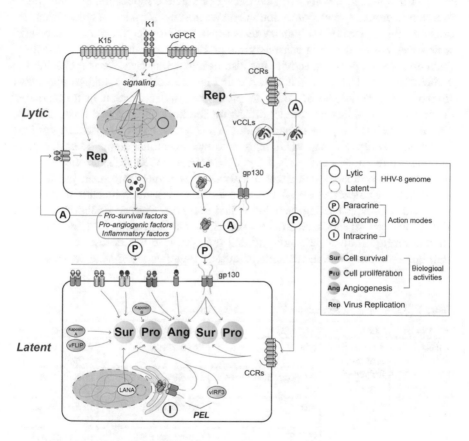

Fig. 4 Overview of potential contributions of HHV-8 proteins to virus-associated neoplasia. The general activities related to HHV-8 malignant pathogenesis are indicated: cell survival (Sur), cell proliferation (Pro), angiogenesis (Ang), and virus replication (Rep). Both latent and lytic proteins have the potential to contribute to disease in an autocrine (also intracrine) and/or paracrine manner. The viral chemokine receptor, vGPCR, and the terminal membrane proteins, K1 and K15, activate cell signaling and contribute to neoplasia via induction of mitogenic, pro-survival, and angiogenic secreted cellular factors. The viral cytokines (lytic) are secreted and can act by both autocrine and paracrine mechanisms to influence cell growth and survival. These activities can promote virus replication (by autocrine signaling) in addition to contributing to viral pathogenesis (via paracrine signaling). Latent expression of vIL-6 in PEL cells is likely to contribute to pathogenesis, mainly via intracrine signaling (refer to Fig. 1 and the text for details). vIRF-3 is also expressed during latency in PEL cells and, like vIL-6, promotes PEL cell viability. The latency proteins have the potential to contribute to HHV-8-associated malignancies by direct autocrine effects on cell proliferation and survival by mechanisms typical of oncogenes and tumor suppressors (see text). Kaposin A and vFLIP have the potential to function as promoters of cell survival

maximally during productive replication, but there is evidence for their expression
during latency as well in some cell types. Furthermore, it is notable that latently
expressed vIL-6, vIRF-2, and vIRF-3 are of demonstrable importance for PEL cell
growth and viability. A third key point is that the virally encoded chemokines
vCCL-1 and vCCL-2, while secreted during productive replication and thought to
function to promote virus production via paracrine effects on the microenvironment
(most notably to evade host immune responses), can also act directly on the cells in
which they are produced to enhance virus production via anti-apoptotic signaling.
Such direct pro-replication activity has also been demonstrated for vGPCR. For the
v-chemokines and vGPCR, it is possible that induced cellular cytokines may serve
similar and/or additional activities to promote virus replication in an autocrine
fashion as well as having broader effects on the host microenvironment. Finally,
several of the HHV-8 proteins have multiple interactions with a broad range of host
factors in the originally and additionally reported subcellular compartments, a point
summarized in Table 2 and exemplified by LANA and vIRF-1. Thus, viral proteins
can have extraordinarily multifaceted activities via numerous protein interactions,
and detailed characterization of these interactions and their functional effects is
important for understanding their individual and combined contributions to virus
biology and pathogenesis. Such characterization can potentially provide the basis
for the development of new antiviral and therapeutic drugs designed to interfere
with specific virus–host interactions of critical importance for viral replication or

Table 2 Virus–host protein interactions and their activities

Protein	Class	Target(s)	Activity/function	References
LANA	Apoptosis Proliferation Epigenetic	p53	Pro-survival	Friborg et al. (1999)
		Angiogenin/annexin A2	Pro-survival	Paudel et al. (2012)
		pRb	Pro-mitogenic	Radkov et al. (2000)
		GSK3β	Pro-mitogenic	Fujimuro and Hayward (2003)
		c-Myc	Pro-mitogenic	Liu et al. (2007b)
		Histones H2A/B	Viral genome–chromosome tethering	Barbera et al. (2006)
		Bub1	Chromosome instability	Lang et al. (2018)
		DNMT3a	Transcriptional repression	Shamay et al. (2006)
		SUV39H1	Transcriptional repression	Sakakibara et al. (2004)
		mSin3/SAP30/CIR	Transcriptional repression	Krithivas et al. (2000)

(continued)

Table 2 (continued)

Protein	Class	Target(s)	Activity/function	References
Cytoplasmic LANA	Innate immune	cGAS	IFN signaling inhibition Promotion of lytic replication	Zhang et al. (2016a)
		Rad50-Mre11-CARD9	NF-κB inhibition Promotion of lytic replication	Mariggio et al. (2017)
vFLIP/K13	Apoptosis Autophagy Signaling	IKKα/β	NF-κB activation Survival Repression of lytic replication	Matta et al. (2007)
		Procaspase-8	Inhibition of caspase activation Pro-survival	Belanger et al. (2001)
		Peroxisomal MAVS	NF-κB activation vFLIP stabilization	Choi et al. (2018)
		ATG3	Anti-autophagy	Lee et al. (2009a)
Kaposin A	Apoptosis Signaling	Cytohesin-1	MAPK activation Promotion of cell adhesion	Kliche et al. (2001)
		Septin 4 variant	Pro-survival	Lin et al. (2007)
Kaposin B	Cytokine production	MK2 kinase	Stabilization of ARE-containing mRNAs (e.g., IL-6 and PROX1) Reprogramming of endothelial cells	McCormick and Ganem (2005, Corcoran et al. (2015)
vIL-6	Ligand Cytokine	gp130/gp80	Pro-mitogenic Pro-survival Pro-inflammatory Pro-angiogenic Promotion of virus replication	Chow et al. (2001), Boulanger et al. (2004), Chen et al. 2009b), Cousins et al. (2014)
		VKORC1v2	Pro-mitogenic Pro-survival	Chen et al. (2012), Chen and Nicholas (2015), Li et al. (2018)
vCCL-1	Ligand Chemokine	CCR8	*Agonist* Th2 polarization Pro-survival Pro-angiogenic Promotion of virus replication	Dairaghi et al. (1999), Choi and Nicholas (2008)

(continued)

Table 2 (continued)

Protein	Class	Target(s)	Activity/function	References
vCCL-2	Ligand Chemokine	CCR3, CCR8	*Agonist* Th2 polarization Pro-survival Pro-angiogenic Promotion of virus replication	Boshoff et al. (1997), Nakano et al. (2003), Choi and Nicholas (2008)
		CCR5, CCR2, CCR10, CXCR4, CX$_3$CR1, XCR1	*Antagonist* Pro-survival Promotion of virus replication	Boshoff et al. (1997), Luttichau et al. (2007), Chen et al. (1998), Kledal et al. (1997), Luttichau et al. (2001)
vCCL-3	Ligand Chemokine	CCR4, XCR1	*Agonist* Th2 polarization Pro-angiogenic	Stine et al. (2000), Luttichau et al. (2007)
vGPCR	Signaling receptor	Gα (i, q, 12/13)	Pro-survival/mitogenic Promotion of virus replication	Couty et al. (2001), Liu et al. (2004), Shepard et al. (2001), Sandford et al. (2009)
		SHP2	Pro-angiogenic	(Philpott et al. (2011)
vIRF-1	Innate response Apoptosis	IRF1, IRF3, p300	IFN signaling inhibition	Burysek et al. 1999a), Li et al. (2000), Lin et al. (2001)
		MAVS	IFN signaling inhibition Promotion of virus replication	Hwang and Choi (2016)
		STING	IFN signaling inhibition	Ma et al. (2015)
		p53, ATM	Pro-survival	Nakamura et al. (2001, Seo et al. (2001), Shin et al. (2006)
		GRIM19	Pro-survival	Seo et al. (2002)
		BOPs (BIM and BID)	Pro-survival Promotion of virus replication	Choi et al. (2012), Choi and Nicholas (2010)

(continued)

Table 2 (continued)

Protein	Class	Target(s)	Activity/function	References
		USP7	Pro-survival Promotion of virus replication	Xiang et al. (2018), Chavoshi et al. (2016)
		NIX	Mitophagy activation Promotion of virus replication	Vo et al. (2019)
		SMAD3, SMAD4	TGFβ signaling inhibition	Seo et al. (2005)
vIRF-2	Innate response	ISGF-3	IFN signaling inhibition	Mutocheluh et al. (2011)
		IRF1, IRF2, p65, p300	IFN signaling inhibition	Burysek et al. 1999b), Fuld et al. (2006)
		PKR	IFN signaling inhibition	Burysek and Pitha (2001)
vIRF-3	Innate response Apoptosis Angiogenesis	IRF3, IRF5, IRF7	IFN signaling inhibition	Joo et al. (2007), Lubyova and Pitha (2000), Wies et al. (2009)
		14-3-3σ, FOXO3a	Inhibition of cell cycle arrest	Munoz-Fontela et al. (2007)
		p53	Pro-survival	Rivas et al. (2001)
		CASP3, CASP7	Pro-survival	Wies et al. (2008)
		PKR	Pro-survival	Esteban et al. (2003)
		USP7	Pro-survival	Xiang et al. (2018)
		HIF-1α,	Pro-angiogenic	Shin et al. (2008)
		HDAC5	Pro-angiogenic	Lee et al. (2018)
vIRF-4	Innate response Apoptosis	MDM2	Pro-survival (via p53 destabilization)	Lee et al. (2009c)
		USP7	Pro-survival (via p53 destabilization)	Lee et al. (2011)
		CSL/CBF1	Unknown	Heinzelmann et al. (2010)
K7/vIAP	Apoptosis	Bcl-2/Caspase-3	Pro-survival	Wang et al. (2002)
		CAML	Pro-survival	Feng et al. (2002)

(continued)

Table 2 (continued)

Protein	Class	Target(s)	Activity/function	References
		PLIC1	Pro-survival	Feng et al. (2004)
K1/VIP	Apoptosis Angiogenesis Signaling	Src kinases, PI3K, SHP2, PLCγ2, RasGAP	Pro-survival/mitogenic Pro-angiogenic Inflammatory cytokine production	Lee et al. 1998b), Tomlinson and Damania (2004), Lee et al. (2005)
K15	Apoptosis Angiogenesis Signaling	Src kinases, TRAF2	Pro-survival/mitogenic	Brinkmann et al. (2003), Pietrek et al. (2010)
		Intersectin-2	Protein trafficking	Lim et al. (2007)
		HAX-1	Pro-survival	Sharp et al. (2002)
		PI3K-C2α	Promotion of lytic replication	Abere et al. (2018)
vBcl-2	Apoptosis Autophagy Lytic replication	Aven	Pro-survival	Chau et al. (2000)
		Beclin-1	Anti-autophagy	Pattingre et al. (2005), Liang et al. (2006)
		ORF55	Promotion of lytic replication	Liang et al. (2018)
Nuclear vBcl-2	Lytic replication	Unknown	Promotion of lytic replication	Gallo et al. (2017)

pathogenesis. This chapter has attempted to provide an overview of the various novel HHV-8–host interactions and related activities that contribute to these processes and that could perhaps be targeted in this way. The interactions described also illustrate the breadth and complexity of virus–host interactions and suggest that similar activities and mechanisms may be operative in other viral systems.

References

Abend JR, Uldrick T, Ziegelbauer JM (2010) Regulation of tumor necrosis factor-like weak inducer of apoptosis receptor protein (TWEAKR) expression by Kaposi's sarcoma-associated herpesvirus microRNA prevents TWEAK-induced apoptosis and inflammatory cytokine expression. J Virol 84(23):12139–12151. https://doi.org/10.1128/JVI.00884-10

Abend JR, Ramalingam D, Kieffer-Kwon P, Uldrick TS, Yarchoan R, Ziegelbauer JM (2012) Kaposi's sarcoma-associated herpesvirus microRNAs target IRAK1 and MYD88, two components of the toll-like receptor/interleukin-1R signaling cascade, to reduce inflammatory-cytokine expression. J Virol 86(21):11663–11674. https://doi.org/10.1128/JVI.01147-12

Abere B, Samarina N, Gramolelli S, Ruckert J, Gerold G, Pich A et al (2018) Kaposi's sarcoma-associated herpesvirus nonstructural membrane protein pK15 recruits the class II phosphatidylinositol 3-Kinase PI3K-C2alpha to activate productive viral replication. J Virol 92 (17). https://doi.org/10.1128/jvi.00544-18

Albrecht JC, Nicholas J, Cameron KR, Newman C, Fleckenstein B, Honess RW (1992) Herpesvirus saimiri has a gene specifying a homologue of the cellular membrane glycoprotein CD59. Virology 190(1):527–530

Alexander L, Denekamp L, Knapp A, Auerbach MR, Damania B, Desrosiers RC (2000) The primary sequence of rhesus monkey rhadinovirus isolate 26-95: sequence similarities to Kaposi's sarcoma-associated herpesvirus and rhesus monkey rhadinovirus isolate 17577. J Virol 74(7):3388–3398

Alonzi T, Gorgoni B, Screpanti I, Gulino A, Poli V (1997) Interleukin-6 and CAAT/enhancer binding protein beta-deficient mice act as tools to dissect the IL-6 signalling pathway and IL-6 regulation. Immunobiology 198(1–3):144–156

Ambrosini G, Adida C, Altieri DC (1997) A novel anti-apoptosis gene, survivin, expressed in cancer and lymphoma. Nat Med 3(8):917–921

An J, Sun Y, Sun R, Rettig MB (2003) Kaposi's sarcoma-associated herpesvirus encoded vFLIP induces cellular IL-6 expression: the role of the NF-kappaB and JNK/AP1 pathways. Oncogene 22(22):3371–3385. https://doi.org/10.1038/sj.onc.1206407

An FQ, Compitello N, Horwitz E, Sramkoski M, Knudsen ES, Renne R (2005) The latency-associated nuclear antigen of Kaposi's sarcoma-associated herpesvirus modulates cellular gene expression and protects lymphoid cells from p16 INK4A-induced cell cycle arrest. J Biol Chem 280(5):3862–3874. https://doi.org/10.1074/jbc.M407435200

Anders PM, Zhang Z, Bhende PM, Giffin L, Damania B (2016) The KSHV K1 protein modulates AMPK function to enhance cell survival. PLoS Pathog 12(11):e1005985. https://doi.org/10.1371/journal.ppat.1005985

Angell JE, Lindner DJ, Shapiro PS, Hofmann ER, Kalvakolanu DV (2000) Identification of GRIM-19, a novel cell death-regulatory gene induced by the interferon-beta and retinoic acid combination, using a genetic approach. J Biol Chem 275(43):33416–33426. https://doi.org/10.1074/jbc.M003929200

Aoki Y, Tosato G (1999) Role of vascular endothelial growth factor/vascular permeability factor in the pathogenesis of Kaposi's sarcoma-associated herpesvirus-infected primary effusion lymphomas. Blood 94(12):4247–4254

Aoki Y, Tosato G (2003) Targeted inhibition of angiogenic factors in AIDS-related disorders. Curr Drug Targets Infect Disord 3(2):115–128

Aoki Y, Jaffe ES, Chang Y, Jones K, Teruya-Feldstein J, Moore PS et al (1999) Angiogenesis and hematopoiesis induced by Kaposi's sarcoma-associated herpesvirus-encoded interleukin-6. Blood 93(12):4034–4043

Aoki Y, Yarchoan R, Braun J, Iwamoto A, Tosato G (2000) Viral and cellular cytokines in AIDS-related malignant lymphomatous effusions. Blood 96(4):1599–1601

Aoki Y, Narazaki M, Kishimoto T, Tosato G (2001) Receptor engagement by viral interleukin-6 encoded by Kaposi sarcoma-associated herpesvirus. Blood 98(10):3042–3049

Aoki Y, Feldman GM, Tosato G (2003) Inhibition of STAT3 signaling induces apoptosis and decreases survivin expression in primary effusion lymphoma. Blood 101(4):1535–1542. https://doi.org/10.1182/blood-2002-07-2130

Areste C, Mutocheluh M, Blackbourn DJ (2009) Identification of caspase-mediated decay of interferon regulatory factor-3, exploited by a Kaposi sarcoma-associated herpesvirus immunoregulatory protein. J Biol Chem 284(35):23272–23285. https://doi.org/10.1074/jbc.M109.033290

Arvanitakis L, Mesri EA, Nador RG, Said JW, Asch AS, Knowles DM et al (1996) Establishment and characterization of a primary effusion (body cavity-based) lymphoma cell line (BC-3) harboring kaposi's sarcoma-associated herpesvirus (KSHV/HHV-8) in the absence of Epstein-Barr virus. Blood 88(7):2648–2654

Bais C, Santomasso B, Coso O, Arvanitakis L, Raaka EG, Gutkind JS et al (1998) G-protein-coupled receptor of Kaposi's sarcoma-associated herpesvirus is a viral oncogene and angiogenesis activator. Nature 391(6662):86–89. https://doi.org/10.1038/34193

Ballon G, Akar G, Cesarman E (2015) Systemic expression of Kaposi sarcoma herpesvirus (KSHV) Vflip in endothelial cells leads to a profound proinflammatory phenotype and myeloid lineage remodeling in vivo. PLoS Pathog 11(1):e1004581. https://doi.org/10.1371/journal.ppat. 1004581

Barbera AJ, Chodaparambil JV, Kelley-Clarke B, Joukov V, Walter JC, Luger K et al (2006) The nucleosomal surface as a docking station for Kaposi's sarcoma herpesvirus LANA. Science 311(5762):856–861. https://doi.org/10.1126/science.1120541

Baresova P, Pitha PM, Lubyova B (2012) Kaposi sarcoma-associated herpesvirus vIRF-3 protein binds to F-box of Skp2 protein and acts as a regulator of c-Myc protein function and stability. J Biol Chem 287(20):16199–16208. https://doi.org/10.1074/jbc.M111.335216

Baresova P, Musilova J, Pitha PM, Lubyova B (2014) p53 tumor suppressor protein stability and transcriptional activity are targeted by Kaposi's sarcoma-associated herpesvirus-encoded viral interferon regulatory factor 3. Mol Cell Biol 34(3):386–399. https://doi.org/10.1128/MCB. 01011-13

Bartel DP (2004) MicroRNAs: genomics, biogenesis, mechanism, and function. Cell 116(2): 281–297

Bartoli M, Gu X, Tsai NT, Venema RC, Brooks SE, Marrero MB et al (2000) Vascular endothelial growth factor activates STAT proteins in aortic endothelial cells. J Biol Chem 275(43):33189–33192. https://doi.org/10.1074/jbc.C000318200

Belanger C, Gravel A, Tomoiu A, Janelle ME, Gosselin J, Tremblay MJ et al (2001) Human herpesvirus 8 viral FLICE-inhibitory protein inhibits Fas-mediated apoptosis through binding and prevention of procaspase-8 maturation. J Hum Virol. 4(2):62–73

Bellare P, Ganem D (2009) Regulation of KSHV lytic switch protein expression by a virus-encoded microRNA: an evolutionary adaptation that fine-tunes lytic reactivation. Cell Host Microbe 6(6):570–575. https://doi.org/10.1016/j.chom.2009.11.008

Bellare P, Dufresne A, Ganem D (2015) Inefficient codon usage impairs mRNA accumulation: the case of the v-FLIP gene of Kaposi's sarcoma-associated herpesvirus. J Virol 89(14):7097–7107. https://doi.org/10.1128/JVI.03390-14

Bertin J, Armstrong RC, Ottilie S, Martin DA, Wang Y, Banks S et al (1997) Death effector domain-containing herpesvirus and poxvirus proteins inhibit both Fas- and TNFR1-induced apoptosis. Proc Natl Acad Sci U S A. 94(4):1172–1176

Berx G, van Roy F (2009) Involvement of members of the cadherin superfamily in cancer. Cold Spring Harb Perspect Biol 1(6):a003129. https://doi.org/10.1101/cshperspect.a003129

Boshoff C, Endo Y, Collins PD, Takeuchi Y, Reeves JD, Schweickart VL et al (1997) Angiogenic and HIV-inhibitory functions of KSHV-encoded chemokines. Science 278(5336):290–294

Boss IW, Nadeau PE, Abbott JR, Yang Y, Mergia A, Renne R (2011) A Kaposi's sarcoma-associated herpesvirus-encoded ortholog of microRNA miR-155 induces human splenic B-cell expansion in NOD/LtSz-scid IL2Rgammanull mice. J Virol 85(19):9877–9886. https://doi.org/10.1128/JVI.05558-11

Botto S, Totonchy JE, Gustin JK, Moses AV (2015) Kaposi sarcoma herpesvirus induces HO-1 during De Novo infection of endothelial cells via viral miRNA-dependent and -independent mechanisms. MBio. 6(3):e00668. https://doi.org/10.1128/mBio.00668-15

Boulanger MJ, Chow DC, Brevnova EE, Garcia KC (2003) Hexameric structure and assembly of the interleukin-6/IL-6 alpha-receptor/gp130 complex. Science 300(5628):2101–2104. https://doi.org/10.1126/science.1083901

Boulanger MJ, Chow DC, Brevnova E, Martick M, Sandford G, Nicholas J et al (2004) Molecular mechanisms for viral mimicry of a human cytokine: activation of gp130 by HHV-8 interleukin-6. J Mol Biol 335(2):641–654

Bram RJ, Crabtree GR (1994) Calcium signalling in T cells stimulated by a cyclophilin B-binding protein. Nature 371(6495):355–358. https://doi.org/10.1038/371355a0

Brinkmann MM, Schulz TF (2006) Regulation of intracellular signalling by the terminal membrane proteins of members of the Gammaherpesvirinae. J Gen Virol 87(Pt 5):1047–1074. https://doi.org/10.1099/vir.0.81598-0

Brinkmann MM, Glenn M, Rainbow L, Kieser A, Henke-Gendo C, Schulz TF (2003) Activation of mitogen-activated protein kinase and NF-kappaB pathways by a Kaposi's sarcoma-associated herpesvirus K15 membrane protein. J Virol 77(17):9346–9358

Brinkmann MM, Pietrek M, Dittrich-Breiholz O, Kracht M, Schulz TF (2007) Modulation of host gene expression by the K15 protein of Kaposi's sarcoma-associated herpesvirus. J Virol 81 (1):42–58. https://doi.org/10.1128/JVI.00648-06

Brown HJ, Song MJ, Deng H, Wu TT, Cheng G, Sun R (2003) NF-kappaB inhibits gammaherpesvirus lytic replication. J Virol 77(15):8532–8540

Bubman D, Guasparri I, Cesarman E (2007) Deregulation of c-Myc in primary effusion lymphoma by Kaposi's sarcoma herpesvirus latency-associated nuclear antigen. Oncogene 26(34):4979–4986. https://doi.org/10.1038/sj.onc.1210299

Burger R, Wendler J, Antoni K, Helm G, Kalden JR, Gramatzki M (1994) Interleukin-6 production in B-cell neoplasias and Castleman's disease: evidence for an additional paracrine loop. Ann Hematol 69(1):25–31

Burger R, Neipel F, Fleckenstein B, Savino R, Ciliberto G, Kalden JR et al (1998) Human herpesvirus type 8 interleukin-6 homologue is functionally active on human myeloma cells. Blood 91(6):1858–1863

Burysek L, Pitha PM (2001) Latently expressed human herpesvirus 8-encoded interferon regulatory factor 2 inhibits double-stranded RNA-activated protein kinase. J Virol 75(5):2345–2352. https://doi.org/10.1128/JVI.75.5.2345-2352.2001

Burysek L, Yeow WS, Lubyova B, Kellum M, Schafer SL, Huang YQ et al (1999a) Functional analysis of human herpesvirus 8-encoded viral interferon regulatory factor 1 and its association with cellular interferon regulatory factors and p300. J Virol 73(9):7334–7342

Burysek L, Yeow WS, Pitha PM (1999b) Unique properties of a second human herpesvirus 8-encoded interferon regulatory factor (vIRF-2). J Hum Virol. 2(1):19–32

Cai X, Lu S, Zhang Z, Gonzalez CM, Damania B, Cullen BR (2005) Kaposi's sarcoma-associated herpesvirus expresses an array of viral microRNAs in latently infected cells. Proc Natl Acad Sci U S A. 102(15):5570–5575. https://doi.org/10.1073/pnas.0408192102

Cannon M (2007) The KSHV and other human herpesviral G protein-coupled receptors. Curr Top Microbiol Immunol 312:137–156

Cannon ML, Cesarman E (2004) The KSHV G protein-coupled receptor signals via multiple pathways to induce transcription factor activation in primary effusion lymphoma cells. Oncogene 23(2):514–523. https://doi.org/10.1038/sj.onc.1207021

Cannon M, Philpott NJ, Cesarman E (2003) The Kaposi's sarcoma-associated herpesvirus G protein-coupled receptor has broad signaling effects in primary effusion lymphoma cells. J Virol 77(1):57–67

Carbone A, Cilia AM, Gloghini A, Capello D, Perin T, Bontempo D et al (2000) Primary effusion lymphoma cell lines harbouring human herpesvirus type-8. Leuk Lymphoma 36(5–6):447–456. https://doi.org/10.3109/10428190009148391

Caselli E, Fiorentini S, Amici C, Di Luca D, Caruso A, Santoro MG (2007) Human herpesvirus 8 acute infection of endothelial cells induces monocyte chemoattractant protein 1-dependent capillary-like structure formation: role of the IKK/NF-kappaB pathway. Blood 109(7):2718–2726. https://doi.org/10.1182/blood-2006-03-012500

Chandriani S, Ganem D (2010) Array-based transcript profiling and limiting-dilution reverse transcription-PCR analysis identify additional latent genes in Kaposi's sarcoma-associated herpesvirus. J Virol 84(11):5565–5573. https://doi.org/10.1128/JVI.02723-09

Chang Y, Moore PS (1996) Kaposi's Sarcoma (KS)-associated herpesvirus and its role in KS. Infect Agents Dis 5(4):215–222

Chang Y, Moore PS, Talbot SJ, Boshoff CH, Zarkowska T, Godden K et al (1996) Cyclin encoded by KS herpesvirus. Nature 382(6590):410. https://doi.org/10.1038/382410a0

Chau BN, Cheng EH, Kerr DA, Hardwick JM (2000) Aven, a novel inhibitor of caspase activation, binds Bcl-xL and Apaf-1. Mol Cell 6(1):31–40

Chaudhary PM, Jasmin A, Eby MT, Hood L (1999) Modulation of the NF-kappa B pathway by virally encoded death effector domains-containing proteins. Oncogene 18(42):5738–5746. https://doi.org/10.1038/sj.onc.1202976

Chavoshi S, Egorova O, Lacdao IK, Farhadi S, Sheng Y, Saridakis V (2016) Identification of Kaposi Sarcoma Herpesvirus (KSHV) vIRF1 Protein as a Novel Interaction Partner of Human Deubiquitinase USP7. J Biol Chem 291(12):6281–6291. https://doi.org/10.1074/jbc.M115. 710632

Chen D, Nicholas J (2006) Structural requirements for gp80 independence of human herpesvirus 8 interleukin-6 (vIL-6) and evidence for gp80 stabilization of gp130 signaling complexes induced by vIL-6. J Virol 80(19):9811–9821. https://doi.org/10.1128/JVI.00872-06

Chen D, Nicholas J (2015) Promotion of endoplasmic reticulum-associated degradation of procathepsin D by human herpesvirus 8-encoded viral interleukin-6. J Virol 89(15):7979–7990. https://doi.org/10.1128/JVI.00375-15

Chen S, Bacon KB, Li L, Garcia GE, Xia Y, Lo D et al (1998) In vivo inhibition of CC and CX3C chemokine-induced leukocyte infiltration and attenuation of glomerulonephritis in Wistar-Kyoto (WKY) rats by vMIP-II. J Exp Med 188(1):193–198

Chen X, Cheng L, Jia X, Zeng Y, Yao S, Lv Z et al (2009a) Human immunodeficiency virus type 1 Tat accelerates Kaposi sarcoma-associated herpesvirus Kaposin A-mediated tumorigenesis of transformed fibroblasts in vitro as well as in nude and immunocompetent mice. Neoplasia. 11 (12):1272–1284

Chen D, Sandford G, Nicholas J (2009b) Intracellular signaling mechanisms and activities of human herpesvirus 8 interleukin-6. J Virol 83(2):722–733. https://doi.org/10.1128/JVI.01517-08

Chen W, Hilton IB, Staudt MR, Burd CE, Dittmer DP (2010) Distinct p53, p53:LANA, and LANA complexes in Kaposi's Sarcoma–associated Herpesvirus Lymphomas. J Virol 84 (8):3898–3908. https://doi.org/10.1128/JVI.01321-09

Chen D, Cousins E, Sandford G, Nicholas J (2012) Human herpesvirus 8 viral interleukin-6 interacts with splice variant 2 of vitamin K epoxide reductase complex subunit 1. J Virol 86 (3):1577–1588. https://doi.org/10.1128/JVI.05782-11

Chen M, Sun F, Han L, Qu Z (2016) Kaposi's sarcoma herpesvirus (KSHV) microRNA K12-1 functions as an oncogene by activating NF-kappaB/IL-6/STAT3 signaling. Oncotarget. 7 (22):33363–33373. https://doi.org/10.18632/oncotarget.9221

Chen D, Xiang Q, Nicholas J (2017) Human herpesvirus 8 interleukin-6 interacts with calnexin cycle components and promotes protein folding. J Virol 91(22). https://doi.org/10.1128/jvi. 00965-17

Cheng EH, Nicholas J, Bellows DS, Hayward GS, Guo HG, Reitz MS et al (1997) A Bcl-2 homolog encoded by Kaposi sarcoma-associated virus, human herpesvirus 8, inhibits apoptosis but does not heterodimerize with Bax or Bak. Proc Natl Acad Sci U S A. 94(2):690–694

Choi YB, Nicholas J (2008) Autocrine and paracrine promotion of cell survival and virus replication by human herpesvirus 8 chemokines. J Virol 82(13):6501–6513. https://doi.org/10. 1128/JVI.02396-07

Choi YB, Nicholas J (2010a) Induction of angiogenic chemokine CCL2 by human herpesvirus 8 chemokine receptor. Virology 397(2):369–378. https://doi.org/10.1016/j.virol.2009.11.024

Choi YB, Nicholas J (2010b) Bim nuclear translocation and inactivation by viral interferon regulatory factor. PLoS Pathog 6(8):e1001031. https://doi.org/10.1371/journal.ppat.1001031

Choi JK, Lee BS, Shim SN, Li M, Jung JU (2000) Identification of the novel K15 gene at the rightmost end of the Kaposi's sarcoma-associated herpesvirus genome. J Virol 74(1):436–446

Choi YB, Sandford G, Nicholas J (2012) Human herpesvirus 8 interferon regulatory factor-mediated BH3-only protein inhibition via Bid BH3-B mimicry. PLoS Pathog 8(6): e1002748. https://doi.org/10.1371/journal.ppat.1002748

Choi YB, Choi Y, Harhaj EW (2018) Peroxisomes support human herpesvirus 8 latency by stabilizing the viral oncogenic protein vFLIP via the MAVS-TRAF complex. PLoS Pathog 14 (5):e1007058. https://doi.org/10.1371/journal.ppat.1007058

Chow D, He X, Snow AL, Rose-John S, Garcia KC (2001) Structure of an extracellular gp130 cytokine receptor signaling complex. Science 291(5511):2150–2155. https://doi.org/10.1126/science.1058308

Chugh P, Matta H, Schamus S, Zachariah S, Kumar A, Richardson JA et al (2005) Constitutive NF-kappaB activation, normal Fas-induced apoptosis, and increased incidence of lymphoma in human herpes virus 8 K13 transgenic mice. Proc Natl Acad Sci U S A. 102(36):12885–12890. https://doi.org/10.1073/pnas.0408577102

Cirone M, Lucania G, Aleandri S, Borgia G, Trivedi P, Cuomo L et al (2008) Suppression of dendritic cell differentiation through cytokines released by Primary Effusion Lymphoma cells. Immunol Lett 120(1–2):37–41. https://doi.org/10.1016/j.imlet.2008.06.011

Corcoran JA, Johnston BP, McCormick C (2015) Viral activation of MK2-hsp27-p115RhoGEF-RhoA signaling axis causes cytoskeletal rearrangements, p-body disruption and ARE-mRNA stabilization. PLoS Pathog 11(1):e1004597. https://doi.org/10.1371/journal.ppat.1004597

Cousins E, Nicholas J (2013) Role of human herpesvirus 8 interleukin-6-activated gp130 signal transducer in primary effusion lymphoma cell growth and viability. J Virol 87(19):10816–10827. https://doi.org/10.1128/JVI.02047-13

Cousins E, Gao Y, Sandford G, Nicholas J (2014) Human herpesvirus 8 viral interleukin-6 signaling through gp130 promotes virus replication in primary effusion lymphoma and endothelial cells. J Virol 88(20):12167–12172. https://doi.org/10.1128/JVI.01751-14

Couty JP, Geras-Raaka E, Weksler BB, Gershengorn MC (2001) Kaposi's sarcoma-associated herpesvirus G protein-coupled receptor signals through multiple pathways in endothelial cells. J Biol Chem 276(36):33805–33811. https://doi.org/10.1074/jbc.M104631200

Crump MP, Elisseeva E, Gong J, Clark-Lewis I, Sykes BD (2001) Structure/function of human herpesvirus-8 MIP-II (1-71) and the antagonist N-terminal segment (1-10). FEBS Lett 489(2–3):171–175

Cunningham C, Barnard S, Blackbourn DJ, Davison AJ (2003) Transcription mapping of human herpesvirus 8 genes encoding viral interferon regulatory factors. J Gen Virol 84(Pt 6):1471–1483. https://doi.org/10.1099/vir.0.19015-0

Dairaghi DJ, Fan RA, McMaster BE, Hanley MR, Schall TJ (1999) HHV8-encoded vMIP-I selectively engages chemokine receptor CCR8. Agonist and antagonist profiles of viral chemokines. J Biol Chem 274(31):21569–21574

Damania B, Li M, Choi JK, Alexander L, Jung JU, Desrosiers RC (1999) Identification of the R1 oncogene and its protein product from the rhadinovirus of rhesus monkeys. J Virol 73 (6):5123–5131

Dittmer DP (2003) Transcription profile of Kaposi's sarcoma-associated herpesvirus in primary Kaposi's sarcoma lesions as determined by real-time PCR arrays. Cancer Res 63(9):2010–2015

Djerbi M, Screpanti V, Catrina AI, Bogen B, Biberfeld P, Grandien A (1999) The inhibitor of death receptor signaling, FLICE-inhibitory protein defines a new class of tumor progression factors. J Exp Med 190(7):1025–1032

Dolken L, Malterer G, Erhard F, Kothe S, Friedel CC, Suffert G et al (2010) Systematic analysis of viral and cellular microRNA targets in cells latently infected with human gamma-herpesviruses by RISC immunoprecipitation assay. Cell Host Microbe 7(4):324–334. https://doi.org/10.1016/j.chom.2010.03.008

Domin J, Gaidarov I, Smith ME, Keen JH, Waterfield MD (2000) The class II phosphoinositide 3-kinase PI3K-C2alpha is concentrated in the trans-Golgi network and present in clathrin-coated vesicles. J Biol Chem 275(16):11943–11950

Drexler HG, Meyer C, Gaidano G, Carbone A (1999) Constitutive cytokine production by primary effusion (body cavity-based) lymphoma-derived cell lines. Leukemia 13(4):634–640

Ellis M, Chew YP, Fallis L, Freddersdorf S, Boshoff C, Weiss RA et al (1999) Degradation of p27 (Kip) cdk inhibitor triggered by Kaposi's sarcoma virus cyclin-cdk6 complex. EMBO J 18 (3):644–653. https://doi.org/10.1093/emboj/18.3.644

El-Shewy HM, Luttrell LM (2009) Insulin-like growth factor-2/mannose-6 phosphate receptors. Vitam Horm 80:667–697. https://doi.org/10.1016/S0083-6729(08)00624-9

Ensoli B, Sturzl M (1998) Kaposi's sarcoma: a result of the interplay among inflammatory cytokines, angiogenic factors and viral agents. Cytokine Growth Factor Rev 9(1):63–83

Ensoli B, Sgadari C, Barillari G, Sirianni MC, Sturzl M, Monini P (2001) Biology of Kaposi's sarcoma. Eur J Cancer 37(10):1251–1269

Esteban M, Garcia MA, Domingo-Gil E, Arroyo J, Nombela C, Rivas C (2003) The latency protein LANA2 from Kaposi's sarcoma-associated herpesvirus inhibits apoptosis induced by dsRNA-activated protein kinase but not RNase L activation. J Gen Virol 84(Pt 6):1463–1470. https://doi.org/10.1099/vir.0.19014-0

Feng P, Park J, Lee BS, Lee SH, Bram RJ, Jung JU (2002) Kaposi's sarcoma-associated herpesvirus mitochondrial K7 protein targets a cellular calcium-modulating cyclophilin ligand to modulate intracellular calcium concentration and inhibit apoptosis. J Virol 76(22):11491–11504

Feng P, Scott CW, Cho NH, Nakamura H, Chung YH, Monteiro MJ et al (2004) Kaposi's sarcoma-associated herpesvirus K7 protein targets a ubiquitin-like/ubiquitin-associated domain-containing protein to promote protein degradation. Mol Cell Biol 24(9):3938–3948

Field N, Low W, Daniels M, Howell S, Daviet L, Boshoff C et al (2003) KSHV vFLIP binds to IKK-gamma to activate IKK. J Cell Sci 116(Pt 18):3721–3728. https://doi.org/10.1242/jcs.00691

Fielding CA, McLoughlin RM, Colmont CS, Kovaleva M, Harris DA, Rose-John S et al (2005) Viral IL-6 blocks neutrophil infiltration during acute inflammation. J Immunol. 175(6):4024–4029

Flanagan AM, Letai A (2008) BH3 domains define selective inhibitory interactions with BHRF-1 and KSHV BCL-2. Cell Death Differ 15(3):580–588. https://doi.org/10.1038/sj.cdd.4402292

Friborg J Jr, Kong W, Hottiger MO, Nabel GJ (1999) p53 inhibition by the LANA protein of KSHV protects against cell death. Nature 402(6764):889–894. https://doi.org/10.1038/47266

Fujimuro M, Hayward SD (2003) The latency-associated nuclear antigen of Kaposi's sarcoma-associated herpesvirus manipulates the activity of glycogen synthase kinase-3beta. J Virol 77(14):8019–8030

Fujimuro M, Wu FY, ApRhys C, Kajumbula H, Young DB, Hayward GS et al (2003) A novel viral mechanism for dysregulation of beta-catenin in Kaposi's sarcoma-associated herpesvirus latency. Nat Med 9(3):300–306. https://doi.org/10.1038/nm829

Fuld S, Cunningham C, Klucher K, Davison AJ, Blackbourn DJ (2006) Inhibition of interferon signaling by the Kaposi's sarcoma-associated herpesvirus full-length viral interferon regulatory factor 2 protein. J Virol 80(6):3092–3097. https://doi.org/10.1128/JVI.80.6.3092-3097.2006

Gaidano G, Pastore C, Gloghini A, Volpe G, Capello D, Polito P et al (1997) Human herpesvirus type-8 (HHV-8) in haematopoietic neoplasia. Leuk Lymphoma 24(3–4):257–266. https://doi.org/10.3109/10428199709039013

Gallo A, Lampe M, Gunther T, Brune W (2017) The viral Bcl-2 homologs of Kaposi's sarcoma-associated herpesvirus and rhesus rhadinovirus share an essential role for viral replication. J Virol 91(6). https://doi.org/10.1128/jvi.01875-16

Garrigues HJ, Howard K, Barcy S, Ikoma M, Moses AV, Deutsch GH et al (2017) Full-length isoforms of Kaposi's sarcoma-associated herpesvirus latency-associated nuclear antigen accumulate in the cytoplasm of cells undergoing the lytic cycle of replication. J Virol 91(24). https://doi.org/10.1128/jvi.01532-17

Gay LA, Sethuraman S, Thomas M, Turner PC, Renne R (2018) Modified cross-linking, ligation, and sequencing of hybrids (qCLASH) identifies Kaposi's sarcoma-associated herpesvirus MicroRNA targets in endothelial cells. J Virol 92(8). https://doi.org/10.1128/jvi.02138-17

Gelgor A, Kalt I, Bergson S, Brulois KF, Jung JU, Sarid R (2015) Viral Bcl-2 encoded by the Kaposi's sarcoma-associated herpesvirus is vital for virus reactivation. J Virol 89(10):5298–5307. https://doi.org/10.1128/JVI.00098-15

Glas R, Franksson L, Une C, Eloranta ML, Ohlen C, Orn A et al (2000) Recruitment and activation of natural killer (NK) cells in vivo determined by the target cell phenotype. An adaptive component of NK cell-mediated responses. J Exp Med 191(1):129–138

Glenn M, Rainbow L, Aurade F, Davison A, Schulz TF (1999) Identification of a spliced gene from Kaposi's sarcoma-associated herpesvirus encoding a protein with similarities to latent membrane proteins 1 and 2A of Epstein-Barr virus. J Virol 73(8):6953–6963

Godfrey A, Anderson J, Papanastasiou A, Takeuchi Y, Boshoff C (2005) Inhibiting primary effusion lymphoma by lentiviral vectors encoding short hairpin RNA. Blood 105(6):2510–2518. https://doi.org/10.1182/blood-2004-08-3052

Goncalves PH, Uldrick TS, Yarchoan R (2017a) HIV-associated Kaposi sarcoma and related diseases. AIDS. 31(14):1903–1916. https://doi.org/10.1097/QAD.0000000000001567

Goncalves PH, Ziegelbauer J, Uldrick TS, Yarchoan R (2017b) Kaposi sarcoma herpesvirus-associated cancers and related diseases. Curr Opin HIV AIDS. 12(1):47–56. https://doi.org/10.1097/COH.0000000000000330

Gopalakrishnan R, Matta H, Chaudhary PM (2013) A purine scaffold HSP90 inhibitor BIIB021 has selective activity against KSHV-associated primary effusion lymphoma and blocks vFLIP K13-induced NF-kappaB. Clin Cancer Res 19(18):5016–5026. https://doi.org/10.1158/1078-0432.CCR-12-3510

Gottwein E, Cullen BR (2010) A human herpesvirus microRNA inhibits p21 expression and attenuates p21-mediated cell cycle arrest. J Virol 84(10):5229–5237. https://doi.org/10.1128/JVI.00202-10

Gottwein E, Mukherjee N, Sachse C, Frenzel C, Majoros WH, Chi JT et al (2007) A viral microRNA functions as an orthologue of cellular miR-155. Nature 450(7172):1096–1099. https://doi.org/10.1038/nature05992

Gottwein E, Corcoran DL, Mukherjee N, Skalsky RL, Hafner M, Nusbaum JD et al (2011) Viral microRNA targetome of KSHV-infected primary effusion lymphoma cell lines. Cell Host Microbe 10(5):515–526. https://doi.org/10.1016/j.chom.2011.09.012

Grossmann C, Ganem D (2008) Effects of NFkappaB activation on KSHV latency and lytic reactivation are complex and context-dependent. Virology 375(1):94–102. https://doi.org/10.1016/j.virol.2007.12.044

Grossmann C, Podgrabinska S, Skobe M, Ganem D (2006) Activation of NF-kappaB by the latent vFLIP gene of Kaposi's sarcoma-associated herpesvirus is required for the spindle shape of virus-infected endothelial cells and contributes to their proinflammatory phenotype. J Virol 80 (14):7179–7185. https://doi.org/10.1128/JVI.01603-05

Guasparri I, Keller SA, Cesarman E (2004) KSHV vFLIP is essential for the survival of infected lymphoma cells. J Exp Med 199(7):993–1003. https://doi.org/10.1084/jem.20031467

Guasparri I, Wu H, Cesarman E (2006) The KSHV oncoprotein vFLIP contains a TRAF-interacting motif and requires TRAF2 and TRAF3 for signalling. EMBO Rep 7 (1):114–119. https://doi.org/10.1038/sj.embor.7400580

Haddad L, El Hajj H, Abou-Merhi R, Kfoury Y, Mahieux R, El-Sabban M et al (2008) KSHV-transformed primary effusion lymphoma cells induce a VEGF-dependent angiogenesis and establish functional gap junctions with endothelial cells. Leukemia 22(4):826–834. https://doi.org/10.1038/sj.leu.2405081

Haecker I, Gay LA, Yang Y, Hu J, Morse AM, McIntyre LM et al (2012) Ago HITS-CLIP expands understanding of Kaposi's sarcoma-associated herpesvirus miRNA function in primary effusion lymphomas. PLoS Pathog 8(8):e1002884. https://doi.org/10.1371/journal.ppat.1002884

Hansen A, Henderson S, Lagos D, Nikitenko L, Coulter E, Roberts S et al (2010) KSHV-encoded miRNAs target MAF to induce endothelial cell reprogramming. Genes Dev 24(2):195–205. https://doi.org/10.1101/gad.553410

Hayward GS, Zong JC (2007) Modern evolutionary history of the human KSHV genome. Curr Top Microbiol Immunol 312:1–42

Heinrich PC, Behrmann I, Muller-Newen G, Schaper F, Graeve L (1998) Interleukin-6-type cytokine signalling through the gp130/Jak/STAT pathway. Biochem J 334(Pt 2):297–314

Heinrich PC, Behrmann I, Haan S, Hermanns HM, Muller-Newen G, Schaper F (2003) Principles of interleukin (IL)-6-type cytokine signalling and its regulation. Biochem J 374(Pt 1):1–20. https://doi.org/10.1042/BJ20030407

Heinzelmann K, Scholz BA, Nowak A, Fossum E, Kremmer E, Haas J et al (2010) Kaposi's sarcoma-associated herpesvirus viral interferon regulatory factor 4 (vIRF4/K10) is a novel interaction partner of CSL/CBF1, the major downstream effector of Notch signaling. J Virol 84 (23):12255–12264. https://doi.org/10.1128/JVI.01484-10

Hew K, Dahlroth SL, Venkatachalam R, Nasertorabi F, Lim BT, Cornvik T et al (2013) The crystal structure of the DNA-binding domain of vIRF-1 from the oncogenic KSHV reveals a conserved fold for DNA binding and reinforces its role as a transcription factor. Nucleic Acids Res 41(7):4295–4306. https://doi.org/10.1093/nar/gkt082

Hu S, Vincenz C, Buller M, Dixit VM (1997) A novel family of viral death effector domain-containing molecules that inhibit both CD-95- and tumor necrosis factor receptor-1-induced apoptosis. J Biol Chem 272(15):9621–9624

Hu M, Wang C, Li W, Lu W, Bai Z, Qin D et al (2015) A KSHV microRNA directly targets G protein-coupled receptor kinase 2 to promote the migration and invasion of endothelial cells by inducing CXCR2 and activating AKT signaling. PLoS Pathog 11(9):e1005171. https://doi.org/10.1371/journal.ppat.1005171

Hu H, Dong J, Liang D, Gao Z, Bai L, Sun R et al (2016) Genome-wide mapping of the binding sites and structural analysis of Kaposi's sarcoma-associated herpesvirus viral interferon regulatory factor 2 reveal that it is a DNA-binding transcription factor. J Virol 90(3):1158–1168. https://doi.org/10.1128/JVI.01392-15

Huang Q, Petros AM, Virgin HW, Fesik SW, Olejniczak ET (2002) Solution structure of a Bcl-2 homolog from Kaposi sarcoma virus. Proc Natl Acad Sci U S A. 99(6):3428–3433. https://doi.org/10.1073/pnas.062525799

Hwang KY, Choi YB (2016) Modulation of Mitochondrial Antiviral Signaling by Human Herpesvirus 8 Interferon Regulatory Factor 1. J Virol 90(1):506–520. https://doi.org/10.1128/JVI.01903-15

Ishiyama T, Nakamura S, Akimoto Y, Koike M, Tomoyasu S, Tsuruoka N et al (1994) Immunodeficiency and IL-6 production by peripheral blood monocytes in multicentric Castleman's disease. Br J Haematol 86(3):483–489

Jacobs SR, Damania B (2011) The viral interferon regulatory factors of KSHV: immunosuppressors or oncogenes? Front Immunol 2:19. https://doi.org/10.3389/fimmu.2011.00019

Jarviluoma A, Child ES, Sarek G, Sirimongkolkasem P, Peters G, Ojala PM et al (2006) Phosphorylation of the cyclin-dependent kinase inhibitor p21Cip1 on serine 130 is essential for viral cyclin-mediated bypass of a p21Cip1-imposed G1 arrest. Mol Cell Biol 26(6):2430–2440. https://doi.org/10.1128/MCB.26.6.2430-2440.2006

Jenner RG, Alba MM, Boshoff C, Kellam P (2001) Kaposi's sarcoma-associated herpesvirus latent and lytic gene expression as revealed by DNA arrays. J Virol 75(2):891–902. https://doi.org/10.1128/JVI.75.2.891-902.2001

Jones KD, Aoki Y, Chang Y, Moore PS, Yarchoan R, Tosato G (1999) Involvement of interleukin-10 (IL-10) and viral IL-6 in the spontaneous growth of Kaposi's sarcoma herpesvirus-associated infected primary effusion lymphoma cells. Blood 94(8):2871–2879

Jones T, Ramos da Silva S, Bedolla R, Ye F, Zhou F, Gao SJ (2014) Viral cyclin promotes KSHV-induced cellular transformation and tumorigenesis by overriding contact inhibition. Cell Cycle 13(5):845–58. https://doi.org/10.4161/cc.27758

Joo CH, Shin YC, Gack M, Wu L, Levy D, Jung JU (2007) Inhibition of interferon regulatory factor 7 (IRF7)-mediated interferon signal transduction by the Kaposi's sarcoma-associated herpesvirus viral IRF homolog vIRF3. J Virol 81(15):8282–8292. https://doi.org/10.1128/JVI. 00235-07

Journo G, Tushinsky C, Shterngas A, Avital N, Eran Y, Karpuj MV et al (2018) Modulation of cellular CpG DNA methylation by Kaposi's sarcoma-associated herpesvirus. J Virol 92(16). https://doi.org/10.1128/jvi.00008-18

Karim R, Tse G, Putti T, Scolyer R, Lee S (2004) The significance of the Wnt pathway in the pathology of human cancers. Pathology. 36(2):120–128. https://doi.org/10.1080/00313020410001671957

Kawa K (2000) Epstein-Barr virus–associated diseases in humans. Int J Hematol 71(2):108–117

Kaye KM, Izumi KM, Kieff E (1993) Epstein-Barr virus latent membrane protein 1 is essential for B-lymphocyte growth transformation. Proc Natl Acad Sci U S A. 90(19):9150–9154

Keller SA, Schattner EJ, Cesarman E (2000) Inhibition of NF-kappaB induces apoptosis of KSHV-infected primary effusion lymphoma cells. Blood 96(7):2537–2542

Kieffer-Kwon P, Happel C, Uldrick TS, Ramalingam D, Ziegelbauer JM (2015) KSHV MicroRNAs repress tropomyosin 1 and increase anchorage-independent growth and endothelial tube formation. PLoS ONE 10(8):e0135560. https://doi.org/10.1371/journal.pone.0135560

Kim Y, Cha S, Seo T (2016) Activation of the phosphatidylinositol 3-kinase/Akt pathway by viral interferon regulatory factor 2 of Kaposi's sarcoma-associated herpesvirus. Biochem Biophys Res Commun 470(3):650–656. https://doi.org/10.1016/j.bbrc.2016.01.087

Kishimoto T, Akira S, Narazaki M, Taga T (1995) Interleukin-6 family of cytokines and gp130. Blood 86(4):1243–1254

Kledal TN, Rosenkilde MM, Coulin F, Simmons G, Johnsen AH, Alouani S et al (1997) A broad-spectrum chemokine antagonist encoded by Kaposi's sarcoma-associated herpesvirus. Science 277(5332):1656–1659

Kleijnen MF, Shih AH, Zhou P, Kumar S, Soccio RE, Kedersha NL et al (2000) The hPLIC proteins may provide a link between the ubiquitination machinery and the proteasome. Mol Cell 6(2):409–419

Kliche S, Nagel W, Kremmer E, Atzler C, Ege A, Knorr T et al (2001) Signaling by human herpesvirus 8 kaposin A through direct membrane recruitment of cytohesin-1. Mol Cell 7 (4):833–843

Kovaleva M, Bussmeyer I, Rabe B, Grotzinger J, Sudarman E, Eichler J et al (2006) Abrogation of viral interleukin-6 (vIL-6)-induced signaling by intracellular retention and neutralization of vIL-6 with an anti-vIL-6 single-chain antibody selected by phage display. J Virol 80(17):8510–8520. https://doi.org/10.1128/JVI.00420-06

Krithivas A, Young DB, Liao G, Greene D, Hayward SD (2000) Human herpesvirus 8 LANA interacts with proteins of the mSin3 corepressor complex and negatively regulates Epstein-Barr virus gene expression in dually infected PEL cells. J Virol 74(20):9637–9645

Kuwana T, Bouchier-Hayes L, Chipuk JE, Bonzon C, Sullivan BA, Green DR et al (2005) BH3 domains of BH3-only proteins differentially regulate Bax-mediated mitochondrial membrane permeabilization both directly and indirectly. Mol Cell 17(4):525–535. https://doi.org/10.1016/j.molcel.2005.02.003

Lagunoff M, Ganem D (1997) The structure and coding organization of the genomic termini of Kaposi's sarcoma-associated herpesvirus. Virology 236(1):147–154

Lagunoff M, Lukac DM, Ganem D (2001) Immunoreceptor tyrosine-based activation motif-dependent signaling by Kaposi's sarcoma-associated herpesvirus K1 protein: effects on lytic viral replication. J Virol 75(13):5891–5898. https://doi.org/10.1128/JVI.75.13.5891-5898.2001

Lang F, Sun Z, Pei Y, Singh RK, Jha HC, Robertson ES (2018) Shugoshin 1 is dislocated by KSHV-encoded LANA inducing aneuploidy. PLoS Pathog 14(9):e1007253. https://doi.org/10.1371/journal.ppat.1007253

Laura MV, de la Cruz-Herrera CF, Ferreiros A, Baz-Martinez M, Lang V, Vidal A et al (2015) KSHV latent protein LANA2 inhibits sumo2 modification of p53. Cell Cycle 14(2):277–282. https://doi.org/10.4161/15384101.2014.980657

Lee H, Veazey R, Williams K, Li M, Guo J, Neipel F et al (1998a) Deregulation of cell growth by the K1 gene of Kaposi's sarcoma-associated herpesvirus. Nat Med 4(4):435–440

Lee H, Guo J, Li M, Choi JK, DeMaria M, Rosenzweig M et al (1998b) Identification of an immunoreceptor tyrosine-based activation motif of K1 transforming protein of Kaposi's sarcoma-associated herpesvirus. Mol Cell Biol 18(9):5219–5228

Lee Y, Ahn C, Han J, Choi H, Kim J, Yim J et al (2003a) The nuclear RNase III Drosha initiates microRNA processing. Nature 425(6956):415–419. https://doi.org/10.1038/nature01957

Lee BJ, Koszinowski UH, Sarawar SR, Adler H (2003b) A gammaherpesvirus G protein-coupled receptor homologue is required for increased viral replication in response to chemokines and efficient reactivation from latency. J Immunol. 170(1):243–251

Lee BS, Connole M, Tang Z, Harris NL, Jung JU (2003c) Structural analysis of the Kaposi's sarcoma-associated herpesvirus K1 protein. J Virol 77(14):8072–8086

Lee BS, Lee SH, Feng P, Chang H, Cho NH, Jung JU (2005) Characterization of the Kaposi's sarcoma-associated herpesvirus K1 signalosome. J Virol 79(19):12173–12184. https://doi.org/10.1128/JVI.79.19.12173-12184.2005

Lee JS, Li Q, Lee JY, Lee SH, Jeong JH, Lee HR et al (2009a) FLIP-mediated autophagy regulation in cell death control. Nat Cell Biol 11(11):1355–1362. https://doi.org/10.1038/ncb1980

Lee HR, Kim MH, Lee JS, Liang C, Jung JU (2009b) Viral interferon regulatory factors. J Interferon Cytokine Res 29(9):621–627. https://doi.org/10.1089/jir.2009.0067

Lee HR, Toth Z, Shin YC, Lee JS, Chang H, Gu W et al (2009c) Kaposi's sarcoma-associated herpesvirus viral interferon regulatory factor 4 targets MDM2 to deregulate the p53 tumor suppressor pathway. J Virol 83(13):6739–6747. https://doi.org/10.1128/JVI.02353-08

Lee HR, Choi WC, Lee S, Hwang J, Hwang E, Guchhait K et al (2011) Bilateral inhibition of HAUSP deubiquitinase by a viral interferon regulatory factor protein. Nat Struct Mol Biol 18 (12):1336–1344. https://doi.org/10.1038/nsmb.2142

Lee HR, Li F, Choi UY, Yu HR, Aldrovandi GM, Feng P et al (2018) Deregulation of HDAC5 by viral interferon regulatory factor 3 plays an essential role in Kaposi's sarcoma-associated herpesvirus-induced lymphangiogenesis. MBio 9(1). https://doi.org/10.1128/mbio.02217-17

Lei X, Bai Z, Ye F, Xie J, Kim CG, Huang Y et al (2010) Regulation of NF-kappaB inhibitor IkappaBalpha and viral replication by a KSHV microRNA. Nat Cell Biol 12(2):193–199. https://doi.org/10.1038/ncb2019

Lei X, Zhu Y, Jones T, Bai Z, Huang Y, Gao SJ (2012) A Kaposi's sarcoma-associated herpesvirus microRNA and its variants target the transforming growth factor beta pathway to promote cell survival. J Virol 86(21):11698–11711. https://doi.org/10.1128/JVI.06855-11

Li M, Lee H, Yoon DW, Albrecht JC, Fleckenstein B, Neipel F et al (1997) Kaposi's sarcoma-associated herpesvirus encodes a functional cyclin. J Virol 71(3):1984–1991

Li M, Damania B, Alvarez X, Ogryzko V, Ozato K, Jung JU (2000) Inhibition of p300 histone acetyltransferase by viral interferon regulatory factor. Mol Cell Biol 20(21):8254–8263

Li H, Komatsu T, Dezube BJ, Kaye KM (2002a) The Kaposi's sarcoma-associated herpesvirus K12 transcript from a primary effusion lymphoma contains complex repeat elements, is spliced, and initiates from a novel promoter. J Virol 76(23):11880–11888

Li M, Chen D, Shiloh A, Luo J, Nikolaev AY, Qin J et al (2002b) Deubiquitination of p53 by HAUSP is an important pathway for p53 stabilization. Nature 416(6881):648–653. https://doi.org/10.1038/nature737

Li M, Brooks CL, Kon N, Gu W (2004) A dynamic role of HAUSP in the p53-Mdm2 pathway. Mol Cell 13(6):879–886

Li J, Wan Y, Ji Q, Fang Y, Wu Y (2013) The role of microRNAs in B-cell development and function. Cell Mol Immunol 10(2):107–112. https://doi.org/10.1038/cmi.2012.62

Li W, Jia X, Shen C, Zhang M, Xu J, Shang Y et al (2016a) A KSHV microRNA enhances viral latency and induces angiogenesis by targeting GRK2 to activate the CXCR2/AKT pathway. Oncotarget 7(22):32286–32305. https://doi.org/10.18632/oncotarget.8591

Li W, Yan Q, Ding X, Shen C, Hu M, Zhu Y et al (2016b) The SH3BGR/STAT3 Pathway Regulates Cell Migration and Angiogenesis Induced by a Gammaherpesvirus MicroRNA. PLoS Pathog 12(4):e1005605. https://doi.org/10.1371/journal.ppat.1005605

Li W, Hu M, Wang C, Lu H, Chen F, Xu J et al (2017) A viral microRNA downregulates metastasis suppressor CD82 and induces cell invasion and angiogenesis by activating the c-Met signaling. Oncogene 36(38):5407–5420. https://doi.org/10.1038/onc.2017.139

Li Q, Chen D, Xiang Q, Nicholas J (2018) Insulin-like growth factor 2 receptor expression is promoted by human herpesvirus 8-encoded interleukin-6 and contributes to viral latency and productive replication. J Virol. https://doi.org/10.1128/JVI.02026-18

Li T, Zhu Y, Cheng F, Lu C, Jung JU, Gao SJ (2019) Oncogenic Kaposi's sarcoma-associated herpesvirus upregulates argininosuccinate synthase 1, a rate-limiting enzyme of the citrulline-nitric oxide cycle, to activate the STAT3 pathway and promote growth transformation. J Virol 93(4). https://doi.org/10.1128/jvi.01599-18

Liang C, Feng P, Ku B, Dotan I, Canaani D, Oh BH et al (2006) Autophagic and tumour suppressor activity of a novel Beclin1-binding protein UVRAG. Nat Cell Biol 8(7):688–699. https://doi.org/10.1038/ncb1426

Liang D, Lin X, Lan K (2011a) Looking at Kaposi's Sarcoma-Associated Herpesvirus-Host Interactions from a microRNA Viewpoint. Front Microbiol. 2:271. https://doi.org/10.3389/fmicb.2011.00271

Liang D, Gao Y, Lin X, He Z, Zhao Q, Deng Q et al (2011b) A human herpesvirus miRNA attenuates interferon signaling and contributes to maintenance of viral latency by targeting IKKepsilon. Cell Res 21(5):793–806. https://doi.org/10.1038/cr.2011.5

Liang Q, Chang B, Brulois KF, Castro K, Min CK, Rodgers MA et al (2013) Kaposi's sarcoma-associated herpesvirus K7 modulates Rubicon-mediated inhibition of autophagosome maturation. J Virol 87(22):12499–12503. https://doi.org/10.1128/JVI.01898-13

Liang Q, Chang B, Lee P, Brulois KF, Ge J, Shi M et al (2015) Identification of the essential role of viral Bcl-2 for Kaposi's sarcoma-associated herpesvirus lytic replication. J Virol 89 (10):5308–5317. https://doi.org/10.1128/JVI.00102-15

Liang Q, Wei D, Chung B, Brulois KF, Guo C, Dong S et al (2018) Novel role of vBcl2 in the virion assembly of Kaposi's sarcoma-associated herpesvirus. J Virol 92(4). https://doi.org/10.1128/jvi.00914-17

Lim CS, Seet BT, Ingham RJ, Gish G, Matskova L, Winberg G et al (2007) The K15 protein of Kaposi's sarcoma-associated herpesvirus recruits the endocytic regulator intersectin 2 through a selective SH3 domain interaction. Biochemistry 46(35):9874–9885. https://doi.org/10.1021/bi700357s

Lin R, Genin P, Mamane Y, Sgarbanti M, Battistini A, Harrington WJ Jr et al (2001) HHV-8 encoded vIRF-1 represses the interferon antiviral response by blocking IRF-3 recruitment of the CBP/p300 coactivators. Oncogene 20(7):800–811. https://doi.org/10.1038/sj.onc.1204163

Lin CW, Tu PF, Hsiao NW, Chang CY, Wan L, Lin YT et al (2007) Identification of a novel septin 4 protein binding to human herpesvirus 8 kaposin A protein using a phage display cDNA library. J Virol Methods 143(1):65–72. https://doi.org/10.1016/j.jviromet.2007.02.010

Lin X, Liang D, He Z, Deng Q, Robertson ES, Lan K (2011) miR-K12-7-5p encoded by Kaposi's sarcoma-associated herpesvirus stabilizes the latent state by targeting viral ORF50/RTA. PLoS ONE 6(1):e16224. https://doi.org/10.1371/journal.pone.0016224

Lindner SE, Sugden B (2007) The plasmid replicon of Epstein-Barr virus: mechanistic insights into efficient, licensed, extrachromosomal replication in human cells. Plasmid 58(1):1–12. https://doi.org/10.1016/j.plasmid.2007.01.003

Liu C, Okruzhnov Y, Li H, Nicholas J (2001) Human herpesvirus 8 (HHV-8)-encoded cytokines induce expression of and autocrine signaling by vascular endothelial growth factor (VEGF) in HHV-8-infected primary-effusion lymphoma cell lines and mediate VEGF-independent antiapoptotic effects. J Virol 75(22):10933–10940. https://doi.org/10.1128/JVI.75.22.10933-10940.2001

Liu L, Eby MT, Rathore N, Sinha SK, Kumar A, Chaudhary PM (2002) The human herpes virus 8-encoded viral FLICE inhibitory protein physically associates with and persistently activates the Ikappa B kinase complex. J Biol Chem 277(16):13745–13751. https://doi.org/10.1074/jbc.M110480200

Liu C, Sandford G, Fei G, Nicholas J (2004) Galpha protein selectivity determinant specified by a viral chemokine receptor-conserved region in the C tail of the human herpesvirus 8 g protein-coupled receptor. J Virol 78(5):2460–2471

Liu J, Martin H, Shamay M, Woodard C, Tang QQ, Hayward SD (2007a) Kaposi's sarcoma-associated herpesvirus LANA protein downregulates nuclear glycogen synthase kinase 3 activity and consequently blocks differentiation. J Virol 81(9):4722–4731. https://doi.org/10.1128/JVI.02548-06

Liu J, Martin HJ, Liao G, Hayward SD (2007b) The Kaposi's sarcoma-associated herpesvirus LANA protein stabilizes and activates c-Myc. J Virol 81(19):10451–10459. https://doi.org/10.1128/JVI.00804-07

Liu Y, Sun R, Lin X, Liang D, Deng Q, Lan K (2012) Kaposi's sarcoma-associated herpesvirus-encoded microRNA miR-K12-11 attenuates transforming growth factor beta signaling through suppression of SMAD5. J Virol 86(3):1372–1381. https://doi.org/10.1128/JVI.06245-11

Liu X, Happel C, Ziegelbauer JM (2017) Kaposi's sarcoma-associated herpesvirus MicroRNAs target GADD45B to protect infected cells from cell cycle arrest and apoptosis. J Virol 91(3). https://doi.org/10.1128/jvi.02045-16

Louafi F, Martinez-Nunez RT, Sanchez-Elsner T (2010) MicroRNA-155 targets SMAD2 and modulates the response of macrophages to transforming growth factor-β. J Biol Chem 285(53):41328–41336. https://doi.org/10.1074/jbc.M110.146852

Louahed J, Struyf S, Demoulin JB, Parmentier M, Van Snick J, Van Damme J et al (2003) CCR8-dependent activation of the RAS/MAPK pathway mediates anti-apoptotic activity of I-309/ CCL1 and vMIP-I. Eur J Immunol 33(2):494–501. https://doi.org/10.1002/immu.200310025

Lu F, Stedman W, Yousef M, Renne R, Lieberman PM (2010a) Epigenetic regulation of Kaposi's sarcoma-associated herpesvirus latency by virus-encoded microRNAs that target Rta and the cellular Rbl2-DNMT pathway. J Virol 84(6):2697–2706. https://doi.org/10.1128/JVI.01997-09

Lu CC, Li Z, Chu CY, Feng J, Feng J, Sun R et al (2010b) MicroRNAs encoded by Kaposi's sarcoma-associated herpesvirus regulate viral life cycle. EMBO Rep 11(10):784–790. https://doi.org/10.1038/embor.2010.132

Lubyova B, Pitha PM (2000) Characterization of a novel human herpesvirus 8-encoded protein, vIRF-3, that shows homology to viral and cellular interferon regulatory factors. J Virol 74(17):8194–8201

Lubyova B, Kellum MJ, Frisancho JA, Pitha PM (2007) Stimulation of c-Myc transcriptional activity by vIRF-3 of Kaposi sarcoma-associated herpesvirus. J Biol Chem 282(44):31944–31953. https://doi.org/10.1074/jbc.M706430200

Lund E, Guttinger S, Calado A, Dahlberg JE, Kutay U (2004) Nuclear export of microRNA precursors. Science 303(5654):95–98. https://doi.org/10.1126/science.1090599

Luttichau HR (2008) The herpesvirus 8 encoded chemokines vCCL2 (vMIP-II) and vCCL3 (vMIP-III) target the human but not the murine lymphotactin receptor. Virol J. 5:50. https://doi.org/10.1186/1743-422X-5-50

Luttichau HR, Lewis IC, Gerstoft J, Schwartz TW (2001) The herpesvirus 8-encoded chemokine vMIP-II, but not the poxvirus-encoded chemokine MC148, inhibits the CCR10 receptor. Eur J Immunol 31(4):1217–1220. https://doi.org/10.1002/1521-4141(200104)31:4%3c1217:AID-IMMU1217%3e3.0.CO;2-S

Luttichau HR, Johnsen AH, Jurlander J, Rosenkilde MM, Schwartz TW (2007) Kaposi sarcoma-associated herpes virus targets the lymphotactin receptor with both a broad spectrum antagonist vCCL2 and a highly selective and potent agonist vCCL3. J Biol Chem 282(24):17794–17805. https://doi.org/10.1074/jbc.M702001200

Ma X, Kalakonda S, Srinivasula SM, Reddy SP, Platanias LC, Kalvakolanu DV (2007) GRIM-19 associates with the serine protease HtrA2 for promoting cell death. Oncogene 26(33):4842–4849. https://doi.org/10.1038/sj.onc.1210287

Ma Z, Jacobs SR, West JA, Stopford C, Zhang Z, Davis Z et al (2015) Modulation of the cGAS-STING DNA sensing pathway by gammaherpesviruses. Proc Natl Acad Sci U S A. 112 (31):E4306–E4315. https://doi.org/10.1073/pnas.1503831112

Mandel-Gutfreund Y, Kosti I, Larisch S (2011) ARTS, the unusual septin: structural and functional aspects. Biol Chem 392(8–9):783–790. https://doi.org/10.1515/BC.2011.089

Mann DJ, Child ES, Swanton C, Laman H, Jones N (1999) Modulation of p27(Kip1) levels by the cyclin encoded by Kaposi's sarcoma-associated herpesvirus. EMBO J 18(3):654–663. https://doi.org/10.1093/emboj/18.3.654

Marcos-Villar L, Lopitz-Otsoa F, Gallego P, Munoz-Fontela C, Gonzalez-Santamaria J, Campagna M et al (2009) Kaposi's sarcoma-associated herpesvirus protein LANA2 disrupts PML oncogenic domains and inhibits PML-mediated transcriptional repression of the survivin gene. J Virol 83(17):8849–8858. https://doi.org/10.1128/JVI.00339-09

Marcos-Villar L, Gallego P, Munoz-Fontela C, de la Cruz-Herrera CF, Campagna M, Gonzalez D et al (2014) Kaposi's sarcoma-associated herpesvirus lana2 protein interacts with the pocket proteins and inhibits their sumoylation. Oncogene 33(4):495–503. https://doi.org/10.1038/onc.2012.603

Mariggio G, Koch S, Zhang G, Weidner-Glunde M, Ruckert J, Kati S et al (2017) Kaposi sarcoma herpesvirus (KSHV) latency-associated nuclear antigen (LANA) recruits components of the MRN (Mre11-Rad50-NBS1) repair complex to modulate an innate immune signaling pathway and viral latency. PLoS Pathog 13(4):e1006335. https://doi.org/10.1371/journal.ppat.1006335

Matta H, Chaudhary PM (2004) Activation of alternative NF-kappa B pathway by human herpes virus 8-encoded Fas-associated death domain-like IL-1 beta-converting enzyme inhibitory protein (vFLIP). Proc Natl Acad Sci U S A 101(25):9399–9404. https://doi.org/10.1073/pnas.0308016101

Matta H, Mazzacurati L, Schamus S, Yang T, Sun Q, Chaudhary PM (2007) Kaposi's sarcoma-associated herpesvirus (KSHV) oncoprotein K13 bypasses TRAFs and directly interacts with the IkappaB kinase complex to selectively activate NF-kappaB without JNK activation. J Biol Chem 282(34):24858–24865. https://doi.org/10.1074/jbc.M700118200

Mbulaiteye S, Marshall V, Bagni RK, Wang CD, Mbisa G, Bakaki PM et al (2006) Molecular evidence for mother-to-child transmission of Kaposi sarcoma-associated herpesvirus in Uganda and K1 gene evolution within the host. J Infect Dis 193(9):1250–1257. https://doi.org/10.1086/503052

McCormick C, Ganem D (2005) The kaposin B protein of KSHV activates the p38/MK2 pathway and stabilizes cytokine mRNAs. Science 307(5710):739–741. https://doi.org/10.1126/science.1105779

McCormick C, Ganem D (2006) Phosphorylation and function of the kaposin B direct repeats of Kaposi's sarcoma-associated herpesvirus. J Virol 80(12):6165–6170. https://doi.org/10.1128/JVI.02331-05

Mesri EA, Cesarman E, Boshoff C (2010) Kaposi's sarcoma and its associated herpesvirus. Nat Rev Cancer 10(10):707–719. https://doi.org/10.1038/nrc2888

Miles SA, Rezai AR, Salazar-Gonzalez JF, Vander Meyden M, Stevens RH, Logan DM et al (1990) AIDS Kaposi sarcoma-derived cells produce and respond to interleukin 6. Proc Natl Acad Sci U S A. 87(11):4068–4072

Montaner S, Sodhi A, Molinolo A, Bugge TH, Sawai ET, He Y et al (2003) Endothelial infection with KSHV genes in vivo reveals that vGPCR initiates Kaposi's sarcomagenesis and can promote the tumorigenic potential of viral latent genes. Cancer Cell 3(1):23–36

Montaner S, Sodhi A, Servitja JM, Ramsdell AK, Barac A, Sawai ET et al (2004) The small GTPase Rac1 links the Kaposi sarcoma-associated herpesvirus vGPCR to cytokine secretion and paracrine neoplasia. Blood 104(9):2903–2911. https://doi.org/10.1182/blood-2003-12-4436

Moore PS, Boshoff C, Weiss RA, Chang Y (1996) Molecular mimicry of human cytokine and cytokine response pathway genes by KSHV. Science 274(5293):1739–1744

Mori Y, Nishimoto N, Ohno M, Inagi R, Dhepakson P, Amou K et al (2000) Human herpesvirus 8-encoded interleukin-6 homologue (viral IL-6) induces endogenous human IL-6 secretion. J Med Virol 61(3):332–335

Munoz-Fontela C, Collado M, Rodriguez E, Garcia MA, Alvarez-Barrientos A, Arroyo J et al (2005) Identification of a nuclear export signal in the KSHV latent protein LANA2 mediating its export from the nucleus. Exp Cell Res 311(1):96–105. https://doi.org/10.1016/j.yexcr.2005.08.022

Munoz-Fontela C, Marcos-Villar L, Gallego P, Arroyo J, Da Costa M, Pomeranz KM et al (2007) Latent protein LANA2 from Kaposi's sarcoma-associated herpesvirus interacts with 14-3-3 proteins and inhibits FOXO3a transcription factor. J Virol 81(3):1511–1516. https://doi.org/10.1128/JVI.01816-06

Muralidhar S, Pumfery AM, Hassani M, Sadaie MR, Kishishita M, Brady JN et al (1998) Identification of kaposin (open reading frame K12) as a human herpesvirus 8 (Kaposi's sarcoma-associated herpesvirus) transforming gene. J Virol 72(6):4980–4988

Muralidhar S, Veytsmann G, Chandran B, Ablashi D, Doniger J, Rosenthal LJ (2000) Characterization of the human herpesvirus 8 (Kaposi's sarcoma-associated herpesvirus) oncogene, kaposin (ORF K12). J Clin Virol 16(3):203–213

Murthy SC, Trimble JJ, Desrosiers RC (1989) Deletion mutants of herpesvirus saimiri define an open reading frame necessary for transformation. J Virol 63(8):3307–3314

Mutocheluh M, Hindle L, Areste C, Chanas SA, Butler LM, Lowry K et al (2011) Kaposi's sarcoma-associated herpesvirus viral interferon regulatory factor-2 inhibits type 1 interferon signalling by targeting interferon-stimulated gene factor-3. J Gen Virol 92(Pt 10):2394–2398. https://doi.org/10.1099/vir.0.034322-0

Nachmani D, Stern-Ginossar N, Sarid R, Mandelboim O (2009) Diverse herpesvirus microRNAs target the stress-induced immune ligand MICB to escape recognition by natural killer cells. Cell Host Microbe 5(4):376–385. https://doi.org/10.1016/j.chom.2009.03.003

Nakamura H, Li M, Zarycki J, Jung JU (2001) Inhibition of p53 tumor suppressor by viral interferon regulatory factor. J Virol 75(16):7572–7582. https://doi.org/10.1128/JVI.75.16.7572-7582.2001

Nakamura H, Lu M, Gwack Y, Souvlis J, Zeichner SL, Jung JU (2003) Global changes in Kaposi's sarcoma-associated virus gene expression patterns following expression of a tetracycline-inducible Rta transactivator. J Virol 77(7):4205–4220

Nakano K, Isegawa Y, Zou P, Tadagaki K, Inagi R, Yamanishi K (2003) Kaposi's sarcoma-associated herpesvirus (KSHV)-encoded vMIP-I and vMIP-II induce signal transduction and chemotaxis in monocytic cells. Arch Virol 148(5):871–890. https://doi.org/10.1007/s00705-002-0971-7

Nayar U, Lu P, Goldstein RL, Vider J, Ballon G, Rodina A et al (2013) Targeting the Hsp90-associated viral oncoproteome in gammaherpesvirus-associated malignancies. Blood 122(16):2837–2847. https://doi.org/10.1182/blood-2013-01-479972

Neipel F, Albrecht JC, Ensser A, Huang YQ, Li JJ, Friedman-Kien AE et al (1997a) Human herpesvirus 8 encodes a homolog of interleukin-6. J Virol 71(1):839–842

Neipel F, Albrecht JC, Fleckenstein B (1997b) Cell-homologous genes in the Kaposi's sarcoma-associated rhadinovirus human herpesvirus 8: determinants of its pathogenicity? J Virol 71(6):4187–4192

Nicholas J (2005) Human gammaherpesvirus cytokines and chemokine receptors. J Interferon Cytokine Res 25(7):373–383. https://doi.org/10.1089/jir.2005.25.373

Nicholas J, Ruvolo VR, Burns WH, Sandford G, Wan X, Ciufo D et al (1997) Kaposi's sarcoma-associated human herpesvirus-8 encodes homologues of macrophage inflammatory protein-1 and interleukin-6. Nat Med 3(3):287–292

Nicholas J, Zong JC, Alcendor DJ, Ciufo DM, Poole LJ, Sarisky RT et al (1998) Novel organizational features, captured cellular genes, and strain variability within the genome of KSHV/HHV8. J Natl Cancer Inst Monogr. 23:79–88

Offermann MK (2007) Kaposi sarcoma herpesvirus-encoded interferon regulator factors. Curr Top Microbiol Immunol 312:185–209

Okano M (2000) Haematological associations of Epstein-Barr virus infection. Baillieres Best Pract Res Clin Haematol. 13(2):199–214. https://doi.org/10.1053/beha.1999.0068

Oksenhendler E, Carcelain G, Aoki Y, Boulanger E, Maillard A, Clauvel JP et al (2000) High levels of human herpesvirus 8 viral load, human interleukin-6, interleukin-10, and C reactive protein correlate with exacerbation of multicentric castleman disease in HIV-infected patients. Blood 96(6):2069–2073

Park J, Lee MS, Yoo SM, Jeong KW, Lee D, Choe J et al (2007) Identification of the DNA sequence interacting with Kaposi's sarcoma-associated herpesvirus viral interferon regulatory factor 1. J Virol 81(22):12680–12684. https://doi.org/10.1128/JVI.00556-07

Pati S, Cavrois M, Guo HG, Foulke JS Jr, Kim J, Feldman RA et al (2001) Activation of NF-kappaB by the human herpesvirus 8 chemokine receptor ORF74: evidence for a paracrine model of Kaposi's sarcoma pathogenesis. J Virol 75(18):8660–8673

Pattingre S, Tassa A, Qu X, Garuti R, Liang XH, Mizushima N et al (2005) Bcl-2 antiapoptotic proteins inhibit Beclin 1-dependent autophagy. Cell 122(6):927–939. https://doi.org/10.1016/j.cell.2005.07.002

Paudel N, Sadagopan S, Balasubramanian S, Chandran B (2012) Kaposi's sarcoma-associated herpesvirus latency-associated nuclear antigen and angiogenin interact with common host proteins, including annexin A2, which is essential for survival of latently infected cells. J Virol 86(3):1589–1607. https://doi.org/10.1128/JVI.05754-11

Paulose-Murphy M, Ha NK, Xiang C, Chen Y, Gillim L, Yarchoan R et al (2001) Transcription program of human herpesvirus 8 (kaposi's sarcoma-associated herpesvirus). J Virol 75 (10):4843–4853. https://doi.org/10.1128/JVI.75.10.4843-4853.2001

Pearce M, Matsumura S, Wilson AC (2005) Transcripts encoding K12, v-FLIP, v-cyclin, and the microRNA cluster of Kaposi's sarcoma-associated herpesvirus originate from a common promoter. J Virol 79(22):14457–14464. https://doi.org/10.1128/JVI.79.22.14457-14464.2005

Petre CE, Sin SH, Dittmer DP (2007) Functional p53 signaling in Kaposi's sarcoma-associated herpesvirus lymphomas: implications for therapy. J Virol 81(4):1912–1922. https://doi.org/10.1128/JVI.01757-06

Pfeffer S, Zavolan M, Grasser FA, Chien M, Russo JJ, Ju J et al (2004) Identification of virus-encoded microRNAs. Science 304(5671):734–736. https://doi.org/10.1126/science.1096781

Pfeffer S, Sewer A, Lagos-Quintana M, Sheridan R, Sander C, Grasser FA et al (2005) Identification of microRNAs of the herpesvirus family. Nat Methods 2(4):269–276. https://doi.org/10.1038/nmeth746

Philpott N, Bakken T, Pennell C, Chen L, Wu J, Cannon M (2011) The Kaposi's sarcoma-associated herpesvirus G protein-coupled receptor contains an immunoreceptor tyrosine-based inhibitory motif that activates Shp2. J Virol 85(2):1140–1144. https://doi.org/10.1128/JVI.01362-10

Pietrek M, Brinkmann MM, Glowacka I, Enlund A, Havemeier A, Dittrich-Breiholz O et al (2010) Role of the Kaposi's sarcoma-associated herpesvirus K15 SH3 binding site in inflammatory signaling and B-cell activation. J Virol 84(16):8231–8240. https://doi.org/10.1128/JVI.01696-09

Plaisance-Bonstaff K, Choi HS, Beals T, Krueger BJ, Boss IW, Gay LA et al (2014) KSHV miRNAs decrease expression of lytic genes in latently infected PEL and endothelial cells by targeting host transcription factors. Viruses. 6(10):4005–4023. https://doi.org/10.3390/v6104005

Platt G, Carbone A, Mittnacht S (2002) p16INK4a loss and sensitivity in KSHV associated primary effusion lymphoma. Oncogene 21(12):1823–1831. https://doi.org/10.1038/sj.onc.1205360

Polizzotto MN, Uldrick TS, Hu D, Yarchoan R (2012) Clinical manifestations of Kaposi sarcoma herpesvirus lytic activation: multicentric castleman disease (KSHV-MCD) and the KSHV inflammatory cytokine syndrome. Front Microbiol 3:73. https://doi.org/10.3389/fmicb.2012. 00073

Poole LJ, Zong JC, Ciufo DM, Alcendor DJ, Cannon JS, Ambinder R et al (1999) Comparison of genetic variability at multiple loci across the genomes of the major subtypes of Kaposi's sarcoma-associated herpesvirus reveals evidence for recombination and for two distinct types of open reading frame K15 alleles at the right-hand end. J Virol 73(8):6646–6660

Pozharskaya VP, Weakland LL, Zimring JC, Krug LT, Unger ER, Neisch A et al (2004) Short duration of elevated vIRF-1 expression during lytic replication of human herpesvirus 8 limits its ability to block antiviral responses induced by alpha interferon in BCBL-1 cells. J Virol 78 (12):6621–6635. https://doi.org/10.1128/JVI.78.12.6621-6635.2004

Prakash O, Tang ZY, Peng X, Coleman R, Gill J, Farr G et al (2002) Tumorigenesis and aberrant signaling in transgenic mice expressing the human herpesvirus-8 K1 gene. J Natl Cancer Inst 94(12):926–935

Punj V, Matta H, Schamus S, Tamewitz A, Anyang B, Chaudhary PM (2010) Kaposi's sarcoma-associated herpesvirus-encoded viral FLICE inhibitory protein (vFLIP) K13 suppresses CXCR4 expression by upregulating miR-146a. Oncogene 29(12):1835–1844. https:// doi.org/10.1038/onc.2009.460

Pyakurel P, Pak F, Mwakigonja AR, Kaaya E, Heiden T, Biberfeld P (2006) Lymphatic and vascular origin of Kaposi's sarcoma spindle cells during tumor development. Int J Cancer 119 (6):1262–1267. https://doi.org/10.1002/ijc.21969

Qin Z, Freitas E, Sullivan R, Mohan S, Bacelieri R, Branch D et al (2010a) Upregulation of xCT by KSHV-encoded microRNAs facilitates KSHV dissemination and persistence in an environment of oxidative stress. PLoS Pathog 6(1):e1000742. https://doi.org/10.1371/ journal.ppat.1000742

Qin Z, Kearney P, Plaisance K, Parsons CH (2010b) Pivotal advance: Kaposi's sarcoma-associated herpesvirus (KSHV)-encoded microRNA specifically induce IL-6 and IL-10 secretion by macrophages and monocytes. J Leukoc Biol 87(1):25–34

Qin J, Li W, Gao SJ, Lu C (2017) KSHV microRNAs: Tricks of the Devil. Trends Microbiol 25 (8):648–661. https://doi.org/10.1016/j.tim.2017.02.002

Radkov SA, Kellam P, Boshoff C (2000) The latent nuclear antigen of Kaposi sarcoma-associated herpesvirus targets the retinoblastoma-E2F pathway and with the oncogene Hras transforms primary rat cells. Nat Med 6(10):1121–1127. https://doi.org/10.1038/80459

Ramalingam D, Happel C, Ziegelbauer JM (2015) Kaposi's sarcoma-associated herpesvirus microRNAs repress breakpoint cluster region protein expression, enhance Rac1 activity, and increase in vitro angiogenesis. J Virol 89(8):4249–4261. https://doi.org/10.1128/JVI.03687-14

Ritthipichai K, Nan Y, Bossis I, Zhang Y (2012) Viral FLICE inhibitory protein of rhesus monkey rhadinovirus inhibits apoptosis by enhancing autophagosome formation. PLoS ONE 7(6): e39438. https://doi.org/10.1371/journal.pone.0039438

Rivas C, Thlick AE, Parravicini C, Moore PS, Chang Y (2001) Kaposi's sarcoma-associated herpesvirus LANA2 is a B-cell-specific latent viral protein that inhibits p53. J Virol 75(1):429–438. https://doi.org/10.1128/JVI.75.1.429-438.2001

Rosenkilde MM, Waldhoer M, Luttichau HR, Schwartz TW (2001) Virally encoded 7TM receptors. Oncogene 20(13):1582–1593. https://doi.org/10.1038/sj.onc.1204191

Roth WK (1991) HIV-associated Kaposi's sarcoma: new developments in epidemiology and molecular pathology. J Cancer Res Clin Oncol 117(3):186–191

Russo JJ, Bohenzky RA, Chien MC, Chen J, Yan M, Maddalena D et al (1996) Nucleotide sequence of the Kaposi sarcoma-associated herpesvirus (HHV8). Proc Natl Acad Sci U S A. 93 (25):14862–14867

Sadagopan S, Sharma-Walia N, Veettil MV, Bottero V, Levine R, Vart RJ et al (2009) Kaposi's sarcoma-associated herpesvirus upregulates angiogenin during infection of human dermal microvascular endothelial cells, which induces 45S rRNA synthesis, antiapoptosis, cell proliferation, migration, and angiogenesis. J Virol 83(7):3342–3364. https://doi.org/10.1128/JVI.02052-08

Sadagopan S, Valiya Veettil M, Paudel N, Bottero V, Chandran B (2011) Kaposi's sarcoma-associated herpesvirus-induced angiogenin plays roles in latency via the phospholipase C gamma pathway: blocking angiogenin inhibits latent gene expression and induces the lytic cycle. J Virol 85(6):2666–2685. https://doi.org/10.1128/JVI.01532-10

Sadagopan S, Veettil MV, Chakraborty S, Sharma-Walia N, Paudel N, Bottero V et al (2012) Angiogenin functionally interacts with p53 and regulates p53-mediated apoptosis and cell survival. Oncogene 31(46):4835–4847. https://doi.org/10.1038/onc.2011.648

Sadler R, Wu L, Forghani B, Renne R, Zhong W, Herndier B et al (1999) A complex translational program generates multiple novel proteins from the latently expressed kaposin (K12) locus of Kaposi's sarcoma-associated herpesvirus. J Virol 73(7):5722–5730

Sakakibara S, Ueda K, Nishimura K, Do E, Ohsaki E, Okuno T et al (2004) Accumulation of heterochromatin components on the terminal repeat sequence of Kaposi's sarcoma-associated herpesvirus mediated by the latency-associated nuclear antigen. J Virol 78(14):7299–7310. https://doi.org/10.1128/JVI.78.14.7299-7310.2004

Samaniego F, Pati S, Karp JE, Prakash O, Bose D (2001) Human herpesvirus 8 K1-associated nuclear factor-kappa B-dependent promoter activity: role in Kaposi's sarcoma inflammation? J Natl Cancer Inst Monogr. 28:15–23

Samols MA, Hu J, Skalsky RL, Renne R (2005) Cloning and identification of a microRNA cluster within the latency-associated region of Kaposi's sarcoma-associated herpesvirus. J Virol 79 (14):9301–9305. https://doi.org/10.1128/JVI.79.14.9301-9305.2005

Samols MA, Skalsky RL, Maldonado AM, Riva A, Lopez MC, Baker HV et al (2007) Identification of cellular genes targeted by KSHV-encoded microRNAs. PLoS Pathog 3(5): e65. https://doi.org/10.1371/journal.ppat.0030065

Sandford G, Choi YB, Nicholas J (2009) Role of ORF74-encoded viral G protein-coupled receptor in human herpesvirus 8 lytic replication. J Virol 83(24):13009–13014. https://doi.org/10.1128/JVI.01399-09

Sarek G, Jarviluoma A, Ojala PM (2006) KSHV viral cyclin inactivates p27KIP1 through Ser10 and Thr187 phosphorylation in proliferating primary effusion lymphomas. Blood 107(2):725–732. https://doi.org/10.1182/blood-2005-06-2534

Sarid R, Sato T, Bohenzky RA, Russo JJ, Chang Y (1997) Kaposi's sarcoma-associated herpesvirus encodes a functional bcl-2 homologue. Nat Med 3(3):293–298

Schmidt K, Wies E, Neipel F (2011) Kaposi's sarcoma-associated herpesvirus viral interferon regulatory factor 3 inhibits gamma interferon and major histocompatibility complex class II expression. J Virol 85(9):4530–4537. https://doi.org/10.1128/JVI.02123-10

Schwarz M, Murphy PM (2001) Kaposi's sarcoma-associated herpesvirus G protein-coupled receptor constitutively activates NF-kappa B and induces proinflammatory cytokine and chemokine production via a C-terminal signaling determinant. J Immunol. 167(1):505–513

Schwarz DS, Hutvagner G, Du T, Xu Z, Aronin N, Zamore PD (2003) Asymmetry in the assembly of the RNAi enzyme complex. Cell 115(2):199–208

Searles RP, Bergquam EP, Axthelm MK, Wong SW (1999) Sequence and genomic analysis of a Rhesus macaque rhadinovirus with similarity to Kaposi's sarcoma-associated herpesvirus/human herpesvirus 8. J Virol 73(4):3040–3053

Sears R, Nuckolls F, Haura E, Taya Y, Tamai K, Nevins JR (2000) Multiple Ras-dependent phosphorylation pathways regulate Myc protein stability. Genes Dev 14(19):2501–2514

Seo T, Lee D, Lee B, Chung JH, Choe J (2000) Viral interferon regulatory factor 1 of Kaposi's sarcoma-associated herpesvirus (human herpesvirus 8) binds to, and inhibits transactivation of CREB-binding protein. Biochem Biophys Res Commun 270(1):23–27. https://doi.org/10.1006/bbrc.2000.2393

Seo T, Park J, Lee D, Hwang SG, Choe J (2001) Viral interferon regulatory factor 1 of Kaposi's sarcoma-associated herpesvirus binds to p53 and represses p53-dependent transcription and apoptosis. J Virol 75(13):6193–6198. https://doi.org/10.1128/JVI.75.13.6193-6198.2001

Seo T, Lee D, Shim YS, Angell JE, Chidambaram NV, Kalvakolanu DV et al (2002) Viral interferon regulatory factor 1 of Kaposi's sarcoma-associated herpesvirus interacts with a cell death regulator, GRIM19, and inhibits interferon/retinoic acid-induced cell death. J Virol 76 (17):8797–8807

Seo T, Park J, Choe J (2005) Kaposi's sarcoma-associated herpesvirus viral IFN regulatory factor 1 inhibits transforming growth factor-beta signaling. Cancer Res 65(5):1738–1747. https://doi.org/10.1158/0008-5472.CAN-04-2374

Shamay M, Krithivas A, Zhang J, Hayward SD (2006) Recruitment of the de novo DNA methyltransferase Dnmt3a by Kaposi's sarcoma-associated herpesvirus LANA. Proc Natl Acad Sci U S A. 103(39):14554–14559. https://doi.org/10.1073/pnas.0604469103

Sharma-Walia N, Paul AG, Bottero V, Sadagopan S, Veettil MV, Kerur N et al (2010) Kaposi's sarcoma associated herpes virus (KSHV) induced COX-2: a key factor in latency, inflammation, angiogenesis, cell survival and invasion. PLoS Pathog 6(2):e1000777. https://doi.org/10.1371/journal.ppat.1000777

Sharp TV, Wang HW, Koumi A, Hollyman D, Endo Y, Ye H et al (2002) K15 protein of Kaposi's sarcoma-associated herpesvirus is latently expressed and binds to HAX-1, a protein with antiapoptotic function. J Virol 76(2):802–816

Shepard LW, Yang M, Xie P, Browning DD, Voyno-Yasenetskaya T, Kozasa T et al (2001) Constitutive activation of NF-kappa B and secretion of interleukin-8 induced by the G protein-coupled receptor of Kaposi's sarcoma-associated herpesvirus involve G alpha(13) and RhoA. J Biol Chem 276(49):45979–45987. https://doi.org/10.1074/jbc.M104783200

Shim WS, Ho IA, Wong PE (2007) Angiopoietin: a TIE(d) balance in tumor angiogenesis. Mol Cancer Res 5(7):655–665. https://doi.org/10.1158/1541-7786.MCR-07-0072

Shin YC, Nakamura H, Liang X, Feng P, Chang H, Kowalik TF et al (2006) Inhibition of the ATM/p53 signal transduction pathway by Kaposi's sarcoma-associated herpesvirus interferon regulatory factor 1. J Virol 80(5):2257–2266. https://doi.org/10.1128/JVI.80.5.2257-2266.2006

Shin YC, Joo CH, Gack MU, Lee HR, Jung JU (2008) Kaposi's sarcoma-associated herpesvirus viral IFN regulatory factor 3 stabilizes hypoxia-inducible factor-1 alpha to induce vascular endothelial growth factor expression. Cancer Res 68(6):1751–1759. https://doi.org/10.1158/0008-5472.CAN-07-2766

Skalsky RL, Samols MA, Plaisance KB, Boss IW, Riva A, Lopez MC et al (2007) Kaposi's sarcoma-associated herpesvirus encodes an ortholog of miR-155. J Virol 81(23):12836–12845. https://doi.org/10.1128/JVI.01804-07

Smith CG, Kharkwal H, Wilson DW (2017) Expression and subcellular localization of the Kaposi's sarcoma-associated herpesvirus K15P protein during latency and lytic reactivation in primary effusion lymphoma cells. J Virol 91(21). https://doi.org/10.1128/jvi.01370-17

Staskus KA, Zhong W, Gebhard K, Herndier B, Wang H, Renne R et al (1997) Kaposi's sarcoma-associated herpesvirus gene expression in endothelial (spindle) tumor cells. J Virol 71 (1):715–719

Stern-Ginossar N, Elefant N, Zimmermann A, Wolf DG, Saleh N, Biton M et al (2007) Host immune system gene targeting by a viral miRNA. Science 317(5836):376–381. https://doi.org/10.1126/science.1140956

Stern-Ginossar N, Gur C, Biton M, Horwitz E, Elboim M, Stanietsky N et al (2008) Human microRNAs regulate stress-induced immune responses mediated by the receptor NKG2D. Nat Immunol 9(9):1065–1073. https://doi.org/10.1038/ni.1642

Stine JT, Wood C, Hill M, Epp A, Raport CJ, Schweickart VL et al (2000) KSHV-encoded CC chemokine vMIP-III is a CCR4 agonist, stimulates angiogenesis, and selectively chemoattracts TH2 cells. Blood 95(4):1151–1157

Sturzl M, Blasig C, Schreier A, Neipel F, Hohenadl C, Cornali E et al (1997) Expression of HHV-8 latency-associated T0.7 RNA in spindle cells and endothelial cells of AIDS-associated, classical and African Kaposi's sarcoma. Int J Cancer 72(1):68–71

Sun R, Lin SF, Staskus K, Gradoville L, Grogan E, Haase A et al (1999) Kinetics of Kaposi's sarcoma-associated herpesvirus gene expression. J Virol 73(3):2232–2242

Sun Q, Matta H, Lu G, Chaudhary PM (2006) Induction of IL-8 expression by human herpesvirus 8 encoded vFLIP K13 via NF-kappaB activation. Oncogene 25(19):2717–2726. https://doi.org/10.1038/sj.onc.1209298

Sun Z, Xiao B, Jha HC, Lu J, Banerjee S, Robertson ES (2014) Kaposi's sarcoma-associated herpesvirus-encoded LANA can induce chromosomal instability through targeted degradation of the mitotic checkpoint kinase Bub1. J Virol 88(13):7367–7378. https://doi.org/10.1128/JVI.00554-14

Szpakowska M, Dupuis N, Baragli A, Counson M, Hanson J, Piette J et al (2016) Human herpesvirus 8-encoded chemokine vCCL2/vMIP-II is an agonist of the atypical chemokine receptor ACKR3/CXCR7. Biochem Pharmacol 114:14–21. https://doi.org/10.1016/j.bcp.2016.05.012

Tachibana K, Hirota S, Iizasa H, Yoshida H, Kawabata K, Kataoka Y et al (1998) The chemokine receptor CXCR4 is essential for vascularization of the gastrointestinal tract. Nature 393 (6685):591–594. https://doi.org/10.1038/31261

Takaoka A, Tamura T, Taniguchi T (2008) Interferon regulatory factor family of transcription factors and regulation of oncogenesis. Cancer Sci 99(3):467–478. https://doi.org/10.1111/j.1349-7006.2007.00720.x

Thakker S, Strahan RC, Scurry AN, Uppal T, Verma SC (2018) KSHV LANA upregulates the expression of epidermal growth factor like domain 7 to promote angiogenesis. Oncotarget. 9 (1):1210–1228. https://doi.org/10.18632/oncotarget.23456

Thomas M, Augustin HG (2009) The role of the Angiopoietins in vascular morphogenesis. Angiogenesis 12(2):125–137. https://doi.org/10.1007/s10456-009-9147-3

Thome M, Schneider P, Hofmann K, Fickenscher H, Meinl E, Neipel F et al (1997) Viral FLICE-inhibitory proteins (FLIPs) prevent apoptosis induced by death receptors. Nature 386 (6624):517–521. https://doi.org/10.1038/386517a0

Tomkowicz B, Singh SP, Cartas M, Srinivasan A (2002) Human herpesvirus-8 encoded Kaposin: subcellular localization using immunofluorescence and biochemical approaches. DNA Cell Biol 21(3):151–162. https://doi.org/10.1089/10445490252925413

Tomkowicz B, Singh SP, Lai D, Singh A, Mahalingham S, Joseph J et al (2005) Mutational analysis reveals an essential role for the LXXLL motif in the transformation function of the human herpesvirus-8 oncoprotein, kaposin. DNA Cell Biol 24(1):10–20. https://doi.org/10.1089/dna.2005.24.10

Tomlinson CC, Damania B (2004) The K1 protein of Kaposi's sarcoma-associated herpesvirus activates the Akt signaling pathway. J Virol 78(4):1918–1927

Tsai YH, Wu MF, Wu YH, Chang SJ, Lin SF, Sharp TV et al (2009) The M type K15 protein of Kaposi's sarcoma-associated herpesvirus regulates microRNA expression via its SH2-binding motif to induce cell migration and invasion. J Virol 83(2):622–632. https://doi.org/10.1128/JVI.00869-08

Umbach JL, Cullen BR (2010) In-depth analysis of Kaposi's sarcoma-associated herpesvirus microRNA expression provides insights into the mammalian microRNA-processing machinery. J Virol 84(2):695–703. https://doi.org/10.1128/JVI.02013-09

Verma SC, Lan K, Robertson E (2007) Structure and function of latency-associated nuclear antigen. Curr Top Microbiol Immunol 312:101–136

Verschuren EW, Klefstrom J, Evan GI, Jones N (2002) The oncogenic potential of Kaposi's sarcoma-associated herpesvirus cyclin is exposed by p53 loss in vitro and in vivo. Cancer Cell 2(3):229–241

Verschuren EW, Jones N, Evan GI (2004a) The cell cycle and how it is steered by Kaposi's sarcoma-associated herpesvirus cyclin. J Gen Virol 85(Pt 6):1347–1361. https://doi.org/10.1099/vir.0.79812-0

Verschuren EW, Hodgson JG, Gray JW, Kogan S, Jones N, Evan GI (2004b) The role of p53 in suppression of KSHV cyclin-induced lymphomagenesis. Cancer Res 64(2):581–589

Verzijl D, Fitzsimons CP, Van Dijk M, Stewart JP, Timmerman H, Smit MJ et al (2004) Differential activation of murine herpesvirus 68- and Kaposi's sarcoma-associated herpesvirus-encoded ORF74 G protein-coupled receptors by human and murine chemokines. J Virol 78(7):3343–3351

Vo MT, Smith BJ, Nicholas J, Choi YB (2019) Activation of NIX-mediated mitophagy by an interferon regulatory factor homologue of human herpesvirus. Nat Commun. 10(1):3203. https://doi.org/10.1038/s41467-019-11164-2

Wang HW, Sharp TV, Koumi A, Koentges G, Boshoff C (2002) Characterization of an anti-apoptotic glycoprotein encoded by Kaposi's sarcoma-associated herpesvirus which resembles a spliced variant of human survivin. EMBO J 21(11):2602–2615. https://doi.org/10.1093/emboj/21.11.2602

Wang HW, Trotter MW, Lagos D, Bourboulia D, Henderson S, Makinen T et al (2004) Kaposi sarcoma herpesvirus-induced cellular reprogramming contributes to the lymphatic endothelial gene expression in Kaposi sarcoma. Nat Genet 36(7):687–693. https://doi.org/10.1038/ng1384

Wang L, Dittmer DP, Tomlinson CC, Fakhari FD, Damania B (2006) Immortalization of primary endothelial cells by the K1 protein of Kaposi's sarcoma-associated herpesvirus. Cancer Res 66 (7):3658–3666. https://doi.org/10.1158/0008-5472.CAN-05-3680

Wang HR, Chen DL, Zhao M, Shu SW, Xiong SX, Gan XD et al (2012) C-reactive protein induces interleukin-6 and thrombospondin-1 protein and mRNA expression through activation of nuclear factor-kB in HK-2 cells. Kidney Blood Press Res. 35(4):211–219. https://doi.org/10.1159/000332402

Weber KS, Grone HJ, Rocken M, Klier C, Gu S, Wank R et al (2001) Selective recruitment of Th2-type cells and evasion from a cytotoxic immune response mediated by viral macrophage inhibitory protein-II. Eur J Immunol 31(8):2458–2466. https://doi.org/10.1002/1521-4141 (200108)31:8%3c2458:AID-IMMU2458%3e3.0.CO;2-L

Whitby D, Marshall VA, Bagni RK, Wang CD, Gamache CJ, Guzman JR et al (2004) Genotypic characterization of Kaposi's sarcoma-associated herpesvirus in asymptomatic infected subjects from isolated populations. J Gen Virol 85(Pt 1):155–163. https://doi.org/10.1099/vir.0.19465-0

Wies E, Mori Y, Hahn A, Kremmer E, Sturzl M, Fleckenstein B et al (2008) The viral interferon-regulatory factor-3 is required for the survival of KSHV-infected primary effusion lymphoma cells. Blood 111(1):320–327. https://doi.org/10.1182/blood-2007-05-092288

Wies E, Hahn AS, Schmidt K, Viebahn C, Rohland N, Lux A et al (2009) The Kaposi's sarcoma-associated herpesvirus-encoded vIRF-3 inhibits cellular IRF-5. J Biol Chem 284 (13):8525–8538. https://doi.org/10.1074/jbc.M809252200

Willis SN, Adams JM (2005) Life in the balance: how BH3-only proteins induce apoptosis. Curr Opin Cell Biol 17(6):617–625. https://doi.org/10.1016/j.ceb.2005.10.001

Wong EL, Damania B (2006) Transcriptional regulation of the Kaposi's sarcoma-associated herpesvirus K15 gene. J Virol 80(3):1385–1392. https://doi.org/10.1128/JVI.80.3.1385-1392.2006

Xi X, Persson LM, O'Brien MW, Mohr I, Wilson AC (2012) Cooperation between viral interferon regulatory factor 4 and RTA to activate a subset of Kaposi's sarcoma-associated herpesvirus lytic promoters. J Virol 86(2):1021–1033. https://doi.org/10.1128/JVI.00694-11

Xiang Q, Ju H, Li Q, Mei SC, Chen D, Choi YB et al (2018) Human herpesvirus 8 interferon regulatory factors 1 and 3 mediate replication and latency activities via interactions with USP7 deubiquitinase. J Virol 92(7). https://doi.org/10.1128/jvi.02003-17

Yang TY, Chen SC, Leach MW, Manfra D, Homey B, Wiekowski M et al (2000) Transgenic expression of the chemokine receptor encoded by human herpesvirus 8 induces an angioproliferative disease resembling Kaposi's sarcoma. J Exp Med 191(3):445–454

Yoo J, Kang J, Lee HN, Aguilar B, Kafka D, Lee S et al (2010) Kaposin-B enhances the PROX1 mRNA stability during lymphatic reprogramming of vascular endothelial cells by Kaposi's sarcoma herpes virus. PLoS Pathog 6(8):e1001046. https://doi.org/10.1371/journal.ppat.1001046

Yoshizaki K, Matsuda T, Nishimoto N, Kuritani T, Taeho L, Aozasa K et al (1989) Pathogenic significance of interleukin-6 (IL-6/BSF-2) in Castleman's disease. Blood 74(4):1360–1367

Young LS, Murray PG (2003) Epstein-Barr virus and oncogenesis: from latent genes to tumours. Oncogene 22(33):5108–5121. https://doi.org/10.1038/sj.onc.1206556

Zeng Y, Yi R, Cullen BR (2003) MicroRNAs and small interfering RNAs can inhibit mRNA expression by similar mechanisms. Proc Natl Acad Sci U S A. 100(17):9779–9784. https://doi.org/10.1073/pnas.1630797100

Zhang YJ, Bonaparte RS, Patel D, Stein DA, Iversen PL (2008) Blockade of viral interleukin-6 expression of Kaposi's sarcoma-associated herpesvirus. Mol Cancer Ther 7(3):712–720. https://doi.org/10.1158/1535-7163.MCT-07-2036

Zhang G, Chan B, Samarina N, Abere B, Weidner-Glunde M, Buch A et al (2016a) Cytoplasmic isoforms of Kaposi sarcoma herpesvirus LANA recruit and antagonize the innate immune DNA sensor cGAS. Proc Natl Acad Sci U S A. 113(8):E1034–E1043. https://doi.org/10.1073/pnas.1516812113

Zhang Z, Chen W, Sanders MK, Brulois KF, Dittmer DP, Damania B (2016b) The K1 protein of Kaposi's sarcoma-associated herpesvirus augments viral lytic replication. J Virol 90(17):7657–7666. https://doi.org/10.1128/JVI.03102-15

Zhao J, Punj V, Matta H, Mazzacurati L, Schamus S, Yang Y et al (2007) K13 blocks KSHV lytic replication and deregulates vIL6 and hIL6 expression: a model of lytic replication induced clonal selection in viral oncogenesis. PLoS ONE 2(10):e1067. https://doi.org/10.1371/journal.pone.0001067

Zhong W, Wang H, Herndier B, Ganem D (1996) Restricted expression of Kaposi sarcoma-associated herpesvirus (human herpesvirus 8) genes in Kaposi sarcoma. Proc Natl Acad Sci U S A. 93(13):6641–6646

Zhu Y, Ramos da Silva S, He M, Liang Q, Lu C, Feng P et al (2016) An oncogenic virus promotes cell survival and cellular transformation by suppressing glycolysis PLoS Pathog 12(5):c1005648. https://doi.org/10.1371/journal.ppat.1005648

Ziech D, Franco R, Pappa A, Malamou-Mitsi V, Georgakila S, Georgakilas AG et al (2010) The role of epigenetics in environmental and occupational carcinogenesis. Chem Biol Interact 188(2):340–349. https://doi.org/10.1016/j.cbi.2010.06.012

Ziegelbauer JM, Sullivan CS, Ganem D (2009) Tandem array-based expression screens identify host mRNA targets of virus-encoded microRNAs. Nat Genet 41(1):130–134. https://doi.org/10.1038/ng.266

Zimring JC, Goodbourn S, Offermann MK (1998) Human herpesvirus 8 encodes an interferon regulatory factor (IRF) homolog that represses IRF-1-mediated transcription. J Virol 72(1):701–707

Zong JC, Ciufo DM, Alcendor DJ, Wan X, Nicholas J, Browning PJ et al (1999) High-level variability in the ORF-K1 membrane protein gene at the left end of the Kaposi's sarcoma-associated herpesvirus genome defines four major virus subtypes and multiple variants or clades in different human populations. J Virol 73(5):4156–4170

Merkel Cell Polyomavirus and Human Merkel Cell Carcinoma

Wei Liu and Jianxin You

1 Introduction

Merkel cell polyomavirus (MCPyV) is a member of the Polyomaviridae. It was first identified in Merkel cell carcinoma (MCC) using digital transcriptome subtraction methodology (Feng et al. 2008). MCPyV is also the first polyomavirus proven to be associated with human cancer. MCPyV-associated MCC typically presents as a type of neuroendocrine cancer. In 1972, it was first described by Dr. Cyril Toker, who named it "trabecular carcinoma of the skin" (Toker 1972). MCC is one of the most aggressive skin cancers, with disease-associated mortality of nearly 46% (Becker 2010; Harms 2017; Agelli et al. 2010), which exceeds the mortality rate of melanoma. It kills more patients than some well-known cancers such as cutaneous T-cell lymphoma and chronic myelogenous leukemia (Lemos and Nghiem 2007; Bhatia et al. 2011).

About 80% of MCC cases can be directly linked to MCPyV infection (Feng et al. 2008; Sihto et al. 2009). Immunosuppression caused by aging (Fitzgerald et al. 2015; Bichakjian et al. 2007), HIV infection (Engels et al. 2002), and organ transplant (Clarke et al. 2015) has been shown to stimulate the development of MCPyV-positive MCC. Sunlight exposure and ultraviolet (UV) radiation are also important risk factors for MCC development (Lunder and Stern 1998; Heath et al. 2008).

Epidemiological surveys for MCPyV seropositivity (Tolstov et al. 2009; Kean et al. 2009) and sequencing analyses (Foulongne et al. 2012) have shown that MCPyV is an abundant virus frequently shed from healthy human skin, suggesting that MCPyV infection is widespread in the general population (Schowalter et al.

W. Liu · J. You (✉)
Department of Microbiology, Perelman School of Medicine, University of Pennsylvania,
Philadelphia, PA 19104, USA
e-mail: jianyou@pennmedicine.upenn.edu

© Springer Nature Switzerland AG 2021
T.-C. Wu et al. (eds.), *Viruses and Human Cancer*, Recent Results
in Cancer Research 217, https://doi.org/10.1007/978-3-030-57362-1_12

2010). Most of the primary MCPyV infection occurs during early childhood. Once acquired, the virus becomes a permanent component of the skin flora (Chen et al. 2011). Integration of MCPyV genome into the host genome has been shown to occur before the clonal expansion of the tumor, in which continued expression of the viral oncogenes drives MCC tumor growth. These findings provide key evidence to support the oncogenic role of MCPyV in MCC tumor development (Feng et al. 2008; Shuda et al. 2008). The incidence of MCC has tripled over the past twenty years (Hodgson 2005; Stang et al. 2018) and increased by >95% in the US since 2000 (Paulson et al. 2017). With the high prevalence of MCPyV infection and the increasing amount of MCC diagnosis (Hodgson 2005), there is a growing concern for MCC (Hodgson 2005). Understanding MCPyV biology and its oncogenic mechanism will provide insights for developing novel prevention and treatment strategies for MCC. In this chapter, we present the recent advancement in MCPyV virology and associated MCC tumors.

2 The Life Cycle of MCPyV

2.1 MCPyV Genome Structure

MCPyV is a small, non-enveloped, icosahedral, double-stranded circular DNA virus (Feng et al. 2008). The 5.4 kb viral genome encodes seven gene products under the control of early and late promoters (Fig. 1). A non-coding regulatory

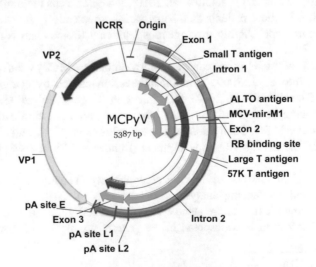

Fig. 1 MCPyV genome. This schematic diagram shows the non-coding regulatory region (NCRR), early genes, late genes, and a microRNA (miR-M1) encoded by the MCPyV genome. Alternate LT ORF (ALTO), Early gene poly A site (pA site E), Late gene poly A site 1 (pA site L1), and Late gene poly A site 2 (pA site L2)

region (NCRR) divides the genome into early and late regions (Gjoerup and Chang 2010). NCRR contains the viral origin (Ori) of replication and bidirectional promoters for viral transcription (Harrison et al. 2011; Kwun et al. 2009).

The MCPyV early region encodes large tumor antigen (LT antigen), small tumor antigen (sT antigen), 57-kilodalton tumor antigen (57KT antigen), and alternate LT ORF (ALTO) (Feng et al. 2008; Schowalter et al. 2010; Carter et al. 2013). Among these early proteins, LT antigen is the largest viral coding protein. It is encoded by T antigen exon 1, exon 2, intron 2 and exon 3 (Shuda et al. 2008). LT antigen not only regulates the viral genome replication but also controls the host cell cycle progression. The N-terminus of the protein contains a conserved region 1 (CR1), a DnaJ domain (for binding heat-shock proteins), and a LxCxE motif that is responsible for binding retinoblastoma protein (RB), which regulates the cell cycle (Feng et al. 2008; Shuda et al. 2008). The C-terminal region of LT antigen contains an Origin Binding Domain (OBD), which binds the MCPyV Ori GAGGC pentameric sequences (Harrison et al. 2011; Kwun et al. 2009), and a helicase domain, which unwinds double-strand MCPyV DNA to initiate replication (Li et al. 2013). The MCPyV sT antigen is encoded by T antigen exon 1 and intron 1 (Shuda et al. 2008). Therefore, it shares the LT N-terminal CR1 and DnaJ domains but carries a unique C-terminal protein phosphatase 2A (PP2A) binding site (Kwun et al. 2015) as well as two highly conserved iron-sulfur clusters, [2Fe–2S] and [4Fe–4S] (Tsang et al. 2016). The MCPyV 57KT antigen is encoded by T antigen exon 1, exon 2 and exon 3 (Cheng et al. 2013). Therefore, it does not have an OBD domain and a complete helicase domain. The MCPyV ALTO protein is translated from an overprinting ORF that is +1 frameshifted relative to the T antigen exon 2 (Carter et al. 2013). In contrast to LT and sT antigens, the function of 57KT antigen and ALTO remains poorly understood.

The MCPyV late region encodes structural proteins VP1 and VP2 (Fig. 1) (Schowalter et al. 2010; Schowalter and Buck 2013), which are the major and minor subunits of the viral capsid, respectively. VP1 and VP2 form the capsids that encapsidate the MCPyV genome (Schowalter and Buck 2013; Schowalter et al. 2011). The major capsid protein VP1 is indispensable and sufficient for producing pseudovirions, whereas the minor protein VP2 confers specificity in host cell targeting (Schowalter and Buck 2013). MCPyV minor capsid protein VP3 is not detectable in either MCPyV-infected cells or native MCPyV virions. Phylogenetic analysis suggests that MCPyV belongs to a unique clade of polyomaviruses that does not encode the conserved VP3 N-terminus (Schowalter and Buck 2013).

Besides the early and late genes, MCPyV also encodes a microRNA, miR-M1 (Seo et al. 2009) (Fig. 1), which has been shown to down-regulate the expression of LT. miR-M1 also appears to be important for long-term MCPyV episome maintenance in cell culture as well as for persistent infection in vivo (Theiss et al. 2015).

2.2 MCPyV Entry into Host Cells

The particle size of MCPyV virions is about 50 nM. The viral capsid is composed of VP1 and VP2 at a ratio of 5:2. Like most polyomaviruses, the major capsid protein VP1 determines antigenicity and receptor specificity. The entry of MCPyV virions into the host cell is mediated by VP1 binding to cellular receptors. Although VP2 knockout does not affect virion assembly, viral DNA packaging, or cell attachment, it reduces native MCPyV infectivity by more than 100-fold (Schowalter and Buck 2013).

MCPyV enters into its target cells through a gradual and asynchronous motion. The initial attachment receptors for MCPyV VP1 are sulfated glycosaminoglycans (GAGs), specifically the N-sulfated and/or 6-O-sulfated forms of heparan sulfate (Schowalter et al. 2011). It has been shown by X-ray structure analysis that a shallow binding site on the apical surface of the VP1 capsomer recognizes the linear sialylated disaccharide Neu5Ac-α2,3-Gal, which is present in ganglioside GT1b containing sialic acids on both arms (Erickson et al. 2009). The study also revealed VP1's interaction with the Neu5Ac motif of GD1a, 3SLN, and DSL oligosaccharides (Neu et al. 2012). This finding indicates that, during MCPyV infectious entry, sialylated glycans serve as post-attachment co-receptors after MCPyV primary attachment through GAGs. MCPyV penetrates cells through caveolar/lipid raft-mediated endocytosis (Becker et al. 2019). The virus internalizes in small endocytic pits, which deliver the virus to endosomes (Becker et al. 2019). From there, the virus moves to the endoplasmic reticulum by taking advantage of microtubule transport, acidification of endosomes, and a functional redox environment. The virus was found to gain a membrane envelope within endosomes, a phenomenon that has not been observed for other viruses (Becker et al. 2019).

2.3 MCPyV Replication

Both MCPyV LT and sT antigens play an important role in replicating viral DNA. After binding to the viral replication Ori through its OBD domain, LT unwinds the Ori using its helicase domain to initiate viral DNA replication (Kwun et al. 2009; Diaz et al. 2014). Several LT phosphorylation sites have been discovered through mass spectrometry analysis (Diaz et al. 2014). Mutagenesis and functional analysis revealed that phosphorylation of LT at these sites dynamically regulates viral replication by controlling Ori recognition, adjusting LT-Ori affinity, as well as initiating viral DNA unwinding (Diaz et al. 2014). As discussed below, MCPyV LT contributes to viral genome replication by recruiting cellular proteins as well. MCPyV sT is also required for efficient viral DNA replication. It does so mostly through increasing LT protein stability (Kwun et al. 2013). It was discovered that LT is normally targeted for proteasomal degradation by the cellular SCF^{Fbw7} E3 ligase. sT can bind and inhibit this E3 ligase to prevent LT degradation (Kwun et al. 2013).

Several host proteins involved in MCPyV replication have been discovered. Vam6p, a vacuolar sorting protein associated with MCPyV LT, is the first cellular factor shown to have an effect on MCPyV replication (Feng et al. 2011). Mutation of the Vam6p binding site on LT enhances MCPyV replication, whereas overexpression of exogenous Vam6p reduces MCPyV virion production by more than 90% (Feng et al. 2011). These studies suggest that Vam6p can inhibit MCPyV replication through its interaction with LT antigen (Feng et al. 2011). Bromodomain protein 4 (BRD4) is another cellular protein that interacts with MCPyV LT antigen and plays a vital role in viral DNA replication. BRD4 colocalizes with the MCPyV LT/replication origin complexes (MCPyV replication center) in the nucleus and recruits replication factor C (RFC) to the viral replication sites (Wang et al. 2012). *BRD4* knockdown inhibits MCPyV replication, which can be rescued by adding purified recombinant BRD4 protein in vitro. Human DNA damage response (DDR) factors are important for MCPyV DNA replication as well. Components of the Ataxia telangiectasia-mutated (ATM)- and Ataxia-telangiectasia-mutated and Rad3-related (ATR)-mediated DDR pathways accumulate in MCPyV replication centers inside the cells infected with recombinant MCPyV virions (Li et al. 2013). This DDR factor recruitment does not happen when a replication-defective LT mutant or an MCPyV Ori mutant was introduced instead of their wild-type counterparts (Li et al. 2013). Components of promyelocytic leukemia nuclear bodies (PML-NB) are another set of host factors that control MCPyV DNA replication. Notably, MCPyV replication was increased in cells depleted of Sp100, one of the key factors of PML-NBs. This observation suggests that Sp100 is a negative regulator of MCPyV DNA replication (Neumann et al. 2016).

2.4 Assembly and Release

The assembly and release processes of MCPyV are largely unexplored. Based on VP1 protein localization during virus infection (Schowalter et al. 2011; Liu et al. 2016), it has been suggested that the virus packages in the nucleus and induces cell lysis events so that it can be released from the infected cells (Liu et al. 2016).

2.5 MCPyV Host Cellular Tropism

Although the MCPyV binding factors, such as heparan sulfate and sialic acid that mediates viral attachment and entry, are ubiquitously expressed, MCPyV infects and replicates poorly in the majority of cell lines tested in a number of studies (Schowalter et al. 2011; Neu et al. 2012). The cells naturally infected by MCPyV have not been discovered until very recently. Several lines of evidence implicate the skin as the major site of MCPyV productive infection in humans. First, multiple deep sequencing studies have detected persistent and asymptomatic infection of MCPyV in adult skin (Foulongne et al. 2012; Schowalter et al. 2010). In addition, cell culture experiments suggest that the cell types conducive for MCPyV replication are either epithelial or fibroblast in origin (Kwun et al. 2009; Feng et al. 2011; Wang et al. 2012). Finally, MCC is a tumor derived from the dermis and the presumed cells of origin for MCC, Merkel cells, reside in the epidermis. Following this line of reasoning, different types of cells in human skin were surveyed for MCPyV infectability. It was discovered that among all of the skin cell types tested, only human dermal fibroblasts (HDFs) could support robust MCPyV propagation (Liu et al. 2016). It was also found that epidermal growth factor (EGF) and basic fibroblast growth factor (bFGF) are essential to support MCPyV infection in HDFs, likely by inducing cellular factors to promote a cellular environment beneficial to MCPyV infection (Liu et al. 2016). Interestingly, in human skin, EGF and FGF are typically stimulated during the wounding and healing process (Quan et al. 2009), suggesting that wounding of human skin may spur MCPyV infection. It was further discovered that the expression of matrix metalloproteinases (MMPs), which can be stimulated by the Wnt/β-catenin signaling pathway, is important for MCPyV infection of HDFs (Liu et al. 2016). Several MCC risk factors, including UV exposure and aging, can upregulate MMPs (Quan et al. 2009; Cho et al. 2009; Fisher et al. 1996; Gill and Parks 2008; Quan and Fisher 2015; Varani et al. 2006), suggesting they may promote MCPyV infection to stimulate MCC development.

Despite the discovery of productive MCPyV replication in HDFs, much remains to be elucidated with respect to MCPyV natural infection and host cellular tropism. For instance, both MCPyV attachment receptors, sialic acid and heparan sulfate are ubiquitously expressed. It is unclear how MCPyV is able to effectively enter HDFs but not many other cell types (Schowalter et al. 2012). MCPyV DNA has also been detected in respiratory, urine, and blood samples (Spurgeon and Lambert 2013). Therefore, the range of tissues in which MCPyV establishes persistent infection remains unclear.

2.6 MCPyV Species Tropism

Mechanistic studies aiming to fully clarify the oncogenic mechanisms of MCPyV have been hampered by the lack of MCPyV infection animal models. To overcome this hurdle, recombinant MCPyV virions and several MCPyV chimeric viruses were used to test the infectivity of dermal fibroblasts isolated from a variety of model animals, including chimpanzee (*Pan troglodytes*), mouse (*Mus musculus*), rabbit (*Oryctolagus cuniculus*), rat (*Rattus norvegicus*), rhesus macaque (*Macaca mulatta*), common woolly monkey (*Lagothrix lagotricha),* patas monkey (*Erythrocebus patas*), red-chested mustached tamarin (*Saguinus labiatus*) and tree shrew (*Tupaia belangeri*). Interestingly, among all of the cells tested, only chimpanzee dermal fibroblasts supported strong MCPyV gene expression and viral replication, and they did so to a much greater extent when compared to HDFs. Therefore, among all of the tested small mammals and non-human primates, chimpanzee represents the only animal type that can support native MCPyV infection (Liu et al. 2018). Since chimpanzee is not available to be used as an animal model for MCPyV research, additional studies are needed to establish more suitable animal models. Chimeric viruses that can overcome species-specific restriction should be constructed to support these studies.

3 The MCPyV Tumorigenic Mechanisms

MCPyV DNA is frequently integrated into MCC genome (Feng et al. 2008; Liu et al. 2016; Krump and You 2018). The MCPyV genome in MCC tumor cells is invariably truncated by the integration event such that it is replication-incompetent, yet the cell growth-promoting functions of viral genes called tumor antigens are preserved (Shuda et al. 2008). MCPyV-positive MCCs typically express intact sT and a tumor-specific truncation mutant of LT that preserves the N-terminal half of LT, referred to as LTT (tumor-derived LT) antigen (Feng et al. 2008; Shuda et al. 2008; Sastre-Garau et al. 2009; Borchert et al. 2014; Houben et al. 2012). MCPyV-positive MCCs harbor very few genetic mutations (Harms et al. 2015; Goh et al. 2016), suggesting that the expression of these viral oncogenes is sufficient to drive tumor development. Indeed, sT and LTT have demonstrated robust oncogenic potential to promote tumorigenesis (Spurgeon and Lambert 2013; Grundhoff and Fischer 2015; Wendzicki et al. 2015; Shuda et al. 2011; Verhaegen et al. 2014). MCPyV-positive MCC cells are addicted to sT/LTT oncogenes and require their continued expression from integrated viral genome to survive (Houben et al. 2010; Shuda et al. 2014). Knockdown of sT/LTT antigens induces growth arrest and cell death in MCPyV-positive MCC cells (Houben et al. 2010; Shuda et al. 2014) and leads to tumor regression in xeno-transplantation (Houben et al. 2012). These key findings demonstrated the important impact of viral oncogene expression in the development of MCPyV-associated MCCs.

A common characteristic of MCPyV genomes integrated into the MCC genome is the selection for mutations in the LT antigen coding sequence that introduce premature stop codons, which delete the LT C-terminal helicase domains (Shuda et al. 2008). The resulting tumor-specific LTT antigen retains the CR1, DnaJ, and RB-binding motifs, allowing the LTT molecules to efficiently disrupt the host cell cycle (Shuda et al. 2008). Phosphorylation of serine 220 of MCPyV LTT is required this viral oncogene to inactivate RB in MCC cells (Schrama et al. 2016). This RB-inhibiting function of MCPyV LTT antigen has also been shown to stimulate cell proliferation by upregulating cyclin E and CDK2 (Richards et al. 2015).

Unlike MCPyV LT, intact MCPyV sT is consistently expressed in MCPyV-positive MCC tumors. Nearly no mutations have been detected in the sT-coding regions integrated into MCC genome (Shuda et al. 2008; Starrett 2017), corroborating a key functional role for this viral oncogene in the development of MCPyV-positive tumors. MCPyV sT has been shown to transform immortalized rat fibroblasts in cell culture (Shuda et al. 2011). Its transforming activity has also been demonstrated in transgenic mouse models (Verhaegen et al. 2017; Verhaegen et al.

2015; Shuda et al. 2015). In line with these observations, co-expression of MCPyV sT antigen with Atonal bHLH transcription factor 1 (ATOH1) induces cell aggregates with morphology and marker expression pattern mimicking MCC (Verhaegen et al. 2017). The oncogenic activity of MCPyV sT antigen is mostly supported by its ability to induce hyperphosphorylated, and thus inactivated, 4E-BP1, causing dysfunction of cap-dependent translation to stimulate cell proliferation and transformation (Shuda et al. 2011; Velasquez et al. 2016; Shuda et al. 2015; Sun et al. 2011). As described above, MCPyV sT antigen can inhibit the E3 ubiquitin ligase SCF^{Fbw7}. This sT function prevents the proteasomal degradation of MCPyV LT antigen as well as several important cellular proliferative proteins, such as c-Myc and cyclin E, which are normally targeted by SCF^{Fbw7} (Kwun et al. 2013). From a large-scale co-immunoprecipitation and proteomic study, MCPyV sT antigen was found to be associated with the MYCL-EP400 complex, which together bind promoters of specific cellular genes to stimulate their expression and cellular transformation (Cheng et al. 2017). In line with this finding, a transcriptome analysis of normal human fibroblasts with inducible expression of MCPyV sT revealed its ability to dynamically change cellular gene expression (Berrios et al. 2016). sT expression leads to upregulation of glycolytic genes, including the monocarboxylate lactate transporter SLC16A1 (MCT1) (Berrios et al. 2016). Additional functional analysis suggested that these gene expression changes lead to elevated aerobic glycolysis, which may also contribute to the MCPyV-dependent cellular transformation (Berrios et al. 2016). In addition, MCPyV sT modulates cellular microtubule network, motility, and migration through upregulation of microtubule- and actin-associated proteins as well as the cellular sheddases, A disintegrin and metalloproteinase (ADAM) 10 and 17. Together, these cellular factors contribute to sT-induced cell dissociation and motility, a feature that may support MCPyV-mediated cellular transformation and metastasis (Nwogu et al. 2018; Stakaityte 2018; Knight et al. 2015).

Clonal integration of MCPyV DNA into the host genome is a key causative factor for MCC development (Houben et al. 2009; Chang and Moore 2012). However, the molecular mechanism that contributes to viral integration remains poorly understood. As described above, both LTT and sT expressed from the integrated viral genome demonstrate strong potential for modulating cellular proteins to drive cell proliferation. The function of these viral oncogenes offers a strong growth advantage for selecting the precancerous cells with the integrated viral genome expressing these viral oncogenes. Another selective pressure may be presented by the loss of viral DNA replication activity caused by the deletion of the LT C-terminal OBD and helicase domains after the integration of the viral DNA into the host genome. As continuous LT-mediated replication from the integrated viral Ori could result in replication fork collisions and double-strand breaks in the host DNA, disrupting the OBD and helicase domains of LT antigen would relieve this genotoxic stress. Finally, other functional activities of the LT antigen C-terminal domain may also need to be negatively selected during tumorigenesis. For example, expression of the C-terminal helicase-containing region of MCPyV LT induces a host cellular DNA damage response, leading to p53 activation, upregulation of its downstream target genes, and cell cycle arrest (Li et al. 2013). Compared to the N-terminal MCPyV LT region normally preserved and expressed in MCC tumors, full-length MCPyV LT shows a significantly decreased potential to support cellular proliferation, focus formation, and anchorage-independent cell growth (Li et al. 2013). It was further discovered that activated ATM phosphorylates the MCPyV LT C-terminal residue serine 816, which functions to promote apoptosis (Li et al. 2015). In line with these observations, an additional study showed that expression of the MCPyV LT antigen C-terminal 100 residues was sufficient to cause growth inhibition in many different cell types (Cheng et al. 2013). Together, the growth-inhibitory activities of the MCPyV LT C-terminal domain revealed in these studies suggest that the truncation mutations that remove the MCPyV LT C-terminal region found in MCC is not only needed to prevent replication of the integrated viral genome but also essential for overcoming the anti-tumorigenic properties intrinsic to the MCPyV LT C-terminus (Li et al. 2013; Cheng et al. 2013; Li et al. 2015). MCPyV viral genome integration, therefore, promotes MCC tumorigenesis by overcoming the obstacles to oncogenesis presented by replicative stress, DNA damage responses, and cell cycle arrest.

Several recent studies have attempted to establish transgenic mouse models for MCPyV oncogenes. Although expression of MCPyV sT and LT can induce hyperplasia and benign lesions in the epidermis of the transgenic mice, they failed to induce lesions that fully recapitulate MCC pathogenesis (Verhaegen et al. 2014; Spurgeon et al. 2015). Therefore, better MCPyV animal models are needed to investigate MCPyV-induced MCC development in vivo.

4   Therapeutic Strategy Targeting MCPyV Infection

An FDA-approved MEK inhibitor, Trametinib, inhibits MCPyV infection in cultured HDFs, making it the first drug capable of blocking the viral infection (Liu et al. 2016). Trametinib could potently reduce the MCPyV viral load in immunocompromised patients. Therefore, it has the potential of preventing the development of MCC tumors in these individuals (Liu et al. 2016).

5   Human Merkel Cell Carcinoma

5.1   MCC Histopathologic Features

MCC is a rare and aggressive cutaneous malignancy of neuroendocrine origin. Based on its histopathologic patterns, it was first named by Dr. Toker as "trabecular carcinoma of the skin" (Toker 1972). Additional names for MCC include the Toker tumor, primary cutaneous neuroendocrine tumor, primary small cell carcinoma of the skin, and malignant trichodiscoma (Schwartz and Lambert 2005). It is the

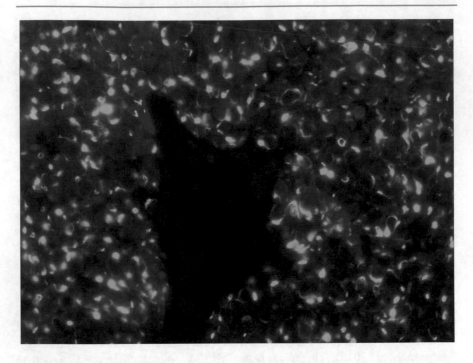

Fig. 2 Cytokeratin 20 (CK20) staining of MCC. Perinuclear dot-like cytokeratin 20 staining (Green). Nuclei (Blue)

second most common cause of skin cancer death after melanoma (Albores-Saavedra et al. 2010). Because of the similarity in histopathologic patterns, MCC was sometimes misdiagnosed as the malignant small blue cell tumors (Xue and Thakuria 2019). The advent of immunohistochemistry staining for Cytokeratin 20 (CK20) has greatly improved the diagnostic accuracy of MCC (Fig. 2) (Scott and Helm 1999).

5.2   Origin of MCC

The origin of tumors remains a central question for MCC research. Historically, MCC tumors were thought to arise from Merkel cells, which are mechanoreceptor cells located in the basal layer of the epidermis and also in hair follicles of the skin (Winkelmann and Breathnach 1973) that share the immunohistochemical marker CK20 with MCC tumors (Table 1). However, Merkel cells are known to be derived from the epidermal lineage (Morrison et al. 2009; Van Keymeulen et al. 2009), whereas MCCs mostly develop within the dermis and subcuitis (Calder and Smoller 2010). In addition, Merkel cells are post-mitotic and have lost the proliferative

Table 1 Prospective cells of origin for MCC and the markers and key characteristics they shared with MCC (Liu et al. 2016)

	Markers/characteristics
Merkel cell carcinoma	CK20+, NSE+, PAX5+, NFP+, TdT+, TTF-1−, LCA−, S100−
Human dermal fibroblasts	Major skin cell types that support productive MCPyV infection, transcription, and replication
Merkel cells	CK20+, electron microscope morphology
Pro/pre-B and pre-B cells	PAX5+, TdT+

CK20: cytokeratin 20; NSE: neuron-specific enolase; NFP: neurofilament protein; PAX5: Paired box gene 5; TdT: terminal deoxynucleotidyl transferase; LCA: leukocyte common antigen; TTF-1: thyroid transcription factor 1

activity, thus are less likely to develop into tumorigenic MCC cells (Vaigot et al. 1987; Moll et al. 1996). These evidence suggest that Merkel cells are not likely to be the cells of origin for MCC.

Identification of HDFs as the target cells of productive MCPyV infection (Liu et al. 2016) is in line with the clinical observation that most MCCs develop in the dermis (Calder and Smoller 2010). It also suggests that MCPyV infection of HDFs in the dermis may ultimately give rise to MCC tumors. For example, it is possible that over-amplification of MCPyV in HDFs may stimulate viral genome integration into the host DNA to induce cellular transformation. Alternatively, MCPyV actively replicating in HDFs located at the boundary between epidermis and dermis may release viral particles that inadvertently infect bystander Merkel cells or their precursor cells present in the immediate vicinity within the basal layer of the epidermis. The dead-end replication environment present in Merkel cells may favor viral integration and cellular transformation.

Recent studies also showed that MCC tumors express Paired box gene 5 (PAX5) and terminal deoxynucleotidyl transferase (TdT), which, under physiological conditions, are specifically expressed in pro/pre-B and pre-B cells (https://pubmed. ncbi.nlm.nih.gov/23576560/). Based on this finding, it was speculated that MCC tumors might originate from pro-B/pre-B cells, although this theory remains to be examined experimentally.

5.3 MCPyV-Positive and -Negative MCCs

The majority of MCC tumors are associated with MCPyV infection, while the remaining can be attributed to UV-induced mutation (Harms et al. 2015; Goh et al. 2016; Starrett 2017; Cohen et al. 2016). In the Northern hemisphere, approximately 80% of MCC tumors carry monoclonally integrated MCPyV genome (Feng et al. 2008; Sihto et al. 2009). However, the percentage of MCPyV-positive MCC is significantly lower in other geographic areas such as Australia ($\sim 30\%$) (Garneski et al. 2009). The fact that MCPyV-positive MCC tumors typically carry an integrated viral genome is reminiscent of papillomavirus-induced cancers (Feng et al. 2008).

However, unlike papillomavirus-associated malignancies, MCPyV-positive MCC tumors grow swiftly with no clear precancerous stage. MCPyV-positive MCC tumors are believed to be derived from monoclonal expansion and have a very low mutation burden (Harms et al. 2015; Goh et al. 2016; Starrett et al. 2017; Cohen et al. 2016). These findings suggest that MCPyV-induced MCC tumors may develop rapidly after MCPyV genome integration.

Whole-genome sequencing has begun to shed light on the differences in the causes of MCPyV-positive and -negative MCC tumors. These genetic studies revealed that UV radiation is the primary cause of MCPyV-negative MCC tumors, which accounts for about 20% of all MCC cases (Harms et al. 2015; Goh et al. 2016; Starrett 2017; Cohen et al. 2016). Compared to MCPyV-positive MCCs, MCPyV-negative tumors revealed a prominent UV-mediated DNA damage signature and displayed a dramatically higher mutation burden (Harms et al. 2015; Goh et al. 2016; Starrett et al. 2017; Cohen et al. 2016). Common cancer activating mutations often observed in MCPyV-negative MCC tumors include mutations in RB1, p53, PIK3CA as well as key components of the host DNA damage response, Notch signaling, and chromatin remodeling pathways (Harms et al. 2015; Goh et al. 2016; Starrett 2017; Cohen et al. 2016). Much lower levels of cancer-promoting mutations were observed in MCPyV-positive MCC tumors, supporting that the MCPyV sT and LTT oncogenes expressed from the integrated viral genomes are the predominant oncogenic drivers for these tumors (Harms et al. 2015; Goh et al. 2016; Starrett et al. 2017; Cohen et al. 2016).

6 Current Therapeutic Strategies for MCC

6.1 Surgery and Radiation Therapy

Early-stage, localized MCC tumors are mostly treated with wide-section surgery and radiation. However, MCC frequently undergoes metastasis, which increases the chance that tumors may be developed in body sites that are harder to reach and fully eradicated with radiotherapy (Bichakjian et al. 2007; Allen et al. 1999). Therefore, chemotherapy has been applied to treat advanced-stage MCCs. Despite the early MCC response to chemotherapy, the duration of the response is usually short-lived and many tumors often develop chemoresistance (Brummer et al. 2016; Saini and Miles 2015; Cassler et al. 2016). Because chemotherapy also has an immunosuppressive effect, which counteracts the cellular immune response to MCC tumors, it offers little overall survival benefit for MCC tumors. Currently, there are very few feasible options for patients with advanced MCCs (Cassler et al. 2016).

6.2 Immunotherapy

MCPyV antigens or ultraviolet-mutation-associated neoantigens expressed in MCC tumors represent ideal targets for anti-tumor immunotherapy. Robust intratumoral

CD8+ T-cell infiltration has been associated with 100% MCC-specific survival, independent of tumor stage (Paulson et al. 2011). This strong correlation between immune function and prognosis reveals the potential for using immunotherapies to treat metastatic MCCs.

6.2.1 Immune Checkpoint Inhibitor Therapy

Targeting the programmed cell death receptor 1/programmed cell death ligand 1 (PD-1/PD-L1) checkpoint has become an attractive treatment option for MCC (Mantripragada and Birnbaum 2015). Both MCPyV-positive and -negative MCCs have been treated with anti-PD-L1 (Kaufman et al. 2016) or anti-PD-1 therapy (Nghiem et al. 2016). In one of the studies, 88 patients with advanced MCCs were treated with the anti-PD-L1 antibody and followed for at least 12 months. An objective response rate of 33.0% was confirmed with 11.4% of the patients showing durable and complete responses. A one-year progression-free survival (PFS) rate of 30% and overall survival (OS) rates of 52% were achieved (Kaufman et al. 2018). In another study with a median follow-up time of 14.9 months (Nghiem 2019), the treatment of 50 patients with the anti-PD-1 antibody, pembrolizumab, resulted in a 56% objective response rate (ORR), including 24% of the patients showing complete response and 32% partial response. Since these studies, the anti-PD-L1 antibody Avelumab and the anti-PD-1 antibody pembrolizumab have been approved by FDA as new treatments for metastatic MCC. Several clinical trials are ongoing to assess the safety and efficacy of anti-PD-1 and anti-PD-L1 immuno-checkpoint therapies for MCC. These early studies using PD-1/PD-L1 immune checkpoint blockade therapies showed promising results but a significant portion of MCC patients does not respond to the treatment (Nghiem et al. 2016; Becker et al. 2017; Terheyden and Becker 2017; Winkler et al. 2017; D'Angelo 2018).

6.2.2 Adoptive Cell Transfer Therapy

MCPyV-encoded T antigens are continuously expressed in MCC to support tumor growth; therefore, they represent an appealing target for viral oncoprotein-directed T-cell therapy. Tumor-infiltrating CD8+ T cells are associated with improved survival of MCC patients. However, CD8+ T-cell infiltration is present in less than 18% of MCC tumors (Miller et al. 2017), suggesting that MCC may benefit from adoptive T-cell transfer therapy. In several recent studies, naturally processed epitopes of MCPyV LT antigen were identified and the T antigen-specific T-cell receptors (TCRs) isolated (Miller et al. 2017; Iyer et al. 2011; Lyngaa et al. 2014). These studies also revealed that intratumoral infiltration of MCPyV-specific T cells is associated with significantly improved MCC-specific survival, demonstrating the therapeutic benefit of MCPyV-specific T cells (Miller et al. 2017). Indeed, tumor-bearing animals treated with engineered T cells expressing MCPyV T antigen-specific TCR leads to tumor regression (Gavvovidis et al. 2018). Recently, Chapuis group treated two patients with advanced MCC tumors with autologous MCPyV-specific CD8+ T cells followed by immune checkpoint inhibitors (Paulson et al. 2018). In both cases, significant tumor regressions were associated with increased CD8+ T-cell infiltration into the regressing tumors (Paulson et al. 2018). However, tumors relapsed and escaped T-cell treatment during the late stage. Single-cell RNA sequencing suggests that treatment failure could be caused by HLA-loss (Paulson et al. 2018). Therefore, genetically engineered T cells with chimeric antigen receptor (CAR), which can recognize cancer cells in an HLA-independent manner, is an alternative approach for overcoming the problem of HLA-loss during MCC treatment.

6.3 DNA Cancer Vaccine

One of the earliest therapeutic approaches explored for treating MCC tumors is an MCPyV DNA vaccine. MCPyV LTT and sT consistently expressed in MCC are attractive foreign antigen targets for vaccine development. In 2012, the Hung laboratory developed a DNA vaccine to specifically target the MCPyV LTT region. When tested in mice injected with the B16/LT murine melanoma cell line stably expressing LTT, this vaccine showed protection against the LTT-expressing tumors in vivo (Zeng et al. 2012). These anti-tumor effects of the DNA vaccine appear to be mediated by CD4 + T-cell stimulation, natural killer cells and CD8+ T cells (Zeng et al. 2012). Because CD8+ T cells are associated with a better outcome, the Hung group went on to produce another DNA vaccine specifically designed to promote MCPyV LT-specific CD8+ T-cell responses. This vaccine encodes LTT antigen fused to a damage-associated molecular pattern protein, calreticulin (CRT), which has the ability to induce CD8+ T cells when fused to other foreign antigens (Zeng et al. 2012; Gomez et al. 2012). This new vaccine, named CRT/LT, showed prolonged survival after tumor challenge in the B16/LT mice model compared to mice vaccinated with the previous MCPyV LT vaccine. It was further demonstrated that this better performance was mediated by the induction of MCPyV LT-specific CD8+ T cells (Gomez et al. 2012). Another MCPyV DNA vaccine developed in the Hung group targeted the sT antigen, which is the main driver of MCC oncogenesis. The DNA vaccine pcDNA3-MCC/sT generated a significant number of sT antigenic peptide-specific CD8+ T cells and demonstrated markedly enhanced protection and treatment, leading to increased survival and decreased tumor volume in vivo (Gomez et al. 2013). These encouraging preliminary results provide a great platform for the development of MCPyV-targeted vaccines for MCC treatment.

6.4 Targeted Therapies

Targeted therapies are necessary for patients with advanced-stage MCCs that don't respond well to immunotherapy. Currently, multiple types of targeted therapies have been evaluated in MCC cell lines as well as in xenograft models, with some of them entering early phase clinical trials. Anti-apoptosis gene *BCL-2* is frequently unregulated in MCC and has become a major target for MCC therapies (Verhaegen et al. 2014). BCL-2 antisense oligonucleotides have been shown to inhibit tumor growth in MCC xenograft models (Schlagbauer-Wadl et al. 2000). However, they were not able to induce an objective response in a phase II trial (Shah et al. 2009). On the other hand, ABT-263, a small-molecule inhibitor of the BCL-2 family members (BCL-2, BCL-XL and BCL-W) could effectively induce apoptosis in most of the MCC cell lines tested (Verhaegen et al. 2014). MCC cell lines are also responsive to inhibitors of PI3K and mTOR pathways (Chteinberg et al. 2018; Kannan et al. 2016; Lin et al. 2015; Nardi et al. 2012; Hafner et al. 2012). In addition, most of the MCPyV-positive MCCs maintain wild-type p53 (Park et al. 2019). Not surprisingly, inhibitors of MDM2 (Mouse double minute 2 homolog or HDM2), which targets p53 for degradation, have been found to be effective in triggering p53-dependent apoptosis and cell cycle arrest in the majority of MCPyV-positive MCC tumors tested (Houben et al. 2013). A clinical trial is ongoing to evaluate a novel MDM2 small molecular inhibitor, KRT-232, for the effectiveness in treating patients with wild-type p53 MCC tumors but have failed anti-PD-1 or PD-L1 immunotherapy (NCT03787602). Since somatostatin receptors are highly expressed in MCC tumors, somatostatin analogues have been explored for their potential to be used in MCC molecular imaging and treatment (Orlova et al. 2018; Sollini et al. 2016; Buder et al. 2014). PEN-221, an inhibitor of somatostatin receptor 2 (SSTR2), is being evaluated in a clinical trial to target MCC and other advanced cancers with highly expressed SSTR2 (NCT02936323). Finally, antiapoptotic factor survivin is highly up-regulated in MCPyV-positive MCCs when compared to MCPyV-negative MCCs (Arora et al. 2012). Consistent with the fact that survivin expression is essential to support the survival of these tumor cells, YM155, a small-molecule inhibitor of survivin, has yielded promising results in inhibiting the growth of MCPyV-positive MCC cell lines both in cell culture and in xenograft models (Arora et al. 2012; Dresang et al. 2013).

7 Remaining Questions and Future Perspectives

MCPyV offers a unique opportunity to explore the oncogenic mechanism of a DNA tumor virus. Although the small viral genome of 5.4 kb DNA encodes just seven gene products, MCPyV successfully infects the skin of most humans and can establish long-lasting infections (Tolstov et al. 2009; Schowalter et al. 2010). While most of the MCPyV infections remain asymptomatic throughout the life of the infected hosts, in rare cases, it can cause extremely lethal skin cancer, MCC (Feng et al. 2008; Gjoerup and Chang 2010; Liu et al. 2016).

Despite recent advancements, much remain to be learned about MCPyV and its role in the development of MCC tumors. For instance, the cells of origin for MCC are currently unknown and the mechanisms by which MCPyV infection leads to cancer also remain enigmatic. MCPyV maintains persistent and latent infection in more than 80% of the general population (Tolstov et al. 2009; Foulongne et al. 2012; Schowalter et al. 2010) but tends to cause MCC in immunocompromised individuals (Heath et al. 2008). This observation suggests that the virus has evolved to exist in a dynamic state of mutual antagonism with the host cells and that changes to host immune status for which MCPyV is not adapted can result in cellular transformation and malignancy. However, the mechanism by which MCPyV escapes host immune eradication and establishes persistent infection remains unexplored. Few studies have examined the immunomodulatory effects of MCPyV-encoded proteins, and none have done so in the context of natural MCPyV infection. It is also unclear how changes to host conditions, such as the decline of immune competency, increase the chance of MCPyV-associated tumorigenesis. Although the uncontrolled proliferation of MCPyV may make viral genome integration more likely, the events that precede MCPyV integration into the host genome have not been elucidated.

Until recently, it was impossible to study biologically relevant host responses to MCPyV as the host cell of MCPyV was unknown. The discovery of HDFs as the host cells supporting productive MCPyV infection allows the development of a physiologically relevant in vitro model system (Liu et al. 2016), thus offering many new opportunities to explore the virus-host interactions in the setting of productive infection. No effective chemotherapies for metastatic MCC are currently available. Recent MCC immunotherapy successes suggest that overcoming MCC immune non-responsiveness is likely to yield improved patient outcomes. The continued discovery of the host restriction mechanisms that normally prevent MCPyV infection and viral oncogenesis could unveil more effective strategies for preventing and treating MCPyV-associated human cancers.

References

Agelli M et al (2010) The etiology and epidemiology of Merkel cell carcinoma. Curr Probl Cancer 34(1):14–37

Albores-Saavedra J et al (2010) Merkel cell carcinoma demographics, morphology, and survival based on 3870 cases: a population based study. J Cutan Pathol 37(1):20–27

Allen PJ et al (1999) Surgical management of Merkel cell carcinoma. Ann Surg 229(1):97–105

Arora R, et al (2012) Survivin is a therapeutic target in Merkel cell carcinoma. Sci Transl Med 4 (133):133ra56

Becker JC (2010) Merkel cell carcinoma. Ann Oncol 21(Suppl 7):vii81-5

Becker JC et al (2017) Merkel cell carcinoma. Nat Rev Dis Primers 3:17077

Becker M, et al (2019) Infectious entry of Merkel cell polyomavirus. J Virol

Berrios C et al (2016) Merkel cell polyomavirus small T antigen promotes pro-glycolytic metabolic perturbations required for transformation. PLoS Pathog 12(11):e1006020

Bhatia S et al (2011) Immunobiology of Merkel cell carcinoma: implications for immunotherapy of a polyomavirus-associated cancer. Curr Oncol Rep 13(6):488–497

Bichakjian CK et al (2007) Merkel cell carcinoma: critical review with guidelines for multidisciplinary management. Cancer 110(1):1–12

Borchert S et al (2014) High-affinity Rb binding, p53 inhibition, subcellular localization, and transformation by wild-type or tumor-derived shortened Merkel cell polyomavirus large T antigens. J Virol 88(6):3144–3160

Brummer GC et al (2016) Merkel cell carcinoma: current issues regarding diagnosis, management, and emerging treatment strategies. Am J Clin Dermatol 17(1):49–62

Buder K et al (2014) Somatostatin receptor expression in Merkel cell carcinoma as target for molecular imaging. BMC Cancer 14:268

Calder KB, Smoller BR (2010) New insights into Merkel cell carcinoma. Adv Anat Pathol 17 (3):155–161

Carter JJ et al (2013) Identification of an overprinting gene in Merkel cell polyomavirus provides evolutionary insight into the birth of viral genes. Proc Natl Acad Sci U S A 110(31):12744–12749

Cassler NM et al (2016) Merkel cell carcinoma therapeutic update. Curr Treat Options Oncol 17 (7):36

Chang Y, Moore PS (2012) Merkel cell carcinoma: a virus-induced human cancer. Annu Rev Pathol 7:123–144

Chen T et al (2011) Serological evidence of Merkel cell polyomavirus primary infections in childhood. J Clin Virol 50(2):125–129

Cheng J et al (2013) Merkel cell polyomavirus large T antigen has growth-promoting and inhibitory activities. J Virol 87(11):6118–6126

Cheng J et al (2017) Merkel cell polyomavirus recruits MYCL to the EP400 complex to promote oncogenesis. PLoS Pathog 13(10):e1006668

Cho S et al (2009) Effects of infrared radiation and heat on human skin aging in vivo. J Investig Dermatol Symp Proc 14(1):15–19

Chteinberg E et al (2018) Phosphatidylinositol 3-kinase p110delta expression in Merkel cell carcinoma. Oncotarget 9(51):29565–29573

Clarke CA, et al (2015) Risk of Merkel cell carcinoma after solid organ transplantation. J Natl Cancer Inst 107(2)

Cohen PR et al (2016) Genomic portfolio of Merkel cell carcinoma as determined by comprehensive genomic profiling: implications for targeted therapeutics. Oncotarget 7 (17):23454–23467

D'Angelo SP, et al (2018) Efficacy and safety of first-line avelumab treatment in patients with stage IV metastatic Merkel cell carcinoma: a preplanned interim analysis of a clinical trial. JAMA Oncol

Diaz J et al (2014) Phosphorylation of large T antigen regulates Merkel cell polyomavirus replication. Cancers (Basel) 6(3):1464–1486

Dresang LR et al (2013) Response of Merkel cell polyomavirus-positive Merkel cell carcinoma xenografts to a survivin inhibitor. PLoS ONE 8(11):e80543

Engels EA et al (2002) Merkel cell carcinoma and HIV infection. Lancet 359(9305):497–498

Erickson KD et al (2009) Ganglioside GT1b is a putative host cell receptor for the Merkel cell polyomavirus. J Virol 83(19):10275–10279

Feng H et al (2008) Clonal integration of a polyomavirus in human Merkel cell carcinoma. Science 319(5866):1096–1100

Feng H et al (2011) Cellular and viral factors regulating Merkel cell polyomavirus replication. PLoS ONE 6(7):e22468

Fisher GJ et al (1996) Molecular basis of sun-induced premature skin ageing and retinoid antagonism. Nature 379(6563):335–339

Fitzgerald TL et al (2015) Dramatic increase in the incidence and mortality from Merkel cell carcinoma in the United States. Am Surg 81(8):802–806

Foulongne V et al (2012) Human skin microbiota: high diversity of DNA viruses identified on the human skin by high throughput sequencing. PLoS ONE 7(6):e38499

Garneski KM et al (2009) Merkel cell polyomavirus is more frequently present in North American than Australian Merkel cell carcinoma tumors. J Invest Dermatol 129(1):246–248

Gavvovidis I et al (2018) Targeting Merkel cell carcinoma by engineered T cells specific to T-antigens of Merkel cell polyomavirus. Clin Cancer Res 24(15):3644–3655

Gill SE, Parks WC (2008) Metalloproteinases and their inhibitors: regulators of wound healing. Int J Biochem Cell Biol 40(6–7):1334–1347

Gjoerup O, Chang Y (2010) Chapter 1—update on human polyomaviruses and cancer. In: George FVW, George K (eds.) Advances in cancer research, pp. 1–51. Academic Press

Goh G et al (2016) Mutational landscape of MCPyV-positive and MCPyV-negative Merkel cell carcinomas with implications for immunotherapy. Oncotarget 7(3):3403–3415

Gomez BP et al (2012) Strategy for eliciting antigen-specific CD8+ T cell-mediated immune response against a cryptic CTL epitope of Merkel cell polyomavirus large T antigen. Cell Biosci 2(1):36

Gomez B et al (2013) Creation of a Merkel cell polyomavirus small T antigen-expressing murine tumor model and a DNA vaccine targeting small T antigen. Cell Biosci 3(1):29

Grundhoff A, Fischer N (2015) Merkel cell polyomavirus, a highly prevalent virus with tumorigenic potential. Curr Opin Virol 14:129–137

Hafner C et al (2012) Activation of the PI3K/AKT pathway in Merkel cell carcinoma. PLoS ONE 7(2):e31255

Harms PW (2017) Update on Merkel cell carcinoma. Clin Lab Med 37(3):485–501

Harms PW et al (2015) The distinctive mutational spectra of polyomavirus-negative Merkel cell carcinoma. Cancer Res 75(18):3720–3727

Harrison CJ et al (2011) Asymmetric assembly of Merkel cell polyomavirus large T-antigen origin binding domains at the viral origin. J Mol Biol 409(4):529–542

Heath M et al (2008) Clinical characteristics of Merkel cell carcinoma at diagnosis in 195 patients: the AEIOU features. J Am Acad Dermatol 58(3):375–381

Hodgson NC (2005) Merkel cell carcinoma: changing incidence trends. J Surg Oncol 89(1):1–4

Houben R et al (2009) Molecular pathogenesis of Merkel cell carcinoma. Exp Dermatol 18 (3):193–198

Houben R et al (2010) Merkel cell polyomavirus-infected Merkel cell carcinoma cells require expression of viral T antigens. J Virol 84(14):7064–7072

Houben R et al (2012) An intact retinoblastoma protein-binding site in Merkel cell polyomavirus large T antigen is required for promoting growth of Merkel cell carcinoma cells. Int J Cancer 130(4):847–856

Houben R et al (2013) Mechanisms of p53 restriction in Merkel cell carcinoma cells are independent of the Merkel cell polyoma virus T antigens. J Invest Dermatol 133(10):2453–2460

Iyer JG et al (2011) Merkel cell polyomavirus-specific CD8(+) and CD4(+) T-cell responses identified in Merkel cell carcinomas and blood. Clin Cancer Res 17(21):6671–6680

Kannan A et al (2016) Dual mTOR inhibitor MLN0128 suppresses Merkel cell carcinoma (MCC) xenograft tumor growth. Oncotarget 7(6):6576–6592

Kaufman HL et al (2016) Avelumab in patients with chemotherapy-refractory metastatic Merkel cell carcinoma: a multicentre, single-group, open-label, phase 2 trial. Lancet Oncol 17 (10):1374–1385

Kaufman HL et al (2018) Updated efficacy of avelumab in patients with previously treated metastatic Merkel cell carcinoma after >/=1 year of follow-up: JAVELIN Merkel 200, a phase 2 clinical trial. J Immunother Cancer 6(1):7

Kean JM et al (2009) Seroepidemiology of human polyomaviruses. PLoS Pathog 5(3):e1000363

Knight LM et al (2015) Merkel cell polyomavirus small T antigen mediates microtubule destabilization to promote cell motility and migration. J Virol 89(1):35–47

Krump NA, You J (2018) Molecular mechanisms of viral oncogenesis in humans. Nat Rev Microbiol 16(11):684–698

Kwun HJ et al (2009) The minimum replication origin of Merkel cell polyomavirus has a unique large T-antigen loading architecture and requires small T-antigen expression for optimal replication. J Virol 83(23):12118–12128

Kwun HJ et al (2013) Merkel cell polyomavirus small T antigen controls viral replication and oncoprotein expression by targeting the cellular ubiquitin ligase SCFFbw7. Cell Host Microbe 14(2):125–135

Kwun HJ et al (2015) Restricted protein phosphatase 2A targeting by Merkel cell polyomavirus small T antigen. J Virol 89(8):4191–4200

Lemos B, Nghiem P (2007) Merkel cell carcinoma: more deaths but still no pathway to blame. J Invest Dermatol 127(9):2100–2103

Li J et al (2013) Merkel cell polyomavirus large T antigen disrupts host genomic integrity and inhibits cellular proliferation. J Virol 87(16):9173–9188

Li J et al (2015) Phosphorylation of Merkel cell polyomavirus large tumor antigen at serine 816 by ATM kinase induces apoptosis in host cells. J Biol Chem 290(3):1874–1884

Lin Z et al (2015) Effect of the dual phosphatidylinositol 3-kinase/mammalian target of rapamycin inhibitor NVP-BEZ235 against human Merkel cell carcinoma MKL-1 cells. Oncol Lett 10 (6):3663–3667

Liu W, et al (2016) Identifying the target cells and mechanisms of Merkel cell polyomavirus infection. Cell Host Microbe

Liu W, et al (2018) Merkel cell polyomavirus infection of animal dermal fibroblasts. J Virol 92(4)

Liu W et al (2016a) Identifying the target cells and mechanisms of Merkel cell polyomavirus infection. Cell Host Microbe 19(6):775–787

Liu W et al (2016b) Merkel cell polyomavirus infection and Merkel cell carcinoma. Curr Opin Virol 20:20–27

Lunder EJ, Stern RS (1998) Merkel-cell carcinomas in patients treated with methoxsalen and ultraviolet a radiation. N Engl J Med 339(17):1247–1248

Lyngaa R et al (2014) T-cell responses to oncogenic Merkel cell polyomavirus proteins distinguish patients with Merkel cell carcinoma from healthy donors. Clin Cancer Res 20(7):1768–1778

Mantripragada K, Birnbaum A (2015) Response to anti-PD-1 therapy in metastatic Merkel cell carcinoma metastatic to the heart and pancreas. Cureus 7(12):e403

Miller NJ et al (2017) Tumor-infiltrating Merkel cell polyomavirus-specific T cells are diverse and associated with improved patient survival. Cancer Immunol Res 5(2):137–147

Moll I et al (1996) Proliferative Merkel cells were not detected in human skin. Arch Dermatol Res 288(4):184–187

Morrison KM et al (2009) Mammalian Merkel cells are descended from the epidermal lineage. Dev Biol 336(1):76–83

Nardi V et al (2012) Activation of PI3K signaling in Merkel cell carcinoma. Clin Cancer Res 18 (5):1227–1236

Neu U et al (2012) Structures of Merkel cell polyomavirus VP1 complexes define a sialic acid binding site required for infection. PLoS Pathog 8(7):e1002738

Neumann F et al (2016) Replication of Merkel cell polyomavirus induces reorganization of promyelocytic leukemia nuclear bodies. J Gen Virol 97(11):2926–2938

Nghiem P, et al (2019) Durable tumor regression and overall survival in patients with advanced Merkel cell carcinoma receiving pembrolizumab as first-line therapy. J Clin Oncol JCO1801896

Nghiem PT et al (2016) PD-1 blockade with pembrolizumab in advanced Merkel-cell carcinoma. N Engl J Med 374(26):2542–2552

Nwogu N et al (2018) Cellular sheddases are induced by Merkel cell polyomavirus small tumour antigen to mediate cell dissociation and invasiveness. PLoS Pathog 14(9):e1007276

Orlova KV et al (2018) Somatostatin receptor type 2 expression in Merkel cell carcinoma as a prognostic factor. J Eur Acad Dermatol Venereol 32(6):e236–e237

Park DE et al (2019) Dual inhibition of MDM2 and MDM4 in virus-positive Merkel cell carcinoma enhances the p53 response. Proc Natl Acad Sci U S A 116(3):1027–1032

Paulson KG et al (2011) Transcriptome-wide studies of Merkel cell carcinoma and validation of intratumoral CD8+ lymphocyte invasion as an independent predictor of survival. J Clin Oncol 29(12):1539–1546

Paulson KG, et al (2017) Merkel cell carcinoma: current United States incidence and projected increases based on changing demographics. J Am Acad Dermatol

Paulson KG et al (2018) Acquired cancer resistance to combination immunotherapy from transcriptional loss of class I HLA. Nat Commun 9(1):3868

Quan T, Fisher GJ (2015) Role of age-associated alterations of the dermal extracellular matrix microenvironment in human skin aging: a mini-review. Gerontology 61(5):427–434

Quan T et al (2009) Matrix-degrading metalloproteinases in photoaging. J Investig Dermatol Symp Proc 14(1):20–24

Richards KF et al (2015) Merkel cell polyomavirus T antigens promote cell proliferation and inflammatory cytokine gene expression. J Gen Virol 96(12):3532–3544

Saini AT, Miles BA (2015) Merkel cell carcinoma of the head and neck: pathogenesis, current and emerging treatment options. Onco Targets Ther 8:2157–2167

Sastre-Garau X et al (2009) Merkel cell carcinoma of the skin: pathological and molecular evidence for a causative role of MCV in oncogenesis. J Pathol 218(1):48–56

Schlagbauer-Wadl H et al (2000) Bcl-2 antisense oligonucleotides (G3139) inhibit Merkel cell carcinoma growth in SCID mice. J Invest Dermatol 114(4):725–730

Schowalter RM, Buck CB (2013) The Merkel cell polyomavirus minor capsid protein. PLoS Pathog 9(8):e1003558

Schowalter RM et al (2010) Merkel cell polyomavirus and two previously unknown polyomaviruses are chronically shed from human skin. Cell Host Microbe 7(6):509–515

Schowalter RM et al (2011) Glycosaminoglycans and sialylated glycans sequentially facilitate Merkel cell polyomavirus infectious entry. PLoS Pathog 7(7):e1002161

Schowalter RM et al (2012) Entry tropism of BK and Merkel cell polyomaviruses in cell culture. PLoS ONE 7(7):e42181

Schrama D et al (2016) Serine 220 phosphorylation of the Merkel cell polyomavirus large T antigen crucially supports growth of Merkel cell carcinoma cells. Int J Cancer 138(5):1153–1162

Schwartz RA, Lambert WC (2005) The Merkel cell carcinoma: a 50-year retrospect. J Surg Oncol 89(1):5

Scott MP, Helm KF (1999) Cytokeratin 20: a marker for diagnosing Merkel cell carcinoma. Am J Dermatopathol 21(1):16–20

Seo GJ et al (2009) Merkel cell polyomavirus encodes a micro RNA with the ability to autoregulate viral gene expression. Virology 383(2):183–187

Shah MH et al (2009) G3139 (Genasense) in patients with advanced Merkel cell carcinoma. Am J Clin Oncol 32(2):174–179

Shuda M et al (2008) T antigen mutations are a human tumor-specific signature for Merkel cell polyomavirus. Proc Natl Acad Sci U S A 105(42):16272–16277

Shuda M et al (2011) Human Merkel cell polyomavirus small T antigen is an oncoprotein targeting the 4E-BP1 translation regulator. J Clin Invest 121(9):3623–3634

Shuda M et al (2014) Merkel cell polyomavirus-positive Merkel cell carcinoma requires viral small T-antigen for cell proliferation. J Invest Dermatol 134(5):1479–1481

Shuda M et al (2015a) Merkel cell polyomavirus small T antigen induces cancer and embryonic Merkel cell proliferation in a transgenic mouse model. PLoS ONE 10(11):e0142329

Shuda M et al (2015b) Mitotic 4E-BP1 hyperphosphorylation and cap-dependent translation. Cell Cycle 14(19):3005–3006

Sihto H et al (2009) Clinical factors associated with Merkel cell polyomavirus infection in Merkel cell carcinoma. J Natl Cancer Inst 101(13):938–945

Sollini M et al (2016) Somatostatin receptor positron emission tomography/computed tomography imaging in Merkel cell carcinoma. J Eur Acad Dermatol Venereol 30(9):1507–1511

Spurgeon ME, Lambert PF (2013) Merkel cell polyomavirus: a newly discovered human virus with oncogenic potential. Virology 435(1):118–130

Spurgeon ME et al (2015) Tumorigenic activity of Merkel cell polyomavirus T antigens expressed in the stratified epithelium of mice. Cancer Res 75(6):1068–1079

Stakaityte G, et al (2018) Merkel cell polyomavirus small T antigen drives cell motility via Rho-GTPase-induced filopodium formation. J Virol 92(2)

Stang A et al (2018) The association between geographic location and incidence of Merkel cell carcinoma in comparison to melanoma: an international assessment. Eur J Cancer 94:47–60

Starrett GJ, et al (2017) Merkel cell polyomavirus exhibits dominant control of the tumor genome and transcriptome in virus-associated Merkel cell carcinoma. MBio 8(1)

Sun CH et al (2011) Activation of the PI3K/Akt/mTOR pathway correlates with tumour progression and reduced survival in patients with urothelial carcinoma of the urinary bladder. Histopathology 58(7):1054–1063

Terheyden P, Becker JC (2017) New developments in the biology and the treatment of metastatic Merkel cell carcinoma. Curr Opin Oncol

Theiss JM et al (2015) A comprehensive analysis of replicating Merkel Cell polyomavirus genomes delineates the viral transcription program and suggests a role for mcv-miR-M1 in episomal persistence. PLoS Pathog 11(7):e1004974

Toker C (1972) Trabecular carcinoma of the skin. Arch Dermatol 105(1):107–110

Tolstov YL et al (2009) Human Merkel cell polyomavirus infection II. MCV is a common human infection that can be detected by conformational capsid epitope immunoassays. Int J Cancer 125(6):1250–1256

Tsang SH et al (2016) The oncogenic small tumor antigen of Merkel cell polyomavirus is an iron-sulfur cluster protein that enhances viral DNA replication. J Virol 90(3):1544–1556

Vaigot P et al (1987) The majority of epidermal Merkel cells are non-proliferative: a quantitative immunofluorescence analysis. Acta Derm Venereol 67(6):517–520

Van Keymeulen A et al (2009) Epidermal progenitors give rise to Merkel cells during embryonic development and adult homeostasis. J Cell Biol 187(1):91–100

Varani J et al (2006) Decreased collagen production in chronologically aged skin: roles of age-dependent alteration in fibroblast function and defective mechanical stimulation. Am J Pathol 168(6):1861–1868

Velasquez C et al (2016) Mitotic protein kinase CDK1 phosphorylation of mRNA translation regulator 4E-BP1 Ser83 may contribute to cell transformation. Proc Natl Acad Sci U S A 113 (30):8466–8471

Verhaegen ME, et al (2014) Merkel cell polyomavirus small T antigen is oncogenic in transgenic mice. J Invest Dermatol

Verhaegen ME et al (2014) Merkel cell carcinoma dependence on bcl-2 family members for survival. J Invest Dermatol 134(8):2241–2250

Verhaegen ME et al (2015) Merkel cell polyomavirus small T antigen is oncogenic in transgenic mice. J Invest Dermatol 135(5):1415–1424

Verhaegen ME et al (2017) Merkel cell polyomavirus small T antigen initiates Merkel cell carcinoma-like tumor development in mice. Cancer Res 77(12):3151–3157

Wang X, et al (2012) Bromodomain protein Brd4 plays a key role in Merkel cell polyomavirus DNA replication. PLoS Pathog 8(11):e1003021. PMCID:PMC3493480

Wang X et al (2012) Bromodomain protein Brd4 plays a key role in Merkel cell polyomavirus DNA replication. PLoS Pathog 8(11):e1003021

Wendzicki JA et al (2015) Large T and small T antigens of Merkel cell polyomavirus. Curr Opin Virol 11:38–43

Winkelmann RK, Breathnach AS (1973) The Merkel cell. J Invest Dermatol 60(1):2–15

Winkler JK et al (2017) PD-1 blockade: a therapeutic option for treatment of metastatic Merkel cell carcinoma. Br J Dermatol 176(1):216–219

Xue Y, Thakuria M (2019) Merkel cell carcinoma review. Hematol Oncol Clin North Am 33 (1):39–52

Zeng Q et al (2012) Development of a DNA vaccine targeting Merkel cell polyomavirus. Vaccine 30(7):1322–1329

Chapter XX Antiviral Treatment and Cancer Control

Wei-Liang Shih, Chi-Tai Fang, and Pei-Jer Chen

1 Introduction

Hepatitis B virus (HBV), hepatitis C virus (HCV), human papilloma virus (HPV), Epstein-Barr virus (EBV), human T-cell lymphotropic virus type 1 (HTLV-1), Kaposi's sarcoma-associated herpesvirus (KSHV), and Merkel cell polyomavirus (MCV) are the seven viruses that are currently known to cause chronic infection and associated with specific cancers in human (Moore and Chang 2010). In total, viruses contributed to the development of 10–15% human cancer cases worldwide. Although only a small proportion of infected individuals actually develop cancers, the clinical prognosis is usually very poor.

Viral factors have long been proposed to play important roles in carcinogenesis. Advances in molecular technologies now allow to rapid and accurate quantification of viral load, a marker of virus replication activity in human body. Accumulated data show that viral load and cancer risk often parallels for virus-associated cancers, such as HBV and HCV for hepatocellular carcinoma (HCC) and EBV for nasopharyngeal carcinoma (NPC). In light of the importance of virus replication activity in the carcinogenesis of virus-associated cancers, antiviral therapy that can suppress or eliminate viruses could be one important strategy for cancer prevention.

P.-J. Chen (✉)
Institute of Clinical Medicine, National Taiwan University, Taipei, Taiwan
e-mail: peijerchen@ntu.edu.tw

W.-L. Shih · C.-T. Fang
Institute of Epidemiology and Preventive Medicine, National Taiwan University, Taipei, Taiwan

© Springer Nature Switzerland AG 2021
T.-C. Wu et al. (eds.), *Viruses and Human Cancer*, Recent Results in Cancer Research 217, https://doi.org/10.1007/978-3-030-57362-1_13

Table 1 Human cancer viruses, associated cancers, and their specific antiviral therapies

Virus[a]	Cancer type[b]	Antiviral therapy[c]	
HBV	HCC	Interferons	(interferon-α and pegylated interferon-α2a)
		Nucleos(t)ide analogues	(lamivudine, adefovir, entecavir, telbivudine, tenofovir disoproxil fumarate, and tenofovir alafenamide)
HCV	HCC	Peglyated interferon plus ribavirin	
		Direct antiviral agents (protease inhibitor)	(Boceprevir, telaprevir, sofosbuvir/velpatasvir, glecaprevie/pibrentasvir)
KSHV	AIDS-KS	Antiviral herpes virus drug	(Ganciclovir and valganciclovir)
		HAART, PI-based	
		HAART, NNRTI-based	
	HIV-negative KS	HIV-protease inhibitor	(Indinavir)
HTLV-1	ATL (acute, chronic, and smoldering forms)	Interferon-α plus zidovudine/zalcitabine alone or combined with chemotherapy	
EBV	PTLD, NPC, HL	Immunotherapy	(EBV-specific cytotoxic T-cells)
	EBV-associate lymphoma	Virus-directed	(Lytic replication inducer plus acyclovir/ganciclovir)
HPV	Cervical cancer	Therapeutic HPV vaccine	
		RNA interference-based therapy	(Antisense oligonucleotides, ribozymes, and siRNAs)
MCV	MCC	Interferon	(Interferon-α and interferon-β)

[a]*HBV* hepatitis B virus, *HCV* hepatitis C virus, *EBV* Epstein-Barr virus, *HPV* human papilloma virus, *HTLV-1* human T-cell lymphotorpic virus type 1, *KSHV* Kaposi's sarcoma-associated herpesvirus, *MCV* Merkel cell polyomavirus
[b]*HCC* hepatocellular carcinoma, *AIDS-KS* AIDS-Kaposi's sarcoma, *ATL* adult T-cell lymphoma, *PTLD* post-transplant lymphoproliferative disorder, *NPC* nasopharyngeal carcinoma, *HL* Hodgkin's lymphoma, *MCC* Merkel cell carcinoma
[c]*HAART* highly active antiretroviral treatment
PI protease inhibitor, *NNRTI* nonnucleoside reverse transcriptase inhibitor

Currently, antiviral therapy has been applied on these seven viruses-associated cancers (Table 1), and clinical benefits have been proven for antiviral therapy for HBV and HCV.

In this chapter, we reviewed and updated the current evidences on the correlation between viral load and clinical outcomes (e.g., cancer risk and survival), and the current understandings on the effect of antiviral treatments for cancer-associated viruses.

2 Hepatitis B Virus and Hepatitis C Virus

Chronic HBV and HCV infections are major etiological factors in 80% of HCC cases and responsible for 96% of all hepatitis mortality worldwide (WHO Global hepatitis 2017; Yang et al. 2019). Universal HBV vaccination on newborns in Taiwan since 1985 has led to a 70% reduction in HBV-related HCC in children or teenagers (Chang et al. 2009). Nevertheless, millions of individuals who already chronically infected by HBV are still under the risk of developing HCC. Unfortunately, no effective vaccine is available for preventing HCV infection now. The proportion of HCV-related HCC has progressively increased, especially in the developed countries (El-Serag 2012) and global burden of HCV-related HCC has increased by 56.7% from 1990 to 2015 (Global Burden of Disease Liver Cancer Collaboration 2017). For these people persistently infected by either HBV or HCV, antiviral therapies target on HBV and HCV can be an effective strategy to reduce risk of HCC. In Taiwan, the national viral hepatitis therapy program launched in 2003 has provided the significant reduction in respect of the incidence and mortality of HCC, chronic liver diseases, and cirrhosis (Chiang et al. 2015).

2.1 Anti-HBV Therapies

HBV replication is the key force to drive the progression of HBV-related diseases (Liaw 2006). HBeAg is a well-known HBV replication marker (Chen et al. 2009; Fang et al. 2003). Epidemiologic studies also consistently show the elevated HBV viral load, which represented a higher HBV replication activity, is associated with high HCC risk, worse progression, and poor survival (Chen et al. 2009). Reducing HBV DNA to undetectable level or induction of HBeAg seroconversion has been the main therapeutic endpoints (Feld et al. 2009).

Conventional interferon-α, pegylated interferon-α2a, and nucleos(t)ide analogues (NUCs), including lamivudine, adefovir, entecavir, telbivudine, tenofovir disoproxil fumarate (TDF) are currently approved treatments for chronic HBV infection. Treatment with newer NUCs, including entecavir, telbivudine, and TDF could suppress HBV DNA by average 6.2–6.9 and 4.6–5.2 $\log_{10}$ IU/mL in HBeAg-positive (Marcellin et al. 2008; Lai et al. 2007; Chang et al. 2006) and HBeAg-negative patients (Marcellin et al. 2008; Lai et al. 2007, 2006), respectively. Undetectable level of HBV DNA can be achieved in 60–80% of HBeAg-positive patients (Marcellin et al. 2008; Lai et al. 2007; Chang et al. 2006; Dienstag 2009) and 88–95% of HBeAg-negative patients (Marcellin et al. 2008; Lai et al. 2007, 2006; Dienstag 2009). Tenofovir alafenamide (TAF), a most recent approved prodrug of tenofoviir, had similar virological responses but with fewer bone and renal adverse effects (Chan et al. 2016; Buti et al. 2016). Treatment with lamivudine and adefovir yields less reduction of HBV DNA and lower proportion of undetectable HBV DNA in both HBeAg-positive and HBeAg-negative patients. In addition, HBeAg seroconversion was observed in 12–23% of NUCs-treated patients (Marcellin et al. 2008; Lai et al. 2007; Chang et al. 2006; Dienstag 2009; Terrault et al. 2018). After one-year course of NUCs treatment, sustained virological response was maintained in relatively small proportion of initial responders (Dienstag 2009). Thus, NUCs were usually used as a long-term therapy. No optimal duration of NUCs therapy is available now, but extending treatment for at least more six months after HBeAg seroconversion to consolidate the sustained response is currently an acceptable practice. For treatment-naïve patients with chronic hepatitis B, long-term NUCs therapy did have the benefits in prevention or delay of the occurrence of hepatic decompensation, HCC, and liver-related death (Wei and Kao 2017). However, the risk of untoward side effects in long-term NUCs-treated patients with compensated cirrhosis remained an issue to be concerned.

Although NUCs are highly potent and safe, emergence of drug resistance after prolonged use of NUCs is a major concern. It has been reported that patients with lamivudine resistance had higher HCC risk than NUCs naïve patients (Papatheodoridis et al. 2010). Lamivudine resistance would accumulate rapidly to 15–25% by 12 months and to 60–65% by four years of treatment (Papatheodoridis et al. 2008). Resistance to adefovir and telbivudine can also reach 25–30% after long-term treatment (Dienstag 2009). Only treatment with entecavir, TDF, and TAF showed negligible resistance (European Association for the Study of the Liver 2017). Thus, due to better resistance profile, excellent safety profile and superior efficacy, entecavir, TDF, and TAF have now been suggested as the first-line NUCs therapy (Dienstag 2009; Terrault et al. 2018; European Association for the Study of the Liver 2017).

In contrast to NUCs, interferon-based therapy was less used for HBV infection due to side effects and poor tolerability. Nevertheless, recent studies have provided more supportive evidence for its role in anti-HBV therapy. Treatment of HBeAg-positive patients with pegylated interferon for 48–56 weeks achieved undetectable HBV DNA in 10–25% of patients and <2–4.5 $\log_{10}$ copies/mL mean reduction of HBV DNA (Lau et al. 2005; Janssen et al. 2005). HBeAg loss and HBeAg seroconversion were durable in 81 and 70% of initial responders with or without concomitant lamivudine therapy, respectively, for a mean of three years follow-up after treatment (Buster et al. 2008). In addition to the benefit of HBV DNA reduction, HBeAg-negative patients with pegylated interferon-based treatment (with or without lamivudine) had higher proportion of undetectable HBV DNA ($\sim 20\%$) after 24 weeks of follow-up (Marcellin et al. 2004) and 46% of these initial responders had the sustained response in suppression of HBV DNA to undetectable level at three years follow-up after end of treatment (Marcellin et al. 2009). Although pegylated interferon resulted in virological response in a less proportion of patients than NUCs therapy, it seemed to have higher probability to achieve sustained off-therapy response. Due to this advantage and the fixed duration of treatment, pegylated interferon still has a therapeutic role in selected patients.

The long-term effect of anti-HBV therapy with NUCs and interferon on survival and incidence of HCC has also been investigated. For interferon-based therapy, two early studies reported that sustained virological responders showed significantly better survival and lower risk of developing HCC (van Zonneveld et al. 2004;

Niederau et al. 1996). Compared with nontreated patients, interferon-treated patients had lower HCC incidence with RR of 0.23–0.66 (Table 2). For NUCs, two early large randomized control trials of treatment for chronic HBV infected patients with advanced liver diseases shown that lamivudine could reduce risk of disease progression including developing HCC in treated patients and in patients with persistent viral suppression (Liaw et al. 2004; Di Marco et al. 2004). Several meta-analyses consistently concluded that NUCs treatment was associated with a lower HCC incidence (Table 2). In a recent multicenter cohort study, long-term entecavir therapy in chronic hepatitis B patients with cirrhosis showed a 60%, 86%, and 85% reduction in HCC risk, liver-related mortality, and all-cause mortality, respectively (Su et al. 2016).

2.2 Anti-HCV Therapies

For HCV, although there is still no vaccine available, current effective anti-HCV therapies have provided great improvement in clinical outcomes for HCV patients.

Interferon-based treatments resulted in sustained virological response (SVR) in about 50% of HCV patients with genotype 1 and 80% of HCV patients with genotypes 2 and 3 (Munir et al. 2010). The achievement of SVR is durable (Hofmann and Zeuzem 2011) and highly associated with good overall clinical outcome, including decreased risk of HCC and improvement of overall and recurrence-free survival (Table 2) and liver-related deaths (Masuzaki et al. 2010). In addition, pegylated interferon combined with ribavirin increased SVR to about 40% in HCV genotype 1b patients with high viral load (Masuzaki et al. 2010).

However, direct acting antiviral (DAA) agents have revolutionized the current standard of care for HCV patients because of the dramatic improvement in SVR, short treatment duration, and overall good tolerability (Sandmann et al. 2019). For typical and pangenotypic combination DAA regimens (ex. sofosbuvir/velpatasvir, glecaprevie/pibrentasvir), the SVR rates were all more than 95% (European Association for the Study of the Liver 2018). The treatment duration of most regimens was 8–12 weeks and some have to extend to 16 weeks depending on HCV genotype, presence of cirrhosis, fibrosis stage, and prior treatment experience (European Association for the Study of the Liver 2018). Most HCV patients tolerate well with current DAA regimens, even a combination use of ribavirin for patients with decompensated liver cirrhosis (European Association for the Study of the Liver 2018). However, HCV patients with prior or present HCC showed lower SVR rate after DAA therapy (Table 2). In regard to the effect of DAA therapy, several large studies had showed that DAA-induced SVR was associated with the reduction of HCC incidence (Ioannou et al. 2018; Calvaruso et al. 2018) and did not increase risk of HCC recurrence (ANRS collaborative study group on hepatocellular carcinoma 2016; Singal et al. 2019). No difference was observed in HCC incidence or recurrence between DAA and interferon therapies in patients with or without cirrhosis based on current studies results (Ioannou et al. 2018; Waziry et al. 2017; Kobayashi et al. 2017). DAA therapy also provided more benefit for HCV patients

Table 2 Meta-analyses of nucleos(t)ide analogues and interferon-based therapy for HBV and HCV infection

Study	Design	No. of subjects	Therapy	Endpoint	Main findings
HBV					
Mommeja-Marin et al. (2003)	26 prospective studies	N = 3428 (2524 HBeAg-positive)	IFN versus CN and NUCs versus CN	Histologic response Biochemical response Serologic response	Treatment-induced HBV DNA reduction correlated with histologic, biochemical, and serologic responses
Sung et al. (2008)	12 studies	N = 2742 1292 treated 1450 untreated	IFN versus CN	HBV-related HCC	RR = 0.66 (0.48–0.89)
	5 studies	N = 2289 1267 treated 1022 untreated	NUCs versus CN	HBV-related HCC	RR = 0.22 (0.10–0.50)
Yang et al. (2009)	11 clinical trials	N = 2082 1006 treated 1076 untreated	IFN versus CN	HBV-related HCC incidence	RR = 0.59 (0.43–0.81)
	5 clinical trials	N = 935 516 treated 419 untreated	IFN versus CN	HBV-related cirrhosis	RR = 0.65 (0.47–0.91)
Wong et al. (2010)	11 studies	N = 2122 975 treated 1147 untreated	IFN versus CN	Overall hepatic events Liver-related mortality	RR = 0.55 (0.43–0.70) RR = 0.63 (0.42–0.96)
Papatheodoridis et al. (2010)	21 studies	N = 4415 3881 treated 534 untreated	NUCs versus CN	HCC incidence	HCC incidence: treated: 2.8% untreated: 6.4%

(continued)

Table 2 (continued)

Study	Design	No. of subjects	Therapy	Endpoint	Main findings
Zhang et al. (2011)	2 RCTs	N = 158 95 treated 63 untreated	Nonmaintenance IFN IFN versus CN	HBV-related HCC incidence	RR = 0.23 (0.05–1.04)
Singal et al. (2013)	6 studies	N = 6877 3306 treated 3571 untreated	NUCs versus CN	HCC risk	OR = 0.48 (0.38–0.61)
Lok et al. (2016)	5 studies	N = 677 219 treated 458 untreated	IFN versus CN	HCC incidence	RR = 0.64 (0.43–0.94)
	4 studies	N = 2681 790 treated 1891 untreated	NUCs versus CN	HCC incidence	RR = 0.61 (0.39–0.96)
HCV					
Cammá et al. (2001)	14 studies	HCV-related cirrhosis N = 3109	IFN treated versus nontreated	HCC incidence	OR = 0.28 (0.22–0.36)
Papatheodoridis et al. (2001)	11 studies	HCV-related cirrhosis N = 2178	CN versus IFN	HCC incidence	OR = 3.02 (2.35–3.89)
	5 studies	HCV-related cirrhosis N = 683	nonSVR versus SVR	HCC incidence	OR = 3.65 (1.71–7.78)
Zhang et al. (2011)	4 RCTs	HCV patients N = 378	Nonmaintenance IFN treated versus nontreated	HCV-related HCC incidence	RR = 0.39 (0.26–0.59)
	2 RCTs	HCV patients N = 223	Nonmaintenance IFN treated versus nontreated	HCV SVR	RR = 0.30 (0.04–2.15)
	2 RCTs	Initial nonresponders of IFN therapy N = 1101	Maintenance IFN treated versus nontreated	HCV-related HCC incidence	RR = 0.96 (0.59–1.56)

(continued)

Table 2 (continued)

Study	Design	No. of subjects	Therapy	Endpoint	Main findings
Singal et al. (2010)	20 studies	HCV-related cirrhosis N = 4700	IFN/IFN ± ribavirin treated versus nontreated	HCC incidence	RR = 0.43 (0.33–0.56)
	14 studies	N = 3310	IFN/IFN ± ribavirin SVR versus nonSVR	HCC incidence	RR = 0.35 (0.26–0.46)
Qu et al. (2012)	8 RCTs	HCV-related cirrhosis N = 1505	IFN treated versus nontreated	HCC incidence	OR = 0.29 (0.10–0.80)
	3 RCTs	HCV-related cirrhosis N = 1155	Maintenance IFN therapy treated versus nontreated	HCC incidence	OR = 0.54 (0.32–0.90)
Morgan et al. (2013)	12 studies	HCV patients at all stages of fibrosis N = 25,497	SVR versus nonSVR	HCC incidence	HR = 0.24 (0.18–0.31)
	8 studies	HCV patients with advanced liver diseases N = 2649	SVR versus nonSVR	HCC incidence	HR = 0.23 (0.16–0.35)
Waziry et al. (2017)	26 studies	N = 11,523 5521 IFN treated 6002 DAA treated	DAA treatment versus IFN treatment	HCC incidence	RR = 0.68 (0.18–2.55)
	17 studies	N = 2352 1485 IFN treated 867 DAA treated	DAA treatment versus IFN treatment	HCC recurrence	RR = 0.62 (0.11–3.45)
Manthravadi et al. (2017)	7 studies	N = 1519 223 SVR 1296 nonSVR/nonIFN	IFN/IFN ± ribavirin SVR versus nonSVR/nonIFN	Overall survival	HR = 0.18 (0.11–0.29)
	8 studies	N = 1241 226 SVR 1015 nonSVR/nonIFN	IFN/IFN ± ribavirin SVR versus nonSVR/nonIFN	Recurrence-free survival	HR = 0.50 (0.40–0.63)

(continued)

Chapter XX Antiviral Treatment and Cancer Control

Table 2 (continued)

Study	Design	No. of subjects	Therapy	Endpoint	Main findings
Ji et al. (2019)	49 studies	N = 39,042 3341 HCC 35,701 non-HCC	DAA regimens HCC versus non-HCC	HCV SVR	Overall SVR rate HCC: 89.6% non-HCC: 93.3%
He et al. (2020)	27 studies	N = 52,264 3126 HCC 49,138 non-HCC	DAA regimens HCC versus non-HCC	HCV SVR	Overall SVR rate HCC: 88.2% non-HCC: 92.4% OR = 0.54 (0.43–0.68)

CN control group (placebo or no treatment), *RR* relative risk, *OR* odds ratio, *HR* hazard ratio, *SVR* sustained virologic response
IFN interferon, *NUCs* nucleos(t)ide analogue, *DAA* direct acting antiviral
RCT randomized controlled trial

with cirrhosis in terms of higher overall survival (hazard ratio = 0.39, compared with DAA-untreated patients) (Cabibbo et al. 2019). Many recent studies further demonstrated that typical and pangenotypic DAA therapy also resulted in high SVR (>95%), safety, and efficacy for HCV patients with moderate to severe renal impairment (Lawitz et al. 2020) and liver or kidney transplant (Reau et al. 2018; Agarwal et al. 2018). However, potential drug–drug interaction between DAAs and concomitant medications should be an important issue to be considered in the era of DAA therapy.

3   HBV-HCV Coinfection

Treatment of HBV-HCV coinfection is important in endemic area because of its fairly high prevalence due to shared routes of transmission and increased risk for liver diseases, including HCC.

Both interferon-based and DAA therapy for HBV-HCV coinfection have the risk of HBV reactivation. However, HBV activation may occur more frequently and easier during DAA therapy because its treatment target is HCV only. This had been issued a warning by the US Food and Drug Administration (FDA) and will be an important concern when DAA therapy gradually became the main option for HCV treatment. Recently, clinical trial and meta-analyses demonstrated that treatment of HBV-HCV coinfected patients with DAA regimens resulted in transient HBV DNA increase and higher HBV activation rate (Liu et al. 2018; Mucke et al. 2018; Jiang et al. 2018).

Before initiation of therapy, determination of replicative status of HBV and HCV is required. For HBsAg-positive HCV patients, DAA therapy was recommended as that for monoinfected HCV patients and anti-HBV treatment should be started concurrently with (or before) DAA therapy (Terrault et al. 2018; European Association for the Study of the Liver 2018). The effect of DAA therapy in SVR was not impaired on the condition of HBV coinfection (Liu et al. 2018). In regard to HCV patients with HBsAg-negative and anti-HBc-positive, close monitoring of alanine transaminase (ALT), HBsAg, and HBV DNA should be performed during DAA therapy. When increased or abnormal ALT is observed and the following HBsAg or HBV DNA is detectable, HBV treatment should be started (Terrault et al. 2018; European Association for the Study of the Liver 2018). In addition, a recent meta-analysis reported that the risk of HBV reactivation in HCV patients with HBsAg-positive during DAA therapy could be reduced by preemptive anti-HBV therapy (Jiang et al. 2018).

In regard to the long-term efficacy, the decreased risk of HCC, liver-related mortality, and all-cause mortality has been reported in dually infected patients with the interferon-based (interferon plus ribavirin) treatment by a large population-based study (Liu et al. 2014). The risk reduction of HCC, liver-related mortality, and all-cause mortality were 35%, 62%, and 59%, respectively. However, the long-term efficacy of DAA therapy for dually infected patients still needs to be studied.

4   HBV-HIV Coinfection and HCV-HIV Coinfection

Coinfection of HBV-HIV and HCV-HIV usually hastens the development of liver diseases, including fibrosis and HCC. Generally speaking, monotherapy of interferons, pegylated interferons, lamivudine, adefovir, entecavir, and telbivudine has yielded much less satisfactory responses in HBV-HIV coinfected patients (Lacombe and Rockstroh 2012). Since highly active antiretroviral treatment (HAART) (also referred as combination antiretroviral therapy, cART) has dramatically improved therapeutic and long-term outcomes, all HBV-HIV coinfection patients should receive tenofovir (TDF or TAF)-based HAART regardless of the number of CD4 cell (Terrault et al. 2018; European Association for the Study of the Liver 2017). Generally, HAART containing tenofovir plus lamivudine/emtricitabine is the preferred regimen. However, intermittent use must be avoided because of the high risk of viral breakthrough-induced deleterious consequences.

For lamivudine-treated patients, adding TDF/TAF (preferred) or entecavir is recommended. Currently, data of TAF using in HBV-HIV-coinfected patients are limited. A recent trial showed that switching from TDF-based to TAF-based regimen maintained or achieved HIV and HBV suppression in 91.7% of 72 HBV-HIV-coinfected patients with improved renal and bone safety (Gallant et al. 2016). Besides, lamivudine-containing HAART has been shown to be able to suppress HBV replication in HBeAg-negative HBV-HIV coinfected patients (Fang et al. 2003). Although the above guidelines have been proposed, more solid evidences are still needed to make evidence-based therapeutic decisions.

In regard to treatment of HCV-HIV coinfected patients, HCV treatment has a priority because of reduced risk due to HCV clearance (AASLD-IDSA HCV Guidance Panel 2018). Interferon-based therapy (pegylated interferon with ribavirin) provided the SVR rates in 27–50% (higher for genotype 2 and 3, lower for genotype 1 and 4) and regimens would be different according to HCV genotypes and virological response (Lacombe and Rockstroh 2012). In the era of DAA therapy, treatment recommendations for HCV-HIV coinfection were similar with those for HCV monoinfection (AASLD-IDSA HCV Guidance Panel 2018). Recent clinical trials showed that the SVR rates of pangenotypic regimens in HCV-HIV coinfection patients were similar with those to HCV-monoinfected patients regardless of HCV genotypes (Rockstroh et al. 2018; Wyles et al. 2017). Nevertheless, potential drug–drug interactions between antiretroviral drugs and DAA should be carefully considered (European Association for the Study of the Liver 2018; AASLD-IDSA HCV Guidance Panel 2018). In addition, based on results from clinical studies, HAART may offer positive impact on control of prognosis of liver damage in HCV-HIV coinfected patients and most of first-line HAART have good fitness in these patients (Jones and Nunez 2011).

Currently, the evaluation of treatment effect on clinical outcomes (e.g., HCC) in HBV-HIV and HCV-HIV coinfected patients remained limited. An observational study from sub-Saharan Africa observed that TDF-based HAART provided a positive effect for HBV-HIV coinfected patients to have similar mortality with

HIV-monoinfected patients, whereas HBV-HIV coinfected patients had higher mortality than HIV-monoinfected patients on the treatment of nonTDF-based HAART (Hawkins et al. 2013). A retrospective cohort study showed low liver-related mortality in HBV-HIV coinfected patients with TDF-based HAART; however, the liver-related deaths were mainly due to HCC (Huang and Nunez 2015). A recent prospective European cohort study observed the relatively stable (rate ratio per addition year = 0.95, 95% CI = 0.85 − 1.06) and steadily increasing (rate ratio per addition year = 1.14, 95% CI = 1.07 − 1.21) HCC incidence pattern in HBV-HIV coinfected patients with and without TDF-based HAART, respectively (Wandeler et al. 2019). In regard to HCV-HIV coinfection, two recent studies from Spain evaluated the effect of DAA treatment on HCC risk. DAA treatment of HCV-HIV coinfected patients with cirrhosis (>90% with ART therapy) resulted in a significant reduction of HCC risk than those with non-DAA regimens and lower HCC risk was observed in patients achieving SVR (Merchante et al. 2018). Compared HCV-HIV coinfected and cirrhotic patients with DAA regimen and those with IFN-based regimen, there was no statistical difference in incidence rate (Merchante et al. 2018) and recurrence (Merchante et al. 2018) of HCC after the achievement of SVR. These findings were based on small sample size and more studies with large sample size from different population will be needed.

5   Kaposi's Sarcoma-Associated Herpesvirus (KSHV)

KSHV, also known as human herpesvirus 8 (HHV-8), is a necessary factor for Kaposi's sarcoma (KS)—the most common malignancy in HIV patients who became immunocompromised. Elevated KSHV viral load was observed more frequently in KS patients than in asymptomatic KSHV-infected patients and was associated with a higher risk of AIDS-KS (Gantt and Casper 2011; Sunil et al. 2010). There is still no effective vaccine for KSHV. Nevertheless, several studies showed that antiherpes virus drugs, such as ganciclovir and valganciclovir, not only reduced KSHV viral load but also prevented AIDS-KS (Gantt and Casper 2011; Coen et al. 2014). Another antiviral drug, foscarnet, also presented the effect in reduction of KS incidence (Gantt and Casper 2011; Coen et al. 2014). Though these small studies still need further confirmation (Gantt and Casper 2011), it will be interesting to examine the clinical effects of these promising agents in combination with DNA synthesis blockers or lytic replication inducer in future clinical trials with large samples.

The replication of KSHV strongly depends on HIV-induced immunodeficiency (Mesri et al. 2010). Early use of highly active antiretroviral treatment (HAART) (also referred as combination antiretroviral therapy, cART) can restore host immunity and decreased the incidence and mortality of AIDS-KS in HAART-treated patients (Mesri et al. 2010; Bower et al. 2006). Compared with that in pre-HAART era, KS incidence in HAART era dropped by sixfold (Sunil et al. 2010). Even in resource-limited regions, early use of HAART can result in reduced KS incidence (Casper 2011).

Large US and Europe cohort studies presented that KS incidence in HIV-positive patients had a similar changing pattern, which showed the highest incidence at the first six months after the initiation of HAART use and then a steep decrease thereafter to bottom line (Yanik et al. 2013; Cancer Project Working Group for the Collaboration of Observational HIV Epidemiological Research Europe (COHERE) study in EuroCoord 2016). In the individual patients level, HAART is also significantly associated with control of KSHV viraemia (Bourboulia et al. 2004). However, for late presenters who already developed KS at the time of initial diagnosis, HAART alone induces complete remission in only half of these patients (Nguyen et al. 2008). For AIDS-KS patients which did not reach complete remission with HAART alone, co-administration of HAART and chemotherapy could improve the response rate to 81.5% (Bower et al. 2006). The combination of HAART and chemotherapy also provided a high survival for advanced KS (Bower et al. 2014). More recently, results from clinical trial showed that immediate HAART initiation for early HIV-infected patients could reduce the risk of KS (Borges et al. 2016). HAART regimens can be classified to protease inhibitor (PI)-based and nonnucleoside reverse transcriptase inhibitor (NNRTI)-based. No difference of KS incidence rate between these two types of regimens was observed in small observational studies, but PI-based HAART seemed to have better efficacy because of more complete KS remission (Gantt and Casper 2011). In addition, more relapse was reported when switching therapy from PI-based to NNRTI-based HAART (Gantt and Casper 2011). A recent retrospective analysis of 25,529 HIV-infected male veterans exhibited that longer boosted PIs use produced a significant decrease of KS incidence (Kowalkowski et al. 2015). Nevertheless, whether PI-based HAART does have a therapeutic advantage for KSHV or not, this superiority still requires more convincing evidence to support and its potential effect in increasing the risk of non-AIDS-defining malignancies with cumulative exposure to PI-based therapy in HIV-positive patients should be monitored (Borges 2017). Controlled trials with large sample size and optimized HAART regimens are urgently needed.

For KS in HIV-negative patients, a 28-patient trial study showed that treatment with individual ART (indinavir)-induced tumor regression and stabilization of disease progression in some patients (Monini et al. 2009). In addition to KS, KSHV was also linked to primary effusion lymphoma (PEL) and multicentric Castleman's disease (MCD) (Chang et al. 2006), which are rare malignancies with very poor prognosis. Recently, few studies provided that treatment with HAART alone or with other therapy (e.g., monoclonal antibodies) for PEL and MCD may prolong resolution of symptoms (Sunil et al. 2010; Uldrick et al. 2011).

6 Human T-Cell Lymphotropic Virus Type 1 (HTLV-1)

HTLV-1 is the etiologic agent of adult T-cell leukemia-lymphoma (ATL) (Poiesz et al. 1980). An update estimation reported there were 5–10 million HTLV-1-infected individuals worldwide (Gessain and Cassar 2012). Although the lifetime risk to develop ATL in HTLV-1 infected people is only 3–5% after long latent period (approximately 30–50 years) (Ishitsuka and Tamura 2014; Hermine et al. 2018), the prognosis for ATL patients is very poor. Median survivals of acute and lymphomatous ATLs are less one year (Goncalves et al. 2010).

Compared with asymptomatic carriers, ATL patients had significantly higher level of HTLV-1 proviral DNA and antibody titer (Manns et al. 1999). The quantity of HTLV-1 proviral DNA is a predictive factor for the development of ATL and is correlated with clinical outcomes (Okayama et al. 2004; Iwanaga et al. 2010; Etoh et al. 1999; Hodson et al. 2013). Based on the accumulated data, antiviral therapy now is one of treatment options for ATL (Ishitsuka and Tamura 2014; Hermine et al. 2018; Tsukasaki et al. 2009). Treatment of ATL with antiviral drugs was investigated in the early 1980s in Japan, but encouraged advance was obtained from two trials which combined interferon-α and zidovudine (AZT) in treating ATL patients reported in 1995 (Hermine et al. 1995; Gill et al. 1995). These two studies described impressive high response rate (more than 50%) and mild toxic effects and the survival time was prolonged to more than one year. After that, many small studies (all less than 30 patients) using interferon-α and AZT/zalcitabine were performed in France (Hermine et al. 2002), United Kingdom (Matutes et al. 2001), Martinique (French West Indies) (Besson et al. 2002), and United States (Ratner et al. 2009). Overall, these studies showed consistent efficacy of antiviral therapy. Recently, a worldwide meta-analysis with 254 ATL patients treated with interferon-α and AZT combination and/or with chemotherapy provided further evidence that combination of interferon-α and AZT treatment resulted in high response and significantly prolonged survival (Bazarbachi et al. 2010). However, the survival advantage was limited to acute, chronic, and smoldering subtypes. For lymphomatous ATL, chemotherapy seemed to be more effective than antiviral therapy. In a subsequent retrospective study, the use of interferon-α/AZT with chemotherapy showed benefit of improved survival in acute ATL as well as in lymphomatous ATL (Hodson et al. 2011). Besides, treatment with combination of interferon-α/AZT and arsenic trioxide also resulted in promising outcome on seven patients with relapsed/refractory acute or lymphomatous ATL (Hermine et al. 2004), and another Phase II study using ten chronic ATL patients showed 100% response to the treatment (Kchour et al. 2009). Although these were preliminary observations, the feasibility of this regimen has been considered.

Improvement of treatment response and survival resulted from antiviral drugs was an important advance in treating ATL. Studies also showed a reduction of HTLV-1 proviral DNA load in ATL patients after antiviral therapy, and the decrease was correlated with the response of treatment (Kchour et al. 2009, 2007). However, all these results came from studies with small sample size and a prospective evaluation of the efficacy of antiviral therapy will be needed.

7 Epstein-Barr Virus (EBV)

As the first identified carcinogenic virus, EBV has been linked to many malignancies, including Burkitt's lymphoma, Hodgkin's lymphoma (HL), immune-suppressed B-cell lymphoma, post-transplant lymphoproliferative disorder (PTLD), gastric carcinoma and NPC (Kutok and Wang 2006; Kimura and Kwong 2019). The quantification of EBV viral load has been used in the diagnosis and management of EBV-associated diseases, such as PTLD and NPC and in evaluating the treatment response (Kimura and Kwong 2019). The World Health Organization International Standard for EBV quantification was released recently (Fryer et al. 2016), which would accelerate the establishment of the guidelines for diagnosis and management of EBV-associated diseases. Despite of different and complex etiopathological and clinical features, the clinical significance of EBV viral load varies across all types of EBV-associated malignancies (Kimura and Kwong 2019; Kimura et al. 2008).

Immunotherapy which uses adoptive EBV-specific cytotoxic T-cells (EBV-CTLs) to kill infected cells through the reconstitution of cellular immunity has been used to treat PTLD (Rooney et al. 1995; Bollard et al. 2012), NPC (Chua et al. 2001), and HL (Bollard et al. 2004). In addition to EBV DNA reduction and stabilization of viral load (Rooney et al. 1995; Gustafsson et al. 2000), application of the T-cell-based therapy in PTLD also exhibited beneficial consequences in prevention of development for PTLD and prolonged activity of infused CTLs against viral reactivation for up to nine years (Heslop et al. 2010). Immunotherapy was considered for the treatment of NPC and HL because of expression of type II latency antigens (EBNA1, LMP1, and LMP2), which were the targets of EBV-CTLs (Dasari et al. 2019). Mainly focused on refractory/relapse/metastatic NPC patients, some encouraging immunotherapeutic results came from achievement of complete remission (Louis et al. 2010; Straathof et al. 2005), lack of significant toxicity, the control of disease progression (Comoli et al. 2005), and increase in overall survival (Smith et al. 2012). An adenoviral vector (AdE1-LMPpoly) was applied to further increase the immunogenicity and antigen specificity of EBV-CTLs for NPC treatment (Smith et al. 2012, 2017). For relapsed HL patients, complete/partial remission, decrease of EBV DNA in PBMC and stable disease progression were observed (Bollard et al. 2004). A trial by administering adenoviral vector-based EBV-CTLs in high-risk or multiple-relapse HL patients showed the induction of durable complete responses without significant toxicity (Bollard et al. 2014). Current data on immunotherapy for the treatment of EBV-associated malignancies were limited by the small numbers of patients and the short-lasting effects and are still under investigation (Dasari et al. 2019; Merlo et al. 2011).

Virus-directed approaches, which use the EBV viral genome in tumor cells as the target, could be divided into two groups. One group includes designs of inhibition of EBV oncoprotein expression, inducing loss of EBV episome, and production of cellular toxins in EBV-infected tumors (Israel and Kenney 2003; Ghosh et al. 2012). The other approach focuses on EBV lytic cycle and includes two strategies, including inhibition of the EBV lytic cycle by anti-EBV natural compounds and the combination therapy which combined the induction of EBV lytic replication and then followed by use of cytotoxic drugs, e.g., acyclovir and ganciclovir (Israel and Kenney 2003; Ghosh et al. 2012; Li et al. 2018). Phase I/II clinical trials for EBV-associated lymphoma found that combination of arginine butyrate (one kind of HDAC inhibitor) and ganciclovir obtained noticeable clinical response in some refractory patients (Perrine et al. 2007; Faller et al. 2001). For lytic replication inducer, many chemical agents such as phorbol esters, DNA methyl transferase inhibitor (ex. 5-Azacytidine), and HDAC inhibitors have been developed and evaluated in in vitro and animal studies (Li et al. 2018). Taken together, studies of antiviral therapies on EBV-associated malignancies provided some encouraging and promising results to pave the way for future large and long-term studies.

8   Human Papillomavirus (HPV)

Of more than 200 identified HPV genotypes, 13 high-risk HPV genotypes have been defined as carcinogenic or probably carcinogenic (Roden and Stern 2018; Bzhalava et al. 2015). The causative role of HPV-16 and HPV-18 in the cervical cancer is the most well documented. Viral load of HPV in general did not correlate well with disease severity, duration of infection, clearance of disease, or prognosis of disease (Boulet et al. 2008; Woodman et al. 2007), probably due to the extremely large variation in oncogenic potential among different HPV genotypes (Josefsson et al. 2000). For the highest risk genotype HPV-16, nevertheless, high viral load appears to be associated with prevalent cervical cancer precursors and may be used to predict the development of incident disease and persistence of infection (Josefsson et al. 2000; Xi et al. 2011; Gravitt et al. 2007).

Prophylactic HPV vaccines are currently available for preventing HPV infection. Although the systematic review and meta-analysis showed strong effectiveness of prophylactic HPV vaccination in decreasing HPV infection and cervical intraepithelial neoplasia grade 2 + (CIN2 +) (Drolet et al. 2019), its population-level impact on control of cervical and other HPV-associated cancers still needs to be evaluated in next decades (Bonanni et al. 2015). In addition, prophylactic HPV vaccines had no efficacy for established HPV infections and associated lesions (Ma et al. 2012). Therapeutic HPV vaccines and RNA interference (RNAi)-based therapies are two emerging antiviral strategies for treating HPV infection and HPV-induced cervical cancer.

Therapeutic HPV vaccines, including proteins/peptides-based, DNA-based, and live attenuated bacteria/viral vectors vaccines, were majorly designed to induce cell-mediated immune response by targeting E6 and E7 oncoproteins and most of the developing vaccines were in Phase I/II stage (Yang et al. 2016; Clark and Trimble 2020; Bharti et al. 2018; Chabeda et al. 2018). Based on the results of

Phase I/II trials, HPV-specific cytotoxic T lymphocyte (CTL) response, HPV DNA clearance, and disease regression were observed in some patients and most side effects were local mild to moderate (Yang et al. 2016; Bharti et al. 2018). A viral vector-based vaccine (MVA E2), which had completed Phase III trial with 1356 patients demonstrated high remission rate (89–100%) of anogenital intraepithelial lesions and high HPV DNA clearance (83%), was observed in patients with treatment (Rosales et al. 2014). In addition, there are two candidate vaccines (a DNA vaccine and a bacterial vector-based vaccine) in on-going Phase III trials (Chabeda et al. 2018). These encouraging progress had shown the potential for the promising therapeutic HPV vaccines in the near future.

RNAi-based therapies used antisense oligonucleotides, ribozymes, and small-interfering RNAs (siRNAs) to inhibit expression of HPV E6 and E7 genes at post-transcriptional level (Jung et al. 2015). Current studies provided that RNAi therapies-induced apoptosis, reduction of growth rate and cell death in cervical cancer cell and mouse models (Zhou et al. 2012; Benitez-Hess et al. 2011; Jonson et al. 2008; Sima et al. 2007) and also could increase sensitivity of cancer cells to some chemotherapy, such as cisplain, to exhibit the synergistic therapeutic effect, which provided the possibility for RNAi-based combination therapeutics (Putral et al. 2005; Koivusalo et al. 2005; Jung et al. 2012). Nevertheless, most of these beneficial results came from pre-clinical studies and Phase I trials, so that practical implication in the clinical setting and the long-term potential as treatment are still unclear. Furthermore, problems of target selection, delivery efficiency, and stability of the RNA molecule itself when using antisense molecules or ribozymes still need to be overcome.

9 Merkel Cell Polyomavirus (MCV)

MCV is a polyomavirus newly discovered in 2008 and associated with human cancer (Feng et al. 2008; MacDonald and You 2017). As a widespread viral infection, MCV can be detected in 25–64% healthy individuals (Tolstov et al. 2009; Kean et al. 2009; Carter et al. 2009) and can be found in many normal tissues such as esophagus, liver, colon, lung, and bladder (Loyo et al. 2010). MCV is a causal factor of Merkel cell carcinoma (MCC), a rare but aggressive and lethal neuroendocrine skin cancer (Feng et al. 2008). MCC had a significantly higher MCV viral load than other tissues (Loyo et al. 2010). Given the high association of MCV with MCC, antiviral therapy may be a new approach to treat MCC beyond the current surgical excision, lymph node surgery, radiation therapy, and chemotherapy (Schrama et al. 2012). Although early limited studies showed the opportunities of antiviral drugs, such as interferon-α and interferon-β (Krasagakis et al. 2008; Biver-Dalle et al. 2011; Nakajima et al. 2009; Willmes et al. 2012), no further advances in antiviral therapy was reported in recent years. Due to limited data, more study is needed to evaluate the effect of antiviral therapy on MCC.

10 Conclusion and Future Perspective

The tremendous advances in antiviral therapy against oncogenic viruses, particularly HBV and HCV, have revolutionized the concepts and practice of cancer prevention. Primary prevention of HCC in chronically HBV- and/or HCV-infected patients is now a realistic goal, along with significant life- and cost- savings (Kim et al. 2007; Toy et al. 2008). The widespread use of HAART also dramatically reduces the incidence of KS in HIV-infected patients. For other oncogenic viruses, there is still a lack of convenient effective antiviral pharmacologic agents that can be used for primary prevention. Nevertheless, antiviral therapy can effectively prolong patients' survival in some types of HTLV-1-induced ATL and has become one of the treatment options. Further studies on experimental antiviral therapy and technology may yield promising results in adjuvant therapy for EBV, HPV, and MCV-associated malignancies.

Furthermore, to facilitate the development of new preventive or therapeutic approaches and agents, future researches are needed to continually improve our understanding in biological mechanism of viral-related carcinogenesis, and the relationship between dynamics of virus and natural history of diseases. These efforts could add to our arsenal against these viruses and yield a more favorable/satisfactory and clinical outcomes in the future.

References

AASLD-IDSA HCV Guidance Panel (2018) Hepatitis C Guidance 2018 Update: AASLD-IDSA recommendations for testing, managing, and treating hepatitis C virus infection. Clin Infect Dis 67(10):1477–1492. https://doi.org/10.1093/cid/ciy585

Agarwal K, Castells L, Müllhaupt B, Rosenberg WMC, McNabb B, Arterburn S, Camus G, McNally J, Stamm LM, Brainard DM, Mani Subramanian G, Mariño Z, Dufour J-F, Forns X (2018) Sofosbuvir/velpatasvir for 12 weeks in genotype 1–4 HCV-infected liver transplant recipients. J Hepatol 69(3):603–607. https://doi.org/10.1016/j.jhep.2018.05.039

ANRS collaborative study group on hepatocellular carcinoma (2016) Lack of evidence of an effect of direct-acting antivirals on the recurrence of hepatocellular carcinoma: data from three ANRS cohorts. J Hepatol 65(4):734–740. https://doi.org/10.1016/j.jhep.2016.05.045

Bazarbachi A, Plumelle Y, Carlos Ramos J, Tortevoye P, Otrock Z, Taylor G, Gessain A, Harrington W, Panelatti G, Hermine O (2010) Meta-analysis on the use of zidovudine and interferon-alfa in adult T-cell leukemia/lymphoma showing improved survival in the leukemic subtypes. J Clin Oncol 28(27):4177–4183. https://doi.org/10.1200/JCO.2010.28.0669

Benitez-Hess ML, Reyes-Gutierrez P, Alvarez-Salas LM (2011) Inhibition of human papillomavirus expression using DNAzymes. Methods Mol Biol 764:317–335. https://doi.org/10.1007/978-1-61779-188-8_21

Besson C, Panelatti G, Delaunay C, Gonin C, Brebion A, Hermine O, Plumelle Y (2002) Treatment of adult T-cell leukemia-lymphoma by CHOP followed by therapy with antinucleosides, alpha interferon and oral etoposide. Leuk Lymphoma 43(12):2275–2279

Bharti AC, Singh T, Bhat A, Pande D, Jadli M (2018) Therapeutic startegies for human papillomavirus infection and associated cancers. Front Biosci (Elite Ed) 10:15–73. https://doi.org/10.2741/e808

Biver-Dalle C, Nguyen T, Touze A, Saccomani C, Penz S, Cunat-Peultier S, Riou-Gotta MO, Humbert P, Coursaget P, Aubin F (2011) Use of interferon-alpha in two patients with Merkel cell carcinoma positive for Merkel cell polyomavirus. Acta Oncol 50(3):479–480. https://doi.org/10.3109/0284186X.2010.512924

Bollard CM, Aguilar L, Straathof KC, Gahn B, Huls MH, Rousseau A, Sixbey J, Gresik MV, Carrum G, Hudson M, Dilloo D, Gee A, Brenner MK, Rooney CM, Heslop HE (2004) Cytotoxic T lymphocyte therapy for Epstein-Barr virus⁺ Hodgkin's disease. J Exp Med 200 (12):1623–1633. https://doi.org/10.1084/jem.20040890

Bollard CM, Rooney CM, Heslop HE (2012) T-cell therapy in the treatment of post-transplant lymphoproliferative disease. Nat Rev Clin Oncol 9(9):510–519. https://doi.org/10.1038/nrclinonc.2012.111

Bollard CM, Gottschalk S, Torrano V, Diouf O, Ku S, Hazrat Y, Carrum G, Ramos C, Fayad L, Shpall EJ, Pro B, Liu H, Wu MF, Lee D, Sheehan AM, Zu Y, Gee AP, Brenner MK, Heslop HE, Rooney CM (2014) Sustained complete responses in patients with lymphoma receiving autologous cytotoxic T lymphocytes targeting Epstein-Barr virus latent membrane proteins. J Clin Oncol 32(8):798–808. https://doi.org/10.1200/JCO.2013.51.5304

Bonanni P, Bechini A, Donato R, Capei R, Sacco C, Levi M, Boccalini S (2015) Human papilloma virus vaccination: impact and recommendations across the world. Ther Adv Vaccines 3(1):3–12. https://doi.org/10.1177/2051013614557476

Borges ÁH (2017) Combination antiretroviral therapy and cancer risk. Curr Opin HIV AIDS 12 (1):12–19. https://doi.org/10.1097/COH.0000000000000334

Borges AH, Neuhaus J, Babiker AG, Henry K, Jain MK, Palfreeman A, Mugyenyi P, Domingo P, Hoffmann C, Read TR, Pujari S, Meulbroek M, Johnson M, Wilkin T, Mitsuyasu R, Group ISS (2016) Immediate antiretroviral therapy reduces risk of infection-related cancer during early HIV infection. Clin Infect Dis 63(12):1668–1676. https://doi.org/10.1093/cid/ciw621

Boulet GA, Horvath CA, Berghmans S, Bogers J (2008) Human papillomavirus in cervical cancer screening: important role as biomarker. Cancer Epidemiol Biomarkers Prev 17(4):810–817. https://doi.org/10.1158/1055-9965.EPI-07-2865

Bourboulia D, Aldam D, Lagos D, Allen E, Williams I, Cornforth D, Copas A, Boshoff C (2004) Short- and long-term effects of highly active antiretroviral therapy on Kaposi sarcoma-associated herpesvirus immune responses and viraemia. AIDS 18(3):485–493. https://doi.org/10.1097/00002030-200402200-00015

Bower M, Palmieri C, Dhillon T (2006) AIDS-related malignancies: changing epidemiology and the impact of highly active antiretroviral therapy. Curr Opin Infect Dis 19(1):14–19. https://doi.org/10.1097/01.qco.0000200295.30285.13

Bower M, Dalla Pria A, Coyle C, Andrews E, Tittle V, Dhoot S, Nelson M (2014) Prospective stage-stratified approach to AIDS-related Kaposi's sarcoma. J Clin Oncol 32(5):409–414. https://doi.org/10.1200/JCO.2013.51.6757

Buster EH, Flink HJ, Cakaloglu Y, Simon K, Trojan J, Tabak F, So TM, Feinman SV, Mach T, Akarca US, Schutten M, Tielemans W, van Vuuren AJ, Hansen BE, Janssen HL (2008) Sustained HBeAg and HBsAg loss after long-term follow-up of HBeAg-positive patients treated with peginterferon alpha-2b. Gastroenterology 135(2):459–467. https://doi.org/10.1053/j.gastro.2008.05.031

Buti M, Gane E, Seto WK, Chan HL, Chuang WL, Stepanova T, Hui AJ, Lim YS, Mehta R, Janssen HL, Acharya SK, Flaherty JF, Massetto B, Cathcart AL, Kim K, Gaggar A, Subramanian GM, McHutchison JG, Pan CQ, Brunetto M, Izumi N, Marcellin P, GS-US-320-0108 Investigators (2016) Tenofovir alafenamide versus tenofovir disoproxil fumarate for the treatment of patients with HBeAg-negative chronic hepatitis B virus infection: a randomised, double-blind, phase 3, non-inferiority trial. Lancet Gastroenterol Hepatol 1 (3):196–206. https://doi.org/10.1016/S2468-1253(16)30107-8

Bzhalava D, Eklund C, Dillner J (2015) International standardization and classification of human papillomavirus types. Virology 476:341–344. https://doi.org/10.1016/j.virol.2014.12.028

Cabibbo G, Celsa C, Calvaruso V, Petta S, Cacciola I, Cannavò MR, Madonia S, Rossi M, Magro B, Rini F, Distefano M, Larocca L, Prestileo T, Malizia G, Bertino G, Benanti F, Licata A, Scalisi I, Mazzola G, Di Rosolini MA, Alaimo G, Averna A, Cartabellotta F, Alessi N, Guastella S, Russello M, Scifo G, Squadrito G, Raimondo G, Trevisani F, Craxì A, Di Marco V, Cammà C (2019) Direct-acting antivirals after successful treatment of early hepatocellular carcinoma improve survival in HCV-cirrhotic patients. J Hepatol 71(2):265–273. https://doi.org/10.1016/j.jhep.2019.03.027

Calvaruso V, Cabibbo G, Cacciola I, Petta S, Madonia S, Bellia A, Tinè F, Distefano M, Licata A, Giannitrapani L, Prestileo T, Mazzola G, Di Rosolini MA, Larocca L, Bertino G, Digiacomo A, Benanti F, Guarneri L, Averna A, Iacobello C, Magro A, Scalisi I, Cartabellotta F, Savalli F, Barbara M, Davì A, Russello M, Scifo G, Squadrito G, Cammà C, Raimondo G, Craxì A, Di Marco V, Craxì A, Di Marco V, Cammà C, Calvaruso V, Petta S, Cabbibo G, Colletti P, Mazzola G, Cascio A, Montalto G, Licata A, Giannitrapani L, Prestileo T, Di Lorenzo F, Sanfilippo A, Ficalora A, Madonia S, Tinè F, Malizia G, Latteri F, Maida M, Cartabellotta F, Vassallo R, Cacciola I, Caccamo G, Maimone S, Saitta C, Squadrito G, Raimondo G, Mondello L, Smedile A, D'Andrea S, Bertino G, Ardiri AL, Frazzetto E, Rigano G, Montineri A, Larocca LN, Cacopardo B, Benanti F, Russello M, Benigno R, Bellia A, Iacobello C, Davì A, Di Rosolini MA, Digiacomo A, Fuduli G, Scifo G, Distefano M, Portelli V, Savalli F, Scalici I, Gioia G, Magro A, Alaimo G, Alinovi MR, Salvo A, Averna A, Lomonaco F, Guarneri L, Maffeo F, Falzone E, Pulvirenti F (2018) Incidence of hepatocellular carcinoma in patients with HCV-associated cirrhosis treated with direct-acting antiviral agents. Gastroenterology 155(2):411–421.e414. https://doi.org/10.1053/j.gastro.2018.04.008

Cammà C, Giunta M, Andreone P, Craxi A (2001) Interferon and prevention of hepatocellular carcinoma in viral cirrhosis: an evidence-based approach. J Hepatol 34(4):593–602. https://doi.org/10.1016/S0168-8278(01)00005-8

Cancer Project Working Group for the Collaboration of Observational HIV Epidemiological Research Europe (COHERE) study in EuroCoord (2016) Changing incidence and risk factors for Kaposi sarcoma by time since starting antiretroviral therapy: collaborative analysis of 21 European cohort studies. Clin Infect Dis 63(10):1373–1379. https://doi.org/10.1093/cid/ciw562

Carter JJ, Paulson KG, Wipf GC, Miranda D, Madeleine MM, Johnson LG, Lemos BD, Lee S, Warcola AH, Iyer JG, Nghiem P, Galloway DA (2009) Association of Merkel cell polyomavirus-specific antibodies with Merkel cell carcinoma. J Natl Cancer Inst 101(21):1510–1522. https://doi.org/10.1093/jnci/djp332

Casper C (2011) The increasing burden of HIV-associated malignancies in resource-limited regions. Annu Rev Med 62:157–170. https://doi.org/10.1146/annurev-med-050409-103711

Chabeda A, Yanez RJR, Lamprecht R, Meyers AE, Rybicki EP, Hitzeroth II (2018) Therapeutic vaccines for high-risk HPV-associated diseases. Papillomavirus Res 5:46–58. https://doi.org/10.1016/j.pvr.2017.12.006

Chan HL, Fung S, Seto WK, Chuang WL, Chen CY, Kim HJ, Hui AJ, Janssen HL, Chowdhury A, Tsang TY, Mehta R, Gane E, Flaherty JF, Massetto B, Gaggar A, Kitrinos KM, Lin L, Subramanian GM, McHutchison JG, Lim YS, Acharya SK, Agarwal K, Gs-Us-Investigators (2016) Tenofovir alafenamide versus tenofovir disoproxil fumarate for the treatment of HBeAg-positive chronic hepatitis B virus infection: a randomised, double-blind, phase 3, non-inferiority trial. Lancet Gastroenterol Hepatol 1(3):185–195. https://doi.org/10.1016/S2468-1253(16)30024-3

Chang MH, You SL, Chen CJ, Liu CJ, Lee CM, Lin SM, Chu HC, Wu TC, Yang SS, Kuo HS, Chen DS, Taiwan Hepatoma Study Group (2009) Decreased incidence of hepatocellular carcinoma in hepatitis B vaccinees: a 20-year follow-up study. J Natl Cancer Inst 101(19):1348–1355. https://doi.org/10.1093/jnci/djp288

Chang TT, Gish RG, de Man R, Gadano A, Sollano J, Chao YC, Lok AS, Han KH, Goodman Z, Zhu J, Cross A, DeHertogh D, Wilber R, Colonno R, Apelian D, B. EHoLD AI463022 Study

Group (2006) A comparison of entecavir and lamivudine for HBeAg-positive chronic hepatitis B. N Engl J Med 354(10):1001–1010. https://doi.org/10.1056/NEJMoa051285

Chen CJ, Yang HI, Iloeje UH (2009) Hepatitis B virus DNA levels and outcomes in chronic hepatitis B. Hepatology 49(5 Suppl):S72–84. https://doi.org/10.1002/hep.22884

Chiang CJ, Yang YW, Chen JD, You SL, Yang HI, Lee MH, Lai MS, Chen CJ (2015) Significant reduction in end-stage liver diseases burden through the national viral hepatitis therapy program in Taiwan. Hepatology 61(4):1154–1162. https://doi.org/10.1002/hep.27630

Chua D, Huang J, Zheng B, Lau SY, Luk W, Kwong DL, Sham JS, Moss D, Yuen KY, Im SW, Ng MH (2001) Adoptive transfer of autologous Epstein-Barr virus-specific cytotoxic T cells for nasopharyngeal carcinoma. Int J Cancer 94(1):73–80. https://doi.org/10.1002/ijc.1430

Clark KT, Trimble CL (2020) Current status of therapeutic HPV vaccines. Gynecol Oncol 156 (2):503–510. https://doi.org/10.1016/j.ygyno.2019.12.017

Coen N, Duraffour S, Snoeck R, Andrei G (2014) KSHV targeted therapy: an update on inhibitors of viral lytic replication. Viruses 6(11):4731–4759. https://doi.org/10.3390/v6114731

Comoli P, Pedrazzoli P, Maccario R, Basso S, Carminati O, Labirio M, Schiavo R, Secondino S, Frasson C, Perotti C, Moroni M, Locatelli F, Siena S (2005) Cell therapy of stage IV nasopharyngeal carcinoma with autologous Epstein-Barr virus-targeted cytotoxic T lympho-cytes. J Clin Oncol 23(35):8942–8949. https://doi.org/10.1200/JCO.2005.02.6195

Dasari V, Sinha D, Neller MA, Smith C, Khanna R (2019) Prophylactic and therapeutic strategies for Epstein-Barr virus-associated diseases: emerging strategies for clinical development. Expert Rev Vaccines 18(5):457–474. https://doi.org/10.1080/14760584.2019.1605906

Di Marco V, Marzano A, Lampertico P, Andreone P, Santantonio T, Almasio PL, Rizzetto M, Craxi A (2004) Clinical outcome of HBeAg-negative chronic hepatitis B in relation to virological response to lamivudine. Hepatology 40(4):883–891. https://doi.org/10.1002/hep.20381

Dienstag JL (2009) Benefits and risks of nucleoside analog therapy for hepatitis B. Hepatology 49 (5 Suppl):S112–121. https://doi.org/10.1002/hep.22920

Drolet M, Benard E, Perez N, Brisson M, HPV Vaccination Impact Study Group (2019) Population-level impact and herd effects following the introduction of human papillomavirus vaccination programmes: updated systematic review and meta-analysis. Lancet 394 (10197):497–509. https://doi.org/10.1016/S0140-6736(19)30298-3

El-Serag HB (2012) Epidemiology of viral hepatitis and hepatocellular carcinoma. Gastroenterol-ogy 142(6):1264–1273 e1261. https://doi.org/10.1053/j.gastro.2011.12.061

Etoh KI, Yamaguchi K, Tokudome S, Watanabe T, Okayama A, Stuver S, Mueller N, Takatsuki K, Matsuoka M (1999) Rapid quantification of HTLV-1 provirus load: detection of monoclonal proliferation of HTLV-1-infected cells among blood donors. Int J Cancer 81 (6):859–864

European Association for the Study of the Liver (2017) EASL 2017 Clinical Practice Guidelines on the management of hepatitis B virus infection. J Hepatol 67(2):370–398. https://doi.org/10.1016/j.jhep.2017.03.021

European Association for the Study of the Liver (2018) EASL recommendations on treatment of hepatitis C 2018. J Hepatol 69(2):461–511. https://doi.org/10.1016/j.jhep.2018.03.026

Faller DV, Mentzer SJ, Perrine SP (2001) Induction of the Epstein-Barr virus thymidine kinase gene with concomitant nucleoside antivirals as a therapeutic strategy for Epstein-Barr virus-associated malignancies. Curr Opin Oncol 13(5):360–367. https://doi.org/10.1097/00001622-200109000-00008

Fang CT, Chen PJ, Chen MY, Hung CC, Chang SC, Chang AL, Chen DS (2003) Dynamics of plasma hepatitis B virus levels after highly active antiretroviral therapy in patients with HIV infection. J Hepatol 39(6):1028–1035. https://doi.org/10.1016/s0168-8278(03)00416-1

Feld JJ, Wong DK, Heathcote EJ (2009) Endpoints of therapy in chronic hepatitis B. Hepatology 49(5 Suppl):S96–S102. https://doi.org/10.1002/hep.22977

Feng H, Shuda M, Chang Y, Moore PS (2008) Clonal integration of a polyomavirus in human Merkel cell carcinoma. Science 319(5866):1096–1100. https://doi.org/10.1126/science.1152586

Fryer JF, Heath AB, Wilkinson DE, Minor PD (2016) A collaborative study to establish the 1st WHO International Standard for Epstein-Barr virus for nucleic acid amplification techniques. Biologicals 44(5):423–433. https://doi.org/10.1016/j.biologicals.2016.04.010

Gallant J, Brunetta J, Crofoot G, Benson P, Mills A, Brinson C, Oka S, Cheng A, Garner W, Fordyce M, Das M, McCallister S, GS-US-292-1249 Study Investigators (2016) Brief report: efficacy and safety of switching to a single-tablet regimen of elvitegravir/cobicistat/emtricitabine/tenofovir alafenamide in HIV-1/hepatitis B-coinfected adults. J Acquir Immune Defic Syndr 73(3):294–298. https://doi.org/10.1097/QAI.0000000000001069

Gantt S, Casper C (2011) Human herpesvirus 8-associated neoplasms: the roles of viral replication and antiviral treatment. Curr Opin Infect Dis 24(4):295–301. https://doi.org/10.1097/QCO.0b013e3283486d04

Gessain A, Cassar O (2012) Epidemiological aspects and world distribution of HTLV-1 infection. Front Microbiol 3:388. https://doi.org/10.3389/fmicb.2012.00388

Ghosh SK, Perrine SP, Faller DV (2012) Advances in virus-directed therapeutics against Epstein-Barr virus-associated malignancies. Adv Virol 2012:509296. https://doi.org/10.1155/2012/509296

Gill PS, Harrington W Jr, Kaplan MH, Ribeiro RC, Bennett JM, Liebman HA, Bernstein-Singer M, Espina BM, Cabral L, Allen S, Kornblau S, Pike MC, Levine AM (1995) Treatment of adult T-cell leukemia-lymphoma with a combination of interferon alfa and zidovudine. N Engl J Med 332(26):1744–1748. https://doi.org/10.1056/NEJM199506293322603

Global Burden of Disease Liver Cancer Collaboration (2017) The burden of primary liver cancer and underlying etiologies from 1990 to 2015 at the global, regional, and national level: results from the Global Burden of Disease Study 2015. JAMA Oncol 3(12):1683–1691. https://doi.org/10.1001/jamaoncol.2017.3055

Goncalves DU, Proietti FA, Ribas JG, Araujo MG, Pinheiro SR, Guedes AC, Carneiro-Proietti AB (2010) Epidemiology, treatment, and prevention of human T-cell leukemia virus type 1-associated diseases. Clin Microbiol Rev 23(3):577–589. https://doi.org/10.1128/CMR.00063-09

Gravitt PE, Kovacic MB, Herrero R, Schiffman M, Bratti C, Hildesheim A, Morales J, Alfaro M, Sherman ME, Wacholder S, Rodriguez AC, Burk RD (2007) High load for most high risk human papillomavirus genotypes is associated with prevalent cervical cancer precursors but only HPV16 load predicts the development of incident disease. Int J Cancer 121(12):2787–2793. https://doi.org/10.1002/ijc.23012

Gustafsson Å, Levitsky V, Zou JZ, Frisan T, Dalianis T, Ljungman P, Ringden O, Winiarski J, Ernberg I, Masucci MG (2000) Epstein-Barr virus (EBV) load in bone marrow transplant recipients at risk todevelop posttransplant llymphoproliferative disease: prophylactic infusion of EBV-specific cytotoxic T cells. Blood 95(3):807–814

Hawkins C, Christian B, Ye JT, Nagu T, Aris E, Chalamilla G, Spiegelman D, Mugusi F, Mehta S, Fawzi W (2013) Prevalence of hepatitis B co-infection and response to antiretroviral therapy among HIV-infected patients in Tanzania. AIDS 27(6):919–927. https://doi.org/10.1097/QAD.0b013e32835cb9c8

He S, Lockart I, Alavi M, Danta M, Hajarizadeh B, Dore GJ (2020) Systematic review with meta-analysis: effectiveness of direct-acting antiviral treatment for hepatitis C in patients with hepatocellular carcinoma. Aliment Pharmacol Ther 51(1):34–52. https://doi.org/10.1111/apt.15598

Hermine O, Bouscary D, Gessain A, Turlure P, Leblond V, Franck N, Buzyn-Veil A, Rio B, Macintyre E, Dreyfus F, Bazarbachi A (1995) Treatment of adult T-cell leukemia-lymphoma with zidovudine and interferon alfa. N Engl J Med 332(26):1749–1751

Hermine O, Allard I, Levy V, Arnulf B, Gessain A, Bazarbachi A, French ATL therapy group (2002) A prospective phase II clinical trial with the use of zidovudine and interferon-alpha in the acute and lymphoma forms of adult T-cell leukemia/lymphoma. Hematol J 3(6):276–282. https://doi.org/10.1038/sj.thj.6200195

Hermine O, Dombret H, Poupon J, Arnulf B, Lefrère F, Rousselot P, Damaj G, Delarue R, Fermand JP, Brouet JC, Degos L, Varet B, de Thé H, Bazarbachi A (2004) Phase II trial of arsenic trioxide and alpha interferon in patients with relapsed/refractory adult T-cell leukemia/lymphoma. Hematol J 5(2):130–134. https://doi.org/10.1038/sj.thj.6200374

Hermine O, Ramos JC, Tobinai K (2018) A review of new findings in adult T-cell leukemia–lymphoma: A focus on current and emerging treatment strategies. Adv Ther 35(2):135–152. https://doi.org/10.1007/s12325-018-0658-4

Heslop HE, Slobod KS, Pule MA, Hale GA, Rousseau A, Smith CA, Bollard CM, Liu H, Wu MF, Rochester RJ, Amrolia PJ, Hurwitz JL, Brenner MK, Rooney CM (2010) Long-term outcome of EBV-specific T-cell infusions to prevent or treat EBV-related lymphoproliferative disease in transplant recipients. Blood 115(5):925–935. https://doi.org/10.1182/blood-2009-08-239186

Hodson A, Crichton S, Montoto S, Mir N, Matutes E, Cwynarski K, Kumaran T, Ardeshna KM, Pagliuca A, Taylor GP, Fields PA (2011) Use of zidovudine and interferon alfa with chemotherapy improves survival in both acute and lymphoma subtypes of adult T-cell leukemia/lymphoma. J Clin Oncol 29(35):4696–4701. https://doi.org/10.1200/JCO.2011.35.5578

Hodson A, Laydon DJ, Bain BJ, Fields PA, Taylor GP (2013) Pre-morbid human T-lymphotropic virus type I proviral load, rather than percentage of abnormal lymphocytes, is associated with an increased risk of aggressive adult T-cell leukemia/lymphoma. Haematologica 98(3):385. https://doi.org/10.3324/haematol.2012.069476

Hofmann WP, Zeuzem S (2011) A new standard of care for the treatment of chronic HCV infection. Nat Rev Gastroenterol Hepatol 8(5):257–264. https://doi.org/10.1038/nrgastro.2011.49

Huang AJ, Nunez M (2015) Outcomes in HIV/HBV-coinfected patients in the tenofovir era are greatly affected by immune suppression. J Int Assoc Provid AIDS Care 14(4):360–368. https://doi.org/10.1177/2325957415586258

Ioannou GN, Green PK, Berry K (2018) HCV eradication induced by direct-acting antiviral agents reduces the risk of hepatocellular carcinoma. J Hepatol 68(1):25–32. https://doi.org/10.1016/j.jhep.2017.08.030

Ishitsuka K, Tamura K (2014) Human T-cell leukaemia virus type I and adult T-cell leukaemia-lymphoma. Lancet Oncol 15(11):e517–526. https://doi.org/10.1016/S1470-2045(14)70202-5

Israel BF, Kenney SC (2003) Virally targeted therapies for EBV-associated malignancies. Oncogene 22(33):5122–5130. https://doi.org/10.1038/sj.onc.1206548

Iwanaga M, Watanabe T, Utsunomiya A, Okayama A, Uchimaru K, Koh KR, Ogata M, Kikuchi H, Sagara Y, Uozumi K, Mochizuki M, Tsukasaki K, Saburi Y, Yamamura M, Tanaka J, Moriuchi Y, Hino S, Kamihira S, Yamaguchi K (2010) Human T-cell leukemia virus type I (HTLV-1) proviral load and disease progression in asymptomatic HTLV-1 carriers: a nationwide prospective study in Japan. Blood 116(8):1211–1219. https://doi.org/10.1182/blood-2009-12-257410

Janssen HL, van Zonneveld M, Senturk H, Zeuzem S, Akarca US, Cakaloglu Y, Simon C, So TM, Gerken G, de Man RA, Niesters HG, Zondervan P, Hansen B, Schalm SW, HBV 99-01 Study Group, Rotterdam Foundation for Liver Research (2005) Pegylated interferon alfa-2b alone or in combination with lamivudine for HBeAg-positive chronic hepatitis B: a randomised trial. Lancet 365(9454):123–129. https://doi.org/10.1016/S0140-6736(05)17701-0

Ji F, Yeo YH, Wei MT, Ogawa E, Enomoto M, Lee DH, Iio E, Lubel J, Wang W, Wei B, Ide T, Preda CM, Conti F, Minami T, Bielen R, Sezaki H, Barone M, Kolly P, Chu PS, Virlogeux V, Eurich D, Henry L, Bass MB, Kanai T, Dang S, Li Z, Dufour JF, Zoulim F, Andreone P, Cheung RC, Tanaka Y, Furusyo N, Toyoda H, Tamori A, Nguyen MH (2019) Sustained virologic response to direct-acting antiviral therapy in patients with chronic hepatitis C and hepatocellular carcinoma: a systematic review and meta-analysis. J Hepatol 71(3):473–485. https://doi.org/10.1016/j.jhep.2019.04.017

Jiang XW, Ye JZ, Li YT, Li LJ (2018) Hepatitis B reactivation in patients receiving direct-acting antiviral therapy or interferon-based therapy for hepatitis C: a systematic review and meta-analysis. World J Gastroenterol 24(28):3181–3191. https://doi.org/10.3748/wjg.v24.i28.3181

Jones M, Nunez M (2011) HIV and hepatitis C co-infection: the role of HAART in HIV/hepatitis C virus management. Curr Opin HIV AIDS 6(6):546–552. https://doi.org/10.1097/COH. 0b013e32834bcbd9

Jonson AL, Rogers LM, Ramakrishnan S, Downs LS Jr (2008) Gene silencing with siRNA targeting E6/E7 as a therapeutic intervention in a mouse model of cervical cancer. Gynecol Oncol 111(2):356–364. https://doi.org/10.1016/j.ygyno.2008.06.033

Josefsson AM, Magnusson PKE, Ylitalo N, Sorensen P, Qwarforth-Tubbin P, Andersen PK, Melbye M, Adami HO, Gyllensten UB (2000) Viral load of human papilloma virus 16 as a determinant for development of cervical carcinoma in situ: a nested case-control study. Lancet 355(9222):2189–2193. https://doi.org/10.1016/S0140-6736(00)02401-6

Jung HS, Erkin OC, Kwon MJ, Kim SH, Jung JI, Oh Y-K, Her SW, Ju W, Choi Y-L, Song SY, Kim JK, Kim YD, Shim GY, Shin YK (2012) The synergistic therapeutic effect of cisplatin with human papillomavirus E6/E7 short interfering RNA on cervical cancer cell lines in vitro and in vivo. Int J Cancer 130(8):1925–1936. https://doi.org/10.1002/ijc.26197

Jung HS, Rajasekaran N, Ju W, Shin YK (2015) Human papillomavirus: current and future RNAi therapeutic strategies for cervical cancer. J Clin Med 4(5):1126–1155. https://doi.org/10.3390/ jcm4051126

Kchour G, Makhoul NJ, Mahmoudi M, Kooshyar MM, Shirdel A, Rastin M, Rafatpanah H, Tarhini M, Zalloua PA, Hermine O, Farid R, Bazarbachi A (2007) Zidovudine and interferon-alpha treatment induces a high response rate and reduces HTLV-1 proviral load and VEGF plasma levels in patients with adult T-cell leukemia from North East Iran. Leuk Lymphoma 48(2):330–336. https://doi.org/10.1080/10428190601071717

Kchour G, Tarhini M, Kooshyar MM, El Hajj H, Wattel E, Mahmoudi M, Hatoum H, Rahimi H, Maleki M, Rafatpanah H, Rezaee SA, Yazdi MT, Shirdel A, de Thé H, Hermine O, Farid R, Bazarbachi A (2009) Phase 2 study of the efficacy and safety of the combination of arsenic trioxide, interferon alpha, and zidovudine in newly diagnosed chronic adult T-cell leukemia/lymphoma (ATL). Blood 113(26):6528–6532. https://doi.org/10.1182/blood-2009-03-211821

Kean JM, Rao S, Wang M, Garcea RL (2009) Seroepidemiology of human polyomaviruses. PLoS Pathog 5(3):e1000363. https://doi.org/10.1371/journal.ppat.1000363

Kim WR, Benson JT, Hindman A, Brosgart C, Fortner-Burton C (2007) Decline in the need for liver transploantation for end stage liver disease secondary to hepatitis B in the US [Abstract]. Hepatology 46(Suppl):238A

Kimura H, Kwong YL (2019) EBV viral loads in diagnosis, monitoring, and response assessment. Front Oncol 9:62. https://doi.org/10.3389/fonc.2019.00062

Kimura H, Ito Y, Suzuki R, Nishiyama Y (2008) Measuring Epstein-Barr virus (EBV) load: the significance and application for each EBV-associated disease. Rev Med Virol 18(5):305–319. https://doi.org/10.1002/rmv.582

Kobayashi M, Suzuki F, Fujiyama S, Kawamura Y, Sezaki H, Hosaka T, Akuta N, Suzuki Y, Saitoh S, Arase Y, Ikeda K, Kumada H (2017) Sustained virologic response by direct antiviral agents reduces the incidence of hepatocellular carcinoma in patients with HCV infection. J Med Virol 89(3):476–483. https://doi.org/10.1002/jmv.24663

Koivusalo R, Krausz E, Helenius H, Hietanen S (2005) Chemotherapy compounds in cervical cancer cells primed by reconstitution of p53 function after short interfering RNA-mediated degradation of human papillomavirus 18 E6 mRNA: opposite effect of siRNA in combination with different drugs. Mol Pharmacol 68(2):372–382. https://doi.org/10.1124/mol.105.011189

Kowalkowski MA, Kramer JR, Richardson PR, Suteria I, Chiao EY (2015) Use of boosted protease inhibitors reduces Kaposi sarcoma incidence among male veterans with HIV infection. Clin Infect Dis 60(9):1405–1414. https://doi.org/10.1093/cid/civ012

Krasagakis K, Kruger-Krasagakis S, Tzanakakis GN, Darivianaki K, Stathopoulos EN, Tosca AD (2008) Interferon-alpha inhibits proliferation and induces apoptosis of merkel cell carcinoma in vitro. Cancer Invest 26(6):562–568. https://doi.org/10.1080/07357900701816477

Kutok JL, Wang F (2006) Spectrum of Epstein-Barr virus-associated diseases. Annu Rev Pathol 1:375–404. https://doi.org/10.1146/annurev.pathol.1.110304.100209

Lacombe K, Rockstroh J (2012) HIV and viral hepatitis coinfections: advances and challenges. Gut 61(Suppl 1):i47–58. https://doi.org/10.1136/gutjnl-2012-302062

Lai CL, Gane E, Liaw YF, Hsu CW, Thongsawat S, Wang Y, Chen Y, Heathcote EJ, Rasenack J, Bzowej N, Naoumov NV, Di Bisceglie AM, Zeuzem S, Moon YM, Goodman Z, Chao G, Constance BF, Brown NA, Globe Study Group (2007) Telbivudine versus lamivudine in patients with chronic hepatitis B. N Engl J Med 357(25):2576–2588. https://doi.org/10.1056/NEJMoa066422

Lai CL, Shouval D, Lok AS, Chang TT, Cheinquer H, Goodman Z, DeHertogh D, Wilber R, Zink RC, Cross A, Colonno R, Fernandes L, B. EHoLD AI463027 Study Group (2006) Entecavir versus lamivudine for patients with HBeAg-negative chronic hepatitis B. N Engl J Med 354(10):1011–1020. https://doi.org/10.1056/NEJMoa051287

Lau GK, Piratvisuth T, Luo KX, Marcellin P, Thongsawat S, Cooksley G, Gane E, Fried MW, Chow WC, Paik SW, Chang WY, Berg T, Flisiak R, McCloud P, Pluck N, Peginterferon Alfa-2a HBeAg-Positive Chronic Hepatitis B Study Group (2005) Peginterferon Alfa-2a, lamivudine, and the combination for HBeAg-positive chronic hepatitis B. N Engl J Med 352 (26):2682–2695. https://doi.org/10.1056/NEJMoa043470

Lawitz E, Flisiak R, Abunimeh M, Sise ME, Park JY, Kaskas M, Bruchfeld A, Worns MA, Aglitti A, Zamor PJ, Xue Z, Schnell G, Jalundhwala YJ, Porcalla A, Mensa FJ, Persico M (2020) Efficacy and safety of glecaprevir/pibrentasvir in renally impaired patients with chronic HCV infection. Liver Int 40(5):1032–1041. https://doi.org/10.1111/liv.14320

Li H, Hu J, Luo X, Bode AM, Dong Z, Cao Y (2018) Therapies based on targeting Epstein-Barr virus lytic replication for EBV-associated malignancies. Cancer Sci 109(7):2101–2108. https://doi.org/10.1111/cas.13634

Liaw YF (2006) Hepatitis B virus replication and liver disease progression: the impact of antiviral therapy. Antivir Ther 11(6):669–679

Liaw YF, Sung JJ, Chow WC, Farrell G, Lee CZ, Yuen H, Tanwandee T, Tao QM, Shue K, Keene ON, Dixon JS, Gray DF, Sabbat J, Cirrhosis Asian Lamivudine Multicentre Study Group (2004) Lamivudine for patients with chronic hepatitis B and advanced liver disease. N Engl J Med 351(15):1521–1531. https://doi.org/10.1056/NEJMoa033364

Liu CJ, Chu YT, Shau WY, Kuo RN, Chen PJ, Lai MS (2014) Treatment of patients with dual hepatitis C and B by peginterferon alpha and ribavirin reduced risk of hepatocellular carcinoma and mortality. Gut 63(3):506–514. https://doi.org/10.1136/gutjnl-2012-304370

Liu CJ, Chuang WL, Sheen IS, Wang HY, Chen CY, Tseng KC, Chang TT, Massetto B, Yang JC, Yun C, Knox SJ, Osinusi A, Camus G, Jiang D, Brainard DM, McHutchison JG, Hu TH, Hsu YC, Lo GH, Chu CJ, Chen JJ, Peng CY, Chien RN, Chen PJ (2018) Efficacy of ledipasvir and sofosbuvir treatment of HCV infection in patients coinfected with HBV. Gastroenterology 154(4):989–997. https://doi.org/10.1053/j.gastro.2017.11.011

Lok ASF, McMahon BJ, Brown RS Jr, Wong JB, Ahmed AT, Farah W, Almasri J, Alahdab F, Benkhadra K, Mouchli MA, Singh S, Mohamed EA, Abu Dabrh AM, Prokop LJ, Wang Z, Murad MH, Mohammed K (2016) Antiviral therapy for chronic hepatitis B viral infection in adults: a systematic review and meta-analysis. Hepatology 63(1):284–306. https://doi.org/10.1002/hep.28280

Louis CU, Straathof K, Bollard CM, Ennamuri S, Gerken C, Lopez TT, Huls MH, Sheehan A, Wu MF, Liu H, Gee A, Brenner MK, Rooney CM, Heslop HE, Gottschalk S (2010) Adoptive transfer of EBV-specific T cells results in sustained clinical responses in patients with locoregional nasopharyngeal carcinoma. J Immunother 33(9):983–990. https://doi.org/10.1097/CJI.0b013e3181f3cbf4

Loyo M, Guerrero-Preston R, Brait M, Hoque MO, Chuang A, Kim MS, Sharma R, Liegeois NJ, Koch WM, Califano JA, Westra WH, Sidransky D (2010) Quantitative detection of Merkel cell virus in human tissues and possible mode of transmission. Int J Cancer 126(12):2991–2996. https://doi.org/10.1002/ijc.24737

Ma B, Maraj B, Tran NP, Knoff J, Chen A, Alvarez RD, Hung CF, Wu TC (2012) Emerging human papillomavirus vaccines. Expert Opin Emerg Drugs 17(4):469–492. https://doi.org/10.1517/14728214.2012.744393

MacDonald M, You J (2017) Merkel cell polyomavirus: a new DNA virus associated with human cancer. In: Cai Q, Yuan Z, Lan K (eds) Infectious agents associated cancers: epidemiology and molecular biology. Springer, Singapore, pp 35–56. https://doi.org/10.1007/978-981-10-5765-6_4

Manns A, Miley WJ, Wilks RJ, Morgan OSC, Hanchard B, Wharfe G, Caranston B, Maloney W, Welles SL, Blattner WA, Waters D (1999) Quantitative proviral DNA and antibody levels in the natural histroy of HTLV-1 infection. J Infect Dis 180(5):1487–1493

Manthravadi S, Paleti S, Pandya P (2017) Impact of sustained viral response postcurative therapy of hepatitis C-related hepatocellular carcinoma: a systematic review and meta-analysis. Int J Cancer 140(5):1042–1049. https://doi.org/10.1002/ijc.30521

Marcellin P, Lau GK, Bonino F, Farci P, Hadziyannis S, Jin R, Lu ZM, Piratvisuth T, Germanidis G, Yurdaydin C, Diago M, Gurel S, Lai MY, Button P, Pluck N, Peginterferon Alfa-2a HBeAg-Negative Chronic Hepatitis B Study Group (2004) Peginterferon alfa-2a alone, lamivudine alone, and the two in combination in patients with HBeAg-negative chronic hepatitis B. N Engl J Med 351(12):1206–1217. https://doi.org/10.1056/NEJMoa040431

Marcellin P, Heathcote EJ, Buti M, Gane E, de Man RA, Krastev Z, Germanidis G, Lee SS, Flisiak R, Kaita K, Manns M, Kotzev I, Tchernev K, Buggisch P, Weilert F, Kurdas OO, Shiffman ML, Trinh H, Washington MK, Sorbel J, Anderson J, Snow-Lampart A, Mondou E, Quinn J, Rousseau F (2008) Tenofovir disoproxil fumarate versus adefovir dipivoxil for chronic hepatitis B. N Engl J Med 359(23):2442–2455. https://doi.org/10.1056/NEJMoa0802878

Marcellin P, Bonino F, Lau GK, Farci P, Yurdaydin C, Piratvisuth T, Jin R, Gurel S, Lu ZM, Wu J, Popescu M, Hadziyannis S (2009) Sustained response of hepatitis B e antigen-negative patients 3 years after treatment with peginterferon alpha-2a. Gastroenterology 136(7):2169–2179 e2161–2164. https://doi.org/10.1053/j.gastro.2009.03.006

Masuzaki R, Yoshida H, Omata M (2010) Interferon reduces the risk of hepatocellular carcinoma in hepatitis C virus-related chronic hepatitis/liver cirrhosis. Oncology 78(Suppl 1):17–23. https://doi.org/10.1159/000315225

Matutes E, Taylor GP, Cavenagh J, Pagliuca A, Bareford D, Domingo A, Hamblin M, Kelsey S, Mir N, Reilly JT (2001) Interferon α and zidovudine therapy in adult T-cell leukaemia lymphoma: response and outcome in 15 patients. Br J Haematol 113(3):779–784. https://doi.org/10.1046/j.1365-2141.2001.02794.x

Merchante N, Rivero-Juarez A, Tellez F, Merino D, Rios-Villegas MJ, Villalobos M, Omar M, Rincon P, Rivero A, Perez-Perez M, Raffo M, Lopez-Montesinos I, Palacios R, Gomez-Vidal MA, Macias J, Pineda JA, on behalf of the HEPAVIR-Cirrhosis Study Group (2018) Sustained virological response to direct-acting antiviral regimens reduces the risk of hepatocellular carcinoma in HIV/HCV-coinfected patients with cirrhosis. J Antimicrob Chemother 73 (9):2435–2443. https://doi.org/10.1093/jac/dky234

Merchante N, Rodriguez-Arrondo F, Revollo B, Merino E, Ibarra S, Galindo MJ, Montero M, Garcia-Deltoro M, Rivero-Juarez A, Tellez F, Delgado-Fernandez M, Rios-Villegas MJ, Garcia MA, Vera-Mendez FJ, Ojeda-Burgos G, Lopez-Ruz MA, Metola L, Omar M, Aleman-Valls MR, Aguirrebengoa K, Portu J, Raffo M, Macias J, Pineda JA, on behalf of the GEHEP-002 Study Group (2018) Hepatocellular carcinoma after sustained virological response with interferon-free regimens in HIV/hepatitis C virus-coinfected patients. AIDS 32(11):1423–1430. https://doi.org/10.1097/QAD.0000000000001809

Merlo A, Turrini R, Dolcetti R, Zanovello P, Rosato A (2011) Immunotherapy for EBV-associated malignancies. Int J Hematol 93(3):281–293. https://doi.org/10.1007/s12185-011-0782-2

Mesri EA, Cesarman E, Boshoff C (2010) Kaposi's sarcoma and its associated herpesvirus. Nat Rev Cancer 10(10):707–719. https://doi.org/10.1038/nrc2888

Mommeja-Marin H, Mondou E, Blum MR, Rousseau F (2003) Serum HBV DNA as a marker of efficacy during therapy for chronic HBV infection: analysis and review of the literature. Hepatology 37(6):1309–1319. https://doi.org/10.1053/jhep.2003.50208

Monini P, Sgadari C, Grosso MG, Bellino S, Di Biagio A, Toschi E, Bacigalupo I, Sabbatucci M, Cencioni G, Salvi E, Leone P, Ensoli B (2009) Clinical course of classic Kaposi's sarcoma in HIV-negative patients treated with the HIV protease inhibitor indinavir. AIDS 23(4):534–538. https://doi.org/10.1097/QAD.0b013e3283262a8d

Moore PS, Chang Y (2010) Why do viruses cause cancer? Highlights of the first century of human tumour virology. Nat Rev Cancer 10(12):878–889. https://doi.org/10.1038/nrc2961

Morgan RL, Baack B, Smith BD, Yartel A, Pitasi M, Falck-Ytter Y (2013) Eradication of hepatitis C virus infection and the development of hepatocellular carcinoma: a meta-analysis of observational studies. Ann Intern Med 158(5 Pt 1):329–337. https://doi.org/10.7326/0003-4819-158-5-201303050-00005

Mucke MM, Backus LI, Mucke VT, Coppola N, Preda CM, Yeh ML, Tang LSY, Belperio PS, Wilson EM, Yu ML, Zeuzem S, Herrmann E, Vermehren J (2018) Hepatitis B virus reactivation during direct-acting antiviral therapy for hepatitis C: a systematic review and meta-analysis. Lancet Gastroenterol Hepatol 3(3):172–180. https://doi.org/10.1016/S2468-1253(18)30002-5

Munir S, Saleem S, Idrees M, Tariq A, Butt S, Rauff B, Hussain A, Badar S, Naudhani M, Fatima Z, Ali M, Ali L, Akram M, Aftab M, Khubaib B, Awan Z (2010) Hepatitis C treatment: current and future perspectives. Virol J 7:296. https://doi.org/10.1186/1743-422X-7-296

Nakajima H, Takaishi M, Yamamoto M, Kamijima R, Kodama H, Tarutani M, Sano S (2009) Screening of the specific polyoma virus as diagnostic and prognostic tools for Merkel cell carcinoma. J Dermatol Sci 56(3):211–213. https://doi.org/10.1016/j.jdermsci.2009.07.013

Nguyen HQ, Magaret AS, Kitahata MM, Van Rompaey SE, Wald A, Casper C (2008) Persistent Kaposi sarcoma in the era of highly active antiretroviral therapy: characterizing the predictors of clinical response. AIDS 22(8):937–945. https://doi.org/10.1097/QAD.0b013e3282ff6275

Niederau C, Heintges T, Lange S, Goldmann G, Niederau CM, Mohr L, Haussinger D (1996) Long-term follow-up of HBeAg-positive patients treated with interferon alfa for chronic hepatitis B. N Engl J Med 334(22):1422–1427. https://doi.org/10.1056/NEJM199605303342202

Okayama A, Stuver S, Matsuoka M, Ishizaki J, Tanaka G, Kubuki Y, Mueller N, Hsieh CC, Tachibana N, Tsubouchi H (2004) Role of HTLV-1 proviral DNA load and clonality in the development of adult T-cell leukemia/lymphoma in asymptomatic carriers. Int J Cancer 110 (4):621–625. https://doi.org/10.1002/ijc.20144

Papatheodoridis GV, Papadimitropoulos VC, Hadziyannis SJ (2001) Effect of interferon therapy on the development of hepatocellular carcinoma in patients with hepatitis C virus-related cirrhosis: a meta-analysis. Aliment Pharmacol Ther 15(5):689–698. https://doi.org/10.1046/j.1365-2036.2001.00979.x

Papatheodoridis GV, Manolakopoulos S, Dusheiko G, Archimandritis AJ (2008) Therapeutic strategies in the management of patients with chronic hepatitis B virus infection. Lancet Infect Dis 8(3):167–178. https://doi.org/10.1016/S1473-3099(07)70264-5

Papatheodoridis GV, Lampertico P, Manolakopoulos S, Lok A (2010) Incidence of hepatocellular carcinoma in chronic hepatitis B patients receiving nucleos(t)ide therapy: a systematic review. J Hepatol 53(2):348–356. https://doi.org/10.1016/j.jhep.2010.02.035

Perrine SP, Hermine O, Small T, Suarez F, O'Reilly R, Boulad F, Fingeroth J, Askin M, Levy A, Mentzer SJ, Di Nicola M, Gianni AM, Klein C, Horwitz S, Faller DV (2007) A phase 1/2 trial of arginine butyrate and ganciclovir in patients with Epstein-Barr virus-associated lymphoid malignancies. Blood 109(6):2571–2578. https://doi.org/10.1182/blood-2006-01-024703

Poiesz BJ, Ruscetti FW, Gazdar AF, Bunn PA, Minna JD, Gallo RC (1980) Detection and isolation of type C retrovirus particles from fresh and cultured lymphcytes of a patient with cutaneous T-cell lymphoma. Proc Natl Acad Sci U S A 77(12):7514–7519

Putral LN, Bywater MJ, Gu W, Saunders NA, Gabrielli BG, Leggatt GR, McMillan NA (2005) RNA interference against human papillomavirus oncogenes in cervical cancer cells results in increased sensitivity to cisplatin. Mol Pharmacol 68(5):1311–1319. https://doi.org/10.1124/mol.105.014191

Qu LS, Chen H, Kuai XL, Xu ZF, Jin F, Zhou GX (2012) Effects of interferon therapy on development of hepatocellular carcinoma in patients with hepatitis C-related cirrhosis:

a meta-analysis of randomized controlled trials. Hepatol Res 42(8):782–789. https://doi.org/
10.1111/j.1872-034X.2012.00984.x

Ratner L, Harrington W, Feng X, Grant C, Jacobson S, Noy A, Sparano J, Lee J, Ambinder R, Campbell N, Lairmore M (2009) Human T cell leukemia virus reactivation with progression of adult T-cell leukemia-lymphoma. PLoS ONE 4(2):e4420. https://doi.org/10.1371/journal.pone.0004420

Reau N, Kwo PY, Rhee S, Brown RS Jr, Agarwal K, Angus P, Gane E, Kao J-H, Mantry PS, Mutimer D, Reddy KR, Tran TT, Hu YB, Gulati A, Krishnan P, Dumas EO, Porcalla A, Shulman NS, Liu W, Samanta S, Trinh R, Forns X (2018) Glecaprevir/pibrentasvir treatment in liver or kidney transplant patients with hepatitis C virus infection. Hepatology 68(4):1298–1307. https://doi.org/10.1002/hep.30046

Rockstroh JK, Lacombe K, Viani RM, Orkin C, Wyles D, Luetkemeyer AF, Soto-Malave R, Flisiak R, Bhagani S, Sherman KE, Shimonova T, Ruane P, Sasadeusz J, Slim J, Zhang Z, Samanta S, Ng TI, Gulati A, Kosloski MP, Shulman NS, Trinh R, Sulkowski M (2018) Efficacy and safety of glecaprevir/pibrentasvir in patients coinfected with hepatitis C virus and human immunodeficiency virus type 1: the EXPEDITION-2 Study. Clin Infect Dis 67(7):1010–1017. https://doi.org/10.1093/cid/ciy220

Roden RBS, Stern PL (2018) Opportunities and challenges for human papillomavirus vaccination in cancer. Nat Rev Cancer 18(4):240–254. https://doi.org/10.1038/nrc.2018.13

Rooney CM, Smith CA, Ng CYC, Loftin S, Li C, Krance RA, Brenner MK, Heslop HE (1995) Use of gene-modified virus-specific T lymphocytes to control Epstein-Barr-virus-related lymphoproliferation. Lancet 345(8941):9–13. https://doi.org/10.1016/S0140-6736(95)91150-2

Rosales R, Lopez-Contreras M, Rosales C, Magallanes-Molina JR, Gonzalez-Vergara R, Arroyo-Cazarez JM, Ricardez-Arenas A, Del Follo-Valencia A, Padilla-Arriaga S, Guerrero MV, Pirez MA, Arellano-Fiore C, Villarreal F (2014) Regression of human papillomavirus intraepithelial lesions is induced by MVA E2 therapeutic vaccine. Hum Gene Ther 25(12):1035–1049. https://doi.org/10.1089/hum.2014.024

Sandmann L, Schulte B, Manns MP, Maasoumy B (2019) Treatment of chronic hepatitis C: efficacy side effects and complications. Visceral Med 35(3):161–170. https://doi.org/10.1159/000500963

Schrama D, Ugurel S, Becker JC (2012) Merkel cell carcinoma: recent insights and new treatment options. Curr Opin Oncol 24(2):141–149. https://doi.org/10.1097/CCO.0b013e32834fc9fe

Sima N, Wang S, Wang W, Kong D, Xu Q, Tian X, Luo A, Zhou J, Xu G, Meng L, Lu Y, Ma D (2007) Antisense targeting human papillomavirus type 16 E6 and E7 genes contributes to apoptosis and senescence in SiHa cervical carcinoma cells. Gynecol Oncol 106(2):299–304. https://doi.org/10.1016/j.ygyno.2007.04.039

Singal AK, Singh A, Jaganmohan S, Guturu P, Mummadi R, Kuo YF, Sood GK (2010) Antiviral therapy reduces risk of hepatocellular carcinoma in patients with hepatitis C virus-related cirrhosis. Clin Gastroenterol Hepatol 8(2):192–199

Singal AK, Salameh H, Kuo YF, Fontana RJ (2013) Meta-analysis: the impact of oral anti-viral agents on the incidence of hepatocellular carcinoma in chronic hepatitis B. Aliment Pharmacol Ther 38(2):98–106. https://doi.org/10.1111/apt.12344

Singal AG, Rich NE, Mehta N, Branch A, Pillai A, Hoteit M, Volk M, Odewole M, Scaglione S, Guy J, Said A, Feld JJ, John BV, Frenette C, Mantry P, Rangnekar AS, Oloruntoba O, Leise M, Jou JH, Bhamidimarri KR, Kulik L, Tran T, Samant H, Dhanasekaran R, Duarte-Rojo A, Salgia R, Eswaran S, Jalal P, Flores A, Satapathy SK, Wong R, Huang A, Misra S, Schwartz M, Mitrani R, Nakka S, Noureddine W, Ho C, Konjeti VR, Dao A, Nelson K, Delarosa K, Rahim U, Mavuram M, Xie JJ, Murphy CC, Parikh ND (2019) Direct-acting antiviral therapy not associated with recurrence of hepatocellular carcinoma in a multicenter north American cohort study. Gastroenterology 156(6):1683–1692.e1681. https://doi.org/10.1053/j.gastro.2019.01.027

Smith C, Tsang J, Beagley L, Chua D, Lee V, Li V, Moss DJ, Coman W, Chan KH, Nicholls J, Kwong D, Khanna R (2012) Effective treatment of metastatic forms of Epstein-Barr virus-associated nasopharyngeal carcinoma with a novel adenovirus-based adoptive immunotherapy. Cancer Res 72(5):1116. https://doi.org/10.1158/0008-5472.CAN-11-3399

Smith C, Lee V, Schuessler A, Beagley L, Rehan S, Tsang J, Li V, Tiu R, Smith D, Neller MA, Matthews KK, Gostick E, Price DA, Burrows J, Boyle GM, Chua D, Panizza B, Porceddu SV, Nicholls J, Kwong D, Khanna R (2017) Pre-emptive and therapeutic adoptive immunotherapy for nasopharyngeal carcinoma: phenotype and effector function of T cells impact on clinical response. Oncoimmunology 6(2):e1273311. https://doi.org/10.1080/2162402X.2016.1273311

Straathof KC, Bollard CM, Popat U, Huls MH, Lopez T, Morriss MC, Gresik MV, Gee AP, Russell HV, Brenner MK, Rooney CM, Heslop HE (2005) Treatment of nasopharyngeal carcinoma with Epstein-Barr virus–specific T lymphocytes. Blood 105(5):1898–1904. https://doi.org/10.1182/blood-2004-07-2975

Su TH, Hu TH, Chen CY, Huang YH, Chuang WL, Lin CC, Wang CC, Su WW, Chen MY, Peng CY, Chien RN, Huang YW, Wang HY, Lin CL, Yang SS, Chen TM, Mo LR, Hsu SJ, Tseng KC, Hsieh TY, Suk FM, Hu CT, Bair MJ, Liang CC, Lei YC, Tseng TC, Chen CL, Kao JH, C-TEAM study group and the Taiwan Liver Diseases Consortium (2016) Four-year entecavir therapy reduces hepatocellular carcinoma, cirrhotic events and mortality in chronic hepatitis B patients. Liver Int 36(12):1755–1764. https://doi.org/10.1111/liv.13253

Sung JJ, Tsoi KK, Wong VW, Li KC, Chan HL (2008) Meta-analysis: treatment of hepatitis B infection reduces risk of hepatocellular carcinoma. Aliment Pharmacol Ther 28(9):1067–1077. https://doi.org/10.1111/j.1365-2036.2008.03816.x

Sunil M, Reid E, Lechowicz MJ (2010) Update on HHV-8-associated malignancies. Curr Infect Dis Rep 12(2):147–154. https://doi.org/10.1007/s11908-010-0092-5

Terrault NA, Lok ASF, McMahon BJ, Chang KM, Hwang JP, Jonas MM, Brown RS Jr, Bzowej NH, Wong JB (2018) Update on prevention, diagnosis, and treatment of chronic hepatitis B: AASLD 2018 hepatitis B guidance. Hepatology 67(4).1560–1599. https://doi.org/10.1002/hep.29800

Tolstov YL, Pastrana DV, Feng H, Becker JC, Jenkins FJ, Moschos S, Chang Y, Buck CB, Moore PS (2009) Human Merkel cell polyomavirus infection II. MCV is a common human infection that can be detected by conformational capsid epitope immunoassays. Int J Cancer 125(6):1250–1256. https://doi.org/10.1002/ijc.24509

Toy M, Veldhuijzen IK, De Man RA, Richardus J, Schalm SW (2008) The potential impact of long-term nucleodise therapy on the mortality and morbidity of high viremic chronic hepatitis B [Abstract]. Hepatology 48(Suppl):717A

Tsukasaki K, Hermine O, Bazarbachi A, Ratner L, Ramos JC, Harrington W Jr, O'Mahony D, Janik JE, Bittencourt AL, Taylor GP, Yamaguchi K, Utsunomiya A, Tobinai K, Watanabe T (2009) Definition, prognostic factors, treatment, and response criteria of adult T-cell leukemia-lymphoma: a proposal from an international consensus meeting. J Clin Oncol 27(3):453–459. https://doi.org/10.1200/JCO.2008.18.2428

Uldrick TS, Polizzotto MN, Aleman K, O'Mahony D, Wyvill KM, Wang V, Marshall V, Pittaluga S, Steinberg SM, Tosato G, Whitby D, Little RF, Yarchoan R (2011) High-dose zidovudine plus valganciclovir for Kaposi sarcoma herpesvirus-associated multicentric Castleman disease: a pilot study of virus-activated cytotoxic therapy. Blood 117(26):6977–6986. https://doi.org/10.1182/blood-2010-11-317610

van Zonneveld M, Honkoop P, Hansen BE, Niesters HG, Darwish Murad S, de Man RA, Schalm SW, Janssen HL (2004) Long-term follow-up of alpha-interferon treatment of patients with chronic hepatitis B. Hepatology 39(3):804–810. https://doi.org/10.1002/hep.20128

Wandeler G, Mauron E, Atkinson A, Dufour JF, Kraus D, Reiss P, Peters L, Dabis F, Fehr J, Bernasconi E, van der Valk M, Smit C, Gjaerde LK, Rockstroh J, Neau D, Bonnet F, Rauch A, Swiss HIV Cohort Study, Athena Observational Cohort Study, EuroSIDA, ANRS CO3 Aquitaine Cohort (2019) Incidence of hepatocellular carcinoma in HIV/HBV-coinfected patients on tenofovir therapy: relevance for screening strategies. J Hepatol 71(2):274–280. https://doi.org/10.1016/j.jhep.2019.03.032

Waziry R, Hajarizadeh B, Grebely J, Amin J, Law M, Danta M, George J, Dore GJ (2017) Hepatocellular carcinoma risk following direct-acting antiviral HCV therapy: a systematic review, meta-analyses, and meta-regression. J Hepatol 67(6):1204–1212. https://doi.org/10.1016/j.jhep.2017.07.025

Wei L, Kao JH (2017) Benefits of long-term therapy with nucleos(t)ide analogues in treatment-naive patients with chronic hepatitis B. Curr Med Res Opin 33(3):495–504. https://doi.org/10.1080/03007995.2016.1264932

WHO Global hepatitis report (2017) April, 2017. https://www.who.int/hepatitis/publications/global-hepatitis-report2017/en/. Accessed 5 Feb 2020

Willmes C, Adam C, Alb M, Volkert L, Houben R, Becker JC, Schrama D (2012) Type I and II IFNs inhibit Merkel cell carcinoma via modulation of the Merkel cell polyomavirus T antigens. Cancer Res 72(8):2120–2128. https://doi.org/10.1158/0008-5472.CAN-11-2651

Wong GL, Yiu KK, Wong VW, Tsoi KK, Chan HL (2010) Meta-analysis: reduction in hepatic events following interferon-alfa therapy of chronic hepatitis B. Aliment Pharmacol Ther 32 (9):1059–1068. https://doi.org/10.1111/j.1365-2036.2010.04447.x

Woodman CB, Collins SI, Young LS (2007) The natural history of cervical HPV infection: unresolved issues. Nat Rev Cancer 7(1):11–22. https://doi.org/10.1038/nrc2050

Wyles D, Brau N, Kottilil S, Daar ES, Ruane P, Workowski K, Luetkemeyer A, Adeyemi O, Kim AY, Doehle B, Huang KC, Mogalian E, Osinusi A, McNally J, Brainard DM, McHutchison JG, Naggie S, Sulkowski M, ASTRAL-5 Investigators (2017) Sofosbuvir and velpatasvir for the treatment of hepatitis C virus in patients coinfected with human immunodeficiency virus type 1: an open-label, phase 3 study. Clin Infect Dis 65(1):6–12. https://doi.org/10.1093/cid/cix260

Xi LF, Hughes JP, Castle PE, Edelstein ZR, Wang C, Galloway DA, Koutsky LA, Kiviat NB, Schiffman M (2011) Viral load in the natural history of human papillomavirus type 16 infection: a nested case-control study. J Infect Dis 203(10):1425–1433. https://doi.org/10.1093/infdis/jir049

Yang YF, Zhao W, Zhong YD, Xia HM, Shen L, Zhang N (2009) Interferon therapy in chronic hepatitis B reduces progression to cirrhosis and hepatocellular carcinoma: a meta-analysis. J Viral Hepat 16(4):265–271. https://doi.org/10.1111/j.1365-2893.2009.01070.x

Yang A, Farmer E, Wu TC, Hung C-F (2016) Perspectives for therapeutic HPV vaccine development. J Biomed Sci 23(1):75. https://doi.org/10.1186/s12929-016-0293-9

Yang JD, Hainaut P, Gores GJ, Amadou A, Plymoth A, Roberts LR (2019) A global view of hepatocellular carcinoma: trends, risk, prevention and management. Nat Rev Gastroenterol Hepatol 16(10):589–604. https://doi.org/10.1038/s41575-019-0186-y

Yanik EL, Napravnik S, Cole SR, Achenbach CJ, Gopal S, Olshan A, Dittmer DP, Kitahata MM, Mugavero MJ, Saag M, Moore RD, Mayer K, Mathews WC, Hunt PW, Rodriguez B, Eron JJ (2013) Incidence and timing of cancer in HIV-infected individuals following initiation of combination antiretroviral therapy. Clin Infect Dis 57(5):756–764. https://doi.org/10.1093/cid/cit369

Zhang CH, Xu GL, Jia WD, Li JS, Ma JL, Ge YS (2011) Effects of interferon treatment on development and progression of hepatocellular carcinoma in patients with chronic virus infection: a meta-analysis of randomized controlled trials. Int J Cancer 129(5):1254–1264. https://doi.org/10.1002/ijc.25767

Zhou J, Peng C, Li B, Wang F, Zhou C, Hong D, Ye F, Cheng X, Lu W, Xie X (2012) Transcriptional gene silencing of HPV16 E6/E7 induces growth inhibition via apoptosis in vitro and in vivo. Gynecol Oncol 124(2):296–302. https://doi.org/10.1016/j.ygyno.2011.10.028

Printed in the United States
by Baker & Taylor Publisher Services